AF251739

ALGEBRAS, RINGS AND THEIR REPRESENTATIONS

Proceedings of the International Conference
on Algebras, Modules and Rings

ALGEBRAS, RINGS AND THEIR REPRESENTATIONS

Proceedings of the International Conference
on Algebras, Modules and Rings

Lisbon, Portugal   14–18 July 2003

editors

Alberto Facchini *(Università di Padova, Italy)*

Kent Fuller *(University of Iowa, USA)*

Claus M Ringel *(Universität Bielefeld, Germany)*

Catarina Santa-Clara *(Universidade de Lisboa, Portugal)*

Cover image by courtesy of João Sotomayor

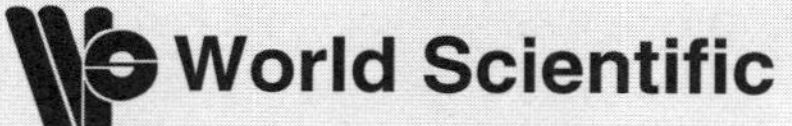

World Scientific

NEW JERSEY · LONDON · SINGAPORE · BEIJING · SHANGHAI · HONG KONG · TAIPEI · CHENNAI

Published by

World Scientific Publishing Co. Pte. Ltd.

5 Toh Tuck Link, Singapore 596224

USA office: 27 Warren Street, Suite 401-402, Hackensack, NJ 07601

UK office: 57 Shelton Street, Covent Garden, London WC2H 9HE

British Library Cataloguing-in-Publication Data
A catalogue record for this book is available from the British Library.

ALGEBRAS, RINGS AND THEIR REPRESENTATIONS
Proceedings of the International Conference on Algebras, Modules and Rings

ISBN 981-256-598-1

Printed in Singapore by B & JO Enterprise

CONTENTS

INTRODUCTION

In memory of Professor António Almeida Costa,
on the centenary of his birth.

Organized by Centro de Álgebra da Universidade de Lisboa
Held at Faculdade de Ciências da Universidade de Lisboa

`http://caul.cii.fc.ul.pt/lisboa2003/`

In homage to Professor António Almeida Costa (1903-1978), Full Professor of the Faculty of Sciences of the University of Lisbon (FCUL), and on occasion of the 100th anniversary of his birth, the Centro de Álgebra da Universidade de Lisboa (CAUL) organized the "International Conference on Algebras, Modules and Rings", held at FCUL, from the 14th to the 18th of July 2003.

This event was widely publicized and attracted 151 participants, from 33 different nationalities, comprising many experts in the areas of Ring and Module Theory, Representation Theory of Algebras, and other closely related fields, especially Non-Commutative Algebraic Geometry. It provided an excellent venue for the communication of recent results and the discussion of new problems in the areas that will surely bring about future progress in research.

The invited speakers consisted of 9 main speakers, each lecturing for 50 minutes, and 9 plenary speakers, speaking for 35 minutes. The main speakers, leaders in their respective areas of research, were headed by the keynote speaker, Michel Van den Bergh, who set the tone for the conference, establishing a connection to Non-Commutative Algebraic Geometry. The plenary speakers are younger mathematicians, with promising work on the different subjects covered. The conference also included 85 short communications (20 minutes each).

The conference was co-funded by CAUL, DMFCUL, FCUL, Fundação para a Ciência e a Tecnologia, Project POCTI/143/2003 of FCT and FEDER, Centro de Matemática da Universidade do Porto, Centro de Estruturas Lineares e Combinatórias and Universidade Lusófona. Grants were awarded to some participants, both by Unesco-Roste and Deutscher Akademische Austausch Dienst. The organization enjoyed some support from Turismo de Lisboa, Lisbon's City Council and Bookshop Escolar Editora. Participants had discounts on flights by TAP and VARIG.

Conference Organization

Scientific Committee

A. Facchini (Padova)
K. Fuller (Iowa)
M. L. Galvão (Lisboa), coordinator
J. L. Gómez Pardo
    (Santiago de Compostela)

J. A. Green (Oxford, Warwick)
C. M. Ringel (Bielefeld)
D. Simson (N. Copernicus)
P. F. Smith (Glasgow)

Organizing Committee

A. P. Alexandre (Nova de Lisboa)
P. Carvalho Lomp (Porto)
A. V. Fonseca (Lusófona, Leicester)
M. L. Galvão (Lisboa)

C. Lomp (Porto)
M. T. Nogueira (Lisboa)
C. Santa-Clara (Lisboa), coordinator
M. E. Simões (Lisboa)

Assistants to the Organization

Staff: M. Manuela Ferreira, Béatrice Huberty.

Students: Catarina Barroca, Patrícia Batista, Marta Cabrita, Eliana Castro, Vidal Delgado, António V. Dias, M. Graça Duffner, Gilda M. Ferreira, Susana Frederique, Jaime Gaspar, Joaquim Graça, Joana M. Matos, Rita Monge, Alexandra Passinhas, Ana Pinheiro, David Raimundo, Filipe Ramos, Andrea Rodrigues, Carina Rodrigues, Ana Isabel Teixeira, Ana Margarida Troncão.

Invited Talks

Keynote

M. Van den Bergh (Limburgs-Diepenbeek): Derived categories in algebraic geometry

Main Talks

S. Koenig (Leicester): Comparing Schur algebras, upwards and downwards
T.-Y. Lam (California, Berkeley): Corner Ring Theory: A Generalization of Peirce Decompositions
L. Levy (Nebraska, Lincoln and Wisconsin, Madison): Representation type of Commutative Noetherian Rings
O. Mathieu (Claude Bernard, Lyon): Connection on stable bundle
B. Osofsky (Rutgers, New Brunswick): Quasideterminants and Roots of Polynomials over Division Rings
C. M. Ringel (Bielefeld): Basic properties of the module category of an artin algebra
M. Saorín (Murcia): Categorical invariance of automorphism groups of rings an algebras
J. Trlifaj (Karlova, Praha): Cotorsion pairs

Plenary Talks

P. Ara (Autònoma, Barcelona): Finitely presented modules over Leavitt algebras
J. A. de la Pena (UNAM, Mexico City): Hochschild cohomology of algebras and epimorphisms
I. Gordon (Glasgow): Rational Cherednik algebras and applications
P. A. Guil Asensio (Murcia): Left Cotorsion Rings
O. Iyama (Himeji): Representation dimension of artin algebras
J. Okniński (Warsaw): Quadratic algebras of skew type
C. Santa-Clara (Lisboa): Goldie dimension applied to Linear Operator Theory
J. Schroer (Leeds): A decomposition theory for irreducible components of module varieties
A. Tonolo (Padova): Cotilting modules versus canonical modules for Cohen-Macaulay rings

Contributed talks

Module Theory

Klaus Robert Aehle: Complexity of degenerations of modules
Khaled Al-Takhman: Comatrix Corings
Lidia Angeleri Hügel: Tilting modules and Gorenstein rings
María José Arroyo Paniagua: Spectral Torsion Theories and a General Theory of Types
Septimiu Crivei: On divisible modules with respect to a torsion theory
Sergio Estrada: Torsion free covers of cotorsion modules
Alberto Facchini: Projective modules and divisor homomorphisms
Grigory Garkusha: Triangulated categories and the Ziegler spectrum
Enrico Gregorio: Cotilting dualities

El-Amin Kaidi: Weakly Noetherian or Artinian modules and rings

Thomas G. Kucera: The shape of indecomposable injective modules over non-commutative noetherian rings

Miroslav Kureš: Weil modules

Christian Lomp: Central closure for ring extensions with additional module structure

Leandro Marín: Topologies and functors between categories of modules for nonunital rings

Federico Raggi: Semiprime preradicals and semiprime modules

Lutz Strüngmann: Generalized E-rings

Lia Vas: Applying Torsion Theories to Finite von Neumann algebras

Indah Emilia Wijayanti: Coprime Coalgebras and Dual Algebras

Non-commutative Algebraic Geometry

Tatiana Gateva-Ivanova: Set-theoretic Solutions of the Yang-Baxter Equations

Dirk Kussin: Projective coordinate algebras of exceptional curves

Wendy Tor Lowen: Deformations of abelian categories

Representation Theory of Algebras

Jan Adrianssens: The modular group and quiver representations

Helena Albuquerque: Quadratic Malcev superalgebras

Daciana Alina Alb: Some coreflective categories of topological modules

Zvi Arad: Classification of integral table algebras via a given subset of their algebra constants

David M. Arnold: Endowild representation type and generic representations of finite posets

Esther Beneish: Lattice invariants, generic matrices and other rationality problems

Grzegorz Bobiński: On simply connected generically finite algebras

Victor Bovdi: Generalized crystallographic groups with indecomposable holonomy group

Aslak Bakke Buan: Tilting in tubes

Antonio J. Calderón Martín: On locally finite split Lie triple systems

Giovanna Carnovale: Quantized universal enveloping algebras at the roots of unity and spherical conjugacy classes

William Chin: Local Theory of Almost Split Sequences for Comodules

Flávio Ulhoa Coelho: Endomorphism algebras of projective modules over laura algebras

Gabriella D'Este: Tilting and cotilting-type modules

Yuriy Drozd: Derived tame and wild algebras

Alice Fialowski: Deformations of three dimensional Lie algebras

André Fonseca: Ringel duality for Harish-Chandra modules

Manuel Forero Piulestán: On infinte dimensional Lie algebras

Angela Holtmann: The s-tame dimension vectors for stars

Andrew Hubery: Representations of a quiver with automorphism

Tibor Juhász: The derived length of Lie soluble group algebras

Mark Kleiner: Abelian categories, almost split sequences, and comodules

Piroska Lakatos: Zeros of Coxeter and reciprocal polynomials

Zbigniew Leszczynski: Locally hereditary tame algebras

Mojgan Mahmoudi: On Injectivity of Projection Algebras

František Marko: Schur superalgebras in characteristic p
Robert Marsh: Tilting modules and cluster algebras
M. Teresa Nogueira: Radicals and socles of an algebra without identity
Tsetska Rashkova: P.I. algebras with involution
Igor K. Redchuk: Generalization of function ρ and representations of quivers in Hilbert spaces
Alexander Retakh: Structure and representations of conformal algebras
Sonia L. Rueda: Finite dimensional representations of invariant differential operators
Ana Paula Santana: Some results on the classification of quasiassociative algebras
José Carlos Santos: Homologically induced modules for Lie superalgebras
Andrzej Skowroński: On Galois coverings of selfinjective algebras
Adrian Williams: Does the Murphy subalgebra of the Hecke algebra have a Grobner-Shirshov basis?
Milos Ziman: On local embeddability of groups and group algebras
Grzegorz Zwara: On the zero set of semi-invariants for quivers

Ring Theory

Elena Aladova: Identities in nil-algebras over a field of a prime characteristic
Pham Ngoc Ánh: On a question of Müller
Javad Asadollahi: Dehaghi Homological dimensions of Modules over Commutative Noetherian rings
Mamadou Barry: On the FGI-Rings
Adalbert Bovdi: Applications of the group identities theory to the group of units
Juan Cuadra: Torsion Classes and F-Noetherian Coalgebras
Ángel del Río: Groups of units of integral group rings of Kleinian type
Vitor O. Ferreira: Invariants of free algebras
Laszlo Fuchs: Primal ideals in commutative rings without finiteness conditions
Franz Halter-Koch: C-monoids and the multiplicative structure of integral domains
Roozbeh Hazrat: Reduced K-theory for Azumaya algebras
Natalia K. Iyudu: About Serre's conjecture in noncommutative case
Algirdas Kaucikas: Strongly prime rings, ideals and their applications
Elena Kireeva: Some extremal varieties of associative algebras
Peter Malcolmson: Expansions of prime ideals in noncommutative rings
Ryszard Mazurek: Pseudo-chain rings and pseudo-uniserial modules
Francesc Perera: The exchange property of Gromov's translation algebras
Robert Raphael: On one-accessible regular algebras
Raquel Reis: On lattice-ordered commutative semigroups
Javier Sánchez Serdà: On a theorem of Ian Hughes about division rings of fractions
Hisa Tsutsui: On rings with the same set of proper ideals
Mihail Ursul: Structure theorems on countably compact rings
Peter Vamos: The matrix type of rings
Paolo Zanardo: Intersections of powers of a principal ideal and primality
Jan Žemlička: Classes of modules over regular semiartinian rings
Rita Zuazua: Standard Noether Normalizations of the Graph Subring

Proceedings Sponsors

Project POCTI/143/2003 of FCT and FEDER

FUNDAÇÃO CALOUSTE GULBENKIAN

Centro de Álgebra da
Universidade de Lisboa

FCT Fundação para a Ciência e a Tecnologia

MINISTÉRIO DA CIÊNCIA E DO ENSINO SUPERIOR   Portugal

ANTÓNIO ALMEIDA COSTA

The International Conference on Algebras, Modules and Rings was dedicated to Professor António Almeida Costa, on the centenary of his birth.

After completing high school in the city of Guarda with distinction, António Almeida Costa attended the first year of the university at this Faculty of Sciences of Lisbon. Then he went to Oporto University, where he got his B.A. in Mathematics, in 1924, with very high marks, having been awarded the prizes "Gomes Teixeira" and "Gomes Ribeiro".

Starting his academic career in Astronomy and Celestial Mechanics in the 1920's, it was he who introduced in Portugal, about 1940, the systematic study of abstract algebra.

In 1928/29 he had assumed responsibility for the courses of Astronomy in the Faculty of Sciences of Oporto and when, in the same year, the Junta de Salvação Nacional—the Portuguese science research council—was set up, he immediately applied for a subsidy to prepare his doctoral dissertation. His proposal was received favourably since he clearly revealed exceptional intelligence and diligence and his area of study (theory of surfaces by vector calculus methods) was recognized to be of interest. However the scholarship was not awarded that year because "the candidate had not yet revealed his talent for research through written work".

The young Almeida Costa was not discouraged and two years later his first publications began to appear: Notas de Cálculo Vectorial (Notes on Vector Calculus) and Sobre a Dinâmica dos Sistemas Holónomos (On the Dynamics of Holonomic Systems), the latter presented with an application for a professorship.

Meanwhile, he trained at the Astronomical Observatory of Lisbon, always thinking about the modernization of the courses he was in charge of.

In 1934 he once more applied for a grant from the Junta de Educação Nacional, this time in order to pursue his studies in astronomy and geometry at the University of Paris, the latter with Elie Cartan, but "with the possibility of replacing, at his own responsibility, the study of geometry by physics under the direction of Louis de Broglie".

The research council was very demanding in its conditions for the grant, particularly with regard to the choice of the specialization. But Almeida Costa did not give up, and in this struggle with the Committee of Scholarships he revealed his determination to fight for what he believed in, a quality that remained with him all his life.

In successive applications he proved to be well informed about science in Europe, and in 1937 (three years after his first application!) he finally got a grant to study theoretical physics, particularly quantum mechanics and theory of relativity, at the Physikalische Institut in Berlin, where he remained for 22 months.

He then began to attend courses on Matrices and Group Theory, as well as Theoretical Electricity and Thermodynamics. When the Portuguese science research council observed that he should not diversify his studies so much, he replied: "I cannot forget that a background in mathematics is absolutely indispensable for the serious study of theoretical physics. Pages and pages of books on Physics are dedicated to matrix theory, series expansions of complete functions, orthogonal or not, group theory, and so on."

The influence of Hilbert in the teaching of mathematics and theoretical physics was obvious at that time, and we know that the last studies of Almeida Costa in Berlin demonstrated his familiarity with methods of mathematical physics according Courant and Hilbert and quantum groups according Hermann Weyl, a disciple of Hilbert.

In conclusion: until 1937, Almeida Costa had lived in the world of applied mathematics, essentially celestial mechanics and astronomy. With the Berlin grant, the Portuguese science research council directed him towards theoretical physics; and, as a result of this, he dropped his interest in astronomy and moved to a completely different scientific world, where he demonstrated a great enthusiasm, transmitting to others what he had learned abroad.

After a remarkable time at the Faculty of Sciences of Oporto, and while he already was a full professor of Celestial Mechanics, Almeida Costa accepted, in 1952, the chair of Algebra at the school where he had been a freshman—this Faculty of Sciences of Lisbon—and here he could give a great contribution to the subjects he had chosen a dozen years before.

In the decade of 1950, Almeida Costa published about 20 papers, among them Para a história dos domínios multiplicativos associativos (key note speech in the XXI Congress of Spanish Association for the Progress of Sciences, Malaga, 1951) and two books Curso de Álgebra Abstracta (1954, 240 pp.) and Elementos de álgebra linear e de geometria linear (1958, 272 pp.); and in 1959, he was selected corresponding member of the Academy of Sciences of Lisbon.

His scientific activity proceeded with enthusiasm and until his retirement (in 1973) he published more than a dozen papers in Anais da Faculdade de Ciências do Porto, Revista da Faculdade de Ciências de Lisboa, Boletim e Memórias da Academia das Ciências de Lisboa, Seminar Dubreil-Pisot (Paris), Publicationes Mathematicae de Debrecen (Hungary) and Mathematische Zeitschrift (Germany) and two volumes of a Course d'Algèbre Générale by the Gulbenkian Foundation. In 1972, he became full member of the Academy of Sciences of Lisbon and five years later he was its president.

His activity in algebra did not finish with his retirement: in 1974, the third volume of the Course d'Algèbre Générale appeared and he presented 5 papers, alone or in collaboration, to the Academy until 1978, the year of his death, on 24 August.

He had work until the end, he who had been a distinguished professor of the Faculties of Sciences of Oporto and of Lisbon, and of the Instituto Superior Técnico; director of the Seminar of Mathematics and of the Center for Mathematics Applied to Nuclear Energy (in Lisbon); member of the Junta Nacional de Educação, of the Research Committee of the Instituto de Alta Cultura and of the Scientific Committee of the Gulbenkian Foundation; member of the Portuguese Mathematical Society, of the American Mathematical Society and honorary of the Real Spanish Mathematical

Society; director of the Faculty of Sciences of Lisbon in the last years of his career and, at last, president of the Academy of Sciences of Lisbon.

He is rightly remembered as the true founder of the Portuguese school of algebra and as an example of a scientist who believes in the importance of Science as a cultural value of a nation and an agent of its progress.

Text based on the Opening Address (Homage to Professor A. Almeida Costa) by Emeritus Professor F. R. Dias Agudo, member of the Academy of Sciences of Lisbon delivered at the Opening Ceremony of the International Conference on Algebras, Modules and Rings

CONFERENCE PARTICIPANTS

Jan Adriaenssens
Univ. Antwerpen, Belgium
jan.adriaenssens@ua.ac.be

Klaus Robert Aehle
ETH-Zurich, Switzerland
aehle@math.ethz.ch

Elena Aladova
Moscow Pedagogical State Univ., Russia
aladovael@mail.ru

Daciana Alina Alb
Univ. Oradea, Romania
dalb@uoradea.ro

Helena Albuquerque
Univ. Coimbra, Portugal
lena@mat.uc.pt

António P. Alexandre
Univ. Nova de Lisboa, Portugal
apa@fct.unl.pt

Khaled Al-Takhman
Birzeit Univ., Palestine
takhman@birzeit.edu

Carlos André
Univ. Lisboa, Portugal
candre@fc.ul.pt

Lidia Angeleri Hügel
Univ. Studi Insubria, Varese, Italy
lidia.angeleri@uninsubria.it

Pere Ara
Univ. Autonoma, Barcelona, Spain
para@mat.uab.es

Zvi Arad
Netanya Academic College, Netanya, Israel
aradtzvi@macs.biu.ac.il

David M. Arnold
Baylor Univ., Waco, Texas, USA
David_Arnold@baylor.edu

María José Arroyo Paniagua
Univ. Autónoma Metropolitana, Iztapalapa, Mexico
mja@xanum.uam.mx

Javad Asadollahi Dehaghi
Shahre-Kord Univ., Iran
Asadollahi@ipm.ir

Mamadou Barry
Cheikh Anta DIOP Univ., Dakar, Senegal
mansabadion1@hotmail.com

Silvana Bazzoni
Univ. Padova, Italy
bazzoni@math.unipd.it

Esther Beneish
Central Michigan Univ., USA
benei1e@cmich.edu

Américo Bento
Univ. Trás-os-Montes e A. Douro, Portugal
abento@utad.pt

Grzegorz Bobiński
Nicholas Copernicus Univ., Torun, Poland
gregbob@mat.uni.torun.pl

Inês M. Borges
Univ. Porto, Portugal
ines_pais@hotmail.com

Maria Fátima Borralho
Univ. Algarve, Portugal
mfborra@ualg.pt

Adalbert Bovdi
Univ. Debrecen, Hungary
bodibela@math.klte.hu

Victor Bovdi
Univ. Debrecen, Hungary
vbovdi@math.klte.hu

Aslak Bakke Buan
*Norwegian Univ. Technology and Science,
Norway*
aslakb@math.ntnu.no

Walter Burgess
Univ. Ottawa, Canada
wburgess@uottawa.ca

Juan Carlos Bustamante
Univ. Sherbroke, Canada
bustaman@dmi.usherb.ca

Antonio J. Calderón Martín
Univ. Cadiz, Spain
ajesus.calderon@uca.es

Giovanna Carnovale
Univ. Padova, Italy
carnoval@math.unipd.it

Paula Carvalho Lomp
Univ. Porto, Portugal
pbcarval@fc.up.pt

Kanokporn Changtong
Univ. Glasgow, UK
kc@maths.gla.ac.uk

William Chin
DePaul Univ., Chicago, USA
wchin@condor.depaul.edu

John Clark
Univ. Otago, Dunedin, New Zealand
jclark@maths.otago.ac.nz

Flávio U. Coelho
Univ. São Paulo, Brazil
fucoelho@ime.usp.br

Riccardo Colpi
Univ. Padova, Italy
colpi@math.unipd.it

Septimiu Crivei
*Babeş-Bolyai Univ., Cluj-Napoca,
Romania*
crivei@math.ubbcluj.ro

Kathi Crow
Univ. California, Santa Barbara, USA
crow@math.ucsb.edu

Juan Cuadra
Univ. Almeria, Spain
jcdiaz@ual.es

José Antonio Cuenca Mira
Univ. Málaga, Spain
cuenra@agt.cie.uma.es

Gabriel Davis
Univ. Leicester, UK
g.davis@mcs.le.ac.uk

José Antonio de la Peña
UNAM, Mexico City, Mexico
jap@penelope.matem.unam.mx

Ángel del Río
Univ. Murcia, Spain
adelrio@um.es

Gabriella D'Este
Univ. Milano, Italy
gabriella.deste@mat.unimi.it

Luca Diracca
Univ. Padova, Italy
diracca@math.unipd.it

Yuriy Drozd
Kyiv Taras Shevchenko Univ., Ukraine
`yuriy@drozd.org`

Sergio Estrada
Univ. Almería, Spain
`sestrada@ual.es`

Alberto Facchini
Univ. Padova, Italy
`facchini@math.unipd.it`

Philipp Fahr
Univ. Bielefeld, Germany
`philfahr@gmx.net`

Vitor O. Ferreira
Univ. São Paulo, Brazil
`vofer@ime.usp.br`

Alice Fialowski
Eotvos Lorand Univ., Budapest, Hungary
`fialowsk@cs.elte.hu`

André V. Fonseca
Univ. Lusófona, Lisboa, Portugal
`A.Fonseca@mcs.le.ac.uk`

Manuel Forero Piulestán
Univ. Cádiz, Spain
`ForeroManuel@hotmail.com`

Pedro Freitas
Univ. Lisboa, Portugal
`pedro@ptmat.fc.ul.pt`

Avital Frumkin
Tel Aviv Univ., Israel
`frumkin@math.tau.ac.il`

Lazlo Fuchs
Tulane Univ., New Orleans, USA
`fuchs@tulane.edu`

Kent Fuller
Univ. Iowa, USA
`kfuller@math.uiowa.edu`

Maria Luísa Galvão
Univ. Lisboa, Portugal
`mlgalvao@ptmat.fc.ul.pt`

Grigory Garkusha
Int. Centre Theoretical Physics, Trieste, Italy
`ggarkusha@mail.ru`

Tatiana Gateva-Ivanova
Bulgarian Academy Sciences, Bulgaria and Univ. Antwerp, Belgium
`tatianagateva@yahoo.com,`
`Tatiana.Ivanova@ua.ac.be`

José Luis Gómez Pardo
Univ. Santiago de Compostela, Spain
`alpardo@uscmail.usc.es`

Iain Gordon
Univ. Glasgow, UK
`i.gordon@maths.gla.ac.uk`

Enrico Gregorio
Univ. Verona, Italy
`Enrico.Gregorio@univr.it`

Pedro A. Guil Asensio
Univ. Murcia, Spain
`paguil@um.es`

Franz Halter-Koch
Univ. Graz, Austria
`franz.halterkoch@uni-graz.at`

Roozbeh Hazrat
Australian National Univ., Canberra, Australia
`hazrat@maths.anu.edu.au`

Angela Holtmann
Univ. Bielefeld, Germany
`aholtman@mathematik.uni-bielefeld.de`

François Huard
Bishop's Univ., Quebec, Canada
`fhuard@ubishops.ca`

Birgit Huber
Ludwig-Maximilians-Univ., Muenchen, Germany
bhuber@crm.es,
Birgit.Huber@mathematik.uni-muenchen.de

Andrew Hubery
Univ. Bielefeld, Germany
hubery@mathematik.uni-bielefeld.de

Laura Iglésias
Inst. Sup. Engenharia Lisboa, Portugal
laura@hermite.cii.fc.ul.pt

Yasuo Iwanaga
Shinshu University, Japan
iwanaga@gipnc.shinshu-u.ac.jp

Osamu Iyama
Himeji Inst.of Technology, Japan
iyama@sci.himeji-tech.ac.jp

Natalia Iyudu
Moscow State Univ., Russia
ioudu@mail.ru

Tibor Juhász
Univ. Debrecen, Hungary
juhaszti@math.klte.hu

El-Amin Kaidi
Univ. Almeria, Spain
elamin@ual.es

Algirdas Kaucikas
Vilnius Pedagogical Univ., Lithuania
algirdasXY@netscape.net

Elena Kireeva
Moscow Pedagogical State Univ., Russia
vvd@cityline.ru

Mark Kleiner
Syracuse Univ., Syracuse, USA
mkleiner@syr.edu

Steffen Koenig
Univ. Leicester, UK
sck5@mcs.le.ac.uk

Toshiko Koyama
Ochanomizu Univ., Tokyo, Japan
koyama@is.ocha.ac.jp

Thomas G. Kucera
Univ., Manitoba, Canada
tkucera@cc.umanitoba.ca

Miroslav Kures
Brno Univ. of Technology, Czech Republic
kures@um.fme.vutbr.cz

Dirk Kussin
Univ. Paderborn, Germany
dirk@math.upb.de

Thomas Laffey
University College Dublin, Ireland
Thomas.Laffey@ucd.ie

Piroska Lakatos
Univ. Debrecen, Hungary
lapi@math.klte.hu

Tsit-Yuen Lam
Univ. California, Berkeley, USA
lam@math.berkeley.edu

Zbigniew Leszczynski
Nicholas Copernicus Univ., Torun, Poland
zbles@mat.uni.torun.pl

Lawrence Levy
Univ. Nebraska, Lincoln and Wisconsin, Madison, USA
levy@math.wisc.edu

Christian Lomp
Univ. Porto, Portugal
clomp@fc.up.pt

Samuel A. Lopes
Univ. Porto, Portugal and Univ. Wisconsin-Madison, USA
lopes@math.wisc.edu

Laszlo Losonczi
Debrecen Univ., Hungary
losi@math.klte.hu

Wendy Tor Lowen
Free Univ. Brussels, Belgium
wlowen@vub.ac.be

Adolf Mader
Univ. Hawaii, USA
adolf@math.hawaii.edu

Dag Madsen
*Norwegian Univ. Technology and Science,
Norway*
dagma@math.ntnu.no

Mojgan Mahmoudi
Shahid Beheshti Univ., Tehran, Iran.
m-mahmoudi@cc.sbu.ac.ir

Peter Malcolmson
Wayne State Univ., Detroit, USA
petem@math.wayne.edu

Francesca Mantese
Univ. Verona, Italy
fmantese@math.unipd.it

Leandro Marín
Univ. Murcia, Spain
leandro@um.es

František Marko
Pennsylvania State Univ., USA
fxm13@psu.edu

Robert Marsh
Univ. Leicester, UK
rjm25@mcs.le.ac.uk

Leandro Martins
Univ. Beira Interior, Portugal
lmartins@noe.ubi.pt

Dragan Masulovic
Univ. Novi Sad, Serbia and Montenegro
masul@im.ns.ac.yu

Olivier Mathieu
Univ. Claude Bernard, Lyon, France
mathieu@desargues.univ-lyon1.fr

Ryszard Mazurek
Technical Univ. Bialystok, Poland
mazurek@pb.bialystok.pl

Ana Neto
Univ. Técnica Lisboa, Portugal
ananeto@ptmat.fc.ul.pt

Maria Teresa Nogueira
Univ. Lisboa, Portugal
tnogueir@cii.fc.ul.pt

Jan Okniński
Univ. Warsaw, Poland
okninski@mimuw.edu.pl

Gabriela Olteanu
*Babeş-Bolyai Univ., Cluj-Napoca,
Romania*
olteanu@math.ubbcluj.ro

Barbara Osofsky
Rutgers University, New Brunswick, USA
osofsky@math.rutgers.edu

Francesc Perera
Queen's Univ., Belfast, UK
perera@qub.ac.uk

Ngoc Ánh Pham
*Alfred Renyi Mathematical Inst.,
Budapest, Hungary*
anh@renyi.hu

Federico Raggi
UNAM, Mexico City, Mexico
fraggi@matem.unam.mx

Margarita Ramalho
Univ. Lisboa, Portugal
mbramalho@fc.ul.pt

Kulumani Rangaswamy
Univ.Colorado, Colorado Springs, USA
Ranga@math.uccs.edu

Robert Raphael
Concordia University, Montreal, Canada
raphael@alcor.concordia.ca

Tsetska Rashkova
Univ. Rousse "Angel Kanchev", Bulgaria
tsrashkova@ru.acad.bg

Igor Redchuk
*National Academy of Science, Kiev,
Ukraine*
red@imath.kiev.ua

Raquel Reis
Univ. Aberta, Portugal
raqreis@univ-ab.pt

Alexander Retakh
Massachusetts Inst. of Technology, USA
retakh@math.mit.edu

Claus M. Ringel
Univ. Bielefeld, Germany
ringel@mathematik.uni-bielefeld.de

José Ríos Montes
UNAM, Mexico City, Mexico
jrios@matem.unam.mx

Nicola Rodinò
Univ. Padova, Italy
rodino@math.unipd.it

Sonia L. Rueda
Univ. Almería, Spain
srueda@ual.es

Pavel Růzička
Karlova Univ., Praha, Czech Republic
ruzicka@karlin.mff.cuni.cz

Javier Sánchez Serdà
Univ. Autònoma Barcelona, Spain
jsanchez@mat.uab.es

Catarina Santa-Clara
Univ. Lisboa, Portugal
cgomes@ptmat.fc.ul.pt

Ana Paula Santana
Univ. Coimbra, Portugal
mamsanghare@hotmail.com

José Carlos Santos
Univ. Porto, Portugal
jcsantos@fc.up.pt

Manuel Saorín
Univ. Murcia, Spain
msaorinc@um.es

Jan Schroer
Univ. Leeds, UK
jschroer@maths.leeds.ac.uk

Maria Elisa Simões
Univ. Lisboa, Portugal
elisa.simoes@fc.ul.pt

Daniel Simson
*Nicholas Copernicus Univ., Torun,
Poland*
simson@mat.uni.torun.pl

Andrzej Skowroński
*Nicholas Copernicus Univ., Torun,
Poland*
skowron@mat.uni.torun.pl

Patrick F. Smith
Univ. Glasgow, UK
pfs@maths.gla.ac.uk

Lutz Strüngmann
Univ. Essen, Germany
lutz.struengmann@uni-essen.de

Peter Thompson
Univ. Glasgow, UK
pt@maths.gla.ac.uk

Noriko Tone
Tokyo Denki Univ., Japan
tone@cck.dendai.ac.jp

Alberto Tonolo
Univ. Padova, Italy
tonolo@math.unipd.it

Jan Trlifaj
Karlova Univ., Praha, Czech Republic
trlifaj@karlin.mff.cuni.cz

Hisaya Tsutsui
Millersville University, USA
Dr.T@math-sci.net

Mihail Ursul
Univ. Oradea, Romania
ursul@uoradea.ro

Peter Vamos
Univ. Exeter, UK
p.vamos@ex.ac.uk

Michel Van den Bergh
Limburgs Univ., Diepenbeek, Belgium
vdbergh@luc.ac.be

Lia Vas
Univ. Sciences Philadelphia, USA
l.vas@usip.edu

Indah Emilia Wijayanti
Gadjah Mada Univ., Jogjakarta,
Indonesia
indah@math2-uni.duesseldorf.de

Adrian Williams
Cornwall, UK
alwleon@hotmail.com

Anatoly Yakovlev
St. Petersburg State Univ., Russia
yakovlev@yak.pdmi.ras.ru

Paolo Zanardo
Univ. Padova, Italy
pzanardo@math.unipd.it

Jan Žemlička
Karlova Univ., Praha, Czech Republic
zemlicka@karlin.mff.cuni.cz

Milos Ziman
Comeniuus Univ., Bratislava, Slovakia
ziman@fmph.uniba.sk

Rita Zuazua
UNAM, Morelia, Mexico
zuazua@matmor.unam.mx

Grzegorz Zwara
Nicholas Copernicus Univ., Torun,
Poland
gzwara@mat.uni.torun.pl

CONTRIBUTORS

ALINA ALB
Mathematics Department, University of Oradea, Romania
dalb@uoradea.ro

A. J. CALDERÓN MARTÍN
Departamento de Matemáticas, Universidad de Cádiz, 11510 Puerto Real,
Cádiz, Spain
ajesus.calderon@uca.es

IULIU CRIVEI
Department of Mathematics, Technical University, Str. C. Daicoviciu 15,
400020 Cluj-Napoca, Romania
crivei@math.utcluj.ro

SEPTIMIU CRIVEI
Faculty of Mathematics and Computer Science, "Babeş-Bolyai" University,
Str. M. Kogălniceanu 1, 400084 Cluj-Napoca, Romania
crivei@math.ubbcluj.ro

ÁNGEL DEL RÍO
Departamento de Matemáticas, Universidad de Murcia, 30071 Murcia,
Spain
adelrio@fcu.um.es

ALBERTO FACCHINI
Dipartimento di Matematica Pura e Applicata, Università di Padova,
Via Belzoni 7, 35131 Padova, Italy
facchini@math.unipd.it

ALICE FIALOWSKI
Eötvös Loránd University, Department of Applied Analysis,
H-1117 Budapest, Pázmány P. sétány. 1/C, Hungary
fialowsk@cs.elte.hu

M. FORERO PIULESTÁN
Departamento de Matemáticas, Universidad de Cádiz, 11510 Puerto Real,
Cádiz, Spain
ForeroManuel@hotmail.com

JEREMY HAEFNER
Department of Mathematics, University of Colorado, Colorado Springs,
CO 60933, USA
haefner@math.uccs.edu

ERIC JESPERS
Department of Mathematics, Vrije Universiteit Brussel, Pleinlaan 2,
1050 Brussel, Belgium
efjesper@vub.ac.be

LEE KLINGLER
Mathematics Department, Florida Atlantic University, Boca Raton,
Florida 33431–0991, USA
klingler@fau.edu

T. Y. LAM
Department of Mathematics, University of California, Berkeley, Ca 94720
lam@math.berkeley.edu

LAWRENCE S. LEVY
Mathematics Department, University of Nebraska, Lincoln, NE 68588–
0323, USA
Mailing address: 2528 Van Hise Ave., Madison, WI 53705-3850, USA
levy@math.wisc.edu

PETER MALCOLMSON
Department of Mathematics, Wayne State University, Detroit, Michigan,
USA
petem@math.wayne.edu

FRANTIŠEK MARKO
Pennsylvania State University, 76 University Drive, Hazleton, PA 18202,
USA and
Mathematical Institute, Slovak Academy of Sciences, Štefánikova 49,
814 38 Bratislava, Slovakia
fxm13@psu.edu

CÁNDIDO MARTÍN GONZÁLEZ
Departamento de Algebra, Geometría y Topología, Universidad de Málaga, Apartado 59, 29080 Málaga, Spain
candido@apncs.cie.uma.es

IZURU MORI
Department of Mathematics, Syracuse University, Syracuse, NY 13244, USA
imori@syr.edu

JAN OKNIŃSKI
Institute of Mathematics, Warsaw University, Banacha 2, 02-097 Warsaw, Poland
okninski@mimuw.edu.pl

FRANK OKOH
Department of Mathematics, Wayne State University, Detroit, Michigan, USA
okoh@math.wayne.edu

BARBARA L. OSOFSKY
Department of Mathematics, Rutgers University, 110 Frelinghuysen Road, Piscataway, NJ 08854-8019, USA
osofsky@math.rutgers.edu

MOHAMED OUKESSOU
Département de Mathématiques, Faculté des Sciences et Techniques, BP523 Beni-Mellal, Maroc
ouk_mohamed@yahoo.fr

TSETSKA GRIGOROVA RASHKOVA
Centre of Mathematics and Informatics, University of Rousse "A. Kanchev", 7017 Rousse, Bulgaria
tcetcka@ami.ru.acad.bg

ALEXANDER RETAKH
Department of Mathematics, Massachusetts Institute of Technology, Cambridge, MA 02139
retakh@math.mit.edu

SONIA L. RUEDA
Departamento de Matemáticas. E.T.S. Arquitectura, Universidad Politécnica de Madrid, Madrid, Spain
srueda@aq.upm.es

xxx

José Carlos de Sousa Oliveira Santos
Departamento de Matemática Pura, Rua do Campo Alegre, 687, 4169–
007 Porto, Portugal
jcsantos@fc.up.pt

Patrick F. Smith
Department of Mathematics, University of Glasgow, Glasgow, G12 8QW,
Scotland, UK
pfs@maths.gla.ac.uk

M. Ursul
Department of Mathematics, University of Oradea, Armatei Române 5,
Oradea, România
ursul@uoradea.ro

SOME COREFLECTIVE CATEGORIES OF TOPOLOGICAL MODULES

ALINA ALB

Mathematics Department
University of Oradea, Romania
E-mail: dalb@uoradea.ro

There are essential differences between the theory of topological algebraic systems with a discrete signature and the theory of topological algebraic systems with a continuous signature (see [C]). A topological module is a classic example of a topological algebraic system with a continuous signature.

It is well known that the class of all topological P-groups (P-rings) is a coreflective subcategory of the category of all topological groups (topological rings). For some results in coreflective subcategories of the category of topological abelian groups see in [GT], [HH].

In this context Professor M. Choban posed the following question: Is the class of all P-modules of *TopMod* a coreflective subcategory of *TopMod*?

We give in this paper a partial answer to this question; the general case remains open.

Notation

All topological rings are assumed associative, Hausdorff and with identity. All topological modules are assumed unitary and Hausdorff. For a topological ring R with identity *TopMod* denotes the category of all topological left R-modules. If M is a left R-module, $X \subseteq R$ and $Y \subseteq M$, set $X \cdot Y = \{rm : r \in X, m \in Y\}$.

Recall that a subcategory $\mathfrak{B}$ of a category $\mathfrak{A}$ is said to be *coreflective* provided for each $X \in \mathfrak{A}$ there is $X' \in \mathfrak{B}$ and a morphism $c_X : X' \to X$ such that for each $Y \in \mathfrak{B}$ and each morphism $\alpha : Y \to X$ there is a unique morphism $\beta : Y \to X'$ in $\mathfrak{B}$ such that the following diagram is

commutative,

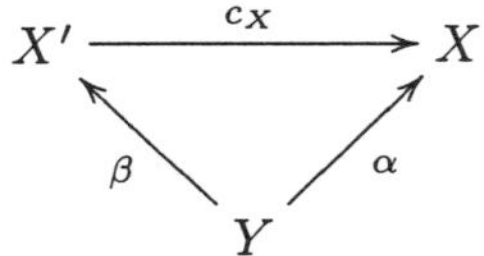

i.e., $\alpha = c_X \circ \beta$.

We will introduce a new cardinal invariant for topological rings.

Definition 1. Let R be a topological ring. Denote by $h(R)$ the minimal infinite regular cardinal number for which there exists a neighborhood V of zero which can be represented as a reunion of a family of cardinality $< h(R)$ of compact subsets.

Remark 2. We note that if $\mathfrak{m}$ is an arbitrary infinite cardinal number then there exist regular cardinal numbers $\mathfrak{m}'$ such that $\mathfrak{m} \leq \mathfrak{m}'$.

Indeed, let $\mathfrak{m} = \aleph_\alpha$, where α is an ordinal. Then $\aleph_\alpha < \aleph_{\alpha+1}$ and $\aleph_{\alpha+1}$ is a regular cardinal number. We obtain that $h(R)$ is defined for any topological ring.

Remark 3. For a topological ring R, $h(R) = \aleph_0$ if and only if R is locally compact.

Definition 4. A topological ring R is called locally σ-compact provided $h(R) = \aleph_1$.

Being locally σ-compact means that there exists a base of neighborhoods of zero consisting of subsets which are σ-compact, i.e., countable unions of compact sets.

Definition 5 (see, e.g., [AGM], p. 275). Let $\mathfrak{m}$ be an infinite cardinal. We will say that a topological space $(X, \mathcal{T})$ is a $\mathfrak{m}$-*space* provided any intersection of a collection of cardinality $< \mathfrak{m}$ of open subsets is open.

We recall the following construction from general topology: Let $(X, \mathcal{T})$ be a topological space and $\mathfrak{m}$ an infinite regular cardinal number. Consider the topology $\mathcal{T}_\mathfrak{m}$ on X having a base consisting of intersections of collections of cardinality $< \mathfrak{m}$ of open sets of $(X, \mathcal{T})$. Then $\mathcal{T}_\mathfrak{m} \geq \mathcal{T}$ and $\mathcal{T}_\mathfrak{m}$ is called the $\mathfrak{m}$-*modification* of $\mathcal{T}$.

If $(X, \mathcal{T})$ is a topological group (ring), then $(X, \mathcal{T}_\mathfrak{m})$ is a topological group (ring), too.

We will outline the proof of this assertion only for topological groups (for topological rings the proof is analogous): Let $x, y \in X$ and $xy^{-1} \in$

$W \in \mathcal{T}_{\mathfrak{m}}$. There exists a collection $\{W_i\}_{i<\mathfrak{m}}$ of open sets of $(X, \mathcal{T})$ such that $W = \cap_{i<\mathfrak{m}} W_i$. Choose for each $i < \mathfrak{m}$, U_i, $V_i \in \mathcal{T}$ such that $x \in U_i$, $y \in V_i$ and $U_i \cdot V_i^{-1} \subseteq W_i$. Set $U = \cap_{i<\mathfrak{m}} U_i$ and $V = \cap_{i<\mathfrak{m}} V_i$ Then $x \in U \in \mathcal{T}_{\mathfrak{m}}$, $y \in V \in \mathcal{T}_{\mathfrak{m}}$ and $U \cdot V^{-1} \subseteq W$.

An important case is when $\mathfrak{m} = \aleph_1$. The spaces $(X, \mathcal{T}_{\aleph_1})$ and $(X, \mathcal{T})$ have the same collections of G_δ-sets.

Recall that a topological space $(X, \mathcal{T})$ is called a *P-space* provided $\mathcal{T} = \mathcal{T}_{\aleph_1}$ ([GJ], p. 62–63). A topological module or ring whose underlying topological space is a P-space is called briefly a *P-module*, or a *P-ring*, respectively.

Lemma 6. *Let $M \in TopMod$ and K be a compact subset of R. Then for every neighborhood W of zero there exists a neighborhood U of zero such that $K \cdot U \subseteq W$.*

Proof. For every $k \in K$ there exist a neighborhood W_k of k and a neighborhood $U^{(k)}$ of zero such that $W_k \cdot U^{(k)} \subseteq W$. By compactness of K there exist $k_0, \ldots, k_n \in K$ such that $K \subseteq W_{k_0} \cup \cdots \cup W_{k_n}$. Then, evidently, $K \cdot U \subseteq W$, where $U = U^{(k_0)} \cap \cdots \cap U^{(k_n)}$. $\square$

Lemma 7. *Let R be a topological ring and $h(R) \leq \mathfrak{m}$, where $\mathfrak{m}$ is an infinite cardinal number. Let $(M, \mathcal{T}) \in TopMod$ and V be a 0_R-neighborhood, $V = \cup_{\alpha \in \Omega} K_\alpha$, $|\Omega| < \mathfrak{m}$, where each K_α is a compact subset. Then for every 0-neighborhood W of $(M, \mathcal{T}_{\mathfrak{m}})$ there exists a 0-neighborhood U of $(M, \mathcal{T}_{\mathfrak{m}})$ such that $V \cdot U \subseteq W$.*

Proof. We can assume without loss of generality that $W = \cap_{\alpha \in \Omega} W_\alpha$, where each W_α is a 0-neighborhood of $(M, \mathcal{T})$. Fix $\alpha \in \Omega$; then by Lemma 6 for every $\beta \in \Omega$ there exists a 0-neighborhood H_β of $(M, \mathcal{T})$ such that $K_\beta \cdot H_\beta \subseteq W_\alpha$. Put $H^{(\alpha)} = \cap_{\beta \in \Omega} H_\beta$; then $V \cdot H^{(\alpha)} \subseteq W_\alpha$. It follows that $V \cdot (\cap_{\alpha \in \Omega} H^{(\alpha)}) \subseteq \cap_{\alpha \in \Omega} W_\alpha$. Then $U = \cap_{\alpha \in \Omega} H^{(\alpha)}$ is an intersection of a collection of cardinality $< \mathfrak{m}$ of 0-neighborhoods for $(M, \mathcal{T})$, hence it is 0-neighborhood of $(M, \mathcal{T}_{\mathfrak{m}})$. $\square$

Lemma 8. *Let R be a topological ring, $h(R) \leq \mathfrak{m}$, where $\mathfrak{m}$ is an infinite regular cardinal number $\geq \aleph_1$ and $M \in TopMod$ is such that the underlying topological space of M is a $\mathfrak{m}$-space. Then for each $m \in M$ there exists a 0_R-neighborhood V such that $Vm = 0$.*

Proof. Let $m \in M$ and V_0 be a 0_R-neighborhood such that $V_0 = \cup_{\alpha \in \Omega} K_\alpha$, $|\Omega| < \mathfrak{m}$, where K_α are compact subspaces. Then for each $\alpha \in \Omega$, $K_\alpha m \subseteq$

M is a compact subspace of M. Since $\mathfrak{m} \geq \aleph_1$, the topological space M is a P-space. Therefore each $K_\alpha m$ is finite ([GJ], Exercise 4K, 1, p. 63).

It follows that $|V_0 m| < \mathfrak{m}$. Fix an enumeration $\{x_\alpha | \alpha \in \Omega\}$ of $V_0 m \backslash \{0\}$ (it is not assumed that $\alpha \neq \beta$ implies $x_\alpha \neq x_\beta$). For each x_α there exists a 0_M-neighborhood U_α such that $x_\alpha \notin U_\alpha$. Then $U = \cap_{\alpha \in \Omega} U_\alpha$ is a 0_M-neighborhood and $V_0 m \cap U = 0$. There exists a 0_R-neighborhood V such that $V \subseteq V_0$, $Vm \subseteq U$. It follows that $Vm \subseteq V_0 m \cap U = 0$, hence $Vm = 0$. $\qquad\qquad\square$

Remark 9. The element x of a topological R-module M for which there exists a neighborhood V of 0 in R such that $Vx = 0$ are precisely those having open annihilator (or left annihilator if the module M is the ring R itself).

Theorem 10. *Let R be a topological ring and $h(R) \leq \mathfrak{m}$, where $\mathfrak{m}$ is a regular cardinal number $\geq \aleph_1$. Then the subcategory of TopMod consisting of all topological R-modules whose topology is a $\mathfrak{m}$-topology is coreflective.*

Proof. Let $(M, \mathcal{T}) \in$ *TopMod*. Consider the submodule M' of M consisting of elements with open annihilator.

We affirm that $(M', \mathcal{T}_\mathfrak{m}|M')$ is a topological R-module.

Obviously, $\mathcal{T}_\mathfrak{m}$ is a group topology on M. By Lemma 7 the mapping $R \times M' \to M'$, $(r, m') \mapsto rm'$ is continuous at $(0,0)$ with respect to the topology $\mathcal{T}_\mathfrak{m}|M'$. By definition of M', for each $m' \in M'$ the mapping $R \to M'$, $r \mapsto rm'$ is continuous with respect to the topology $\mathcal{T}_\mathfrak{m}|M'$. It is obviously that for each $r \in R$ the mapping $M' \to M'$, $m' \mapsto rm'$ is continuous with respect to the topology $\mathcal{T}_\mathfrak{m}|M'$. Therefore $(M', \mathcal{T}_\mathfrak{m}|M')$ is a topological R-module.

Denote by $i : M' \to M$, $i(m') = m'$, $m' \in M'$. Let $(M'', \mathcal{T}'')$ be a R-module whose topology is a $\mathfrak{m}$-topology and $\alpha : M'' \to M$ a continuous homomorphism. We affirm that $\alpha(M'') \subseteq M'$.

By Lemma 8 for each $m'' \in M''$ there exists a 0_R-neighborhood V' such that $V'm'' = 0$. Then $V'\alpha(m'') = \alpha(V'm'') = 0$, i.e., $\alpha(m'') \in M'$.

Put $\widehat{\alpha}(m'') = \alpha(m'')$ for any $m'' \in M''$. Then $\widehat{\alpha}$ is a continuous homomorphism of M'' in M and $\alpha = i \circ \widehat{\alpha}$.

The uniqueness of $\widehat{\alpha}$ is evident. $\qquad\qquad\square$

Corollary 11. *Let R be a fixed locally σ-compact ring. Then the subcategory of all P-modules is coreflective in the category of all topological R-modules.*

Remark 12. If R is a P-ring, then the subcategory of *TopMod* consisting of all P-modules over R is coreflective.

Indeed, if $(M, \mathcal{T}) \in$ *TopMod*, consider the abelian topological group $(M', \mathcal{T}^\delta)$, where $M' = M$ and $\mathcal{T}^\delta$ is the $\aleph_1$-modification of $\mathcal{T}$. Since R is a P-ring, $\mathcal{T}^\delta$ is a R-module topology. Consider the mapping $i : M' \to M$, $m' \mapsto m'$. We affirm that $(M', \mathcal{T}^\delta)$ is the coreflection of $(M, \mathcal{T})$ in the category of all P-modules. Indeed, let $(M'', \mathcal{T}'')$ be a P-module and $\alpha : M'' \to M$ a continuous homomorphism. Put $\widehat{\alpha} : M'' \to M'$, $m'' \mapsto \alpha(m'')$. Then $\widehat{\alpha}$ is a continuous homomorphism of M'' in M and $\alpha = i \circ \widehat{\alpha}$. The uniqueness of $\widehat{\alpha}$ is evidently.

Remark 13 ([GJ], Exercise 4K, 8, p. 63). Every P-space X is zero-dimensional, i.e., has a base consisting closed and open subsets.

We note that if R is a connected topological ring and $_R M = M$ is a P-module, then $M = 0$. Indeed, by Remark 13, M is zero-dimensional. If $x \in M$, then $Rx \subseteq M$ and Rx is connected. Therefore $Rx = 0$ and so $M = 0$.

We obtained that if R is a connected topological ring then the subcategory of *TopMod* consisting of all P-modules is coreflective.

We noted that the $\aleph_1$-modification of a topology of a topological group (ring) is a group (ring) topology. We give here examples of topological modules $(M, \mathcal{T})$ for which the $\aleph_1$-modification of $\mathcal{T}$ is not a module topology:

Example 14. Let R be any nonzero connected topological ring. Consider R as a left topological R-module with the multiplication as a module operation. Then the $\aleph_1$-modification is not a R-module topology.

Another example of this kind is the following.

Example 15. Let p be any prime number and $\mathbb{Z}_p$ be the ring of p-adic integers with the natural compact topology $\mathcal{T}$, $M = \mathbb{Z}_p$. Then the pair $((\mathbb{Z}_p, \mathcal{T}), (M, \mathcal{T}))$ is a compact left $\mathbb{Z}_p$-module. The $\aleph_1$-modification of $\mathcal{T}$ is the discrete topology $\mathcal{T}_d$ on M. Obviously, the pair $((\mathbb{Z}_p, \mathcal{T}), (M, \mathcal{T}_d))$ is not a topological $\mathbb{Z}_p$-module.

We give here an example of a non-discrete P-module over a compact ring.

We will identify the set of all natural numbers with all ordinals $< \omega$, where ω is the first infinite ordinal. As usual, ω_1 denotes the first uncountable ordinal.

Example 16. Let $\mathbb{F}_2$ be the field consisting of two elements and $R = \mathbb{F}_2^\omega$ be the topological product of ω copies of $\mathbb{F}_2$. Put S the subset of R consisting of elements with open annihilator.

Claim. S is a dense ideal of R.

Indeed, let $r_1, r_2 \in S$, then there exist neighborhoods V_1, V_2 of zero such that $V_1 r_1 = 0$, $V_2 r_2 = 0$. Therefore $(V_1 \cap V_2)(r_1 - r_2) = 0 \to r_1 - r_2 \in S$. Let $r \in S$ and $m \in R$, then there exists V a neighborhood of zero such that $Vr = 0$. We have $Vrm = 0 \to rm \in S$. Evidently, $S \supseteq \oplus_{i \in \omega} (\mathbb{F}_2)_i$, hence S is dense.

Now, we consider the group $M = \oplus_{\alpha \in \omega_1} S_\alpha$, where $S_\alpha = S$ (a direct sum of ω_1 copies of R-module S).

For any $\beta \in \omega_1$ put $M_\beta = \{x : x \in M, pr_\alpha(x) = 0 \text{ for every } \alpha \leq \beta\}$. Then $M = M_0 \supseteq M_1 \supseteq \cdots \supseteq M_\alpha \supseteq M_{\alpha+1} \supseteq \cdots$, and $\cap_{\alpha \in \omega_1} M_\alpha = 0$.

Evidently, the family $\{M_\alpha\}_{\alpha \in \omega_1}$ gives a group topology $\mathcal{T}$ on M and that $(M, \mathcal{T})$ is a P-space.

We note that $(M, \mathcal{T})$ is a topological R-module.

Indeed, $RM_\alpha \subseteq M_\alpha$ for each $\alpha \in \omega_1$. Let $m \in M$; then there exists a neighborhood V of zero of R such that $Vm = 0$. We have proved that $(M, \mathcal{T})$ is a topological R-module.

Acknowledgment

The author is grateful to professors Mitrofan Choban and Mihail Ursul for their constant interest and valuable indications. I am also grateful to the referee for her/his very careful reading and substantial helpful suggestions.

References

AGM. V. I. Arnautov, S. T. Glavatsky, A. V. Mikhalev, Introduction to the theory of topological rings and modules, Marcel Dekker Inc., 1996.

C. M. M. Choban, On the theory of topological algebraic systems, Trans. Moscow. Math. Soc., 1986, 115–159.

GJ. L. Gillman, M. Jerison, Rings of Continuous Functions, Van Nostrand, 1960.

GT. M. L. Gramellini, A. Tozzi, Final topological groups and coreflections, Rend. Mat. et Appl., 1(4) (1981), 139–145.

HH. H. Herrlich, M. Husek, Productivity of coreflective classes of topological groups, Comment. Math. Univ. Carolinae, 40 (1999), 551–560.

INHERITANCE OF PRIMENESS BY IDEALS IN LIE TRIPLE SYSTEMS

A. J. CALDERÓN MARTÍN AND M. FORERO PIULESTÁN*

Departamento de Matemáticas
Universidad de Cádiz
11510 Puerto Real, Cádiz (Spain)
E-mail: ajesus.calderon@uca.es
E-mail: ForeroManuel@hotmail.com

We show that semiprime ideals in a Lie triple system inherit primeness. As a tool, we also study the transfer of regularity conditions between a Lie triple system T and its standard algebra envelope L, stating that T is prime (resp. semiprime, with zero annihilator) if and only if L is gr-prime (resp. gr-semiprime, with zero annihilator).

1. Introduction

1.1. The inheritance of semiprimeness and primeness by ideals in associative algebras is an easy consequence of Andrunakievich's Lemma, which is based in the way ideals are generated in associative algebras. The generation of ideals in Jordan and Lie algebras require more complex calculations and, in fact, the Jordan version of Andrunakievich's Lemma is false even for linear Jordan algebras (cf. [6]) and unknown for Lie algebras. Thus, the study of the inheritance of regularity by ideals in Jordan and Lie algebras requires the use of different techniques. In 1984 McCrimmon began the study of ideals in strongly prime Jordan systems (see [4]). Subsequently, he improved this work in [5] where the hypothesis of nondegenerancy was removed and only semiprimeness of the ideals was required for the inheritance of primeness. To do so, he introduced the notion of the eventual annihilator of an ideal I as the union of the annihilators of the powers of I. The eventual annihilator turns out to be an ideal and plays a key role in

*The authors were supported by the PCI of the UCA 'Teoría de Lie y Teoría de Espacios de Banach' and by the PAI of the Spanish Junta de Andalucía with project number FQM-298

8

McCrimmon's work. By following McCrimmon's arguments E. García got in [2] an analogue in Lie algebras. In the framework of Lie triple systems, the eventual annihilator of an ideal J of an ideal I of a Lie triple system T is just a left ideal of T and therefore we need to refine McCrimmon's arguments in the study of the inheritance of regularity by ideals in Lie triple systems. The key tool to do this shall be the study of the behaviour of an adequate ideal of the standard algebra envelope of T with respect to certain eventual annihilator.

1.2. In section 3, we characterize regularity conditions on a Lie triple system T through its standard algebra envelope L, by showing that T is prime (resp. semiprime, with zero annihilator) if and only if L is gr-prime (resp. gr-semiprime, with zero annihilator).

2. Preliminaries

2.1. Let $\mathbb{K}$ be a field of characteristic not two and let T be a vector space over $\mathbb{K}$. We say that T is a *triple system* if it is endowed with a trilinear map

$$\langle \cdot, \cdot, \cdot \rangle : T \times T \times T \to T,$$

called the *triple product* of T.

A triple system T is called a *Lie triple system* if its triple product, denoted by $[\cdot, \cdot, \cdot]$, satisfies

(1) $[x, x, y] = 0$
(2) $[x, y, z] + [y, z, x] + [z, x, y] = 0$ (Jacobi identity)
(3) $[x, y, [a, b, c]] - [a, b, [x, y, c]] = [[x, y, a], b, c] + [a, [x, y, b], c]$

for any $x, y, z, a, b, c \in T$.

Given a Lie algebra L with product $[\cdot, \cdot]$, a Lie triple system called the *underlying triple system* of L can be defined on the linear structure of L by taking the triple product given by $[x, y, z] = [[x, y], z]$.

2.2. A *left ideal* of a Lie triple system T is a subspace I for which $[T, T, I] \subseteq I$. If the subspace I satisfies $[I, T, T] \subseteq I$, we shall say that I is an *ideal* of T. Notice that $[I, T, T] \subseteq I$ implies that $[T, I, T] \subseteq I$ and $[T, T, I] \subseteq I$. A Lie triple system T is called *semiprime* if, for an ideal I of T, $[I, T, I] = 0$ implies $I = 0$. We say that T is *prime* if, for any ideals I, J of T, $[I, T, J] + [J, T, I] = 0$ implies either $I = 0$ or $J = 0$. Notice that $[I, T, J] + [J, T, I] = 0$ implies $[I, J, T] + [J, I, T] + [T, I, J] + [T, J, I] = 0$ by 2.1.1 and 2.1.2. Clearly, any prime Lie triple system is semiprime.

2.3. We recall that the *annihilator* of a subset S of a Lie triple system T is defined as the set of elements x in T such that $[x, S, T] + [T, S, x] = 0$ and is denoted by $\mathrm{Ann}_T(S)$. Notice that the definition also implies $[S, x, T] + [S, T, x] + [x, T, S] + [T, x, S] = 0$ by 2.1.1 and 2.1.2. It is not difficult to check that $\mathrm{Ann}_T(S)$ is an ideal of T when S is an ideal of T, and that a Lie triple system T is prime if and only if the annihilator of every nonzero ideal of T is zero.

2.4. A *two-graded Lie $\mathbb{K}$-algebra* L is a *Lie $\mathbb{K}$-algebra* which splits into the direct sum $L = L_0 \oplus L_1$ of subspaces (called the even and the odd part respectively) satisfying $[L_\alpha, L_\beta] \subset L_{\alpha+\beta}$ for any α, β in $\mathbb{Z}_2$. A *graded ideal*, (gr-ideal), of a two-graded algebra $L = L_0 \oplus L_1$ is an ideal I of L such that there exist two subspaces I_0 and I_1 with $I = I_0 \oplus I_1$ and $I_\alpha \subseteq L_\alpha$ for any $\alpha \in \mathbb{Z}_2$. A two-graded algebra L is *graded-semiprime*, (gr-semiprime), if $[I, I] = 0$ implies $I = 0$ for any gr-ideal I of L, and L is gr-prime if $[I, J] = 0$ implies either $I = 0$ or $J = 0$ for any gr-ideals I, J of L.

2.5. The *annihilator* of a subset S of a two-graded Lie algebra L is defined as $\mathrm{Ann}_L(S) := \{x \in L : [x, S] = 0\}$. As in 2.3., it is not difficult to check that $\mathrm{Ann}_L(S)$ is a gr-ideal of L when S is a gr-ideal of L, and that L is gr-prime if and only if the annihilator of every nonzero gr-ideal of L is zero. We point out that the annihilator is just the classical centralizer of a Lie algebra, i.e., $\mathrm{Ann}_L(S) = C_L(S)$.

2.6. The *standard algebra envelope* of a Lie triple system T, (see for instance [3]), is the two-graded Lie algebra $L = L_0 \oplus L_1$, L_0 being the $\mathbb{K}$-span of $\{\mathcal{L}(x, y) : x, y \in T\}$, where $\mathcal{L}(x, y)$ denotes the left multiplication operator in T, $\mathcal{L}(x, y)(z) := [x, y, z]$; $L_1 := T$ and where the product is given by

$$[(\mathcal{L}(x, y), z), (\mathcal{L}(u, v), w)] :=$$
$$(\mathcal{L}([u, v, y], x) - \mathcal{L}([u, v, x], y) + \mathcal{L}(z, w), [x, y, w] - [u, v, z]).$$

Notice that L_0 is a subalgebra of L, while T is just L_1 as a subtriple system of the underlying triple system of L.

2.7. Let us observe that if T denotes a Lie triple system and $L = L_0 \oplus L_1$ its standard algebra envelope, x_0 in L_0 belongs to $\mathrm{Ann}_L(T)$ if and only if $x_0 = 0$.

2.8. From now on, given $A, B \subset L$, where L denotes a Lie algebra, we shall denote by $[A, B]$ the $\mathbb{K}$-linear span of the set $\{[a, b] : a \in A, b \in B\}$.

3. Characterization of Regularity Conditions on Lie Triple Systems through the Standard Algebra Envelope

The first result on this subject was given by Lister in [3, Theorem 2.13], who proved that a finite dimensional Lie triple system over an algebraically closed field of characteristic zero is simple if and only if its standard algebra envelope is gr-simple. By arguing in the same way, we showed in [1, Proposition 2.1] that Lister's result holds in arbitrary dimension. The following theorem states analogous results concerning primeness, semiprimeness and annihilators.

Theorem 3.1. *Let T be a Lie triple system and let $L = L_0 \oplus L_1$ be its standard algebra envelope. Then the following assertions hold*

(1) T is prime if and only if L is gr-prime.

(2) T is semiprime if and only if L is gr-semiprime.

(3) $\mathrm{Ann}_T(T) = \mathrm{Ann}_L(L)$.

Proof. 1. Let us suppose T is prime and let $I = I_0 \oplus I_1$, $J = J_0 \oplus J_1$ be two gr-ideals of L satisfying $[I, J] = 0$. Let us observe that I_1, (and J_1), is an ideal of T, since by gradedness $[I_1, T, T] = [[I_1, T], T] \subset [I_0, T] \subset I_1$. Let us also observe that $[I_1, T, J_1] = 0$ (and similarly $[J_1, T, I_1] = 0$). Indeed, $[I_1, T, J_1] = [[I_1, T], J_1] \subset [I_0, J_1] = 0$. Primeness of T gives either $I_1 = 0$ or $J_1 = 0$. In the first case, $I = I_0$ and then $[I, T] = [I_0, T] \subset I_1 = 0$, by 2.7 we conclude $I_0 = 0$ and therefore $I = 0$. In the second case we similarly obtain $J = 0$. Hence L is gr-prime.

Conversely, let us now suppose L is gr-prime and let I, J be two ideals of T satisfying $[I, T, J] + [J, T, I] = 0$. It is easy to check that $[I, T] \oplus I$ and $[J, T] \oplus J$ are gr-ideals of L. We assert $[[I, T] \oplus I, [J, T] \oplus J] = 0$. Indeed, for any $x \in [I, T] \oplus I$ we have $x = (\sum_{i=1}^{n_x} \mathcal{L}(x_i, x_i'), z)$ with $z, x_i \in I$, $x_i' \in T$ for $i = 1, \ldots, n_x$, and for any $y \in [J, T] \oplus J$ we can write $y = (\sum_{j=1}^{n_y} \mathcal{L}(y_j, y_j'), t)$ with $t, y_j \in J$, $y_j' \in T$ for $j = 1, \ldots, n_y$. Then

$$[x, y] = \left(\left[\sum_{i=1}^{n_x} \mathcal{L}(x_i, x_i'), \sum_{j=1}^{n_y} \mathcal{L}(y_j, y_j') \right] + \mathcal{L}(z, t), \right.$$
$$\left. \left[\sum_{i=1}^{n_x} \mathcal{L}(x_i, x_i'), t \right] + \left[z, \sum_{j=1}^{n_y} \mathcal{L}(y_j, y_j') \right] \right). \tag{1}$$

Let us study the following products in equation (1):

$$[\sum_{i=1}^{n_x} \mathcal{L}(x_i, x_i'), \sum_{j=1}^{n_y} \mathcal{L}(y_j, y_j')] = \sum_{i,j=1}^{i=n_x, j=n_y} [\mathcal{L}(x_i, x_i'), \mathcal{L}(y_j, y_j')]$$

$$= \sum_{i,j=1}^{i=n_x, j=n_y} (\mathcal{L}([y_j, y_j', x_i'], x_i) - \mathcal{L}([y_j, y_j', x_i], x_i')). \quad (2)$$

We have, for any $u \in T$, $i \in \{1, \ldots, n_x\}$ and $j \in \{1, \ldots, n_y\}$,

$$\mathcal{L}([y_j, y_j', x_i'], x_i)(u) := [[y_j, y_j', x_i'], x_i, u] \in [J, I, T] = 0.$$

Last equality follows from

$$[J, I, T] \subseteq [T, J, I] + [I, T, J] \overset{(by\ 2.1.2)}{\subseteq} [J, T, I] + [I, T, J] \overset{(by\ 2.1.1)}{=} 0.$$

In the same way we obtain $\mathcal{L}([y_j, y_j', x_i], x_i') = 0$ and so (2)=0.

We also have $\mathcal{L}(z, t)(u) = [z, t, u] \in [I, J, T] = 0$ for any $u \in T$ and therefore $\mathcal{L}(z, t) = 0$.

Finally, $[\mathcal{L}(x_i, x_i'), t] := [x_i, x_i', t] \in [I, T, J] = 0$ and

$$[z, \mathcal{L}(y_j, y_j')] = -[y_j, y_j', z] \in [J, T, I] = 0$$

for any $i \in \{1, \ldots, n_x\}$ and $j \in \{1, \ldots, n_y\}$.

We conclude (1)=0, hence either $[I, T] \oplus I = 0$ or $[J, T] \oplus J = 0$ by gr-primeness of L. Therefore T is prime.

2. We just need to repeat the proof of 1 for $J = I$.

3. We claim

$$\mathrm{Ann}_T(T) \subset \mathrm{Ann}_L(L). \quad (3)$$

Indeed, let $x \in \mathrm{Ann}_T(T)$. Since $[x, T, T] = 0$, (and so $[T, x, T] = [T, T, x] = 0$), $[[x, L_1], L_1] = 0$, but $[x, L_1]$ is a subset of L_0 and thus 2.7 yields $[x, L_1] = 0$. Since we also have for any $y \in L_0$, ($y = \sum[z_i, t_i]$, $z_i, t_i \in T$), that $[x, y] = -\sum[z_i, t_i, x] = 0$ we conclude (3).

If we now take $x = x_0 + x_1 \in L$ such that $x \in \mathrm{Ann}_L(L)$, we have by gradedness that $[x_1, T] = [x_0, T] = 0$. These facts respectively imply $x_1 \in \mathrm{Ann}_T(T)$ and, by 2.7, $x_0 = 0$. Hence

$$\mathrm{Ann}_L(L) \subset \mathrm{Ann}_T(T). \quad (4)$$

Equations (3) and (4) complete the proof of 3. $\qquad\square$

4. Inheritance of Primeness by Ideals

4.1. The key in the study of the inheritance of primeness by an ideal I of a Lie triple system T shall be the study of the gr-ideal of its standard algebra envelope $L_I := [T, I] \oplus I$, and its behaviour respect to certain eventual annihilators.

Proposition 4.1. *Let T be a prime Lie triple system, let $L = L_0 \oplus L_1$ be its standard algebra envelope and let I be a semiprime ideal of T. Then $L_I := [T, I] \oplus I$ is a gr-semiprime ideal of L.*

Proof. As in 3.1, L_I is a graded ideal of L. Let $J = J_0 \oplus J_1$ be now a gr-ideal of L_I satisfying $[J, J] = 0$. By gradedness,

$$[[J_0, I], I, I] = [[[J_0, I], I], I] \subset [[J_1, I], I] \subset [J_0, I]$$

and so $[J_0, I]$ is an ideal of I. Since

$$[[J_0, I], I, [J_0, I]] \subset [[[J_0, I], I], [J_0, I]] \subset [[J_1, I], [J_0, I]]$$
$$\subset [J_0, J_1] \subset [J, J] = 0,$$

semiprimeness of I gives us

$$[J_0, I] = 0. \tag{5}$$

We also have

$$[J_0, T] \subset [[T, I], T] \subset [T, I, T] \subset I. \tag{6}$$

By the Jacobi identity and (5)

$$[[J_0, T], I] \subset [[T, I], J_0] + [[I, J_0], T] \subset [[T, I], J_0] \subset J_0. \tag{7}$$

By applying (7) and (5) we get

$$[[J_0, T], I, I] = [[[J_0, T], I], I] \subset [J_0, I] = 0. \tag{8}$$

By (6) and (8), $[J_0, T]$ is an ideal of I. By applying again (6) and (8),

$$[[J_0, T], I, [J_0, T]] = 0.$$

Therefore the semiprimeness of I gives $[J_0, T] = 0$ and so $J_0 = 0$ by 2.7. We then have $J = J_1$. Thus, $[J, I] \subset J_0 = 0$ and so $[J, I, I] = 0$, hence $[J, I, J] = 0$. By applying again the semiprimeness of I, we obtain $J = 0$. $\qquad\square$

By arguing as in [2], we can state the following.

Theorem 4.2. *Let L be a gr-prime two-graded Lie algebra and let I be a gr-semiprime ideal of L. Then I is gr-prime.*

Corollary 4.3. *Let T be a prime Lie triple system, let $L = L_0 \oplus L_1$ be its standard algebra envelope and let I be a semiprime ideal of T. Then $L_I := [T, I] \oplus I$ is a gr-prime ideal of L.*

Proof. Apply 3.1.1, 4.1, and 4.2. $\qquad\qquad\qquad\qquad\qquad\qquad\qquad\square$

4.2. Let I be an ideal of T. We define the *eventual annihilator* of I in T as

$$\mathrm{Evann}_T(I) := \bigcup_{n \geq 1} \mathrm{Ann}_T(I^n),$$

where the I^n is defined inductively by

$$\begin{aligned}
I^1 &= I \\
I^2 &= [T, I, I] \\
I^3 &= [T, I^2, I^2] \\
I^n &= [T, I^{n-1}, I^{n-1}], \qquad n \geq 2
\end{aligned}$$

Following the proof of [3, p. 220] for the finite dimensional case, we can prove, in a general setting, the following.

Lemma 4.4. *For any ideal I of T and $n \geq 1$ we have I^n is also an ideal of T that satisfies $I^n \subset I^{n-1} \subset \cdots \subset I^2 \subset I$.*

Lemma 4.5. *Let T be a Lie triple system and let I be an ideal of T. Then $\mathrm{Evann}_T(I)$ is an ideal of T.*

Proof. As I^n is an ideal of T, we have $\mathrm{Ann}_T(I^n)$ is also an ideal of T (see 2.3). Let $x \in \mathrm{Evann}_T(I)$ be, so there exists $n \in \mathbb{N}$ such that $x \in \mathrm{Ann}_T(I^n)$. By the above remark, $[x, T, T] \subset \mathrm{Ann}_T(I^n)$ and so $[x, T, T] \subset \mathrm{Evann}_T(I)$. $\qquad\qquad\square$

Proposition 4.6. *Let T be a Lie triple system, let I be an ideal of T and let J be an ideal of I. Then the eventual annihilator $\mathrm{Evann}_I(J)$ of J in I is a left-ideal of T.*

Proof. Recall that $\mathrm{Evann}_I(J) = \bigcup_{n \geq 1} \mathrm{Ann}_I(J^n)$. We know that

(i) J^n is an ideal of I for all $n \in \mathbb{N}$.

Moreover,

14

(ii) $[T, T, J^{n+1}] \subset J^n$. Indeed, taking into account (i),

$$[T, T, J^{n+1}] = [T, T, [I, J^n, J^n]] \subset$$
$$[[T, T, I], J^n, J^n] + [I, [T, T, J^n], J^n] + [I, J^n, [T, T, J^n]] \subset$$
$$[I, J^n, J^n] + [I, I, J^n] + [I, J^n, I] \subset J^n.$$

Now take $x \in \mathrm{Evann}_I(J)$, so there exists $n \in \mathbb{N}$ such that $x \in \mathrm{Ann}_I(J^n)$. Let t, w be two elements in T. To show that $[t, w, x] \in \mathrm{Evann}_I(J)$ we will see that $[t, w, x] \in \mathrm{Ann}_I(J^{n+1})$. This is equivalent to establish $[y, j, [t, w, x]] + [[t, w, x], j, y] = 0$ for any $j \in J^{n+1}$ and $y \in I$. As $J^{n+1} \subset J^n$ (see Lemma 4.4), we have $0 = [[t, w, y], j, x] = [t, w, [y, j, x]] - [y, j, [t, w, x]] - [y, [t, w, j], x]$. Since $[y, j, x] = 0$ and, by (ii), $[t, w, j] \in J^n$, we have $[t, w, [y, j, x]] = [y, [t, w, j], x] = 0$ and so

$$[y, j, [t, w, x]] = 0. \tag{9}$$

By arguing in a similar way on $0 = [x, [t, w, y], j]$, we obtain $[[t, w, x], y, j] = 0$. Then Jacobi identity gives us $[y, j, [t, w, x]] + [j, [t, w, x], y] = 0$. By (9) we have $[j, [t, w, x], y] = 0$ and by 2.1.1 $[[t, w, x], j, y] = 0$. Hence $[t, w, x] \in \mathrm{Ann}_I(J^{n+1})$. $\qquad\square$

Proposition 4.7. *Let T be a Lie triple system, let I be an ideal of T and let J be an ideal of I. Then $[I, \mathrm{Evann}_I(J)] \oplus \mathrm{Evann}_I(J)$ is a gr-ideal of $L_I := [I, T] \oplus I$.*

Proof. By applying respectively Lemma 4.5 and Proposition 4.6, we have

$$[[I, \mathrm{Evann}_I(J)], I] = [I, \mathrm{Evann}_I(J), I] \subset \mathrm{Evann}_I(J) \tag{10}$$

and

$$[\mathrm{Evann}_I(J), [I, T]] = [I, T, \mathrm{Evann}_I(J)] \subset \mathrm{Evann}_I(J). \tag{11}$$

Then by applying (9) and (10) we have

$$[[I, \mathrm{Evann}_I(J)] \oplus \mathrm{Evann}_I(J), [I, T] \oplus I] \subset$$
$$\subset [[I, \mathrm{Evann}_I(J)], [I, T]] + \mathrm{Evann}_I(J) + [I, \mathrm{Evann}_I(J)]. \tag{12}$$

By using Jacobi identity and (10), we also have

$$[[I, \mathrm{Evann}_I(J)], [I, T]] \subset [I, \mathrm{Evann}_I(J)]. \tag{13}$$

Now, equations (11) and (12) finally give

$$[[I, \mathrm{Evann}_I(J)] \oplus \mathrm{Evann}_I(J), [I, T] \oplus I] \subset [I, \mathrm{Evann}_I(J)] \oplus \mathrm{Evann}_I(J). \quad\square$$

Lemma 4.8. *Let T be a Lie triple system, let $L = L_0 \oplus L_1$ be its standard algebra envelope, let I be a semiprime ideal of T and let J, K be ideals of I satisfying $[J, K, I] = 0$. Then $[J, K] = 0$.*

Proof. We have

$$[[J, K, T], I, I] \subset [J, K, [T, I, I]] + [T, [J, K, I], I] + [T, I, [J, K, I]] = 0.$$

Since $[J, K, T] \subset I$, the semiprimeness of I gives $[J, K, T] = 0$ and hence, by 2.7, we complete the proof. $\qquad\square$

Lemma 4.9. *Let T be a Lie triple system, let I be a semiprime ideal of T and let J be an ideal of I. Then the following assertions hold:*

(1) $J \cap \mathrm{Evann}_I(J) = 0$
(2) $[J, I, \mathrm{Evann}_I(J)] = [\mathrm{Evann}_I(J), I, J] = 0$
(3) $[J, [I, \mathrm{Evann}_I(J)] \oplus \mathrm{Evann}_I(J)] = 0.$

Proof. 1. Let $S_n := J \cap \mathrm{Ann}_I(J^n)$, which is an ideal of I since J^n is an ideal of I, so both J and $\mathrm{Ann}_I(J^n)$ are ideals of I. Moreover, by Lemma 4.4 $S_n^n \subset S_n \subset \mathrm{Ann}_I(J^n)$ and $S_n^n \subset J^n$, hence we have $S_n^{n+1} = [I, S_n^n, S_n^n] \subset [I, J^n, \mathrm{Ann}_I(J^n)] = 0$. We have proved that S_n is a nilpotent ideal of I and I is semiprime, hence $S_n = 0$. Now $J \cap \mathrm{Evann}_I(J) = J \cap (\bigcup_{n \geq 1} \mathrm{Ann}_I(J^n)) = \bigcup_{n \geq 1} (J \cap \mathrm{Ann}_I(J^n)) = \bigcup_{n \geq 1} S_n = 0$.

2. Taking into account Lemma 4.5, both ideals $[J, I, \mathrm{Evann}_I(J)]$ and $[\mathrm{Evann}_I(J), I, J]$ belong to $J \cap \mathrm{Evann}_I(J)$. Then 1. completes the proof.

3. By 2., J and $\mathrm{Evann}_I(J)$ are ideals of I satisfying $[J, \mathrm{Evann}_I(J), I] = 0$, then Lemma 4.8 implies

$$[J, \mathrm{Evann}_I(J)] = 0. \tag{14}$$

Finally, Equation (13) and 2. complete the proof of 3. $\qquad\square$

We can finally state our main result.

Theorem 4.10. *Let T be a prime Lie triple system and let I be a semiprime ideal of T. Then I is prime.*

Proof. To see that I is prime, we will take a nonzero ideal J of I and show $\mathrm{Ann}_I(J) = 0$ (see 2.3). By Corollary 4.3, $L_I := [T, I] \oplus I$ is a gr-prime ideal of the standard algebra envelope L of T. By Proposition 4.7, $[I, \mathrm{Evann}_I(J)] \oplus \mathrm{Evann}_I(J)$ is a gr-ideal of L_I. By applying 2.5 we have two possibilities, either it is zero or its annihilator is zero: In the first case, $[I, \mathrm{Evann}_I(J)] \oplus \mathrm{Evann}_I(J) = 0$ and so $\mathrm{Ann}_I(J) = 0$, and in the second case $\mathrm{Ann}_{L_I}([I, \mathrm{Evann}_I(J)] \oplus \mathrm{Evann}_I(J)) = 0$. By Lemma 4.9-3, $J \subset \mathrm{Ann}_{L_I}([I, \mathrm{Evann}_I(J)] \oplus \mathrm{Evann}_I(J))$, hence we have $J = 0$, which is a contradiction, so $\mathrm{Ann}_I(J) = 0$. $\qquad\square$

Acknowledgment

The authors are grateful to the referee for his valuable suggestions.

References

1. Calderón, A. J. and Forero M. On locally finite split Lie triple systems. Comm. Algebra. In press.
2. García, E. Inheritance of Primeness by Ideals in Lie Algebras. Int. J. Math. Game Theory Algebra. In press.
3. Lister, W. G. A Structure Theory of Lie Triple Systems. Trans. Amer. Math. Soc. **1952**, 72, 217–242.
4. McCrimmon, K. Strong Prime Inheritance in Jordan Systems. Algebras, Groups and Geom. **1984** 1, 217–232.
5. McCrimmon, K. Prime Inheritance in Jordan Systems. Algebras, Groups and Geom. **1989** 6, 153–237.
6. Medvedev, Yu. A. An Analogue of Andrunakievich's Lemma for Jordan Algebras. Siberian Math. J. **1987** 28, 928–936.

A GEOMETRIC APPROACH TO FOUR-DIMENSIONAL ABSOLUTE VALUED TRIPLE SYSTEMS*

ANTONIO J. CALDERÓN MARTÍN

Departamento de Matemáticas
Universidad de Cádiz
11510 Puerto Real, Cádiz, Spain
E-mail: ajesus.calderon@uca.es

CÁNDIDO MARTÍN GONZÁLEZ

Departamento de Algebra, Geometría y Topología
Universidad de Málaga
Apartado 59, 29080 Málaga, Spain
E-mail: candido@apncs.cie.uma.es

We give a complete description, up to isomorphism, of four-dimensional absolute valued triple systems by using the geometry of the quaternions as the main tool.

1. Introduction and Preliminaries

1.1.

Let $\mathbb{K}$ denote the field of real or complex numbers. An absolute valued algebra over $\mathbb{K}$ is a non-zero algebra A over $\mathbb{K}$, endowed with a norm $|\cdot|$ satisfying $|xy| = |x|\,|y|$ for all $x, y \in A$. The most natural examples of absolute valued algebras are $\mathbb{R}$, $\mathbb{C}$, $\mathbb{H}$ (the algebra of Hamilton quaternions), and $\mathbb{O}$ (the algebra of Cayley numbers), with norms equal to their usual absolute values. In the early paper of A. Albert ([1]) it is proved that the only finite dimensional absolute valued algebra with a unit is $\mathbb{C}$ in the complex case and $\mathbb{R}$, $\mathbb{C}$, $\mathbb{H}$ and $\mathbb{O}$ in the real one, that any finite dimensional

*The first author is supported by the PCI of the UCA 'Teoría de Lie y Teoría de Espacios de Banach', by the PAI of the Spanish Junta de Andalucía with project number FQM-298 and by the Spanish DGICYT with project number BFM 2001-1886. The second author is supported by the Spanish DGICYT with project number BFM 2001-1886, by the Junta de Andalucía projects: FQM-0125, and by the PCI of the UCA 'Teoría de Lie y Teoría de Espacios de Banach'.

absolute valued algebra has dimension 1 in the complex case, 1, 2, 4 or 8 in the real one, and that the absolute values are the usual euclidean norms. Since [1] absolute valued algebras have been intensively studied by many authors, (see for instance the excellent survey [22] and [1, 2, 10, 11, 12, 13, 14, 20, 21, 25]).

Clearly, any finite dimensional absolute valued algebra is a division algebra, conversely, absolute valued division algebras are finite dimensional ([26]). It is easy to see that if two norms on a finite dimensional algebra convert it into an absolute valued algebra, then they must coincide (see for instance [10]). From here, it is also clear that any isomorphism between two finite dimensional absolute valued algebras $f : A \to A'$ is isometric. Indeed, we can define a new norm on A by $|x|'_A := |f(x)|_{A'}$ that makes A an absolute valued algebra, finally the uniqueness of the absolute value gives us $|x|_A = |x|'_A = |f(x)|_{A'}$. The precise determination of isomorphism classes for absolute valued real algebras of dimensions 1 and 2, (non-unital), is contained in ([21]), where the number of classes reduces to 1 and 4 respectively, while a detailed determination for the four-dimensional ones, (non-unital), appears in ([20]).

1.2.

Let T be a vector space over $\mathbb{K}$. We say that T is a *triple system* if it is endowed with a trilinear map

$$\langle \ \ \rangle : T \times T \times T \to T,$$

called the *triple product* of T. Let T, T' be triple systems, a bijective linear map $f : T \to T'$ is called an *isomorphism* of triple systems if it satisfies

$$f(\langle x, y, z \rangle) = \langle f(x), f(y), f(z) \rangle$$

for any $x, y, z \in T$. Triple systems appear in the literature as the natural ternary extension of algebras and have been studied in the associative ([7, 23, 24]), non associative ([3, 4, 9, 15, 16, 18]) and general context ([8]). An absolute valued triple system is defined as follows.

Definition 1.1. An absolute valued triple system, (a.v.t.s.), is a non-zero triple system T over $\mathbb{K}$, $\mathbb{K} = \mathbb{R}$ or $\mathbb{C}$, endowed with a norm $|\cdot|$ that satisfies

$$|\langle x, y, z \rangle| = |x|\,|y|\,|z|$$

for any $x, y, z \in T$.

If T is some finite-dimensional a.v.t.s. and we fix any $v \in T$ such that $|v| = 1$, the algebra whose underlying vector space is T, with the product $x \cdot y := \langle xvy \rangle$ is absolute valued. By Albert's results in 1.1 we conclude that its absolute value comes from an inner product, so the reference [19] is fundamental in this framework and gives the first approach to the classification of finite dimensional a.v.t.s.

Clearly, if we fix $x, y \in T$ with $|x| = |y| = 1$, then the *left, middle and right product operators* $L(x, y), R(x, y), M(x, y) : T \to T$, defined by $L(x, y)z = R(z, y)x = M(x, z)y := \langle x, y, z \rangle$, are isometric.

Any absolute valued algebra A can be seen as an a.v.t.s., with the same norm, by defining for instance the triple product as $\langle x, y, z \rangle := (xy)z$. Then we have that the class of absolute valued algebras is related to the class of a.v.t.s. Moreover, as we said above, given $u \in T$ with $|u| = 1$, we can construct an absolute valued algebra, denoted by T^u, by defining $xy := \langle x, u, y \rangle$ and with the same norm as T. Since $\dim(T) = \dim(T^u)$, Albert's result in 1.1 gives us that any finite dimensional a.v.t.s. has dimension 1 in the complex case and 1, 2, 4 or 8 in the real one,[a] and that the absolute values are the usual euclidean norms. By considering T^u and taking into account the observations in 1.1, we also obtain that if we have two norms on T converting it into an a.v.t.s., then they must coincide and, that any isomorphism between two finite dimensional a.v.t.s. is isometric.

1.3.

The following lemmas will be useful in Section 2. Let us denote by $\mathcal{O}(4)$ the group of all isometries in $\mathbb{R}^4$ and by $S^3 := \{x \in \mathbb{R}^4 : |x| = 1\}$.

Lemma 1.2. *If $f, g \in \mathcal{O}(4)$ satisfy $\frac{1}{\sqrt{2}}(f + g) \in \mathcal{O}(4)$, then $fg^{-1} = -gf^{-1}$.*

Proof. From the equality of transposition and inversion of orthogonal matrices we have $(\frac{1}{\sqrt{2}}(f + g))^{-1} = \frac{1}{\sqrt{2}}(f^{-1} + g^{-1})$. Hence, id $= \frac{1}{2}(2\mathrm{id} + fg^{-1} + gf^{-1})$, and then $fg^{-1} + gf^{-1} = 0$. $\square$

In the following lemma, we shall denote by $\mathrm{Lin}(S)$ the linear span of a set $S \subset \mathbb{H}$.

Lemma 1.3. *Let $f \in \mathcal{O}(4)$ be satisfying $f^{-1} = -f$. Let $u, v \in S^3$ such that $(u \mid v) = 0$ and $f(u) = v$, and let be $s, t \in \mathrm{Lin}(u, v)^{\perp}$ satisfying $s, t \in S^3$ and $(s \mid t) = 0$. Then $f(v) = -u$, $f(s) = \epsilon t$ and $f(t) = -\epsilon s$, where $\epsilon = \pm 1$.*

---

[a]The referee noticed that the dimensions 1, 2, 4, 8 can even be deduced from a 1898 Hurwitz's Theorem, [17].

Proof. Let consider the orthonormal basis of $\mathbb{H}$, $\mathcal{B} = \{u, v, s, t\}$. It is clear that $f(u) = v$ yields $f(v) = -u$. Also, $(s \mid f(s)) = (f(s) \mid f^2(s)) = -(f(s) \mid s)$, so that $(s \mid f(s)) = 0$. Consequently, $f(s) = \epsilon t$, and $f(t) = -\epsilon s$. $\qquad\square$

2. Main results

2.1.

In [5] we study some aspects of the theory of a.v.t.s. and in [6] relate this theory to the one of two-graded absolute valued algebras. In [5, Section 2] we show that the isomorphism classes of a.v.t.s. in dimension 1 reduces to one class in the complex case and two classes in the real one, and that in dimension two, (real), there are an infinity of them. We also give in the same reference an approach to the four-dimensional case as a consequence of the ideas in [19]. The aim of this paper is to develop intrinsic techniques to the geometry of the quaternions to obtain a classification in the four-dimensional case.

2.2.

If T is any four-dimensional (real) a.v.t.s. we will identify T with $\mathbb{H}$ as euclidean vector spaces. From now on, given $x, y \in \mathbb{H}$, the juxtaposition xy will mean the usual product in $\mathbb{H}$. Let us introduce some terminology.

Definition 2.1. Let $\mu \in S_3$ be a permutation of the set $\{1, 2, 3\}$. The triple system obtained from T by maintaining the same vector space structure but by defining a new triple product as $\langle x_1, x_2, x_3 \rangle_\mu := \langle x_{\mu(1)}, x_{\mu(2)}, x_{\mu(3)} \rangle$ for any $x_1, x_2, x_3 \in T$, will be called the *μ-permutation* triple system of T.

Definition 2.2. Let $T = (\mathbb{H}, \langle\ \rangle)$ be a four-dimensional a.v.t.s. The new a.v.t.s. $T' = (\mathbb{H}, \langle\ \rangle')$ obtained by defining the new triple product as $\langle\ \rangle' = v\langle\ \rangle$ for some $v \in \mathbb{H}$, $|v| = 1$, shall be called a *factor* of T.

Definition 2.3. Let $T = (\mathbb{H}, \langle\ \rangle)$ be a four-dimensional a.v.t.s. For any

$$\tau_1, \tau_2, \tau_3 \in \mathcal{O}(4),$$

we shall call a *perturbation* of T (with respect to τ_1, τ_2, τ_3) to the a.v.t.s.

$$T' = (\mathbb{H}, \langle\ \rangle')$$

with the new triple product given by

$$\langle x, y, z \rangle' := \langle \tau_1(x), \tau_2(y), \tau_3(z) \rangle.$$

Lemma 2.4. *Let $T = (\mathbb{H}, \langle\ \rangle)$ be a four-dimensional a.v.t.s. We have*

$$L(1,1) = R(1,1) = M(1,1) = \mathrm{id}$$

up to perturbations and factors.

Proof. Let be $v := \langle 1,1,1 \rangle \neq 0$. The factor $(\mathbb{H}, \langle\ \rangle')$, where

$$\langle\ \rangle' := v^{-1}\langle\ \rangle,$$

satisfies $\langle 1,1,1 \rangle' = 1$ and therefore we may assume $\langle 1,1,1 \rangle = 1$. Let be $\sigma^{-1} := L(1,1)$, $\tau^{-1} := R(1,1)$ and $\rho^{-1} := M(1,1)$. Since $\sigma, \tau, \rho \in \mathcal{O}(4)$ and $\sigma(1) = \tau(1) = \rho(1) = 1$, the triple system $(\mathbb{H}, \langle\ \rangle')$, where the triple product is defined by $\langle x, y, z \rangle' := \langle \tau(x), \rho(y), \sigma(z) \rangle$, is a perturbation of T that satisfies $\langle 1, 1, x \rangle' = \langle 1, 1, \sigma(x) \rangle = \sigma^{-1}\sigma(x) = x$ and, in a similar way, $\langle x, 1, 1 \rangle' = x = \langle 1, x, 1 \rangle'$ for any $x \in \mathbb{H}$. $\qquad\square$

Theorem 2.5. *Let T be a four-dimensional a.v.t.s. Then:*

(1) Up to factors, perturbations and permutations T is isomorphic to $(\mathbb{H}, \langle\ \rangle)$ with triple product $\langle x, y, z \rangle := xyz$ for any $x, y, z \in \mathbb{H}$.

(2) Up to permutations, the triple product of T is

$$\langle x, y, z \rangle = a(\triangle x)b(\square y)c((z)d,$$

where $a, b, c, d \in S^3$ and any $\triangle, \square, (: \mathbb{H} \to \mathbb{H}$ is either the identity or the conjugation on $\mathbb{H}$

Proof. In view of Lemma 2.4 we may assume that $L(1,1) = R(1,1) = M(1,1) = \mathrm{id}$. Let us determine the matrix expression, with respect to the canonical orthonormal basis $\mathcal{B} = \{1, i, j, k\}$, of the operator $L(1,i)$. Since $\left|\frac{1+i}{\sqrt{2}}\right| = 1$, we have $L(1,i), L(1, \frac{1+i}{\sqrt{2}}) \in \mathcal{O}(4)$ and so, if we denote $f := L(1,i)$ we obtain $\frac{1}{\sqrt{2}}(1+f) \in \mathcal{O}(4)$. By applying Lemma 1.2 we obtain $f^{-1} = -f$. Since $f(1) = i$, Lemma 1.3 shows $f(i) = -1$, $f(j) = \epsilon k$ and $f(k) = -\epsilon j$, hence

$$L(1,i) \equiv \begin{pmatrix} 0 & 1 & 0 & 0 \\ -1 & 0 & 0 & 0 \\ 0 & 0 & 0 & \epsilon \\ 0 & 0 & -\epsilon & 0 \end{pmatrix},$$

with $\epsilon \in \pm 1$. The same argument shows $L(1,j)^{-1} = -L(1,j)$ and that the matrix expression of $L(1,j)$ (with respect to $\mathcal{B}$) is

$$L(1,j) \equiv \begin{pmatrix} 0 & 0 & 1 & 0 \\ 0 & 0 & 0 & -\delta \\ -1 & 0 & 0 & 0 \\ 0 & \delta & 0 & 0 \end{pmatrix},$$

with $\delta \in \pm 1$. As $L(1, \frac{i+j}{\sqrt{2}}) \in \mathcal{O}(4)$ then $\frac{1}{\sqrt{2}}(L(1,i) + L(1,j)) \in \mathcal{O}(4)$. By applying Lemma 1.2 we have $L(1,i)L(1,j)^{-1} = -L(1,j)L(1,i)^{-1}$. Since $L(1,i)^{-1} = -L(1,i)$ and $L(1,j)^{-1} = -L(1,j)$, we get $L(1,i)L(1,j) = -L(1,j)L(1,i)$. Finally, as $L(1,i)L(1,j)i = \delta L(1,i)k = -\delta\epsilon j$ and $-L(1,j)L(1,i)i = L(1,j)1 = j$ we conclude that $\delta = -\epsilon$. Therefore

$$L(1,j) \equiv \begin{pmatrix} 0 & 0 & 1 & 0 \\ 0 & 0 & 0 & \epsilon \\ -1 & 0 & 0 & 0 \\ 0 & -\epsilon & 0 & 0 \end{pmatrix}.$$

By arguing in this fashion we obtain that the table

$$\begin{vmatrix} L(1,1) & L(1,i) & L(1,j) & L(1,k) \\ L(i,1) & L(i,i) & L(i,j) & L(i,k) \\ L(j,1) & L(j,i) & L(j,j) & L(j,k) \\ L(k,1) & L(k,i) & L(k,j) & L(k,k) \end{vmatrix}$$

can be written, with respect to $\mathcal{B}$, as

$$\begin{vmatrix}
1 &
\begin{pmatrix} 0 & 1 & 0 & 0 \\ -1 & 0 & 0 & 0 \\ 0 & 0 & 0 & \epsilon \\ 0 & 0 & -\epsilon & 0 \end{pmatrix} &
\begin{pmatrix} 0 & 0 & 1 & 0 \\ 0 & 0 & 0 & \epsilon \\ -1 & 0 & 0 & 0 \\ 0 & -\epsilon & 0 & 0 \end{pmatrix} &
\begin{pmatrix} 0 & 0 & 0 & 1 \\ 0 & 0 & \epsilon & 0 \\ 0 & -\epsilon & 0 & 0 \\ -1 & 0 & 0 & 0 \end{pmatrix} \\
\begin{pmatrix} 0 & 1 & 0 & 0 \\ -1 & 0 & 0 & 0 \\ 0 & 0 & 0 & \epsilon \\ 0 & 0 & -\epsilon & 0 \end{pmatrix} &
-1 &
\begin{pmatrix} 0 & 0 & 0 & -\delta' \\ 0 & 0 & -\epsilon\delta' & 0 \\ 0 & \epsilon\delta' & 0 & 0 \\ \delta' & 0 & 0 & 0 \end{pmatrix} &
\begin{pmatrix} 0 & 0 & \delta' & 0 \\ 0 & 0 & 0 & -\epsilon\delta' \\ -\delta' & 0 & 0 & 0 \\ 0 & \epsilon\delta' & 0 & 0 \end{pmatrix} \\
\begin{pmatrix} 0 & 0 & 1 & 0 \\ 0 & 0 & 0 & -\epsilon \\ -1 & 0 & 0 & 0 \\ 0 & \epsilon & 0 & 0 \end{pmatrix} &
\begin{pmatrix} 0 & 0 & 0 & \delta' \\ 0 & 0 & \delta'\epsilon & 0 \\ 0 & -\delta'\epsilon & 0 & 0 \\ -\delta' & 0 & 0 & 0 \end{pmatrix} &
-1 &
\begin{pmatrix} 0 & -\delta' & 0 & 0 \\ \delta' & 0 & 0 & 0 \\ 0 & 0 & 0 & -\epsilon\delta' \\ 0 & 0 & \epsilon\delta' & 0 \end{pmatrix} \\
\begin{pmatrix} 0 & 0 & 0 & 1 \\ 0 & 0 & \epsilon & 0 \\ 0 & -\epsilon & 0 & 0 \\ -1 & 0 & 0 & 0 \end{pmatrix} &
\begin{pmatrix} 0 & 0 & -\delta' & 0 \\ 0 & 0 & 0 & \epsilon\delta' \\ \delta' & 0 & 0 & 0 \\ 0 & -\epsilon\delta' & 0 & 0 \end{pmatrix} &
\begin{pmatrix} 0 & \delta' & 0 & 0 \\ -\delta' & 0 & 0 & 0 \\ 0 & 0 & 0 & \epsilon\delta' \\ 0 & 0 & -\epsilon\delta' & 0 \end{pmatrix} &
-1
\end{vmatrix}$$

with $\epsilon, \delta' \in \{-1, 1\}$. Let study *all four cases*.

(1) $\epsilon = 1$, $\delta' = -1$. We have that $\langle x, y, z \rangle = xyz$. This can be verified by comparing the above matrices with those arising in the triple product $\langle x, y, z \rangle = xyz$. In a similar way we will obtain (2), (3) and (4).

(2) $\epsilon = 1$, $\delta' = 1$. Then $\langle x, y, z \rangle = yxz$.

(3) $\epsilon = -1$, $\delta' = 1$. We have $\langle x, y, z \rangle = zyx$.

(4) $\epsilon = -1$, $\delta' = -1$. In this case $\langle x, y, z \rangle = zxy$.

We observe that any of the above four cases is a permutation of T with the triple product $\langle x, y, z \rangle = xyz$ and then we complete assertion 1 of the theorem.

Assertion 2 is an easy consequence of the following well known facts. The isometries of $\mathbb{H}$ are of the form $x \mapsto v\sigma(x)$ with $|v| = 1$ and σ being either an automorphism or an antiautomorphism of the algebra $\mathbb{H}$, and all these automorphisms can be written as $x \mapsto qxq^{-1}$ where $q \in S^3$ and the antiautomorphism as $x \mapsto q\bar{x}q^{-1}$ with q as above and where $-$ is the conjugation of $\mathbb{H}$. $\qquad\square$

Acknowledgment

The authors are grateful to the referee for his valuable suggestions.

References

1. Albert, A. A. *Absolute valued real algebras*, Ann. Math. **48** (1947), 495–501.
2. Albert, A. A. *Absolute valued algebraic algebras.* Bull. Amer. Math. Soc. **55** (1949), 763–768.
3. Calderón, A. J. and Martín, C. *On L^*-triples and Jordan H^*-pairs*, Ring theory and Algebraic Geometry, Marcel Dekker, Inc., 2001, 87–94.
4. Calderón, A. J. and Martín, C. *Hilbert space methods in the theory of Lie triple systems*, Recent Progress in Functional Analysis, North-Holland Math. Studies, 2001, 309–319.
5. Calderón, A. J. and Martín, C. *Absolute valued triple systems*, submitted to International Mathematical Journal.
6. Calderón, A. J. and Martín, C. *On two-graded absolute valued algebras*, preprint.
7. Castellón, A. and Cuenca, J. A. *Associative H^*-triple Systems*, Workshop on Nonassociative Algebraic Models, Nova Science Publishers New York, 1992, 45–67.
8. Castellón, A. and Cuenca, J. A. *The Centroid and Metacentroid of an H^*-triple system*, Bull. Soc. Math. Belg. **45** (1993), 85–93.
9. Castellón, A., Cuenca, J. A. and Martín, C. *Special Jordan H^*-triple systems*, Comm. Algebra **28** (2000), 4699–4706.
10. Cuenca, J. A. and Rodríguez, A. *Absolute values on H^*-algebras*, Comm. Algebra **23** (1995), 1709–1740.
11. El-Amin, K., Ramírez, M. I. and Rodríguez, A. *Absolute-Valued Algebraic Algebras are Finite-Dimensional*, J. Algebra **195** (1997), 295–307.

12. El-Mallah, M. L. *Absolute valued algebras containing a central idempotent*, J. Algebra **128** (1990), 180–187.

13. El-Mallah, M. L. *Absolute valued algebras containing a central element*, Italian J. Pure Appl. Math. **3** (1998), 103–105.

14. El-Mallah, M. L. *Absolute valued algebras satisfying* $(x, x, x^2) = 0$, Arch. Math. **77** (2001), 378–382.

15. Faulkner, J. R. *Dynkin diagrams for Lie triple systems*, J. Algebra **62** (1980), 384–392.

16. Hopkins, N. C. *Some structure theory for a class of triple systems*, Trans. Amer. Math. Soc. **1** (1985), 203–212.

17. Hurwitz, A. *Über die Komposition der quadratischen Formen von beliebig vielen Variabeln*, Nachr. Ges. Wiss. Göttingen (1898), 309–316.

18. Lister, W. G. *A structure theory of Lie triple systems*, Trans. Amer. Math. Soc. **72** (1952), 217–242.

19. McCrimmon, K., *Quadratic forms permitting triple composition*, Trans. Amer. Math. Soc. **275** (1983), 107–130.

20. Ramírez, M. I. *On four dimensional absolute-valued algebras*, Proceedings of the International Conference on Jordan Structures, Univ. Málaga, Málaga, 1999, 169–173.

21. Rodríguez, A. *Absolute valued algebras of degree two*, Non-Associative Algebra and Its Applications, Kluwer Academic Publishers, 1994, 350–357.

22. Rodríguez, A. *Absolute-valued algebras and absolute-valuable Banach spaces*, Proceedings of the First International Course of Mathematical Analysis in Andalucía. In press.

23. Shaw, R. *Ternary composition algebras. I. Structure theorems: definite and neutral signatures*, Proc. Soc. London Ser. A **431** (1990), 1–9.

24. Shaw, R. *Ternary composition algebras. II. Automorphism groups and subgroups*, Proc. Soc. London Ser. A. **431** (1990), 21–36.

25. Urbanik, K. and Wright, F. B. *Absolute valued algebras*, Proc. Amer. Math. Soc. **11** (1960), 861–866.

26. Wright, F. B. *Absolute valued algebras*, Proc. Nat. Acad. Sci. U.S.A. **39** (1953), 330–332.

DIVISIBLE MODULES WITH RESPECT TO A TORSION THEORY*

IULIU CRIVEI

*Department of Mathematics, Technical University,
Str. C. Daicoviciu 15, 400020 Cluj-Napoca, Romania
E-mail: crivei@math.utcluj.ro*

SEPTIMIU CRIVEI

*Faculty of Mathematics and Computer Science, "Babeş-Bolyai" University,
Str. M. Kogălniceanu 1, 400084 Cluj-Napoca, Romania
E-mail: crivei@math.ubbcluj.ro*

For a hereditary torsion theory τ, a module is called τ-divisible if it is injective with respect to every monomorphism having a τ-torsionfree cokernel. Alongside of some properties of the class of τ-divisible modules, we are interested in self-τ-divisible modules, that generalize τ-complemented modules. We show that if $A_1, \ldots, A_n$ are relatively injective modules, then $A_1 \oplus \cdots \oplus A_n$ is self-τ-divisible if and only if each A_i is self-τ-divisible.

1. Introduction

Divisible modules with respect to a hereditary torsion theory τ, i.e., modules D such that $\operatorname{Ext}_R^1(F, D) = 0$ for every τ-torsionfree module F, generalize to the torsion-theoretic context injective modules, as τ-injective modules do, but also (Warfield) cotorsion modules, i.e., modules D such that $\operatorname{Ext}_R^1(F, D) = 0$ for every torsionfree module F (for instance, see [4]). They have been studied, sometimes under different names like τ-cotorsion or even τ-pure-injective modules, by J. L. Bueso and B. Torrecillas [1], L. Fuchs [3], J. Henderson and M. Orzech [7], R. Mines [8], A. P. Mishina and L. A. Skornjakov [9], B. Stenström [14] or C. P. Walker [17].

---

*Part of this paper was written during the second author's stay at the University of Murcia between February and June 2003, supported by an AECI fellowship of the Spanish Ministry of Foreign Affairs. He wishes to thank the Department of Mathematics for the kind hospitality.

In the present paper, we are interested in studying (self-)τ-divisible modules mainly in relationship with some relatively recent notions. Thus, we establish some further properties of the class of τ-divisible modules and discuss connections between, on the one hand, (self-)τ-divisible modules and, on the other hand, τ-complemented, extending and (self-)c-injective modules. The final part of the paper is motivated by the following problem. It is known that if $A_1, \ldots, A_n$ are relatively injective modules, then $A_1 \oplus \cdots \oplus A_n$ is extending if and only if each A_i is extending [2, Proposition 7.10]. This characterization is no longer true for τ-complemented modules, that can be seen as the torsion-theoretic generalization of extending modules, but we shall show that it holds for self-τ-divisible modules, that generalize τ-complemented modules.

Some further interest in studying τ-divisible modules is motivated by their similarities with (Warfield) cotorsion modules, that intervene in the context of cotilting theory and cotorsion theories. We give a couple of arguments to support this statement, following the terminology from J. Trlifaj [15].

It is known that, over a domain R, every (Warfield) cotorsion module M has injective dimension i. d.$(M) \leq 1$ [4, Theorem 8.3]. Moreover, a torsionfree module M is (Warfield) cotorsion if and only if it is pure-injective and i. d.$(M) \leq 1$ ([16, Theorem 2.6]). Considering the torsion-theoretic case, for a faithful τ, every τ-divisible module D has i. d.$(D) \leq 1$ [9]. If D is also τ-torsionfree, then clearly $\mathrm{Ext}_R^1(D^\kappa, D) = 0$ for every cardinal κ. Consequently, if τ is faithful, then every τ-torsionfree τ-divisible module is partial cotilting, in the sense that it satisfies the first two conditions of the definition of a cotilting module (for instance, see [15, Definition 4.10]).

For a class of modules $\mathcal{A}$, denote $^\perp \mathcal{A} = \{X \mid \mathrm{Ext}_R^1(X, A) = 0, \forall A \in \mathcal{A}\}$ and $\mathcal{A}^\perp = \{Y \mid \mathrm{Ext}_R^1(A, Y) = 0, \forall A \in \mathcal{A}\}$. A pair $(\mathcal{A}, \mathcal{B})$ of classes of modules is called a *cotorsion theory* if $\mathcal{A} = {}^\perp \mathcal{B}$ and $\mathcal{B} = \mathcal{A}^\perp$. An important example of cotorsion theory is the pair $(\mathcal{TF}, \mathcal{WC})$, where $\mathcal{TF}$ is the class of torsionfree modules and $\mathcal{WC}$ is the class of (Warfield) cotorsion modules [15, p. 9]. Let us now consider the pair $(\mathcal{F}, \mathcal{D})$, where $\mathcal{F}$ is the class of τ-torsionfree modules and $\mathcal{D}$ is the class of τ-divisible modules. Then $\mathcal{F} \subseteq {}^\perp \mathcal{D}$ and $\mathcal{D} = \mathcal{F}^\perp$, but in general $(\mathcal{F}, \mathcal{D})$ is not a cotorsion theory. Clearly, a necessary condition is that τ is faithful. Nevertheless, for any hereditary torsion theory τ, there exists a cotorsion theory involving the class of τ-divisible modules, namely the cotorsion theory $({}^\perp(\mathcal{F}^\perp), \mathcal{F}^\perp)$ cogenerated by $\mathcal{F}$, that is, $({}^\perp \mathcal{D}, \mathcal{D})$.

Let us now set some basic notation and definitions. Throughout, R will be an associative ring with non-zero identity and all modules will be left unital R-modules. For any module A, $E(A)$ denotes the injective hull of A, whereas i. d.(A) and p. d.(A) denote respectively the injective and the projective dimension of A.

A module A is called *extending* if every submodule is essential in a direct summand of A or, equivalently, if every closed submodule is a direct summand of A (for instance, see [2]).

We shall denote by τ a hereditary torsion theory on the category R-Mod of left R-modules and by $T_\tau(A)$ the unique maximal τ-torsion submodule of a module A.

A submodule B of a module A is called τ-*dense* (respectively, τ-*closed*) *in* A if A/B is τ-torsion (respectively, τ-torsionfree). A non-zero module A is called τ-*cocritical* if A is τ-torsionfree and each of its non-zero submodules is τ-dense in A.

A module A is called τ-*complemented* if every submodule is τ-dense in a direct summand of A or, equivalently, if every τ-closed submodule is a direct summand of A [12].

A module is said to be τ-*injective* (respectively, τ-*divisible*) if it is injective with respect to every monomorphism having a τ-torsion (respectively, τ-torsionfree) cokernel. We should emphasize that there are various concepts of τ-divisibility present in literature, the one used in this paper being that given for instance in J. S. Golan [5]. A module is said to be τ-*projective* (respectively, τ-*codivisible*) if it is projective with respect to every epimorphism having a τ-torsion (respectively, τ-torsionfree) kernel.

For further terminology on modules and torsion theories the reader is referred to [18] and [5].

2. On the class of τ-divisible modules

As we have seen above, a τ-divisible module is a module injective with respect to every monomorphism having a τ-torsionfree cokernel or, equivalently, a module D such that $\mathrm{Ext}_R^1(F, D) = 0$ for every τ-torsionfree module F. A basic characterization of τ-divisible modules is the following one.

Proposition 2.1 ([5, Proposition 12.1]). *A module D is τ-divisible if and only if D is a direct summand of every module containing it as a τ-closed submodule.*

Therefore a module is τ-divisible if and only if it cannot be properly extended by any τ-torsionfree module. Actually, this was the basic idea for introducing cotorsion abelian groups by D. K. Harrison [6].

The following refined characterization will be useful.

Lemma 2.2. *A module D is τ-divisible if and only if for every module A and every τ-closed submodule B of A, every monomorphism $B \to D$ extends to a homomorphism $A \to D$.*

Proof. The "only if" part is obvious. For the converse, let A be a module, B be a τ-closed submodule of A and $u\colon B \to D$ be a monomorphism. Write $u = jp$, where $p\colon B \to B/\operatorname{Ker} u$ is the natural homomorphism and $j\colon B/\operatorname{Ker} u \to D$ is the canonical monomorphism. Applying the hypothesis for the monomorphism $f'\colon B/\operatorname{Ker} u \to A/\operatorname{Ker} u$ induced by f, we get a homomorphism $w\colon A/\operatorname{Ker} u \to D$ such that $wf' = j$. If $q\colon A \to A/\operatorname{Ker} u$ denotes the natural homomorphism, then wq extends u. Therefore D is τ-divisible. $\qquad\square$

Let us denote by $\mathcal{D}$ the class of τ-divisible modules. Then $\mathcal{D}$ is closed under extensions, direct products and direct summands [3, p. 34–35]. It is also closed under direct sums provided R is left noetherian and under homomorphic images provided R is left hereditary [3, p. 35]. More generally, we have:

Proposition 2.3. *The following statements are equivalent:*

(i) *The class $\mathcal{D}$ is closed under homomorphic images.*
(ii) *Every τ-torsionfree module has projective dimension at most 1.*
(iii) *Every τ-closed submodule of a projective module is projective.*

Proof. (i) $\Rightarrow$ (ii) Let F be a τ-torsionfree module and let A be any module. Then the exact sequence $0 \to A \to E(A) \to E(A)/A \to 0$ induces the exact sequence

$$\operatorname{Ext}_R^1(F, E(A)/A) \to \operatorname{Ext}_R^2(F, A) \to \operatorname{Ext}_R^2(F, E(A))$$

By hypothesis we have the first Ext zero and clearly the last one is zero, hence $\operatorname{Ext}_R^2(F, A) = 0$. Thus p. d.$(F) \le 1$.

(ii) $\Rightarrow$ (i) Let D be a τ-divisible module and let $f\colon D \to C$ be an epimorphism with kernel B. Also, let F be a τ-torsionfree module. Then the exact sequence $0 \to B \to D \to C \to 0$ induces the exact sequence

$$\operatorname{Ext}_R^1(F, D) \to \operatorname{Ext}_R^1(F, C) \to \operatorname{Ext}_R^2(F, B)$$

Since D is τ-divisible and p. d.$(F) \leq 1$, we have the first and the last Ext zero, hence $\operatorname{Ext}_R^1(F, C) = 0$. Thus C is τ-divisible.

(ii) $\Rightarrow$ (iii) Let P be a projective module and let B be a τ-closed submodule of P. Also, let A be any module. Then the exact sequence $0 \to B \to P \to P/B \to 0$ induces the exact sequence

$$\operatorname{Ext}_R^1(P, A) \to \operatorname{Ext}_R^1(B, A) \to \operatorname{Ext}_R^2(P/B, A)$$

The first Ext is clearly zero and the last one is zero by hypothesis, so $\operatorname{Ext}_R^1(B, A) = 0$. Thus B is projective.

(iii) $\Rightarrow$ (ii) Let F be a τ-torsionfree module and let A be any module. Then, for some projective module P, there exists an epimorphism $P \to F$ with kernel B. Then the exact sequence $0 \to B \to P \to F \to 0$ induces the exact sequence

$$\operatorname{Ext}_R^1(B, A) \to \operatorname{Ext}_R^2(F, A) \to \operatorname{Ext}_R^2(P, A)$$

The last Ext is clearly zero and the first one is zero because B is projective by hypothesis. Hence $\operatorname{Ext}_R^2(F, A) = 0$. Thus p. d.$(F) \leq 1$. $\qquad\square$

We now see when every (τ-torsion, τ-torsionfree) module is τ-divisible.

Theorem 2.4. (a) *The following statements are equivalent:*

    (i) *$\mathcal{D} = R$-Mod.*
    (ii) *Every τ-torsionfree module is projective.*
    (iii) *Every module is τ-complemented.*

(b) *The following statements are equivalent:*

    (i) *Every τ-torsion module is τ-divisible.*
    (ii) *Every τ-torsionfree module is τ-projective.*
    (iii) *τ is splitting.*

(c) *The following statements are equivalent:*

    (i) *Every τ-torsionfree module is τ-divisible.*
    (ii) *Every τ-torsionfree module is τ-codivisible.*
    (iii) *Every τ-torsionfree module is τ-complemented.*

Proof. (a) (i)$\Leftrightarrow$ (ii) This is immediate by the Ext definition of τ-divisible modules. In the view of the proofs of parts (b) and (c), we give another argument, namely: every module is τ-divisible if and only if any exact sequence $0 \to D \to A \to F \to 0$, where F is τ-torsionfree, splits if and only if every τ-torsionfree module is projective.

(ii)$\Leftrightarrow$ (iii) By [13, Proposition 10].

(b) (i) $\Leftrightarrow$ (ii) Adapt the proof of part (a) for D τ-torsion.

(i) $\Rightarrow$ (iii) Assume that every τ-torsion module is τ-divisible. For any module A, $T_\tau(A)$ is τ-divisible by hypothesis, whence $T_\tau(A)$ is a direct summand of A by Proposition 2.1. Thus τ is splitting.

(iii) $\Rightarrow$ (i) Suppose that τ is splitting. Let D be a τ-torsion module. Also, let A be a module, B be a τ-closed submodule of A and $u\colon B \to D$ be a monomorphism. Since τ is splitting, $B = T_\tau(A)$ is a direct summand of A. Now clearly there exists a homomorphism $A \to D$ that extends u. By Lemma 2.2, D is τ-divisible.

(c) (i) $\Leftrightarrow$ (ii) Adapt the proof of part (a) for D τ-torsionfree.

(i) $\Rightarrow$ (iii) Let A be a τ-torsionfree module. Also, let B be a τ-closed submodule of A. Then B is τ-divisible by hypothesis, hence B is a direct summand of A by Proposition 2.1. Therefore, A is τ-complemented.

(iii)$\Rightarrow$ (i) Let D be a τ-torsionfree module. Also, let A be a module, B be a τ-closed submodule of A and $u\colon B \to D$ be a monomorphism. Then B is τ-torsionfree and, consequently, A is τ-torsionfree as well. Then A is τ-complemented, whence it follows that B is a direct summand of A. Now clearly u extends to a homomorphism $A \to D$. Thus D is τ-divisible by Lemma 2.2. $\qquad\square$

Let us now consider the natural notion of relative τ-divisibility. Let A be a module. A module D is called A-τ-*divisible* if for every τ-closed submodule B of A, every homomorphism $B \to D$ extends to a homomorphism $A \to D$. A module D is called *self-τ-divisible* if it is D-τ-divisible.

Lemma 2.5. *Let A be a module and $(D_i)_{i\in I}$ be a family of modules. Then $\prod_{i\in I} D_i$ is A-τ-divisible if and only if D_i is A-τ-divisible for every $i \in I$.*

Proof. Similar to the proof for A-injective modules (for instance, see [18, 16.1]). $\qquad\square$

The following result gives a connection between (self-)τ-divisible modules and τ-complemented modules.

Proposition 2.6. *The following statements are equivalent for a module A:*

(i) *A is τ-complemented.*
(ii) *Every module is A-τ-divisible.*
(iii) *Every τ-closed submodule of A is A-τ-divisible.*

Proof. Clear. $\qquad\square$

In particular, every τ-complemented module is self-τ-divisible. But there exist self-τ-divisible modules that are not τ-complemented, as one can see in the following example.

Example 2.7. Let τ be the trivial torsion theory where every module is τ-torsionfree. Then τ-complemented and self-τ-divisible mean respectively semisimple and quasi-injective. Clearly, there are quasi-injective, and even injective modules, that are not semisimple.

Let us now recall the notion of c-injectivity, introduced in [10]. Let A be a module. A module M is said to be *A-c-injective* if for every closed submodule B of A, every homomorphism $B \to M$ extends to a homomorphism $A \to M$. A module M is called *self-c-injective* if it is M-c-injective.

Every τ-torsion module is τ-complemented and, consequently, self-τ-divisible by Proposition 2.6. We are going to see that τ-torsion modules are even τ-divisible under some extra hypotheses.

Proposition 2.8. *Let τ be stable. Then every τ-torsion c-injective module is τ-divisible.*

Proof. Let D be a τ-torsion c-injective module. Also, let A be a module, B be a τ-closed submodule of A and $u\colon B \to D$ be a non-zero monomorphism. It follows that B is τ-torsion, whence $B = T_\tau(A)$. We may suppose that $T_\tau(A) \neq A$. Since τ is stable, B must be closed in A. Hence u extends to a homomorphism $A \to D$. Now D is τ-divisible by Lemma 2.2. $\qquad\square$

3. Self-τ-divisible modules and direct sums

It is known that every τ-torsionfree τ-complemented module is extending [12, Lemma 1.7] and every extending module is γ-complemented [12, Example 1.3], where γ is the Goldie torsion theory. We establish similar connections between self-τ-divisible modules, that generalize τ-complemented modules, and self-c-injective modules, that generalize extending modules [10, Proposition 2.2].

Proposition 3.1. (i) *Every τ-torsionfree self-τ-divisible module is self-c-injective.*

(ii) *Every self-c-injective module is self-γ-divisible.*

Proof. (i) Let A be a τ-torsionfree self-τ-divisible module and let B be a closed submodule of A. Also, let M be such that $M/B = T_\tau(A/B)$.

We claim that B is essential in M. Indeed, if L is a non-zero submodule of M, then $L/(L \cap B) \cong (L+B)/B$ is τ-torsion. But since L is τ-torsionfree, we have $L \cap B \neq 0$.

Since B is closed in A, it follows that $B = M$. Thus B is τ-closed in A. Now the conclusion is immediate.

(ii) Let A be a self-c-injective module and let B be a submodule of A such that A/B is γ-torsionfree, i.e., non-singular. Then B is a closed submodule of A [2, 4.1]. Now the conclusion follows. $\square$

Therefore we have the following implications.

$$
\begin{array}{ccc}
\tau\text{-torsionfree } \tau\text{-complemented} & \Longrightarrow & \tau\text{-torsionfree self-}\tau\text{-divisible} \\
\Downarrow & & \Downarrow \\
\text{extending} & \Longrightarrow & \text{self-}c\text{-injective} \\
\Downarrow & & \Downarrow \\
\gamma\text{-complemented} & \Longrightarrow & \text{self-}\gamma\text{-divisible}
\end{array}
$$

We would like to thank Professor Patrick F. Smith for communicating us the following example (i).

Example 3.2. (i) We have seen in Example 2.7 that there exist self-τ-divisible modules that are not τ-complemented.

Let us now see another example for the Goldie torsion theory γ. The $\mathbb{Z}$-module $M = \prod_{p \text{ prime}} \mathbb{Z}/p\mathbb{Z}$ is self-c-injective [11, Theorem 6], hence self-γ-divisible. But its torsion submodule is not a direct summand [11, Corollary 3], so that M is not γ-complemented [12, Theorem 1.8].

(ii) By Proposition 3.1, every self-c-injective module is self-γ-divisible. But there exists self-γ-divisible modules that are not self-c-injective.

Indeed, let $M = \mathbb{Z}/p\mathbb{Z} \oplus \mathbb{Z}/p^2\mathbb{Z} \oplus \mathbb{Z}/p^3\mathbb{Z}$ for some prime p. Then the $\mathbb{Z}$-module M is γ-torsion, hence self-γ-divisible, but it is not self-c-injective [10, Example 4.2].

(iii) We have seen that every extending module is γ-complemented, hence self-γ-divisible. But this is no longer true for an arbitrary torsion theory τ.

Indeed, let τ be cogenerated by the Prüfer group $\mathbb{Z}_{p^\infty}$ for some prime p. Then the $\mathbb{Z}$-module $\mathbb{Z}$ is clearly extending. It is easy to see that the natural homomorphism from the closed $\mathbb{Z}$-submodule $p\mathbb{Z}$ of $\mathbb{Z}$ to $\mathbb{Z}$ cannot be extended to an endomorphism of $\mathbb{Z}$, hence $\mathbb{Z}$ is not a self-τ-divisible $\mathbb{Z}$-module.

In what follows, we shall be interested in when a finite direct sum of self-τ-divisible modules is self-τ-divisible. Towards such a result we give some preliminary technical properties.

Proposition 3.3. *Let A_1 and A_2 be modules, $A = A_1 \oplus A_2$ and let p_1, p_2 be the canonical projections on A_1 and A_2 respectively. Then the following statements are equivalent:*

(i) *A_1 is A_2-τ-divisible.*

(ii) *For every submodule B of A such that $B \cap A_1 = 0$ and $p_2(B)$ is a τ-closed submodule of A_2, there exists a submodule C of A such that $A = A_1 \oplus C$ and $B \subseteq C$.*

Proof. (i) $\Rightarrow$ (ii) Suppose first that A_1 is A_2-τ-divisible. Let B be a submodule of A such that $B \cap A_1 = 0$ and $p_2(B)$ is a τ-closed submodule of A_2. The homomorphism $f: B \to p_2(B)$ defined by $f(b) = p_2(b)$ for every $b \in B$ is an isomorphism, because $B \cap A_1 = 0$. By hypothesis, the homomorphism $p_1 f^{-1}: p_2(B) \to A_1$ extends to a homomorphism $v: A_2 \to A_1$. Define $C = \{v(a) + a \mid a \in A_2\}$. It is easy to check that C is a submodule of A and $A = A_1 \oplus C$. Also, we get $B \subseteq C$, because for every $b \in B$ we have $b = p_1(b) + p_2(b) = p_1 f^{-1} p_2(b) + p_2(b) = v p_2(b) + p_2(b) \in C$.

(ii) $\Rightarrow$ (i) Suppose now that (ii) holds. Let M be a τ-closed submodule of A_2 and let $u: M \to A_1$ be a homomorphism. Put $B = \{m - u(m) \mid m \in M\}$. It is easy to check that B is a submodule of A and $B \cap A_1 = 0$. Also, we have $p_2(B) = M$, hence $p_2(B)$ is a τ-closed submodule of A_2. Now there exists a submodule C of A such that $A = A_1 \oplus C$ and $B \subseteq C$. Let $p: A \to A_1$ denote the projection with kernel C and let $q: A_2 \to A_1$ be the restriction of p to A_2. Then for every $m \in M$, $q(m) = p(m) = p(m - u(m)) + u(m) = u(m)$. Hence q extends u and, consequently, A_1 is A_2-τ-divisible. $\qquad\square$

We need the following definition. Two modules A_1 and A_2 are said to be *relatively τ-divisible* if A_1 is A_2-τ-divisible and A_2 is A_1-τ-divisible.

Proposition 3.4. *Let A be a module, let D be an A-τ-divisible module and let B be a submodule of A. Then:*

(i) *D is A/B-τ-divisible.*

(ii) *If B is τ-closed in A, then D is B-τ-divisible.*

Proof. (i) Let C/B be a τ-closed submodule of A/B and let $u: C/B \to D$ be a homomorphism. Denote by $i: C \to A$ and $j: C/B \to A/B$ the inclusions and by $p: C \to C/B$ and $q: A \to A/B$ the natural homomorphisms. Since

C is τ-closed in A and D is A-τ-divisible, there exists a homomorphism $v\colon A \to D$ such that $vi = up$. Thus we obtain the commutative diagram

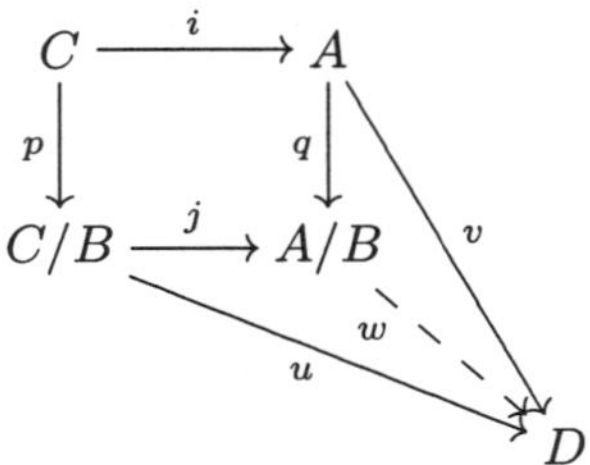

Since B is a submodule of $\mathrm{Ker}\,v$, there exists a homomorphism $w\colon A/B \to D$ such that $wq = v$. For every $c \in C$, we have $wj(c+B) = wqi(c) = vi(c) = up(c) = u(c+B)$, hence $wj = u$. Thus D is A/B-τ-divisible.

(ii) Assume further that B is τ-closed in A. Let M be a τ-closed submodule of B and $u\colon M \to D$ be a homomorphism. Since M is τ-closed in A, u extends to a homomorphism $v\colon A \to D$. Then the restriction of v to B extends u and, consequently, D is B-τ-divisible. $\qquad\square$

Corollary 3.5. *Let A_1 and A_2 be modules such that $A_1 \oplus A_2$ is self-τ-divisible. Then A_1 and A_2 are both self-τ-divisible and relatively τ-divisible.*

Proof. By Proposition 3.4 and Lemma 2.5. $\qquad\square$

Theorem 3.6. *Let A_1 and A_2 be modules. If a module is A_1-τ-divisible and A_2-injective, then it is $(A_1 \oplus A_2)$-τ-divisible.*

Proof. Denote $A = A_1 \oplus A_2$ and let D be an A_1-τ-divisible and A_2-injective module. Let B be a τ-closed submodule of A and $u\colon B \to D$ be a homomorphism. Let f be the restriction of u to $B \cap A_1$, $i\colon B \to A$ and $j\colon B \cap A_1 \to A_1$ be the inclusions and $i_1\colon A_1 \to A$ be the canonical injection. Then $B \cap A_1$ is τ-closed in A_1, because $A_1/(B \cap A_1) \cong (B + A_1)/B \subseteq A/B$. Since D is A_1-τ-divisible, there exists a homomorphism $v\colon A_1 \to D$ such that $vj = f$. Denote $w = vp_1$. Then $wi_1 j = vp_1 i_1 j = vj = f$. Denote by g the restriction of w to B. Since $B \cap A_1 \subseteq \mathrm{Ker}(u - g)$, we can define a homomorphism $h\colon B/(B \cap A_1) \to D$ by $h(b + B \cap A_1) = u(b) - g(b)$ for every $b \in B$. Since $B \cap A_1 \subseteq A_1 = \mathrm{Ker}\,p_2$, we can define a homomorphism $\alpha\colon B/(B \cap A_1) \to A_2$ by $\alpha(b + B \cap A_1) = p_2(b)$ for every $b \in B$. Clearly, α is a monomorphism. Since D is A_2-injective, there exists a homomorphism $\beta\colon A_2 \to D$ such that $\beta\alpha = h$. Denote $\gamma = w + \beta p_2\colon A \to D$. Then for every $b \in B$ we have

$$\gamma i(b) = \gamma(b) = w(b) + \beta p_2(b) = g(b) + \beta \alpha(b + B \cap A_1)$$
$$= g(b) + h(b + B \cap A_1) = g(b) + u(b) - g(b) = u(b),$$

hence $\gamma i = u$. Therefore D is A-τ-divisible. $\qquad\square$

Now we obtain the main result on self-τ-divisible modules.

Corollary 3.7. *Let* $A_1, \ldots, A_n$ *be relatively injective modules. Then* $A_1 \oplus \cdots \oplus A_n$ *is self-τ-divisible if and only if* A_i *is self-τ-divisible for each* i.

Proof. It follows by Lemma 2.5 and Theorem 3.6. $\qquad\square$

Note that such a result holds for extending modules [2, Proposition 7.10] and even for self-c-injective modules [10, Theorem 2.8]. As for the torsion-theoretic case, a weaker property holds for τ-complemented modules, i.e., a finite direct sum of τ-complemented relatively injective modules is again τ-complemented [12, Theorem 3.6]. As we have just seen in Corollary 3.7, we recover the full result for self-τ-divisible modules.

References

1. J. L. Bueso and B. Torrecillas, *On the cotorsion hull in torsion theory*, Comm. Algebra **13** (1985), 1627–1642.
2. N. V. Dung, D. V. Huynh, P. F. Smith and R. Wisbauer, Extending modules, Pitman Research Notes in Mathematics Series, **313**, Longman Scientific & Technical, 1994.
3. L. Fuchs, *Cotorsion modules over noetherian hereditary rings*, Houston J. Math. **3** (1977), 33–46.
4. L. Fuchs and L. Salce, Modules over non-noetherian domains, Mathematical Surveys and Monographs, **84**, Amer. Math. Soc., Providence, 2001.
5. J. S. Golan, Torsion theories, Pitman Monographs and Surveys in Pure and Applied Mathematics, **29**, Longman Scientific & Technical, 1986.
6. D. K. Harrison, *Infinite abelian groups and homological methods*, Ann. Math. **69** (1959), 366–391.
7. J. Henderson and M. Orzech, *Cotorsion modules for torsion theories*, Portugal Math. **36** (1977), 33–40.
8. R. Mines, *Cotorsion modules over noetherian hereditary rings*, in "Abelian Group Theory" (Oberwolfach, 1981), pp. 242–250, Lecture Notes in Math., **874**, Springer, Berlin-New York, 1981.
9. A. P. Mishina and L. A. Skornjakov, *Abelian groups and modules*, Amer. Math. Soc. Transl., Ser. 2, **107** (1976).
10. C. Santa Clara and P. F. Smith, *Modules which are self-injective relative to closed submodules*, in "Algebra and its Applications" (Athens, Ohio, 1999), pp. 487–499, Contemp. Math., **259**, Amer. Math. Soc., Providence, 2000.

11. C. Santa Clara and P. F. Smith, *Direct products of simple modules over Dedekind domains*, Arch. Math. **82** (2004), 8–12.
12. P. F. Smith, A. M. Viola-Prioli and J. E. Viola-Prioli, *Modules complemented with respect to a torsion theory*, Comm. Algebra **25** (1997), 1307–1326.
13. P. F. Smith, A. M. Viola-Prioli and J. E. Viola-Prioli, *On μ-complemented and singular modules*, Comm. Algebra **25** (1997), 1327–1339.
14. B. Stenström, *Pure submodules*, Arkiv for Math. **7** (1967), 159–171.
15. J. Trlifaj, *Covers, envelopes and cotorsion theories*, Lecture notes for the workshop "Homological Methods in Module Theory", Cortona, 2000.
16. J. Trlifaj, *Cotorsion theories induced by tilting and cotilting modules*, in "Abelian Groups, Rings and Modules" (Perth, 2000), pp. 285–300, Contemp. Math., **273**, Amer. Math. Soc., Providence, 2001.
17. C. P. Walker, *Relative homological algebra and abelian groups*, Illinois J. Math. **10** (1966), 186–209.
18. R. Wisbauer, Foundations of module and ring theory, Gordon and Breach, Reading, 1991.

THE GLOBALIZATION PROBLEM FOR INNER AUTOMORPHISMS AND SKOLEM-NOETHER THEOREMS

JEREMY HAEFNER

Department of Mathematics, University of Colorado
Colorado Springs CO, 60933 U.S.A.
E-mail: haefner@math.uccs.edu

ÁNGEL DEL RÍO*

Departamento de Matemáticas
Universidad de Murcia, 30071 Murcia, Spain
E-mail: adelrio@fcu.um.es

We investigate the circumstances under which an inner automorphism of a ring with local units can be built from "local" information. Specifically, we consider three natural "inner" type properties for an automorphism of a ring with local units. We show that every inner automorphism is locally inner but the converse is false, even if the automorphism is "piecewise" inner. On the other hand, we construct a large class of rings for which every locally inner automorphism is actually inner. Finally we obtain some Skolem-Noether type Theorems for infinite matrix and triangular matrix rings.

1. Introduction and preliminaries

We recall (see, e.g., [2]) that a ring A is said to have *local units* provided there exists a set of idempotents E of A such that, for every finite subset $\mathcal{F}$ of A, there is an idempotent $e \in E$ such that $\mathcal{F} \subset eAe$. The set E is said to be a set of local units for A. A natural technique in the study of rings with local units is to try to obtain global information of A from the local unital rings eAe ($e \in E$) of A. In this paper, we investigate the relationship between an inner automorphism and its behavior on the local unital rings eAe ($e \in E$).

Inner automorphisms for rings with local units were introduced in [4] but the following definition comes from [4] and [1].

---

*Research partially supported by DGES and Comunidad Autónoma de Murcia

Definition and Theorem 1.1 ([4], [1]). *Let A be a ring with local units and let $\alpha \in \mathrm{Aut}(A)$. We order the set of idempotents of A via $e \leq f$ if and only if $eAe \subset fAf$. We say that α is* inner *if any of the following equivalent conditions holds:*

(1) *for all $e = e^2 \in A$, (alternatively, for every $e \in E$, being E a set of local units) there exist $u(e) \in \alpha(e)Ae$, and $v(e) \in eA\alpha(e)$, such that for $x \in eAe$, $\alpha(x) = u(e)xv(e)$ and any one of the following equivalent compatibility conditions holds for every two idempotents e, f (for every $e, f \in E$) such that $e \leq f$:*

    (a) $u(f)e = u(e)$.
    (b) $ev(f) = v(e)$.

(2) *α is in the kernel of the natural group homomorphism $\mathrm{Aut}(A) \to \mathrm{Pic}(A)$.*

(3) *α is the restriction of an inner automorphism of $Q(A)$, the ring of multipliers of A in $\mathrm{End}(A_A)$.*

Having defined inner automorphisms globally, we turn our attention to "local" definitions of inner automorphisms. In particular, by considering the behavior of an automorphism when it is restricted to eAe, we can define automorphisms that are "piecewise inner" and "locally inner".

Definition 1.2. Let $\alpha \in \mathrm{Aut}\, A$. For every idempotent e of A, denote by $\alpha_e \colon eAe \to \alpha(e)A\alpha(e)$ the restriction of α to an isomorphism between eAe and $\alpha(e)A\alpha(e)$.

- We say that an idempotent e is *α-invariant* provided $\alpha(e) = e$ and we say that a set of local units E of A is *α-invariant* provided each $e \in E$ is α-invariant.
- The automorphism $\alpha \in \mathrm{Aut}\, A$ is *piecewise constructed* (or *piecewise defined*) if A has a set of α-invariant local units E.
- The automorphism α is *piecewise inner* if A has a set of α-invariant local units E so that for every $e \in E$, α_e is an inner automorphism (in the classical sense) of eAe.
- If e is an idempotent of A, we say that α is *inner at e* provided there are elements $u, v \in A$ such that for all $x \in eAe$, $\alpha(x) = uxv$. Note that if $\alpha(e) = e$, where $e \in E$, then α_e is inner (in the classical sense) if and only if α is inner at e.
- We say that α is *locally inner* provided α is inner at each idempotent e of A.

The main focus of this paper is to determine if every locally inner (piecewise inner) automorphism is inner. While we find that there are locally inner automorphisms that are *not* inner, we also find a large class of rings with local units for which all locally inner automorphisms (and hence all piecewise inner automorphisms) are inner. See Theorem 2.4.

The global/local results of inner automorphisms discussed above also have relevance for unital rings. Assume that A is the ring of multipliers (also known as the translation hull) for a ring I with local units; see [1] or [13] for more details. The idea is to obtain properties of A from properties of the unital rings eIe, where e ranges over a set of local units of I. In this paper, we use this technique to extend the Skolem-Noether result to certain infinitely generated (unital) algebras. See Theorems 4.1, 4.3, 4.4, and 4.5.

We close this section with some comments about some immediate relations between inner, locally inner and piecewise inner automorphisms.

Remark 1.3. Let α denote an automorphism of a ring with local units A.

(1) By definition, every inner automorphism is locally inner.

(2) If the automorphism α is inner at e and $f \le e$, then α is inner at f. This shows that α is locally inner if α is inner at all the elements of a set of local units.

(3) An automorphism α is piecewise inner if and only if it is locally inner and piecewise defined. To see this, assume that α is inner at e and let $u, v \in A$, such that $\alpha(x) = uxv$ for every $x \in eAe$. By replacing u and v by $\alpha(e)ue$ and $ev\alpha(e)$, respectively, we may assume that $u \in \alpha(e)Ae$ and $eA\alpha(e)$. If, in addition, e is α-invariant, then $uv = uev = \alpha(e) = e$ and $vu = \alpha^{-1}(\alpha(vu)) = \alpha^{-1}(uvuv) = \alpha^{-1}(e) = e$, and so α_e is inner. This shows that if α is locally inner and piecewise defined, then it is piecewise inner. The converse is straightforward.

(4) If A is unital then all definitions (inner, locally inner, and piecewise inner) coincide with the classical one; this follows from the previous arguments.

2. Rings for which every locally inner automorphism is inner

In this section, we give a large class of rings with local units for which every locally inner automorphism is inner. Our method relies on transfinite induction on well founded sets (see, e.g., [12]). An ordered set is said to be *well founded* if every nonempty subset contains a minimal element. We recall the formal set theoretical statement of the Recursion Theorem for Well Founded Sets ([12, Theorem III.5.6]): Let $\mathcal{V}$ denote the class of all the ZFC objects, $(A, \leq)$ a well founded set, and $\mathcal{L}: A \times \mathcal{V} \to \mathcal{V}$ a class map. Then there exists a unique map $l: A \to \mathcal{V}$ such that $l(a) = \mathcal{L}(a, \{(x, l(x)) \mid x < a\})$ for every $a \in A$. Well founded sets appear in Set Theory in connection with the Axiom of Foundation which is useful in proofs of relative consistence results (see Chapters III and IV of [12]). We are not aware of any use in Ring Theory of induction of recursion techniques in well founded set. The existence of well founded sublattices in Boolean algebras has been investigated in [5] and [6].

Lemma 2.1. *Every set of local units contains a well founded set of local units.*

Proof. Let E be a set of local units for the ring A with the usual ordering. By the Axiom of Choice, there is a map $f: \mathcal{F} \to E$, where $\mathcal{F}$ is the set of finite subsets of A and $F \subseteq f(F)Af(F)$ for every $F \in \mathcal{F}$. Since $\mathcal{F}$ is well founded by inclusion, there is a unique map $e: \mathcal{F} \to A$ so that $e(F) = f(F \cup \{e(F_1) \mid F_1 \subset F\})$ for every $F \in \mathcal{F}$ by the Recursion Theorem for Well Founded Sets. (Here the class map $\mathcal{L}: \mathcal{F} \times \mathcal{V} \to \mathcal{V}$ which induces e is given by $\mathcal{L}(F, X) = f(F \cup \operatorname{im} X)$, if X is a map whose image is a finite subset of A and $\emptyset$ otherwise.) Since e preserves the order, the image of e is a well founded set of local units. Note that "$\subset$" means properly contained. $\square$

Let e and f be two idempotents of A. Then $\operatorname{Hom}_A(Ae, Af) \simeq eAf$ and we consider this isomorphism as an identification. In particular eAe and fAf are identified with the endomorphism ring of $_AAe$ and $_AAf$ respectively. Then any module isomorphism $_AAe \simeq {}_AAf$ induces a ring isomorphism $eAe \simeq fAf$. Moreover, the isomorphisms $Ae \simeq Af$ correspond, by the previous identification, to elements $v \in eAf$ such that there exists $u \in fAe$ so that $uv = f$ and $vu = e$. More concretely, for such u and v, the map $\phi: Ae \to Af$ given by $\phi(x) = xv$ is an isomorphism of left A-modules with inverse $\phi^{-1}(x) = xu$. In this case, the isomorphism $\alpha: eAe \to fAf$ induced by ϕ_u is given by $\alpha(x) = uxv$. It follows that an automorphism α of A is inner at e if and only if the isomorphism $\alpha_e: eAe \simeq \alpha(e)A\alpha(e)$ is induced by a module isomorphism $\phi_e: Ae \to A\alpha(e)$.

Suppose now that α is locally inner so that for every idempotent e of A, α_e is induced by an isomorphism $\phi_e: Ae \to A\alpha(e)$. If E is a set of local units of A, then we have that $A = \lim_{e \in E} Ae = \lim_{e \in E} A\alpha(e)$. If it happens that $\Phi = (\phi_e \colon Ae \to A\alpha(e) \mid e \in E)$ is a direct system of module isomorphisms, then Φ induces an isomorphism $\phi \colon {}_A A \to {}_A A$ which, in turn, induces α. Then considering A as a subring of its multiplier ring $Q(A)$ which itself is considered as a subring of $\operatorname{End}({}_A A)$, we have that α is inner. (See [1].)

In general, however, Φ is not a direct system. In fact this only happens when the compatibility conditions (either (1) or (2)) of Definition 1.2 hold. But the system Φ that induces α_e at every e is not unique. Our method consists of modifying Φ so that it becomes a direct system. In order to do this, we need to understand how we can modify Φ without changing the automorphism associated to it. This is the task of next lemma.

Lemma 2.2. *Let e and f be two idempotents of A and let $u, u' \in fAe$ and $v, v' \in eAf$ so that $uv = u'v' = f$ and $vu = v'u' = e$. Then the following are equivalent:*

(1) For every $x \in eAe$, $uxv = u'xv'$.

(2) vu' is a central unit of eAe with inverse $v'u$.

(3) There is a central unit c of eAe so that $u' = uc$ and $v' = c^{-1}v$.

Therefore two module isomorphisms $\phi, \psi \colon Ae \to Af$ induce the same ring isomorphism $eAe \to fAf$ if and only if there is a central unit c of eAe so that $\psi(x) = \phi(xc)$ for every $x \in Ae$.

Proof. 1 implies 2. Assume that $uxv = u'xv'$ for every $x \in eAe$. Then $xvu' = vuxvu' = vu'xv'u' = vu'x$. Moreover $vu'v'u = vfu = vu = e$ and similarly $v'uvu' = e$. Therefore vu' is a central unit of eAe.

2 implies 3. Notice that $u' = uvu'$ and $v' = v'uv$, so that $c = vu'$ suffices.

3 implies 1. Let c be a central unit of eAe so that $u' = uc$ and $v' = c^{-1}v$. Then $u'xv' = ucxc^{-1}v = uxv$. The final statement follows from the observation made above. $\qquad\square$

Before giving the main result of this section we adopt the following notation.

Notation 2.3. For every $e = e^2 \in A$, $Z(e)$ denotes the center of eAe and $Z(e)^*$ the group of units of $Z(e)$. For every set E of idempotents of A, let

$$Z(E) = \{(x_e)_{e \in E} \in \prod_{e \in E} Z(e) \mid x_e f = x_f \text{ for every } f \le e\}$$

and $Z(E)^* = Z(E) \cap \prod_{e \in E} Z(e)^*$. If E is a set of idempotents of A and $e \in E$, then set $E_e = \{f \in E \mid f < e\}$ and let $\lambda_e \colon Z(e) \to Z(E_e)$ be the map given by $\lambda_e(x) = (xf)_{f \in E_e}$. Note that if e is a minimal element of E, then $E_e = \emptyset$; in this case we consider $Z(E_e)$ as a singleton set. The map λ_e restricts to a map $Z(e)^* \to Z(E_e)^*$ that we denote also by λ_e.

For every $f \in E$, let $\pi_f \colon \prod_{e \in E} Z(e) \to Z(f)$ be the projection map. Note that if $f \in E_e$, then $xf = \pi_f \lambda_e(x)$ for every $x \in Z(e)$.

We now present the crux of our argument.

Theorem 2.4. *Assume A is a ring with local units such that there is a collection of local units E for which*

1. *$\leq$ is well founded in E, and*
2. *for every non minimal element $e \in E$, $\lambda_e \colon Z(e)^* \to Z(E_e)^*$ is surjective.*

Then any locally inner automorphism of A is inner.

Proof. Let α be a locally inner automorphism of A and let $u, v \colon E \to A$ be maps so that $u(e) \in \alpha(e)Ae$, $v(e) \in eA\alpha(e)$ and $\alpha(x) = u(e)xv(e)$ for every $e \in E$ and $x \in eAe$. By Axiom of Choice and (2), for every non-minimal $e \in E$ there is a map $\mu_e \colon Z(E_e)^* \to Z(e)^*$ such that $\lambda_e \mu_e = 1_{Z(E_e)^*}$. If e is a minimal element, we consider $Z(E_e)^*$ as a singleton, say $Z(E_e)^* = \{\delta\}$, and we choose μ_e by $\mu(\delta) = e$. Consider the class map $\mathcal{L} \colon E \times \mathcal{V} \to \mathcal{V}$ given by

(A) $\mathcal{L}(e, \emptyset) = (u(e), v(e)))$;
(B) $\mathcal{L}(e, X) = (u(e)\mu_e((v(e)U(f))_{f \in E_e}), \mu_e((v(e)U(f))_{f \in E_e})^{-1}v(e))$ if

$$X = \{(f, (U(f), V(f)) \mid f \in E_e\}$$

and the following conditions hold for every $f \in E_e$

    B1. $(v(e)U(f))_{f \in E_e} \in Z(E_e)^*$
    B2. $U(f) \in \alpha(f)Af$, $V(f) \in fA\alpha(f)$ and
    B3. $\alpha(x) = U(f)xV(f)$ for every $x \in fAf$ and

(C) $\mathcal{L}_1(e, X) = \mathcal{L}_2(e, X) = \emptyset$ otherwise.

By the Recursion Theorem for Well Founded Sets there is a unique map $w \colon E \to \mathcal{V}$ such that

$$w(e) = \mathcal{L}(e, \{(f, (w(f)) \mid f \in E_e\})$$

for every $e \in E$.

We claim that there are maps $u', v' \colon E \to A$ such that the following statements hold for every $e \in E$

(1) $w(e) = (u'(e), v'(e))$,
(2) $(v(e)u'(f))_{f \in E_e} \in Z(E_e)^*$ if e is not minimal,
(3) $u'(f) \in \alpha(f)Af$, $v'(f) \in fA\alpha(f)$ and
(4) $\alpha(x) = u'(f)xv'(f)$ for every $x \in fAf$.

We prove this by induction on e for well founded sets. If e is minimal in E, then $w(e) = \mathcal{L}(e, \emptyset) = (u(e), v(e))$, so setting $u'(e) = u(e)$ and $v'(e) = v(e)$ conditions 3 and 4 hold and condition 2 vanishes. Now let e be a non-minimal element of E and assume that $w(f) = (u'(f), v'(f))$ for every $f \in E_e$ and conditions 1-4 hold for all the elements in E_e. If $f \in E_e$ then setting $u = \alpha(f)u(e)f$, $v = fv(e)\alpha(f)$, $u' = u'(f)$ and $v' = v'(f)$, it is easy to see that the following conditions hold: $uv = u'v' = \alpha(f)$, $vu = v'u' = f$ and $\alpha(x) = uxv = u'xv'$ for every $x \in fAf$. By Lemma 2.2 $v(e)u'(f)$ is a central unit of fAf. Moreover for every $f_1 < f < e$, one has $v(e)u'(f)f_1 = v(e)u'(f_1)$. We conclude that $(v(e)u'(f))_{f \in E_e} \in Z(E_e)^*$. Therefore $\{(f, w(f)) \mid f \in E_e\} = \{(f, (u'(f), v'(f))) \mid f \in E_e\}$ is of the form B and satisfies conditions B1-B3 for $U = u'$ and $V = v'$. Therefore

$$w(e) = \mathcal{L}(e, \{(f, w(f)) \mid f \in E_e\})$$
$$= (u(e)\mu_e((v(e)u'(f))_{f \in E_e}), \mu_e((v(e)u'(f))_{f \in E_e})^{-1}v(e)).$$

Now set

$$u'(e) = u(e)\mu_e((v(e)u'(f))_{f \in E_e}) \quad \text{and}$$
$$v'(e) = \mu_e((v)u'(f))_{f \in E_e})^{-1}v(e).$$

Conditions 1, 3 and 4 follow immediately. Finally, 2 follows from the fact that for every $f \in E_E$, $v(e)u'(f) \in Z(f)$ and the following computation

$$u'(e)f = u(e)\mu_e((v(e)u'(g))_{g \in E_e})f$$
$$= u(e)\pi_f(\lambda_e\mu_e((v(e)u'(g))_{g \in E_e}))$$
$$= u(e)\pi_f((v(e)u'(g))_{g \in E_e})$$
$$= u(e)v(e)u'(f) = u'(f),$$

where the second equality is a consequence of applying $\pi_f(\lambda_e(x)) = xf$ to $x = \mu_e((v(e)u'(g))_{g \in E_e})$. So u' and v' satisfy the conditions of Definition and Theorem 1.1 (recall that conditions (a) and (b) of the first statement in Definition and Theorem 1.1 are equivalent) and we conclude that α is inner. $\qquad\square$

It is desirable to have a sufficient condition that guarantees every locally inner automorphism is inner without the condition that a set of local units E is well founded. If every set of local units which satisfies the condition (2) of Theorem 2.4 also contains a well founded set of local units satisfying condition (2), then condition (1) of Theorem 2.4 would be unnecessary. Unfortunately, when passing from a set of local units E to a well founded set of local units E' contained in E, one has that $Z(E_e)^*$ can be strictly contained in $Z(E'_e)^*$ and hence condition (2) does not pass from E to E'. The following Corollary gives an alternative sufficient condition for every locally inner automorphism being inner.

Note that if E is a set of local units and $e \in E$, then E contains a set of local units E' so that e is the unique minimal element of E'.

Corollary 2.5. *If A has a set of local units E with a unique minimal element e so that for every $f \in E$, the map $\rho_f \colon Z(f) \to Z(e)$, $\rho_f(x) = xe$, is an isomorphism, then every locally inner automorphism of A is inner.*

Proof. If E_1 is a well founded set of local units contained in E, then $E_1 \cup \{e\}$ is a well founded set of local units satisfying the hypothesis. Thus by Lemma 2.1 we can assume that E is well founded.

Let $e \neq f \in E$ and let $\pi \colon Z(E_f) \to Z(e)$ denote the projection on the e-th coordinate. Then $\rho_f = \pi \circ \lambda_f$ and therefore π is surjective. If $\pi((a_{f'})_{f' \in E_e}) = 0$, then $\rho_{f'}(a_{f'}) = 0$ for every $f' \in E_f$. Since ρ_f is an isomorphism,$(a_{f'})_{f' \in E_e} = 0$. This shows that π is a bijection and hence so is λ_f Therefore the restriction of λ_f to $Z(e)^* \to Z(E_e)^*$ is also a bijection. $\square$

There are many examples of rings with local units that satisfy the hypotheses of either Theorem 2.4 or Corollary 2.5. We indicate some of these examples next.

Notation 2.6. *Matrix rings.* Given a set X and a ring R, we adopt the following notation:

$$\mathrm{FM}_X(R) = \text{Ring of finite matrices over } R \text{ indexed by } X,$$
$$\mathrm{RFM}_X(R) = \text{Ring of row finite matrices over } R \text{ indexed by } X,$$
$$\mathrm{RCFM}_X(R) = \text{Ring of row and column finite matrices}$$
$$\text{over } R \text{ indexed by } X.$$

Incidence rings. Given a preordered set P, a unital ring R and a two-sided ideal J of R, let

$$\mathrm{FI}(P, R, J) = \{\alpha \in \mathrm{FM}_P(R) \mid x \not\leq y \implies \alpha(x, y) \in J\}$$
$$\mathrm{RCFI}(P, R, J) = \{\alpha \in \mathrm{RCFM}_P(R) \mid x \not\leq y \implies \alpha(x, y) \in J\}.$$

Then $\mathrm{FI}(P, R, J)$ is a subring with local units of $\mathrm{FM}_P(R)$. If $J = 0$, then $FI(P, R) = FI(P, R, 0)$ is the incidence ring of P over R.

Let X be a set and $i, j \in X$. Then e_{ij} is going to denote the matrix indexed by X having 1 at the (i, j)-entry and 0 elsewhere and $e_i = e_{ii}$. The set X will be always clear from the context. If Y is a subset of X, then $e_Y = \sum_{i \in Y} e_i$. Note that the sum makes sense even if Y is infinite.

If we are working with a direct product $\prod_{i \in X} R_i$ of rings, the meaning of e_i and e_Y, for $i \in X$ and $Y \subseteq X$ will be different but without possible confusion. In that situation, e_i is the X-tuple having 1 at the i-coordinate and 0 elsewhere and $e_Y = \sum_{i \in Y} e_i$.

We refer the reader to [3] for the definition and notation about smash product rings.

Corollary 2.7. *For each of the following rings, every locally inner automorphism is inner, and so every piecewise inner automorphism is inner.*

(1) The direct sum of rings with identity.

(2) The ring $\mathrm{BRFM}_P(R)$ of row finite matrices indexed by an infinite set P, having entries from a unital ring such that the cardinality of the set of their nonzero entries is strictly smaller than the cardinality of P.

(3) $\mathrm{FI}(P, R, J)$ where R is a ring with unit, P is a preordered set and J is an ideal of R. This includes the ring $\mathrm{FM}_P(R)$ of finite matrices over R indexed by a set P.

(4) The path algebra of a quiver with infinitely many vertices. That is the non unital algebra generated by the oriented paths of the quiver. (Note that this algebra is unital if and only if the set of vertices of the quiver is finite).

(5) The smash product of a strongly graded ring.

(6) The smash product of a polynomial ring in several (even infinite) variables with coefficients on a unital ring R (graded by the free abelian group generated by the variables).

Proof. (1) Is a particular case of (3) taking I as a partially ordered set with no non trivial relations. However we include a short and less technical proof for this. Let $A = \oplus_{i \in I} R_i$, a direct sum of unital rings. The set $E = \{e_X \mid X$

finite subset of I} is a well founded set of local units. Moreover, if X is a finite subset of I, then $E_{e_X} = \{e_Y \mid Y \subset X\}$ and $Z(e_X) = \oplus_{i \in X} Z(R_i)$. If X has just one element, then $Z(E_{e_X}) = \{0\}$ and λ_{e_X} is surjective. Assume now that X has more than one element and let $a = (a_Y)_{Y \subset X} \in Z(E_{e_X})$. Then the element $b = (b_i) \in e_X A e_X$, defined via $b_i = 0$ if $i \notin X$ and $b_i = a_{\{i\}}$ if $i \in X$, belongs to $Z(e_X)$ and $\lambda_{e_X}(b) = a$. So the hypothesis of Corollary 2.5 holds.

(2) Let $A = \mathrm{BRFM}_P(R)$ and fix an element $x \in X$. Then $\{e_Y \mid x \in Y, |Y| < |X|\}$ is a set of local units of A which satisfies the conditions of Corollary 2.5.

(3) For every subset X of P we define the following equivalent relation $\sim_X$ on X.

$$x \sim_X y \iff (\exists a_0, a_1, \ldots, a_n \in X) x = a_0, y = a_n,$$
$$(\forall i = 1, \ldots n)(a_{i-1} \le a_i \text{ or } a_{i-1} \ge a_i)$$

(Note that $\sim_X$ depends on the set X). Then,

$$Z(e_X) = \{\alpha \in A \mid \alpha(x, y) \ne 0 \implies x = y \in X,$$
$$\alpha(x, x) \in Z(R),$$
$$\alpha(x, x) - \alpha(y, y) \in \mathrm{Ann}_{Z(R)}(J),$$
$$\text{and } x \sim_X y \implies \alpha(x, x) = \alpha(y, y)\}$$

(where $\mathrm{Ann}_{Z(R)}(J)$ denotes the annihilator of J in $Z(R)$. We leave it to the reader to check the details.

Let $\sim$ denote the equivalence relation on P given by $\sim = \cup_{|X| < \infty} \sim_X$. Let Y be a set of representatives of $X/\sim$. Then

$$E = \{e_X \mid X \subseteq P, |X| < \infty, (\forall x \in X)(\exists a_0, a_1, a_2, \ldots, a_n \in X) x = a_0,$$
$$a_n \in Y, \text{ and } (\forall i = 1, \ldots n)(a_{i-1} \le a_i \text{ or } a_{i-1} \ge a_i)\}.$$

is a set of local units of A. We prove that E satisfies the conditions of Theorem 2.4. Plainly E is well founded. Note that for every finite subset X of P, there exist $x = y_0^x, y_1^x, \ldots, y_{n_x}^x \in P$ such that $y_{n_x}^x \in Y$ and for every $i = 1, \ldots, n_x$, either $y_{i-1}^x \le y_i^x$ or $y_{i-1}^x \ge y_i^x$. Thus $X \subseteq \hat{X} = \cup_{x \in X}\{y_0^x, y_1^x, \ldots, y_{n_x}^x\}$ and $e_{\hat{X}} \in E$. Now we check condition (2) of Theorem 2.4. If $e_X \in E$, then the map $\rho_X \colon Z(e_X) \to Z(e_{Y \cap X})$, given by $\rho_X(a) = a e_Y$, is a ring isomorphism. Moreover, $\rho_X = \pi_{Y \cap X} \circ \lambda_e$ where $\pi_{Y \cap X} \colon Z(E_e) \to Z(e_{Y \cap X})$ is the projection. As in the proof of Corollary 2.5 one can prove that π_Y is an isomorphism and hence λ_e is an isomorphism for every $e \in E$. It only remains to apply Theorem 2.4.

Statement (4) follows by similar arguments as those found in (3).

(5) (See [3] for the notation). Let G be a group with identity 1 and R a strongly graded ring. Set $E = \{e_X \mid 1 \in X \subseteq G, |X| < \infty\} \subseteq R\#P_G$. Let X be a finite subset of G which contains 1. For every g in G let $x_g^1, x_g^2, \ldots, x_g^{n_g} \in R_g$ and $y_{g^{-1}}^1, y_{g^{-1}}^2, \ldots, y_{g^{-1}}^{n_g} \in R_{g^{-1}}$ such that $\sum_{i=1}^{n_g} x_g^i y_{g^{-1}}^i = 1$. By straightforward arguments one can show that $Z(e_X)$ is the set of all $\alpha \in e_X R\#P_G e_X$ such that

$$\alpha(g, g') = \begin{cases} 0 & \text{if } g \neq g', \\ \sum_{i=1}^{n_g} x_g^i \alpha(1, 1) y_{g^{-1}}^i & \text{if } g = g' \in X. \end{cases}$$

Then it is clear that the map $\rho_X \colon Z(e_X) \to Z(e_1)$ is an isomorphism and Corollary 2.4 applies.

(6) This can be seen as a particular case of (3) because such a smash product is isomorphic to the incidence ring $\mathrm{FI}(G, R)$, where G is the group generated by the variables with the order given by $x \leq y$ if and only $x^{-1}y$ belongs to the monoid generated by the variables. $\qquad\square$

3. Counterexamples

Having found a class of rings for which inner and locally inner automorphisms are the same, we turn our attention to proving that this is not the case in general. As well as giving one example of a piecewise inner automorphism that is not inner, we give an example of an inner automorphism which is not piecewise defined.

Example 3.1. *An inner (and so locally inner) automorphism which is not piecewise defined (and so not piecewise inner).* Let R be an arbitrary unital ring and $A = \mathrm{FM}_{\mathbf{Z}}(R)$. Let $u \in \mathrm{RCFM}_{\mathbf{Z}}(R)$ be given by $u(x, y) = \delta_{x+1, y}$. Then u is invertible and conjugation by u restricts to an inner automorphism of A. But it is not difficult to see that 0 is the only α-invariant element of A.

Example 3.2. *A piecewise inner automorphism that is not inner.* Let K be any unital ring which admits a non-inner automorphism and let ϕ be such. Let

$$A_1 = K[X_1, X_1^{-1}, X_2, X_2^{-1}, \ldots]$$

be the ring of skew Laurent polynomials in countably many commuting variables, so that $X_n k = \phi(k) X_n$ for every $n \in \mathbf{N}$ and $k \in K$. For every

$n \in \mathbf{N}$, let $A_n = K[X_n, X_n^{-1}, \ldots]$ be the subring of A_1 generated by $K \cup \{X_m, X_m^{-1} \mid m \geq n\}$.

Let A be the ring of upper triangular matrices α indexed by $\mathbf{N}$, such that for every $n, m \in \mathbf{N}$, the (n, m)-entry $\alpha(n, m)$ of α belongs to A_n and $\alpha(n, m) = 0$ for almost all $(n, m) \in \mathbf{N}^2$. For every $n \in \mathbf{N}$, let $f_n = \sum_{i=1}^{n} e_i$. Then $\{f_n \mid n \in N\}$ is a set of local units of A. Let Φ be the automorphism of A given by the following condition: For every $n \in \mathbf{N}$, the restriction $\Phi|_{Af_n}$ of Φ to Af_n is conjugation by $u_n = X_n f_n$. (Note that $Af_n = f_n Af_n$). Φ is well-defined because $X_n^{-1} f_n \cdot X_{n+1} f_{n+1}$ is central in Af_n for every $n \in \mathbf{N}$. And by definition, Φ is piecewise inner.

Assume that Φ is inner. By Definition 1.1, for every $n \in \mathbf{N}$, there are $u_n' \in Af_n$ and $v_n' \in Af_n$ such that if $x \in Af_n$, $\Phi(x) = u_n' x v_n'$ and $u_{n+1}' f_n = u_n'$. By Lemma 2.2, for every $n \in \mathbf{N}$ there are $c_n, d_n \in Z(f_n)$ such that $u_n' = u_n d_n$, $u_n' = c_n u_n$ and $c_n d_n = f_n = d_n c_n$. Then $d_n = \lambda_n f_n$ for some invertible element $\lambda_n \in A_n$. Thus, $u_1' = u_n' f_1 = u_n d_n f_1 = \lambda_n X_n f_1$ and $v_1' = v_n' f_1 = c_n v_n f_1 = \lambda_n^{-1} X_n f_1 \in A_n f_1$ for every $n \in \mathbf{N}$. Therefore $\lambda_1 \in X_1^{-1} A_n$ for every n and hence $u_1' = a f_1$ and $v_1 = a^{-1} f_1$ for some $a \in K$. We conclude that $\phi(x) = a x a^{-1}$ for every $x \in K$ and this yields to a contradiction with the assumption on ϕ.

4. Skolem-Noether results

Several versions of the Skolem-Noether Theorem exist for finitely generated algebras (see [10], [7] and [8]). In this section we show how Theorem 2.4 can be applied to obtain a Skolem-Noether result for certain classes of infinitely generated algebras. As we mentioned in the introduction, this is an example of how non-unital results can be used to obtain results about unital rings.

Let R be a unital ring, P a partially ordered (not necessarily locally finite) and $A = \mathrm{FI}(P, R)$, the incidence ring. Then the ring of multipliers $Q(A)$ of A can be identified with $\mathrm{RCFI}(P, R)$. For example if the order in P is given by $x \leq y$ for every $x, y \in P$, then $\mathrm{FI}(P, R, J) = \mathrm{FM}_P(R)$ and its ring of multipliers is $\mathrm{RCFM}_P(R)$. If P is totally ordered, then $\mathrm{RCFI}(P, R)$ can be viewed as an upper triangular matrix ring.

Recall from [1] that every automorphism of a ring with local units A extends uniquely to an automorphism of $Q(A)$ and an automorphism of A is inner if and only if it is the restriction of a inner automorphism of $Q(A)$. Moreover in some cases every automorphism of $Q(A)$ restricts to an automorphism of A; see [9] and [1].

Given a unital ring A, a subring R of A and a set X we consider R

embedded in $\mathrm{RCFM}_X(A)$ diagonally. In particular, if Y is a finite subset of X and R is a unital subring of A, then $R + M_Y(A)$ is a unital subring of $\mathrm{RCFM}_X(A)$.

Theorem 4.1. *Let A be a ring and R a subring of A. Assume that for every pair of finite sets X and $Y \subseteq X$, every injective R-homomorphism of algebras $R + M_Y(A) \to M_X(A)$ extends to an inner automorphism of $M_X(A)$. Then all the R-automorphisms of the rings $\mathrm{RCFM}_X(A)$ and all the R-automorphisms of $\mathrm{RFM}_X(A)$ are inner.*

Proof. Let α be an R-automorphism of $B = \mathrm{RCFM}_X(A)$. By [9] α restricts to an automorphism of $I = \mathrm{FM}_X(A)$. Let α' be the restriction of α to an automorphism of I. We claim that α' is locally inner. The set $Y = \{e_F \mid F \text{ finite subset of } X\}$ is a set of local units of I. For every finite subset F of X let $I_F = R + e_F I e_F \simeq R + M_F(A)$ and F' be a finite subset of X such that $e_F, \alpha(e_F) \leq F'$. Then α restricts to an injective R-homomorphism $\alpha_F \colon I_F \to I_{F'}$. By the hypothesis and the arguments given prior to the theorem, there exists an invertible element w of $I_{F'}$, such that $\alpha'(x) = \alpha_F(x) = w^{-1}xw$, for every $x \in I_F$. Thus α' is inner at e_F for every finite subset F of X and hence α' is locally inner. By Corollary 2.7, α' is inner. Since B is the ring of multipliers of I, α' is the restriction of an inner automorphism β of B and every automorphism of I extends uniquely to an automorphism of B (see [1]). Thus $\alpha = \beta$ is inner.

Now assume that α is an R-automorphism of $\mathrm{RFM}_X(A)$. By [9, Proposition 6], there is an inner automorphism τ of $\mathrm{RFM}_X(A)$ such that $\alpha\tau$ restricts to an automorphism of I. Thus $\alpha\tau$ restricts to an automorphism of B. By the previous paragraph this automorphism is inner over B. Since every automorphism of I extends uniquely to an automorphism of $\mathrm{RFM}_X(R)$, $\alpha\tau$ is inner over $\mathrm{RFM}_X(A)$ and hence α is inner over $\mathrm{RFM}_X(A)$. $\square$

Note that by the classical Skolem-Noether Theorem, every central simple algebra satisfies the conditions of Theorem 4.1. More generally, if S is a semilocal ring and A is a separable finitely generated projective S-algebra such that the center R of A has no idempotents other than 0 and 1, then the conditions of Theorem 4.1 holds. This is a direct consequence of a Theorem of Childs and DeMeyer [8, Theorem 1.2]. Thus combining Theorem 4.1 and the Childs and DeMeyer Theorem one deduces the following corollary.

Corollary 4.2. *Let S be a semilocal ring and A a separable finitely generated projective S-algebra whose center R of A has no idempotents other than 0 and 1. Then every K-automorphism of the ring $\mathrm{RCFM}_X(A)$ of row*

and column finite matrices (resp. the ring $\mathrm{RFM}_X(A)$ *of row finite matrices)
over A indexed by a set X is inner.*

Jøndrup has given several Skolem-Noether type Theorems for triangular
matrix rings [10] [11]. It is natural to expect to obtain results similar to
Corollary 4.2 for infinite triangular matrix rings by using Jøndrup's results
and arguments as above. Unfortunately the more general statement does
not hold. Indeed, let R be a unital ring, $X = \mathbf{Z}$ and α the automorphism of
the $\mathrm{RCFI}(\mathbf{Z}, R)$ induced by the the automorphism of X given by $n \mapsto n+1$.
Then α is a noninner R-automorphism. However, using an argument similar
to the one found in the proof of Theorem 4.1 together with the above
mentioned results from [10] and [11], we can prove the final three results of
this paper. We leave the details to the reader.

Theorem 4.3. *Let R be a artinian simple algebra finite dimensional over
its center K, X a totally ordered set and $A = \mathrm{RCFI}(X, R)$ the incidence
ring of X over R. Let α be a K-automorphism of A such that the set
$\{e_F \mid F$ is finite$\}$ contains a set of α-invariant local units of $\mathrm{FI}(X, R)$.
Then α is inner.*

Theorem 4.4. *Let K be a commutative ring, R a prime K-algebra and
X a totally ordered set. Let α be an R-automorphism of $\mathrm{RCFI}(X, R)$, such
that $\{e_F \mid F$ is finite$\}$ contains a set of α-invariant local units of $\mathrm{FI}(X, R)$.
Then α is inner.*

Theorem 4.5. *Let R-be a commutative factorial domain, P a prime ideal
of R and X be a totally ordered set. Then every R-automorphism of
$\mathrm{RCFI}(X, R, P)$ such that $\{e_F \mid F$ is finite$\}$ contains a set of α-invariant
local units of $\mathrm{FI}(X, R, P)$ is inner.*

References

1. G. Abrams, J. Haefner and Á. del Río, Approximating rings with local units
   via automorphisms, Acta Math. Hungarica **82**(3) (1999), 207–226.
2. P. N. Ánh and L. Márki, "Morita equivalence for rings without identity",
   Tsukuba J. Math. **11**(1) (1987), 1–16.
3. M. Beattie, A generalization of the smash product of a graded ring, J. Pure
   Appl. Algebra **52** (1988), 219–226.
4. M. Beattie and Á. del Río, The Picard group of a category of graded modules,
   Comm. Algebra **24** (1996), 4397–4414.
5. R. Bonnet and M. Rubin, On well-generated Boolean algebras. Ann. Pure
   Appl. Logic **105** (2000), 1–50.

6. R. Bonnet and M. Rubin, On essentially low, canonically well-generated Boolean algebras. J. Symbolic Logic **67** (2002), 369–396.

7. S.U. Chase, Two remarks on central simple algebras, Comm. Algebra, **12**(18), (1984) 2279–2289.

8. L. N. Childs and F. R. DeMeyer, On automorphisms of separable algebras, Pacific J. Math. **23**, (1967) 25–34.

9. J. Haefner, Á. del Río, J. J. Simón, Isomorphisms of row and column finite matrix rings, Proc. Amer. Math. Soc., **125** (1997), 1651–1658.

10. S. Jøndrup, Automorphisms of upper triangular matrix rings, Arch. Math. **49** (1987), 497–502.

11. S. Jøndrup, The group of automorphisms of certain subalgebras of matrix algebras, J. Algebra **141** (1991), 106–114.

12. K. Kunen, *Set Theory. An Introduction to Independence Proofs*, Studies in Logic and the Foundations of Mathematics 102, North-Holland, Amsterdam, 1980.

13. M. Petrich, Ideal extensions of rings, Acta. Math. Hungarica **45** (1985), 263–285.

KRULL MONOIDS AND THEIR APPLICATION IN MODULE THEORY[*]

ALBERTO FACCHINI

*Dipartimento di Matematica Pura e Applicata,
Università di Padova, I-35131 Padova, Italy*

In this paper, we propose a hierarchy for the classification of the monoids $V(R)$ that describe direct-sum decompositions of finitely generated projective modules over a ring R, and we underline that these direct-sum decompositions are very regular when $V(R)$ is a Krull monoid. We describe the archimedean components, the idempotents and the subgroups of $V(R)$. A well-known characterization of Krull domains also holds for Krull monoids (Theorem 3.4). We prove that the monoid $V(R)$ is free whenever $V(R)$ is a Krull monoid and R is a commutative ring (Theorem 4.2). The submonoid $V(\mathcal{S}_R)$ studied in [15] is always free for a commutative ring R (Theorem 4.4).

Introduction

The algebraic structure that describes direct-sum decompositions of finitely generated projective right modules over a fixed ring R is the reduced commutative monoid $V(R)$ defined as follows. Let proj-R denote the class of all finitely generated projective right R-modules. Fix a set $V(R)$ of representatives of the finitely generated projective right R-modules up to isomorphism. Then for every module $A_R \in$ proj-R there is a unique module $\langle A_R \rangle \in V(R)$ isomorphic to A_R. In particular, for all $A_R, B_R \in$ proj-R, $A_R \cong B_R$ if and only if $\langle A_R \rangle = \langle B_R \rangle$. The set $V(R)$ becomes a monoid, that is, a set with an associative operation and a two-sided identity, if we define the operation (addition) in $V(R)$ by $\langle A_R \rangle + \langle B_R \rangle = \langle A_R \oplus B_R \rangle$ for all $\langle A_R \rangle, \langle B_R \rangle \in V(R)$. In this paper, *all the monoid we shall consider will be commutative monoids*, so that when we say "a monoid" we mean "a commutative monoid". For a monoid M, we shall denote by $U(M)$ the set of all

---

[*]Partially supported by Ministero dell'Istruzione, dell'Università e della Ricerca (Italy) and by Departament d'Universitats, Recerca i Societat de la Informació (Generalitat de Catalunya). This paper was written during a sabbatical year of the author at the Centre de Recerca Matemàtica (Barcelona). He acknowledges the kind hospitality received.

elements $a \in M$ with an opposite $-a \in M$, and we shall say that M is *reduced* if $U(M) = \{0\}$. For every monoid M, the monoid $M_{\mathrm{red}} = M/U(M)$ is reduced. The monoid $V(R)$ is always a reduced commutative monoid. It faithfully describes the behavior of direct-sum decompositions of finitely generated projective R-modules up to isomorphism, in the sense that to every decomposition of a projective module $A_R \in$ proj-R as a direct sum of finitely many submodules there corresponds a decomposition of the element $\langle A_R \rangle$ of the monoid $V(R)$ as a sum of elements of $V(R)$, and two direct-sum decompositions of A_R are isomorphic in the sense of the Krull-Schmidt theorem if and only if they correspond to the same sum decomposition of $\langle A_R \rangle$ in the monoid $V(R)$ up to the order of the summands. More generally, the monoids $V(R)$ faithfully describe the behavior of direct-sum decompositions of *any* right S-module M_S with R isomorphic to the endomorphism ring of M_S [14, p. 186], because the category proj-R turns out to be equivalent to the full subcategory add(M_S) of Mod-S whose objects are all S-modules isomorphic to direct summands of finite direct sums M_S^n of copies of M_S. As the category of finitely generated projective right R-modules is equivalent to the dual of the category of finitely generated projective left R-modules, $V(R)$ describes the behavior of direct-sum decompositions of finitely generated projective left R-modules as well.

There is a natural pre-order (= reflexive and transitive relation) on any commutative additive monoid M, called the *algebraic pre-order* on M, defined by $x \leq y$ if there exists $z \in M$ such that $x + z = y$. An element u of M is an *order-unit* if for every $x \in M$ there exists an integer $n \geq 0$ such that $x \leq nu$. The element $\langle R_R \rangle$ is an order-unit in $V(R)$. More generally, the order-units of $V(R)$ are exactly the elements $\langle A_R \rangle$ of $V(R)$ with A_R a *progenerator*, that is, a finitely generated projective generator

The following is a natural hierarchy that describes how good or bad the structure of $V(R)$ can be, that is, how much the uniqueness of factorization in $V(R)$ can hold or fail.

(1) The best case is that of *the rings R for which $V(R)$ is free*, that is, $V(R)$ is isomorphic to a direct sum $\mathbb{N}^{(I)}$ for some set I. Here $\mathbb{N}$ denotes the additive monoid of non-negative integers, and $\mathbb{N}^{(I)}$ denotes the monoid of all sequences of non-negative integers indexed in I that are zero almost everywhere. As $V(R)$ has an order-unit, the set I must be necessarily a finite set. The monoid $V(R)$ is free if and only if the Krull-Schmidt Theorem holds for finitely generated projective R-modules, that is, every finitely generated projective R-module A_R is a direct sum $A_R = A_1 \oplus \cdots \oplus A_n$ of (necessarily finitely many, finitely generated and projective)

indecomposable R-modules, and if $A_R = A_1 \oplus \cdots \oplus A_n = A'_1 \oplus \cdots \oplus A'_m$ are any two such direct-sum decompositions, then $n = m$ and there is a permutation σ of $\{1, 2, \ldots, n\}$ such that $A_i \cong A'_{\sigma(i)}$ for every $i = 1, 2, \ldots, n$.

(2) The monoid $V(R)$ *is a Krull monoid*. Recall that a *divisor homomorphism* is a monoid homomorphism φ of a monoid M into a monoid N such that $\varphi(x) \le \varphi(y)$ implies $x \le y$ for every $x, y \in M$. A commutative monoid M is a *Krull monoid* if there exists a divisor homomorphism $\varphi\colon M \to F$ into a free monoid F [19, Theorem 23.4]. Recall that the *Grothendieck group* $G(M)$ of a monoid M is the abelian group having the same generators and relations as the monoid M. The Grothendieck group $G(V(R))$ of the monoid $V(R)$ is usually denoted by $K_0(R)$. When M is cancellative, $G(M)$ is the smallest abelian group containing M. Every divisor homomorphism $\varphi\colon M \to F$ of a commutative monoid M into a free monoid F induces an isomorphism $M_{\mathrm{red}} \to \varphi(M)$. As $V(R)$ is reduced, it follows that if $V(R)$ is a Krull monoid, then $V(R)$ is cancellative and the Grothendieck group $K_0(R)$ is a free abelian group. A *valuation* of a monoid M is a non-zero monoid homomorphism $v\colon M \to \mathbb{N}$. Every valuation $M \to \mathbb{N}$ induces a non-zero group homomorphism $G(M) \to \mathbb{Z}$. Conversely, every non-zero group homomorphism $G(M) \to \mathbb{Z}$ that maps M into $\mathbb{N}$ induces a valuation $M \to \mathbb{N}$. Thus valuations can be also seen as non-zero group homomorphisms $G(M) \to \mathbb{Z}$ that map M into $\mathbb{N}$, i.e., non-zero homomorphisms of pre-ordered groups, where $G(M)$ is the pre-ordered group whose positive cone is the image of M in $G(M)$. Valuations are the right tool to study Krull monoids, because one of the characterizations of cancellative Krull monoids is that a cancellative monoid M is a Krull monoid if and only if there are a family of valuations $v_i\colon G(M) \to \mathbb{Z}$ $(i \in I)$ such that, for every $x \in M$, $v_i(x) = 0$ for almost $i \in I$, and M is exactly the set of all $d \in G(M)$ with $v_i(d) \ge 0$ for all $i \in I$. When $V(R)$ is a Krull monoid, valuations separate finitely generated projective R-modules, that is, if $A_R, B_R \in \text{proj-}R$ are not isomorphic, there is a valuation $v\colon V(R) \to \mathbb{N}$ such that $v(\langle A_R \rangle) \ne v(\langle B_R \rangle)$.

For instance, for any ring R, let $J(R)$ denote the Jacobson radical. Recall that a ring R is *semilocal* if $R/J(R)$ is semisimple artinian. The monoid $V(R)$ is a finitely generated reduced Krull monoid for every semilocal ring R. Conversely, for every finitely generated reduced Krull monoid M there exists a semilocal ring R such that $M \cong V(R)$ [13].

(3) *Projective rank functions separate finitely generated projective R-modules.* Recall that a projective rank function on a ring R is a homomorphism of pre-ordered groups $\rho\colon K_0(R) \to \mathbb{R}$ that maps R_R to 1. Equiva-

56

lently, a projective rank function is a monoid homomorphism $V(R) \to \mathbb{R}_{\geq 0}$ that maps $\langle R_R \rangle$ to 1. Here $\mathbb{R}_{\geq 0}$ denotes the set of non-negative real numbers. Thus projective rank functions separate finitely generated projective R-modules if and only if $V(R)$ is isomorphic to a submonoid of a direct product of copies of $\mathbb{R}_{\geq 0}$. In particular, in this case $V(R)$ is cancellative and its Grothendieck group is torsion-free. It follows that there is uniqueness of n-th roots in proj-R: if $A_R, B_R \in$ proj-R, $n \geq 1$ and $A_R^n \cong B_R^n$, then $A_R \cong B_R$. This case was studied by Schofield, Bergman and Cohn, in particular for hereditary R.

(4) *The monoid $V(R)$ is cancellative.* This is the case in which the morphisms $V(R) \to \mathbb{T} = \mathbb{R}/\mathbb{Z}$ separate finitely generated projective modules. For instance, $V(R)$ is cancellative for every commutative Dedekind domain R; see Example 2.

(5) The monoid $V(R)$ is *strongly separative.* A monoid M is called *strongly separative* [2, p. 125] if one of the following equivalent conditions holds:

(a) If $a, b, c \in M$, $a + c = b + c$ and $c \leq na$ for some $n \in \mathbb{N}$, then $a = b$.
(b) If $a, b \in M$ and $2a = a + b$, then $a = b$.
(c) If $a, b, c \in M$ and $a + 2c = b + c$, then $a + c = b$.

The exchange rings R with $V(R)$ strongly separative have been the object of study in [2, §5]. Strong separativity also appears in the study of lattices over orders (Jacobinski's cancellation theorem, [24, Theorem VIII.5.1]).

(6) The monoid $V(R)$ is *separative.* A monoid M is called *separative* if $a, b \in M$ and $a + b = 2a = 2b$ implies $a = b$ [9, p. 131]. A monoid M is separative if and only if the characters of M, that is, the monoid homomorphisms of M into the multiplicative monoid $\mathbb{C}$ of complex numbers, separate elements of M. A monoid M is separative if and only if its archimedean components (see Section 1) are cancellative semigroups [9, p. 131]. The rings for which $V(R)$ is a separative monoid, in particular, the exchange rings with $V(R)$ separative, have been studied in [2]. It is an outstanding open question whether $V(R)$ is separative for all exchange rings R; cf. [2, p. 107]. For a commutative ring R, the monoid $V(R)$ is separative if and only if $V(R)$ is cancellative.

(7) *The general case.* Recall that a monoid M is isomorphic to the monoid $V(R)$ for some ring R if and only if it is a commutative monoid that is reduced and has an order-unit (Bergman and Dicks [4, Theorems 6.2 and 6.4] and [5, p. 315]).

In this hierarchy, the best case after the case in which the Krull-Schmidt Theorem holds (case (1), that is, the case in which $V(R)$ is a free commutative monoid), is when $V(R)$ is a Krull monoid, necessarily reduced and cancellative. A monoid is a reduced Krull monoid if and only if it is isomorphic to $\mathbb{N}^{(I)} \cap G$, where I is a set, $\mathbb{N}^{(I)}$ is the positive cone of the free abelian group $\mathbb{Z}^{(I)}$ with the component-wise order, and G is a subgroup of $\mathbb{Z}^{(I)}$ [8]. This implies that if $V(R)$ is a Krull monoid, then direct-sum decompositions in proj-R have a very regular pattern, because, in the language of Minkowski's Geometry of Numbers, they are represented by a lattice G, that is, a finitely generated free abelian group. If $V(R)$ is a Krull monoid, then $V(R) \cong \mathbb{N}^t \cap G$ has a very regular pattern because it is the intersection of the lattice $G \subseteq \mathbb{Z}^t$ with the positive cone $\mathbb{N}^t$, so that the failure of the Krull-Schmidt theorem is minimal, due only to the presence of the border of $\mathbb{N}^t \cap G$. In other words, when $V(R)$ is a Krull monoid, Krull-Schmidt uniqueness does not necessary hold, but direct-sum decompositions still have a very good pattern.

This has a further extension from the class proj-R of finitely generated projective modules to more general classes $\mathcal{C}$ of modules. Namely, let $\mathcal{C}$ be a class of arbitrary right modules over a fixed ring R. Suppose that

(a) $\mathcal{C}$ is closed under finite direct sums, under direct summands and under isomorphisms;

(b) $\mathcal{C}$ has a set of representatives, that is, a set $V(\mathcal{C}) \subseteq \mathcal{C}$ with the property that every $A_R \in \mathcal{C}$ is isomorphic to a unique element $\langle A_R \rangle \in V(\mathcal{C})$; and

(c) the ring $\mathrm{End}_R(A)$ is semilocal for each $A \in \mathcal{C}$.

Under these hypotheses, $V(\mathcal{C})$ becomes a reduced Krull monoid with the operation $\langle A_R \rangle + \langle B_R \rangle = \langle A_R \oplus B_R \rangle$ [11]. Conversely, let M be a reduced Krull monoid. Then there exists a ring R for which $V(\mathcal{S}_R) \cong M$, where $\mathcal{S}_R$ denotes the class of all finitely generated projective right R-modules with semilocal endomorphism ring [15]. Notice that if $V(R)$ is a Krull monoid, then $V(R)$ is necessarily a finitely generated monoid, but, in the more general case of a class $\mathcal{C}$, we can find Krull monoids $V(\mathcal{C})$ that are not finitely generated monoids.

There are several examples of classes $\mathcal{C}$ of modules satisfying the three conditions (a), (b), (c) above. For example, the class of all finitely generated modules over a commutative Noetherian local ring; the class of all artinian modules over a fixed ring; the class of finite-rank torsion-free modules over a commutative semilocal principal ideal domain or over a valuation domain; the class of all linearly compact modules; and, finally, the class of modules of finite Goldie dimension and finite dual Goldie dimension [10, §4.3].

This paper is divided into four sections. In the first two sections we describe the archimedean components, the idempotents and the subgroups of $V(R)$. In the third one, we study the structure of commutative cancellative Krull monoids and its similarity with the structure of commutative Krull domains. This has already been done, from points of view different form the point of view we study here, in various previous articles, from Chouinard's pioneering work [8] and Krause [21] to more recent articles written by Lettl [23], Halter-Koch [17, 18], Geroldinger [16], Kainrath [20], Chapman, Halter-Koch and Krause [7], and Facchini and Halter-Koch [12]. The point of view we study here is the relationship between prime ideals of height one and essential valuations (Lemma 3.2) and the fact that a cancellative monoid is a Krull monoid if and only if its localizations at prime ideals of height one are valuation monoids (for the precise statement, see Theorem 3.4). This is a classical result for Krull domains, and had been proved for finitely generated Krull monoids in [12]. In the fourth section of the paper, we show that case (2) ($V(R)$ a Krull monoid) is in some sense typically non-commutative, that is, if R is a commutative ring and $V(R)$ is a Krull monoid, then $V(R)$ is free (Theorem 4.2). Moreover, the submonoid $V(\mathcal{S}_R)$ of $V(R)$ is always free for any commutative ring R (Theorem 4.4).

I would like to thank Pere Ara, who allowed me to put in this paper some unpublished results of ours.

1. Archimedean components of $V(R)$

A submonoid N of a commutative monoid M is said to be *divisor-closed* if $x \in M$, $y \in N$ and $x \leq y$ in M implies $x \in N$. For each $x \in M$ we shall denote by $[\![x]\!]$ the smallest divisor-closed submonoid of M containing x. It is the set of all $y \in M$ with $y \leq nx$ for some $n \geq 0$. The order-units of a non-zero monoid M are exactly the elements $u \in M$ such that $M = [\![u]\!]$.

Recall that a commutative semigroup S is called *archimedean* if for every pair (x, y) of elements of S there exists a positive integer n with $x \leq ny$. More generally, let M be a commutative monoid. For $x, y \in M$, define $x \asymp y$ if there exist positive integers n and m such that $x \leq ny$ and $y \leq mx$. Thus $x \asymp y$ if and only if $[\![x]\!] = [\![y]\!]$. It is possible to prove that $\asymp$ is the least congruence on M such that every element in the quotient monoid $M/\asymp$ is idempotent [9, Theorem 4.12]. Any monoid isomorphic to $M/\asymp$ is called a *maximal semilattice homomorphic image* of M [9, pp. 131-132]. The equivalence classes of M modulo $\asymp$ are additively closed subsets of M, called the *archimedean components* of M.

Let R be a ring. For any subclass $\mathcal{U}$ of proj-R, $\mathrm{Tr}_R(\mathcal{U})$ will denote the *trace* of $\mathcal{U}$ in R, that is, the sum of all images $f(A_R)$ where A_R ranges in the modules $A_R \in \mathcal{U}$ and f ranges in the homomorphisms of A_R into R_R. If $\mathcal{U}$ has a unique element A_R, we shall usually write $\mathrm{Tr}_R(A_R)$ instead of $\mathrm{Tr}_R(\mathcal{U})$. The trace $\mathrm{Tr}_R(\mathcal{U})$ is characterized as the smallest two-sided ideal I of R such that $A_R I = A_R$ for every $A_R \in \mathcal{U}$. We shall call *trace ideals* of R all two-sided ideals of R equal to $\mathrm{Tr}_R(\mathcal{U})$ for some subclass $\mathcal{U}$ of proj-R, *finitely generated trace ideals* the ideals equal to $\mathrm{Tr}_R(\mathcal{U})$ for some finite subset $\mathcal{U}$ of proj-R (equivalently, equal to $\mathrm{Tr}_R(A_R)$ for some $A_R \in$ proj-R), and *maximal trace ideals* the trace ideals of R that are maximal in the set of all proper trace ideals of R partially ordered by set inclusion. Every trace ideal is contained in a maximal trace ideal. Recall that a *prime ideal* of a commutative monoid M is a proper subset P of M such that, for any $x, y \in M$, $x + y \in P$ if and only if either $x \in P$ or $y \in P$. The empty set $\emptyset$ is the smallest prime ideal of every commutative monoid. A prime ideal P of a commutative monoid M will be called a prime ideal *of height one* if it is minimal among non-empty prime ideals of M. The set of all finitely generated trace ideals of R and all trace ideals of R will be denoted by $\mathcal{T}_{\mathrm{fg}}(R)$ and $\mathcal{T}(R)$ respectively. In the order-reversing one-to-one correspondence between the set $\mathcal{T}(R)$ of all trace ideals of R and the set $\mathrm{Spec}(R)$ of all prime ideals of $V(R)$ [12, Theorem 2.2(c)], maximal trace ideals of R correspond to prime ideals of height one of $V(R)$. A module $A_R \in$ proj-R is a progenerator if and only if its trace is R.

Proposition 1.1. *Let R be a ring and $A_R, B_R \in$ proj-R. The following conditions are equivalent:*

(1) $\langle A_R \rangle$ *and* $\langle B_R \rangle$ *belong to the same archimedean component of* $V(R)$.
(2) $\mathrm{add}(A_R) = \mathrm{add}(B_R)$.
(3) $\mathrm{Tr}_R(A_R) = \mathrm{Tr}_R(B_R)$.

In particular, there is a one-to-one correspondence between the set of all archimedean components of $V(R)$ and the set $\mathcal{T}_{\mathrm{fg}}(R)$ of all finitely generated trace ideals of R. Moreover, the archimedean component of $\langle R_R \rangle$ in $V(R)$ consists of all the modules belonging to $V(R)$ that are progenerators, or, equivalently, the order-units of $V(R)$.

Proof. (1) $\Rightarrow$ (2) and (2) $\Rightarrow$ (3) follow easily from the definitions. For (3) $\Rightarrow$ (1), notice that $\mathrm{Tr}_R(A_R) = \mathrm{Tr}_R(\langle A_R \rangle) = \mathrm{Tr}_R(\llbracket \langle A_R \rangle \rrbracket)$ for every $A_R \in$ proj-R. In view of the antiisomorphism between the trace ideals of

R and the prime ideals of $V(R)$, we see that $\mathrm{Tr}_R(A_R) = \mathrm{Tr}_R(B_R)$ if and only if the prime ideals $V(R) \setminus [\![\langle A_R \rangle]\!]$ and $V(R) \setminus [\![\langle B_R \rangle]\!]$ of $V(R)$ coincide, that is, if and only if $[\![\langle A_R \rangle]\!] = [\![\langle B_R \rangle]\!]$, i.e., if and only if $\langle A_R \rangle \asymp \langle B_R \rangle$. This proves that $(3) \Rightarrow (1)$. The rest is trivial. $\qquad\square$

Thus there is a one-to-one correspondence between (1) the set of all archimedean components of $V(R)$, (2) the set $\mathcal{T}_{\mathrm{fg}}(R)$ of all finitely generated trace ideals of R, and (3) the set of all prime ideals P of $V(R)$ of the type $P = V(R) \setminus [\![\langle A_R \rangle]\!]$ for some $\langle A_R \rangle \in V(R)$.

The set $\mathcal{T}_{\mathrm{fg}}(R)$ is a commutative additive monoid with respect to the sum of two ideals, and every element of the additive monoid $\mathcal{T}_{\mathrm{fg}}(R)$ is idempotent.

Corollary 1.2. *For any ring R, the additive monoid $\mathcal{T}_{\mathrm{fg}}(R)$ is the maximal semilattice homomorphic image of $V(R)$.*

Proof. The mapping $V(R) \to \mathcal{T}_{\mathrm{fg}}(R)$, $\langle A_R \rangle \in V(R) \mapsto \mathrm{Tr}_R(A_R)$, is a surjective monoid homomorphism whose kernel is $\asymp$. $\qquad\square$

Example 1. Let $\sim_{n,m}$ be the congruence on $\mathbb{N}$ generated by the pair (n, m), that is, the smallest congruence for which n and m are congruent. Equivalently, $\sim_{n,m}$ is defined, for all $s, t \in N$, by $s \sim_{n,m} t$ if and only if either $s = t$, or $s \geq n$, $t \geq n$ and $m - n$ divides $s - t$ in $\mathbb{Z}$. The congruences on the additive monoid $\mathbb{N}$ are only the equality and the congruences $\sim_{n,m}$ with $0 \leq n < m$ integers.

Let F be a field and let $0 < n < m$ be integers. Recall that the *Leavitt algebra* $L(F, n, m)$ is the factor ring $F\langle x_{ij}, y_{ji} \mid i = 1, 2, \ldots, n, \ j = 1, 2, \ldots, m \rangle / I$, where x_{ij}, y_{ji} ($i = 1, 2, \ldots, n$, $j = 1, 2, \ldots, m$) are $2nm$ distinct non-commutative indeterminates, $F\langle x_{ij}, y_{ji} \mid i = 1, 2, \ldots, n, \ j = 1, 2, \ldots, m \rangle$ is the free algebra, and the two-sided ideal I is defined as follows. If X is the $n \times m$ matrix with (i, j) entry x_{ij} and Y is the $m \times n$ matrix with (j, i) entry y_{ji}, then the $n \times n$ matrix $XY - 1_n$ has its n^2 entries in the ring $F\langle x_{ij}, y_{ji} \mid i = 1, 2, \ldots, n, \ j = 1, 2, \ldots, m \rangle$, and the $m \times m$ matrix $YX - 1_m$ has m^2 entries in the same ring. Let I denote the two-sided ideal of $F\langle x_{ij}, y_{ji} \mid i = 1, 2, \ldots, n, \ j = 1, 2, \ldots, m \rangle$ generated by these $n^2 + m^2$ entries. Then $L(F, n, m)$ is the factor ring $F\langle x_{ij}, y_{ji} \mid i = 1, 2, \ldots, n, \ j = 1, 2, \ldots, m \rangle / I$. It is possible to prove [4, Theorem 6.1] that all projective $L(F, n, m)$-modules are free and that, for all $s, t \geq 0$, $R_R^s \cong R_R^t$ if and only if $s \sim_{n,m} t$.

Now let R be any ring over which every finitely generated projective module is free. For instance, R could be (1) a local ring, or (2) a commutative semilocal ring with no non-trivial idempotents, or (3) a free algebra $F\langle X \rangle$, or (4) a Leavitt algebra $L(F, n, m)$. Then $V(R)$ is a cyclic reduced monoid, that is, it is either isomorphic to $\mathbb{N}$ or to the quotient monoid $\mathbb{N}/\sim_{n,m}$ for some $0 < n < m$. It is very easy to check that every such monoid M, that is, every cyclic reduced monoid M, has only the two trivial prime ideals $\emptyset$ and $M \setminus U(M) = M \setminus \{0\}$. The ring R only has the two trivial trace ideals R and $\{0\}$. The monoid $V(R)$ has exactly two archimedean components, one with only the zero element, and the other with all the non-zero elements of $V(R)$.

Example 2. Let R be a commutative Dedekind domain. Then every finitely generated projective R-module is isomorphic to a direct sum $I_1 \oplus I_2 \oplus \cdots \oplus I_m$ of $m \geq 0$ non-zero ideals $I_1, I_2, \ldots, I_m$ of R [25]. Moreover, $I_1 \oplus I_2 \oplus \cdots \oplus I_m \cong I_1' \oplus I_2' \oplus \cdots \oplus I_{m'}'$ for non-zero ideals I_i, I_j' if and only if $m = m'$ and $I_1 I_2 \ldots I_m \cong I_1' I_2' \ldots I_m'$. Recall that the *class group* $\mathrm{Cl}(R)$ of R is the factor semigroup of the multiplicative semigroup of all non-zero fractional ideals of R modulo non-zero principal fractional ideals. Equivalently, $\mathrm{Cl}(R)$ is the multiplicative group of all isomorphism classes of non-zero ideals of R. If we map a non-zero element $\langle A_R \rangle$ of $V(R)$, with $A_R \cong I_1 \oplus I_2 \oplus \cdots \oplus I_m$ and $I_1, I_2, \ldots, I_m$ non-zero ideals of R, to the pair $(m, I_1 I_2 \ldots I_m)$, we get an isomorphism of the semigroup of non-zero elements of $V(R)$ onto the direct product $\mathbb{N}_0 \times \mathrm{Cl}(R)$ of the semigroup $\mathbb{N}_0$ of positive integers and the group $\mathrm{Cl}(R)$. Thus $V(R)$ is isomorphic to the submonoid $M = (\mathbb{N}_0 \times \mathrm{Cl}(R)) \cup \{(0, 0)\}$ of $\mathbb{N} \times \mathrm{Cl}(R)$. In particular, $V(R)$ is cancellative. Notice that $\mathbb{N}_0 \times \mathrm{Cl}(R)$ is the maximal prime ideal of $\mathbb{N} \times \mathrm{Cl}(R)$ consisting of all non-invertible elements of $\mathbb{N} \times \mathrm{Cl}(R)$, that is, $\mathbb{N}_0 \times \mathrm{Cl}(R) = \mathbb{N} \times \mathrm{Cl}(R) \setminus U(\mathbb{N} \times \mathrm{Cl}(R))$. The two subsets $\mathbb{N}_0 \times \mathrm{Cl}(R)$ and $\{(0, 0)\}$ are the two archimedean components of M.

2. Idempotents, archimedean components, subgroups of $V(R)$, and $K_0(R)$

We are particularly interested in the idempotent elements of $V(R)$, that is, the finitely generated projective right R-modules P_R with $P_R^2 \cong P_R$. In the next proposition, we show that there is a strong relation between idempotents of $V(R)$, archimedean components, and subgroups of $V(R)$. *In the rest of this paper, by a subgroup G of a commutative additive monoid M we mean an additively closed subset G of M that is a group with respect to the addition induced on it. In particular, the identity of G does not necessarily coincide with the identity of M. Also, the maximal subgroups* of M are the subgroups of M that are not properly contained in any other subgroup of M.

Recall that for any pre-order $\leq$ on a set S, the *equivalence relation associated to* $\leq$ is defined, for all $x, y \in S$, by $x \sim y$ if $x \leq y$ and $y \leq x$. If M is a commutative monoid and $\leq$ is its algebraic pre-order, then $\sim$ turns out to be a congruence on M. Moreover, the congruence $\sim$ is less or equal to the congruence $\asymp$, that is, every congruence class modulo $\sim$ is contained in exactly one archimedean component.

Proposition 2.1. *Let M be a commutative additive monoid. Then:*

(a) *Every subgroup of M is contained in a unique maximal subgroup of M.*

(b) *The maximal subgroups of M are exactly the equivalence classes of the idempotents of M modulo the congruence $\sim$ associated to the algebraic pre-order. In particular, every subgroup is contained in an archimedean component.*

(c) *There is a one-to-one correspondence between the set of all idempotents of M and the set of all maximal subgroups of M. It associates to every idempotent $e \in M$ the equivalence class of e modulo $\sim$.*

(d) *There is a one-to-one correspondence between the set of all idempotents of M and the set of all archimedean components of M containing an idempotent. It associates to every idempotent $e \in M$ the archimedean component of e.*

Proof. (a) and (b) Any subgroup G of M contains one idempotent element e (the identity of G), and G is contained in the equivalence class of e modulo $\sim$. It is easy to see that the equivalence classes of the idempotents of M modulo $\sim$ are subgroups of M. From this, (a) and (b) follow.

(c) follows from (b).

(d) By (c), it suffices to show that every archimedean component contains at most one idempotent. Let e, f be two idempotents belonging to the same archimedean component C. Then $e \leq nf$ for some positive n, so that $e \leq f$. Similarly, $f \leq e$, hence $e \sim f$. Thus e and f are two idempotent elements of a group. Therefore they are equal. $\qquad\square$

Example 3. Let R be a ring over which every finitely generated projective module is free (cf. Example 1). For such a ring R, $V(R)$ is a cyclic monoid, so that either $V(R) \cong \mathbb{N}$ or $V(R) \cong \mathbb{N}/\!\sim_{n,m}$ for suitable integers $0 < n < m$ (notation as in Example 1). The only idempotent in $\mathbb{N}$ is 0, and the equivalence relation $\sim$ associated to the (pre-)order of $\mathbb{N}$ is the equality. The unique subgroup contained in $\mathbb{N}$ is the subgroup $\{0\}$. This subgroup is one of the two archimedean components of $\mathbb{N}$. The other archimedean component is $\mathbb{N} \setminus \{0\}$.

Let $0 < n < m$ be integers. The factor monoid $\mathbb{N}/\!\sim_{n,m}$ has exactly two idempotents, $\overline{0}$ and $\overline{n(m-n)}$. The equivalence relation associated to the algebraic pre-order is defined, for all $s, t \in \mathbb{N}$, by $s \sim t$ if either $s = t$, or $s \geq n$ and $t \geq n$. The maximal subgroups of $\mathbb{N}/\!\sim_{n,m}$ are two, the subgroup $\{\overline{0}\}$ and the subgroup $\{\,\overline{s} \mid s \in \mathbb{N},\ s \geq n\,\}$, which is a cyclic group of order $m - n$. There are two archimedean components, $\{\overline{0}\}$ and $\mathbb{N}/\!\sim_{n,m} \setminus \{\overline{0}\}$.

Example 4. Let R be a commutative Dedekind domain, so that $V(R) \cong (\mathbb{N}_0 \times \mathrm{Cl}(R)) \cup \{(0,0)\}$ (Example 2). Then $V(R)$ has only the trivial idempotent $\langle 0 \rangle$, and $\{\langle 0 \rangle\}$ is the unique subgroup of $V(R)$.

Example 5. Recall that a module is *directly infinite* if it is isomorphic to a proper direct summand of itself. An idempotent e of a ring R is *infinite* if the the right ideal eR is a directly infinite right R-module [3, Definitions 1.2]. Equivalently, e is infinite if there exist orthogonal idempotents $f, g \in R$ such that $e = f + g$, $eR \cong fR$ and $g \neq 0$. A simple ring R is *purely infinite* if every nonzero right ideal of R contains an infinite idempotent. This concept is left-right symmetric [3]. If R is a purely infinite simple ring and $A_R, B_R \in \mathrm{proj}\text{-}R$ are non-zero, then there exists a non-zero $C_R \in \mathrm{proj}\text{-}R$ such that $A_R \cong B_R \oplus C_R$ [3, Proposition 1.5]. It follows that $V(R) \setminus \{0\}$ is an abelian group. Thus $V(R)$ has exactly two idempotents, one of which is 0. The equivalence relation associated to the algebraic pre-order is defined, for all $A_R, B_R \in \mathrm{proj}\text{-}R$, by $A_R \sim B_R$ if either $A_R \cong B_R$, or $A_R \neq 0$ and $B_R \neq 0$. There are two maximal subgroups, and they coincide with the archimedean components. They are $\{0\}$ and $V(R) \setminus \{0\}$. For every countable abelian group G there exists a purely infinite simple Von Neumann regular ring R with $V(R) \setminus \{0\} \cong G$ [3, Theorem 8.4].

3. Krull monoids

For a prime ideal P of a commutative monoid M, the *localization M_P of M at P* is the monoid whose elements are all differences $x - s$, where $x \in M$ and $s \in M \setminus P$, and where $x - s = x' - s'$ if and only if there exists $t \in M \setminus P$ such that $x + s' + t = x' + s + t$. For example, the Grothendieck group $G(M)$ of M is the localization $M_\emptyset$ of M at its prime ideal $\emptyset$, and, if M is cancellative, $M_P \subseteq M_\emptyset$ for every prime ideal P of M. Let M be an arbitrary commutative monoid, not necessarily cancellative, and let P be a prime ideal of M. The monoid $(M_P)_{\mathrm{red}}$ is called the *reduced localization*

64

of M at P. If $x, x' \in M$ and $s, s' \in M \setminus P$, then $x - s + U(M_P) = x' - s' + U(M_P)$ if and only if there exist elements $t, t' \in M \setminus P$ such that $x + t = x' + t'$. In particular, the canonical homomorphism $M \to (M_P)_{\mathrm{red}}$, defined by $x \mapsto x - 0 + U(M_P)$, is surjective.

Proposition 3.1. *In a commutative monoid with an order-unit, every non-empty prime ideal contains a prime ideal of height one.*

Proof. If u is an order-unit in a commutative monoid M, then every non-empty prime ideal of M contains u. If P is a non-empty prime ideal of M, apply Zorn's Lemma to the set of all non-empty prime ideals of M contained in P, partially ordered by reverse inclusion. $\square$

It is easy to construct examples of reduced cancellative commutative monoids without prime ideals of height one. A trivial example is the monoid with one element. A less trivial example is the following. Let α be a limit ordinal, and let G be the free abelian group of all functions $\alpha \to \mathbb{Z}$ that are zero almost everywhere, so that G is free of rank equal to the cardinality of α. Let M be the submonoid of G whose non-zero elements are all $g \in G$, $g \neq 0$, with $g(\lambda) > 0$, where λ is the greatest element of the finite set $\mathrm{supp}(g) = \{\, \mu \mid \mu < \alpha, \ g(\mu) \neq 0 \,\}$. Then: (1) the prime ideals of this monoid M form a descending chain, well ordered with respect to reverse inclusion; (2) there is exactly one prime ideal P_λ of M for each ordinal $\lambda \leq \alpha$; and (3) $P_\lambda = \{\, g \in M \mid$ there exists an ordinal $\mu < \alpha$ with $g(\mu) \neq 0$ and $\mu \geq \lambda \,\}$. In particular, $P_0 = M \setminus \{0\}$ and $P_\alpha = \emptyset$. As α is a limit ordinal, there is no prime ideal of M of height one. The interest of this example lies in the fact that M is cancellative, reduced, and its Grothendieck group is the free abelian group G, hence it is very close to being a cancellative reduced Krull monoid (cf. the hierarchy in the Introduction), but it does not have prime ideals of height one, while in cancellative Krull monoids prime ideals of height one always exist and play a prominent role (Lemma 3.2 and Theorem 3.4).

A valuation $v \colon M \to \mathbb{N}$ of a commutative monoid M is *essential* if for every $x, y \in M$ with $v(x) \leq v(y)$, there exists $s \in M$ such that $x \leq y + s$ and $v(s) = 0$. If M is a monoid and $v \colon M \to \mathbb{N}$ is a valuation, then $P_v = \{x \in M \mid v(x) > 0\}$ is a prime ideal of M. If v is essential, P_v is a prime ideal of height one [12, Lemma 4.1(c)]. The *index* $e(v)$ of a valuation v is the greatest common divisor of all the elements of $v(M)$. Two valuations $v, v' \colon M \to \mathbb{N}$ are said to be *equivalent* if $e(v)^{-1} v = e(v')^{-1} v'$.

A divisor homomorphism $\varphi \colon M \to F$ of M into a free monoid F is a *divisor theory* if for every $u \in F$ there exist $x_1, \ldots, x_m \in M$ such that $u = \varphi(x_1) \wedge \ldots \wedge \varphi(x_m)$ (notice that a free monoid F is a lattice with respect to its algebraic pre-order $\leq$). Every cancellative Krull monoid M has a divisor theory, and for any two divisor theories $\varphi \colon M \to F$ and $\varphi' \colon M \to F'$ of M, there is a unique isomorphism $\Phi \colon F \to F'$ such that $\Phi \circ \varphi = \varphi'$ [19, Theorems 23.4 and 20.4]. More precisely, if $\varphi = (\varphi_i)_{i \in I} \colon M \to \mathbb{N}^{(I)}$ is a divisor homomorphism of a Krull monoid M such that the $\varphi_i \neq 0$ are pairwise non-equivalent valuations, and J is the set of all indices $j \in I$ with φ_j essential, then $\varphi^* = (e(\varphi_j)^{-1} \varphi_j)_{j \in J} \colon M \to \mathbb{N}^{(J)}$ turns out to be a divisor theory, and φ is a divisor theory if and only if $I = J$ and $e(\varphi_i) = 1$ for all $i \in I$. Every essential valuation of M is equivalent to v_j for some $j \in J$ [12, Satz 1, Satz 2 and Korollar].

The next Lemma is well known (cf. Remark 3.3), but we give an elementary proof for the sake of completeness. We shall denote the set of all prime ideals of height one of a commutative monoid M by $X^{(1)}(M)$.

Lemma 3.2. *Let M be a cancellative Krull monoid. Then:*

(a) *Every non-empty prime ideal of M contains a prime ideal of height one.*

(b) *The mapping $v \mapsto P_v$ is a one-to-one correspondence between the set of essential valuations of M up to equivalence and the set $X^{(1)}(M)$ of prime ideals of M of height one.*

Proof. Let $\varphi = (\varphi_j)_{j \in J} \colon M \to \mathbb{N}^{(J)}$ be a divisor theory for M, so that every essential valuation of M is equivalent to φ_j for some $j \in J$. If P is a prime ideal of M and $P \not\supseteq P_{\varphi_j}$ for every $j \in J$, then for every $j \in J$ there exists $x_j \in M \setminus P$ with $\varphi_j(x_j) > 0$. Let p be a fixed element of P, so that $\varphi_j(p) > 0$ if and only if j belongs to a finite subset F of J. Then $p \leq \sum_{j \in F} \varphi_j(p) x_j \notin P$, so that $p \notin P$. This contradiction shows that $P = \emptyset$. Thus every non-empty prime ideal of M contains a prime ideal of the type P_{φ_j}. In particular, all prime ideals of M of height one are of the type P_{φ_j}. We already know that for any monoid M (not necessarily Krull), P_v is a prime ideal of height one for every essential valuation v. This concludes the proof of (a) and shows that the prime ideals of M of height one are exactly the P_{φ_j}'s.

For (b), it is obvious that two equivalent valuations v, v' yield the same prime ideal $P_v = P_{v'}$. Moreover, the φ_j's, $j \in J$, form a set of representatives of the essential valuations of M up to equivalence. Also, as we have seen in the previous paragraph, the P_{φ_j}'s are exactly the prime ideals of height one. Finally, if $j \neq j'$, then there exists $x \in M$ with $\varphi_j(x) = 0$ and $\varphi_j(x') \neq 0$ because $\varphi = (\varphi_j)_{j \in J} \colon M \to \mathbb{N}^{(J)}$ is a divisor theory. Thus $P_{\varphi_j} \neq P_{\varphi_{j'}}$. $\qquad\square$

A monoid M is a *discrete valuation monoid* if $M_{\mathrm{red}} \cong \mathbb{N}$.

Remark 3.3. In the terminology of [19], a cancellative Krull monoid M is the same as a t-Dedekind monoid, and $X^{(1)}(M)$ is the set of all non-empty prime t-ideals, so that M_P is a discrete valuation monoid for every $P \in X^{(1)}(M)$ [19, Theorem 23.3]. Moreover, every non-empty prime ideal of M contains a non-empty prime t-ideal [19, Proposition 11.6], and P_v is a prime t-ideal for any essential valuation v of M.

The next Theorem also is essentially the definition given in [19, Definition 22.1] in combination with [19, Theorem 23.3].

Theorem 3.4. *The following conditions are equivalent for a cancellative monoid M:*

(a) *M is a Krull monoid.*
(b) *The localization M_P is a discrete valuation monoid for every $P \in X^{(1)}(M)$, every $x \in M$ is contained in at most finitely many prime ideals of height one, and $M = \bigcap_{P \in X^{(1)}(M)} M_P$.*

Proof. (a) $\Rightarrow$ (b) Let M be a Krull monoid and let $\varphi = (\varphi_i)_{i \in I} \colon M \to \mathbb{N}^{(I)}$ be a divisor homomorphism. If $P \in X^{(1)}(M)$, we know from Lemma 3.2 that $P = P_v$ for some essential valuation v of M. Then $M_P = M_{P_v}$ is a discrete valuation monoid by [12, Lemma 4.1(b)], and v induces an isomorphism $\overline{v} \colon (M_P)_{\mathrm{red}} \to e(v)\mathbb{N}$. As M is cancellative, $M \subseteq M_P \subseteq M_\emptyset$ for every $P \in X^{(1)}(M)$, so that $M \subseteq \bigcap_{P \in X^{(1)}(M)} M_P$. For the opposite inclusion, suppose $x - y \in \bigcap_{P \in X^{(1)}(M)} M_P \subseteq M_\emptyset$ with $x, y \in M$. Then $v(x - y) \geq 0$ for every essential valuation v of M, so that $v(x) \geq v(y)$ for every essential valuation v. Thus $\varphi^*(x) \geq \varphi^*(y)$. As $\varphi^* \colon M \to \mathbb{N}^{(J)}$ is a divisor homomorphism, it follows that $x \geq y$ in M. Therefore $x - y$ belongs to the cancellative monoid M. Finally, for every $x \in M$ we have that $\varphi^*(x) \in \mathbb{N}^{(J)}$, so that $v_j(x) \neq 0$ for at most finitely many $j \in J$. Thus x is contained in at most finitely many prime ideals P_j of height one.

(b) $\Rightarrow$ (a) Let M be a cancellative monoid satisfying the conditions stated in (b). Then the canonical homomorphisms $M \to (M_P)_{\mathrm{red}}$ ($P \in X^{(1)}(M)$) have the property that every $x \in M$ is mapped to zero for almost all primes P, and each $(M_P)_{\mathrm{red}}$ is isomorphic to $\mathbb{N}$. Thus these canonical homomorphisms define a monoid homomorphism $\varphi \colon M \to \mathbb{N}^{(X^{(1)}(M))}$. In order to show that φ is a divisor homomorphism, let $x, y \in M$ be such that $\varphi(x) \leq \varphi(y)$. Then, for every P, there exists $s_P \in (M_P)_{\mathrm{red}}$ with $y = x + s_P$ in $(M_P)_{\mathrm{red}}$. Thus for every P there exist $u_P \in M$ and $t_P \in M \setminus P$ with $y + t_P = x + u_P$. Then $y - x = u_P - t_P \in M_P$ for every $P \in X^{(1)}(M)$, from which $y - x \in M$, that is, $x \leq y$ in M. $\qquad\qquad\square$

4. The example of commutative rings

The boolean algebra of all central idempotents of a ring R will be denoted by $B(R)$. Here $e \wedge f = ef$, $e \vee f = e + f - ef$ and $e' = 1 - e$ for all $e, f \in B(R)$.

Recall that an *ideal* in a boolean algebra L is a subset I of L such that $0 \in I$, and $x \vee y, x \wedge e \in I$ for every $x, y \in I$ and every $e \in L$. The set $\mathcal{L}(L)$ of all ideals of a boolean algebra L, partially ordered by set inclusion, is a lattice.

Lemma 4.1. *For a ring R, let $g \colon B(R) \to \mathcal{T}_{\mathrm{fg}}(R)$ and $h \colon \mathcal{L}(B(R)) \to \mathcal{T}(R)$ be the mappings that associate to each element $e \in B(R)$ and each ideal $I \in \mathcal{L}(B(R))$ the two-sided ideals eR and IR of R generated by e and I respectively. Then:*

(a) *The mappings g and h are injective and order-preserving.*
(b) *If R is commutative, the mappings g and h are bijective.*

Proof. (a) By [1, Exercise 7.2], if $e, f \in B(R)$ and $eR \cong fR$, then $e = f$. The injectivity of g follows immediately. In order to prove that h also is injective, it suffices to show that $I = IR \cap B(R)$ for every $I \in \mathcal{L}(B(R))$. The inclusion $\subseteq$ is obvious. To prove the inclusion $\supseteq$, let $i = \sum_{t=1}^{n} i_t r_t$ be a central idempotent of R with $i_t \in I$ and $r_t \in R$ for every $t = 1, 2, \ldots, n$. Multiplying each i_t with the identity $1 = \prod_{\ell=1}^{n} [i_\ell + (1 - i_\ell)]$, we may suppose without loss of generality that, in the sum $\sum_{t=1}^{n} i_t r_t$, the i_t's are pair-wise orthogonal. As i is a central idempotent, it follows that the $i_t r_t$'s also are central idempotents. As $i_t r_t \leq i_t$, we have that each $i_t r_t$ is in I, so that $i = \sum_{t=1}^{n} i_t r_t = \bigvee_{t=1}^{n} i_t r_t$ is in I as well.

(b) Suppose R commutative. In order to see that g is surjective, it suffices to remark that the trace ideal of a finitely generated projective R-module is generated by an idempotent [22, Theorem 2.44]. In order to see that h is surjective, it suffices to show that if I is an ideal of R generated by idempotents, then $I \cap B(R)$ is an ideal of the lattice $B(R)$ and $I = (I \cap B(R))R$. Both these facts are easily checked. $\qquad\square$

Now let R be a commutative ring. In this case, the trace of a finitely generated projective module is generated by an idempotent of R [22, Theorem 2.44], so that for a commutative ring R there is a one-to-one correspondence among the following sets (Proposition 1.1): (1) The set $\mathcal{T}_{\text{fg}}(R)$ of all finitely generated trace ideals; (2) The set of all prime ideals of $V(R)$ of the type $\{\, \langle X_R \rangle \in V(R) \mid X_R \text{ is not isomorphic to a direct summand of } A_R^n \text{ for every } n \geq 0 \,\}$ for some $A_R \in \text{proj-}R$; (3) The set of all archimedean components of $V(R)$; (4) The boolean algebra $B(R)$; (5) The set of all direct summands eR of R_R, $e \in R$ idempotent. In the next Proposition we consider the case in which $V(R)$ is a Krull monoid.

Theorem 4.2. *Let R be a commutative ring and suppose that $V(R)$ is a Krull monoid. Then $V(R)$ is a free monoid.*

Proof. Let $\varphi = (\varphi_j)_{j \in I} \colon V(R) \to \mathbb{N}^{(I)}$ be a divisor theory of the Krull monoid $V(R)$. Then each $\varphi_j \colon V(R) \to \mathbb{N}$ is a surjective essential valuation and, conversely, each essential valuation of $V(R)$ is equivalent to some φ_j. Since $V(R)$ has an order-unit, $V(R)$ is a finitely generated Krull monoid and I is a finite set. By Lemma 3.2(b), $V(R)$ has only finitely many prime ideals of height one. These are the prime ideals $P_j = \{\langle A_R \rangle \in V(R) \mid \varphi_j(\langle A_R \rangle) > 0\}$. It follows that R has only finitely many maximal trace ideals I_j [12, Theorem 2.2(c)]. By Lemma 4.1(b), the boolean algebra $B(R)$ has only finitely many maximal ideals. Thus the Stone space of the boolean algebra $B(R)$ [6, p. 227] is a topological space with finitely many points and the discrete topology, so that the boolean algebra $B(R)$ is necessarily finite. It follows that R has a ring direct-product decomposition $R = e_1 R \times \cdots \times e_n R$, where each factor $e_i R$ has no non-trivial idempotents. The trace ideals of R are generated by some of these idempotents, and the maximal trace ideals are the $(1 - e_i)R$, $i = 1, \ldots, n$. Therefore the prime ideals of $V(R)$ of height one are the ideals $P_i = \{\, \langle A \rangle \in V(R) \mid A(1 - e_i) \neq A_R \,\} = \{\, \langle A \rangle \in V(R) \mid Ae_i \neq 0 \,\}$, $i = 1, \ldots, n$. Let $v_i \colon V(R) \to \mathbb{N}$ be the essential valuation corresponding to P_i, and let $(v_i)_{i=1}^n \colon V(R) \to \mathbb{N}^n$ denote the corresponding divisor theory. Then for every $i = 1, \ldots, n$ there exists a projective R-module A_i with $v_i(A_i) = 1$. As $\langle A_i(1 - e_i) \rangle \notin P_i$, it follows that $v_i(\langle A_i e_i \rangle) = 1$. Moreover $Ae_i e_j = 0$ for $j \neq i$, so that $\langle A_i e_i \rangle \notin P_j$, hence $v_j(\langle A_i e_i \rangle) = 0$ for every $j \neq i$. It follows that the divisor theory is surjective as well, and thus $V(R)$ is free. $\qquad\square$

For a not necessarily commutative ring R, let $\mathcal{S}_R$ be the class of all $A_R \in \text{proj-}R$ with $\text{End}(A_R)$ semilocal. As $\mathcal{S}_R$ is closed under finite direct sums, $V(\mathcal{S}_R)$ turns out to be a submonoid of $V(R)$, and the canonical mapping $V(R) \to V(R/J(R))$ induces by restriction an injective monoid homomorphism $\tau\colon V(\mathcal{S}_R) \to V(\mathcal{S}_{R/J(R)})$ [15]. We will describe the homomorphism $\tau\colon V(\mathcal{S}_R) \to V(\mathcal{S}_{R/J(R)})$ for a commutative ring R (cf. the last paragraph of [15]), showing that it has a very simply structure. Namely, we will see that both $V(\mathcal{S}_R)$ and $V(\mathcal{S}_{R/J(R)})$ are free commutative monoids, and τ maps each canonical generator of $V(\mathcal{S}_R)$ to a sum of distinct canonical generators of $V(\mathcal{S}_{R/J(R)})$, so that τ is completely described by a partition of the canonical free set of generators of $V(\mathcal{S}_{R/J(R)})$ into finite blocks indexed in the canonical free set of generators of $V(\mathcal{S}_R)$. Here by a canonical set of generators of a free commutative monoid we mean its least set of generators, which exists and is unique in every free commutative monoid F and coincides with the set of all atoms of the lattice F with respect to the algebraic pre-order $\le$.

Let R be a commutative ring. Recall that an idempotent $e \in R$ is *indecomposable* if it is non-zero and it is not the sum of two non-zero orthogonal idempotents of R. We shall say that that an idempotent $e \in R$ is *semilocal* if it is contained in all maximal ideals of the ring R except for finitely many maximal ideals. Let $B_{is}(R)$ denote the set of all indecomposable semilocal idempotents of the commutative ring R. Thus an idempotent $e \in R$ belongs to $B_{is}(R)$ if and only if the ring eR is indecomposable and semilocal. As commutative indecomposable semilocal rings with zero Jacobson radical are exactly fields, we see that in the particular case of the ring $\overline{R} = R/J(R)$, an idempotent $\overline{e} \in \overline{R}$ belongs to $B_{is}(\overline{R})$ if and only if the ring $\overline{e}\overline{R}$ is a field.

Lemma 4.3. *Let R be a commutative ring. Then every indecomposable module $A_R \in \mathcal{S}_R$ is isomorphic to eR for some $e \in B_{is}(R)$.*

Proof. Let $A_R \in \mathcal{S}_R$ be an indecomposable module. If I is the trace ideal of A_R, then I is generated by an idempotent e of R [22, Theorem 2.44], and the trace ideal of $I = eR$ is eR itself. Thus A_R and eR have the same trace ideal, so that eR is isomorphic to a direct summand of A_R^n for some n (Proposition 1.1). It follows that $\mathrm{End}(A_R)$ semilocal implies $\mathrm{End}(eR_R)$ semilocal. Thus eR is a semilocal ring, i.e., e is a semilocal idempotent. Since $I = eR$ is the trace ideal of A_R, we have that $A_R e = A_R$. If e is not indecomposable, $e = e' + e''$ with e', e'' non-zero orthogonal idempotents of R, then $A_R = A_R e = A_R e' \oplus A_R e''$ is decomposable, contradiction. Thus $A_R e = A_R$ is a finitely generated projective module over the ring eR. But eR is a semilocal indecomposable ring, and every projective module over a semilocal indecomposable commutative ring is free. Thus $A_R e = A_R \cong eR$. $\qquad\square$

Theorem 4.4. *Let R be a commutative ring. Then the monoid $V(\mathcal{S}_R)$ is a free commutative monoid having the set $\{\, \langle fR \rangle \mid f \in B_{is}(R) \,\}$ as a free set of generators.*

Proof. The monoid $V(\mathcal{S}_R)$ is a Krull monoid [11, Theorem 3.4], in particular every module in $\mathcal{S}_R$ is a direct sum of indecomposables. In view of Lemma 4.3, the set of all $\langle eR \rangle$, with $e \in B_{is}(R)$, generates $V(\mathcal{S}_R)$. Suppose now that $e_1, \ldots, e_n$ are distinct semilocal indecomposable idempotents, $t_1, \ldots, t_n, t_1', \ldots, t_n' \in \mathbb{N}$ and $\oplus_{i=1}^n e_i R^{t_i} \cong \oplus_{i=1}^n e_i R^{t_i'}$. Multiplying by e_j for each $j = 1, \ldots, n$ we see that $e_j R^{t_j} \cong e_j R^{t_j'}$. As the ring $e_j R$ is commutative, hence IBN, we obtain that $t_j = t_j'$. Thus the set of generators is a free set of generators. $\square$

If $\overline{R} = R/J(R)$, a free set of generators of $V(\mathcal{S}_{\overline{R}})$ is given by the set of all $\langle \overline{e}\overline{R} \rangle$, where $\overline{e}$ ranges in the set of all idempotents of $\overline{R}$ with $\overline{e}\overline{R}$ a field, or, equivalently, $\overline{e}\overline{R}$ a simple $\overline{R}$-module [15, Proposition 2.5]. Therefore the canonical mapping $\tau \colon V(\mathcal{S}_R) \to V(\mathcal{S}_{\overline{R}})$, defined by $\langle A_R \rangle \mapsto \langle A \otimes_R \overline{R} \rangle$ and studied in [15], sends the free generator $\langle eR \rangle$ of $V(R)$, e a semilocal indecomposable idempotent of R, to $\langle \overline{e}\overline{R} \rangle$, and $\overline{e}\overline{R}$ is a sum of pair-wise non-isomorphic simple $\overline{R}$-modules. Namely, if e, f are distinct semilocal indecomposable idempotents of R, then $ef = 0$, so that $\overline{e}\overline{f} = \overline{0}$, that is, the simple modules that appear in the sums of simple modules $\overline{e}\overline{R}$ and $\overline{f}\overline{R}$ are pair-wise non-isomorphic. Thus τ maps each generator $\langle fR \rangle$ $(f \in B_{is}(R))$ of $V(\mathcal{S}_R)$ to $\langle \overline{f}\overline{R} \rangle$, which is the sum of the distinct canonical generators of $V(\mathcal{S}_{\overline{R}})$ that correspond to the decomposition of $\overline{f}$ into indecomposable idempotents, so that τ is completely described by the partition $\{\, X_f \mid f \in B_{is}(R) \,\}$ of the free set of generators $\{\, \langle \overline{e}\overline{R} \rangle \mid \overline{e} \in B_{is}(\overline{R}) \,\}$ of $V(\mathcal{S}_{\overline{R}})$ into finite blocks $X_f = \{\, \langle \overline{e}\overline{R} \rangle \mid \overline{e} \in B_{is}(\overline{R}), \ \overline{f}\overline{e} = \overline{e} \,\}$ $(f \in B_{is}(R))$.

References

1. F. W. Anderson and K. R. Fuller, "Rings and Categories of Modules", Second Edition, Springer-Verlag, New York, 1992.
2. P. Ara, K. R. Goodearl, K. C. O'Meara and E. Pardo, *Separative cancellation for projective modules over exchange rings*, Israel J. Math. **105** (1998), 105–137.
3. P. Ara, K. R. Goodearl and E. Pardo, K_0 *of Purely Infinite Simple Regular Rings*, K-Theory **26** (2002), 69–100.
4. G. M. Bergman, *Coproducts and some universal ring constructions*, Trans. Amer. Math. Soc. **200** (1974), 33–88.
5. G. M. Bergman and W. Dicks, *Universal derivations and universal ring constructions*, Pacific J. Math. **79** (1978), 293–337.

6. G. Birkhoff, "Lattice Theory", Third Edition, Amer. Math. Soc, Providence, Rhode Island, 1967.

7. S. T. Chapman, F. Halter-Koch and U. Krause, *Inside factorial monoids and integral domains*, J. Algebra **252** (2002), 350–375.

8. L. G. Chouinard, *Krull semigroups and divisor class groups*, Canad. J. Math. **33** (1981), 1459–1468.

9. A. H. Clifford and G. B. Preston, "The Algebraic Theory of Semigroups", Vol. I, Math. Surveys **7**, Amer. Math. Soc., Providence, R.I., 1961.

10. A. Facchini, "Module Theory. Endomorphism rings and direct sum decompositions in some classes of modules", Birkhäuser Verlag, Basel, 1998.

11. A. Facchini, *Direct sum decompositions of modules, semilocal endomorphism rings, and Krull monoids*, J. Algebra **256** (2002), 280–307.

12. A. Facchini and F. Halter-Koch, *Projective modules and divisor homomorphisms*, J. Algebra Appl. **2**(4) (2003), 435–449.

13. A. Facchini and D. Herbera, K_0 *of a semilocal ring*, J. Algebra **225** (2000), 47–69.

14. A. Facchini and D. Herbera, *Projective modules over semilocal rings*, in "Algebra and its Applications", D. V. Huynh, S. K. Jain, S. R. López-Permouth Eds., Contemporary Math. **259**, Amer. Math. Soc., Providence, 2000, pp. 181–198.

15. A. Facchini and R. Wiegand, *Direct-sum decompositions of modules with semilocal endomorphism rings*, to appear in J. Algebra, 2003.

16. A. Geroldinger *A structure theorem for sets of length*, Colloq. Math. **78** (1998), 225–259.

17. F. Halter-Koch, *Halbgruppen mit Divisorentheorie*, Expo. Math. **8** (1990), 27–66.

18. F. Halter-Koch, *Ein Approximationssatz für Halbgruppen mit Divisorentheorie*, Results Math. **19** (1991), 74–82.

19. F. Halter-Koch, "Ideal Systems. An Introduction to Multiplicative Ideal Theory", Marcel Dekker, New York, 1998.

20. F. Kainrath, *Factorization in Krull monoids with infinite class group*, Colloq. Math. **80** (1999), 23–30.

21. U. Krause, *On monoids of finite real character*, Proc. Amer. Math. Soc. **105** (1989), 546–554.

22. T. Y. Lam, "Lectures on Modules and Rings", Graduate Texts in Math. **189**, Springer-Verlag, New York, 1999.

23. G. Lettl, *Subsemigroups of finitely generated groups with divisor-theory*, Monatsh. Math. **106** (1988), 205–210.

24. K. W. Roggenkamp, "Lattices over Orders II", Lecture Notes in Math. **142**, Springer-Verlag, Berlin, 1970.

25. D. W. Sharpe and P. Vámos, "Injective modules", Cambridge University Press, Cambridge, 1972.

INFINITE PROGENERATOR SUMS

ALBERTO FACCHINI*

Dipartimento di Matematica Pura e Applicata, Università di Padova
Via Belzoni 7, 35131 Padova, Italy
E-mail: facchini@math.unipd.it

LAWRENCE S. LEVY[†]

Mathematics Department, University of Nebraska,
Lincoln, NE 68588–0323 USA
Mailing address: 2528 Van Hise Ave., Madison, WI 53705-3850 USA
E-mail: levy@math.wisc.edu

1. Introduction

Let A, B be modules with semilocal endomorphism rings. If $A^{(\aleph)} \cong B^{(\aleph)}$ (direct sum of $\aleph$ copies) does it follow that $A \cong B$? The answer is known to be "yes" when $\aleph$ is a finite cardinal [F, Proposition 4.8]. This note was motivated by the question of whether the answer is still "yes" when $\aleph = \aleph_0$ and (to make the question nontrivial) A and B are indecomposable. The answer to this $\aleph_0$-*root* question is "no", even if we require A and B to be progenerators [Example 3.1], but is "yes" if we require A and B to be uniserial [P].

The starting point in the construction of our counterexample is a lemma, attributed to Eilenberg, stating that for every progenerator P, the module $P^{(\aleph_0)}$ is free. This led us to the related question of whether *every* infinite direct sum of progenerators is free. The answer is "yes" if the ring is noetherian [B, Corollary 3.2], if "most" of the summands are isomorphic to each other [Corollary 2.2], or if the number of terms is "very large" [Corollary 2.3]. But the answer is "no" in general [Example 4.1].

<hr>

*Research partially supported by Ministero dell'Istruzione, dell'Università e della Ricerca (Italy)

[†]Research was partially supported by an NSA grant (USA)

2. General Progenerator Sums

We learned the following lemma and its proof from [C], where it is used to prove the interesting result that two rings are Morita equivalent if and only the infinite matrix rings over those rings are isomorphic.

Lemma 2.1 (Eilenberg). *Let P, Q be progenerators over any ring, and $\aleph$ any infinite cardinal. Then $P^{(\aleph)} \cong Q^{(\aleph)}$; equivalently, $P^{(\aleph)}$ is free of rank $\aleph$.*

Proof. As P is a progenerator and Q is finitely generated, there is an epimorphism $P^{(n)} \to Q$ for some n. Therefore, since Q is projective, we have $Q \oplus Q' \cong P^{(n)}$ for some Q'. Taking the direct sum of $\aleph$ copies of this isomorphism yields

$$Q^{(\aleph)} \oplus Q'^{(\aleph)} \cong P^{(\aleph)} \tag{2.1.1}$$

Substituting $Q^{(\aleph)} \oplus Q^{(\aleph)}$ for $Q^{(\aleph)}$ in (2.1.1) gives $Q^{(\aleph)} \oplus Q^{(\aleph)} \oplus Q'^{(\aleph)} \cong P^{(\aleph)}$. Therefore the isomorphism in (2.1.1) gives $Q^{(\aleph)} \oplus P^{(\aleph)} \cong P^{(\aleph)}$.

Reversing the roles of P and Q shows that $Q^{(\aleph)} \oplus P^{(\aleph)} \cong Q^{(\aleph)}$, and hence $P^{(\aleph)} \cong Q^{(\aleph)}$. $\square$

The next Corollary follows from the so-called Eilenberg trick: if P is finitely generated projective, we have $P \oplus Q \cong R^{(n)}$ for some finite n, therefore

$$P \oplus R^{(\aleph_0)} \cong P \oplus (Q \oplus P) \oplus (Q \oplus P) \oplus \dots$$
$$\cong (P \oplus Q) \oplus (P \oplus Q) \oplus \dots \cong R^{(\aleph_0)}$$

and therefore $P \oplus R^{(\aleph)} \cong R^{(\aleph)}$ for every infinite cardinal $\aleph$.

Corollary 2.2. *Let $P = \bigoplus_i P_i$ be the direct sum of some infinite number $\aleph$ of finitely generated projective modules P_i over some ring R. Suppose that at least $\aleph$ of the summands P_i are isomorphic to some fixed progenerator Q. Then $P \cong R^{(\aleph)}$.*

Proof. Let $P = \bigoplus_{i \in X} P_i$, and let Y be the subset of X of all indices $i \in X$ with $P_i \cong Q$, so that X and Y have the same cardinality $\aleph$. The case $X = Y$ is Eilenberg's Lemma, so that we can suppose $X \neq Y$. By Eilenberg's Lemma, the direct sum $\bigoplus_{i \in Y} P_i$ is free, i.e., $\bigoplus_{i \in Y} P_i \cong R^{(\aleph)}$. Thus

$$P = \left(\bigoplus_{i \in Y} P_i \right) \oplus \left(\bigoplus_{i \in X \setminus Y} P_i \right) \cong R^{(\aleph)} \oplus \left(\bigoplus_{i \in X \setminus Y} P_i \right)$$
$$\cong \bigoplus_{i \in X \setminus Y} (P_i \oplus R^{(\aleph)})$$

because $X \times (X \setminus Y)$ and X have the same cardinality $\aleph$. But $P_i \oplus R^{(\aleph)} \cong R^{(\aleph)}$ by Eilenberg's trick. Thus P is free. $\qquad\square$

Corollary 2.3. *Let $P = \bigoplus_i P_i$ be the direct sum of $\aleph$ progenerators P_i over some ring R, where $\aleph > \max\{|R|, \aleph_0\}$. Then $P \cong R^{(\aleph)}$.*

Proof. Let $\alpha = \max\{|R|, \aleph_0\}$. Then $R^{(\aleph_0)}$ has cardinality α, so that it has at most α finitely generated submodules. Thus there are at most α finitely generated projective right R-modules up to isomorphism.

Therefore some isomorphism class must be repeated at least $\aleph$ times among the P_i. Therefore Corollary 2.2, implies $P \cong R^{(\aleph)}$. $\qquad\square$

3. Nonunique $\aleph_0$-roots

Example 3.1. Progenerators $P \not\cong Q$ that are indecomposable modules with semilocal endomorphism rings and satisfy $P^{(\aleph_0)} \cong Q^{(\aleph_0)}$.

Proof. Let e be a positive integer ≥ 2. Then there is a (noncommutative) integral domain Λ that is a module-finite algebra over a discrete valuation ring V and such that Λ has an indecomposable progenerator P such that $P^{(e)}$ maps onto Λ, but $P^{(d)}$ does not map onto Λ for $d < e$ [L, Theorem 3.1].

Since $e \geq 2$ we have $P \not\cong \Lambda$; and by Eilenberg's Lemma we have $P^{(\aleph_0)} \cong \Lambda^{(\aleph_0)}$. Thus it suffices to show: (i) Λ is an indecomposable left Λ-module, and (ii) the left Λ-modules Λ and P both have semilocal endomorphism rings.

Since Λ is a domain, Λ has no non-trivial idempotents. Thus (i) follows from the fact that Λ is anti-isomorphic to the endomorphism ring of Λ as a Λ-module. Let M be any finitely generated Λ-module. Then the endomorphism ring of M is a module-finite V-algebra, and is therefore a semilocal ring. This applies, in particular, to the left Λ-modules Λ and P, and therefore proves (ii). $\qquad\square$

4. Non-free Progenerator Sum

Example 4.1. A ring R with progenerators P_n, $n \geq 1$, such that the direct sum $G = \bigoplus_{n=1}^{\infty} P_n$ is not free.

Let $\mathbb{R}_{\geq 0}$ denote the nonnegative real numbers. There is a right self-injective ring R whose monoid $V(R)$ of isomorphism classes of finitely generated projective modules is isomorphic to the additive monoid $\mathbb{R}_{\geq 0}$, and such that the isomorphism $V(R) \cong \mathbb{R}_{\geq 0}$ takes the isomorphism class of R_R to the real number 1. This is a special case of [GW, Corollary 5–3.15] (which characterizes which monoids occur as $V(R)$ for regular right self-injective rings) and [GW, Proposition 3–1.11].

In more detail, there is, for every $\alpha \in \mathbb{R}_{\geq 0}$, a finitely generated projective right R-module Q_α such that the following properties hold.

(i) For every finitely generated projective R-module Q there exists a unique $\alpha \in \mathbb{R}_{\geq 0}$ such that $Q \cong Q_\alpha$.

(ii) $Q_1 \cong R_R$.

(iii) $Q_\alpha \oplus Q_\beta \cong Q_{\alpha+\beta}$ for every α, β.

(iv) Q_α is isomorphic to a direct summand of Q_β if and only if $\alpha \leq \beta$.

We claim: Q_α *is a progenerator whenever* $\alpha \neq 0$. We have $n\alpha \geq 1$ for some positive integer n. Therefore, by (ii)–(iv), $R = Q_1$ is isomorphic to a direct summand of $(Q_\alpha)^{(n)} \cong Q_{n\alpha}$, proving the claim.

Set $G = \bigoplus_{n=1}^{\infty} Q_{1/2^n}$ where summation extends over all positive integers n. We claim that there is a monomorphism of G into R because $\sum_{i=1}^{\infty} 1/2^n = 1$. In more detail, there is a decomposition $R = A_1 \oplus B_1$ with $A_1 \cong B_1 \cong Q_{1/2}$, by (ii) and (iii). Similarly, there is a decomposition $B_1 = A_2 \oplus B_2$ with $A_2 \cong B_2 \cong Q_{1/2^2}$. Thus $R = A_1 \oplus A_2 \oplus B_2$. Continuing in this way we build a submodule $A = \bigoplus_{n=1}^{\infty} A_n$ of R, such that every $A_n \cong Q_{1/2^n}$ and hence $A \cong G$. ($A \neq R$ because R_R is finitely generated while the infinite direct sum A is not.)

Let us prove that G is not free. Assume the contrary, that is, $G \cong R^{(\aleph)}$ for some cardinal $\aleph$. Since G is an infinite direct sum of nonzero modules, G is not cyclic. Therefore $\aleph \geq 2$. Thus there is a monomorphism of $R_R^{(2)}$ into G, and hence into R_R. Since R_R is injective, this implies that $R_R^{(2)} = Q_2$ is isomorphic to a direct summand of $R = Q_1$. This contradiction of property (iv) completes the proof.

References

B.   H. Bass, "Big projective modules are free", *Israel J. Math.* **7** (1963), 24–31.

C.   V. Camillo, "Morita equivalence and infinite matrix rings", *Proc. Amer. Math. Soc.* **90** (1984), 186–188.

F.   A. Facchini, *Module Theory: Endomorphism Rings and Direct Sum Decompositions in some classes of Modules*, Birkhäuser, Basel, 1998.

GW. K. R. Goodearl and F. Wehrung, "The complete dimension theory of partially ordered systems with equivalence and orthogonality", (preprint, presently available at `http://www.math.ucsb.edu/~goodearl/preprints.html`).

L.   L. S. Levy, "Projectives of large uniform-rank, in Krull dimension 1", *Bulletin London Math. Soc.* **21** (1989), 57–64.

P.   P. Prihoda, "On uniserial modules that are not quasi-small", manuscript, 2004.

THE MODULI SPACE AND VERSAL DEFORMATIONS OF THREE DIMENSIONAL LIE ALGEBRAS*

ALICE FIALOWSKI

*Eötvös Loránd University, Department of Applied Analysis
H-1117 Budapest, Pázmány P. sétány. 1/C, Hungary
E-mail: fialowsk@cs.elte.hu*

We consider versal deformations of ordinary (non-graded) three dimensional Lie algebras as special strongly homotopy Lie algebras. They correspond precisely to the 0 even and 3 odd dimensional case. The classification of such Lie algebras is well known. As the symmetric algebra of a three dimensional odd vector space contains terms only of exterior degree less than or equal to three, the construction of versal deformations of these special L_∞ algebras can be carried out completely. We give a characterization of the moduli space of Lie algebras using L_∞ algebra deformation theory as a guide to understanding the picture.

*I dedicate this paper to Izrail Moiseevich Gelfand
on the occasion of his 90-th birthday*

1. Introduction

The classification of low dimensional Lie algebras has been known for a long time. For example, the classification of ordinary Lie algebras of dimension 3 appears in textbooks such as [10]. More recently, the moduli space of three dimensional Lie algebras was studied in [1, 12]. The problem of finding a versal deformation of a given object is a basic question in deformation theory because such a deformation induces all other deformations. This problem turns out to be very difficult. Versal deformation theory was first worked out for the case of Lie algebras in [2, 3, 5] and then extended to strongly homotopy Lie algebras—also called L_∞ algebras—in [6]. We apply these general results to construct versal deformations of three dimensional ordinary Lie algebras, treating them as examples of L_∞ algebras.

*The research of the author was partially supported by grants OTKA T043641 and T043034.

L_∞ algebras are natural generalizations of Lie algebras and superalgebras if one considers $\mathbb{Z}_2$-graded vector spaces. They were discovered in [11] and have recently been the focus of attention. An ordinary 3-dimensional Lie algebra is the same as an L_∞ algebra structure on a $0|3$ (0 even and 3 odd) dimensional $\mathbb{Z}_2$-graded vector space. The advantage of considering Lie algebras as L_∞ algebras is that the deformation problem becomes simpler and we get a clearer insight into the moduli space of the variety of Lie algebras in a given dimension.

For simplicity we will suppose that the underlying vector space is defined over $\mathbb{C}$.

A detailed version of this lecture will appear in [8].

2. Definitions

2.1. *Strongly Homotopy Lie Algebras*

Let W be a $\mathbb{Z}_2$-graded vector space over a field $\mathfrak{K}$ and $T(W)$ the reduced tensor algebra $T(W) = \bigoplus_{n=1}^\infty W^{\otimes n}$. For an element $v = v_1 \otimes \ldots \otimes v_n$ in $T(W)$, define its parity $|v| = |v_1| + \cdots + |v_n|$, and its degree by $deg(v) = n$. With parity $T(W)$ is a $\mathbb{Z}_2$-graded space. The reduced symmetric algebra $S(W)$ is the quotient of the tensor algebra by the graded ideal generated by $u \otimes v - (-1)^{uv} v \otimes u$ for elements $u, v \in W$.

The symmetric algebra $S(W)$ has a natural coalgebra structure, which occurs as a subalgebra of the tensor coalgebra, given by the diagonal mapping

$$\Delta(w_1 \ldots w_n) = \sum_{k=1}^{n-1} \sum_{\sigma \in \mathrm{Sh}(k,n-k)} \epsilon(\sigma) w_{\sigma(1)} \ldots w_{\sigma(k)} \otimes w_{\sigma(k+1)} \ldots w_{\sigma(n)},$$

where we denote the induced product in $S(W)$ by juxtaposition. Here $\mathrm{Sh}(k, n - k)$ is the set of *unshuffles* of type $(k, n - k)$; that is the subset of permutations σ of n elements such that $\sigma(i) < \sigma(i + 1)$ when $i \neq k$, and $\epsilon(\sigma)$ is a sign determined by σ (and $w_1 \ldots w_n$) given by

$$w_{\sigma(1)} \ldots w_{\sigma(n)} = \epsilon(\sigma) w_1 \ldots w_n.$$

Thus if σ interchanges p and $p + 1$, then $\epsilon(\sigma) = (-1)^{v_p v_{p+1}}$.

Obviously the kernel of Δ is W. This mapping is clearly coassociative. The grading on $S(W)$ is compatible with the coalgebra structure, as for homogeneous $c \in S(W)$, $\Delta(c) = \sum_i u_i \otimes v_i$ where $|u_i| + |v_i| = |c|$ for all i, that is Δ has degree 0. With this coalgebra structure, and the $\mathbb{Z}_2$ grading, $S(W)$ is a cocommutatice, coassociative coalgebra (without a counit).

A *coderivation* on the graded coalgebra $S(W)$ is a map $\delta : S(W) \to S(W)$ satisfying

$$\Delta \circ \delta = (\delta \otimes I + I \otimes \delta) \circ \Delta.$$

Let us suppose that the even part of W has basis $e_1 \ldots e_m$, and the odd part has basis $f_1 \ldots f_n$, so that W is an $m|n$ dimensional space. Then a basis of $S(W)$ is given by all vectors of the form $e_1^{k_1} \ldots e_m^{k_m} f_1^{l_1} \ldots f_n^{l_n}$, where k_i is any nonnegative integer, and $l_i \in \mathbb{Z}_2$. An L_∞ structure on W is simply an *odd codifferential* on $S(W)$, that is to say, an odd coderivation whose square is zero. The $\mathfrak{K}$-module $\mathrm{Coder}(S(W))$ of graded coderivations has a natural structure of a graded Lie algebra with the bracket $[\varphi, \psi] = \varphi \circ \psi - (-1)^{\varphi \psi} \psi \circ \varphi$. On the other hand, this space can be identified with $\mathrm{Hom}(S(W), W)$, and the Lie superalgebra structure on $\mathrm{Coder}(S(W))$ determines a Lie bracket on $\mathrm{Hom}(S(W), W)$ as follows. Let

$$L_m = \mathrm{Hom}(S^m(W), W)$$

so that $L = \mathrm{Hom}(S(W), W)$ is the direct product of the spaces L_i. If $\alpha \in L_m$ and $\beta \in L_n$, then $[\alpha, \beta]$ is the element in L_{m+n-1} determined by

$$
\begin{aligned}
[\alpha, \beta](w_1 \ldots w_{m+n-1}) = {} & \\
\sum_{\sigma \in \mathrm{Sh}(n,m-1)} & \epsilon(\sigma) \alpha(\beta(w_{\sigma(1)} \ldots w_{\sigma(n)}) w_{\sigma(n+1)} \ldots w_{\sigma(m+n-1)}) \\
- (-1)^{\alpha\beta} \sum_{\sigma \in \mathrm{Sh}(m,n-1)} & \epsilon(\sigma) \beta(\alpha(w_{\sigma(1)} \ldots w_{\sigma(m)}) w_{\sigma(m+1)} \ldots w_{\sigma(m+n-1)}).
\end{aligned}
\tag{1}
$$

If W is completely odd, and $d \in L_2$, then d determines an ordinary Lie algebra on W, or rather on its parity reversion which is the same space with the parity of elements reversed. This is the case we consider in this talk.

Suppose that $\tilde{g} : S(W) \to S(W')$ is a coalgebra morphism, that is a map satisfying

$$\Delta' \circ \tilde{g} = (\tilde{g} \otimes \tilde{g}) \circ \Delta.$$

If d and d' are L_∞ algebra structures on W and W', resp., then $\tilde{g}$ is a homomorphism between these structures if $\tilde{g} \circ d = d' \circ \tilde{g}$. Two L_∞ structures d and d' on W are equivalent, and we write $d' \sim d$ when there is a coalgebra automorphism $\tilde{g}$ of $S(W)$ such that $d' = \tilde{g}^*(d) = \tilde{g}^{-1} \circ d \circ \tilde{g}$. Furthermore, if $d = d'$, then $\tilde{g}$ is said to be an automorphism of the L_∞ algebra.

2.2. *Versal Deformations*

For the classical theory of formal deformations we refer to [9].

Here we need a more general concept of deformation. A deformation with a base for a Lie algebra was introduced in [2] and worked out in [3]. An augmented local ring $\mathcal{A}$ with maximal ideal $\mathfrak{m}$ will be called an *infinitesimal base* if $\mathfrak{m}^2 = 0$, and a *formal base* if $\mathcal{A} = \varprojlim_n \mathcal{A}/\mathfrak{m}^n$. A deformation of an L_∞ algebra structure d on W with base given by a local ring $\mathcal{A}$ with augmentation $\epsilon : \mathcal{A} \to \mathfrak{K}$ is an $\mathcal{A}\text{-}L_\infty$ structure $\tilde{d}$ on $W \hat{\otimes} \mathcal{A}$ such that the morphism of $\mathcal{A}\text{-}L_\infty$ algebras $\epsilon_* = 1 \otimes \epsilon : L_\mathcal{A} = L \otimes \mathcal{A} \to L \otimes \mathfrak{K} = L$ satisfies $\epsilon_*(\tilde{d}) = d$. (Here $W \hat{\otimes} \mathcal{A}$ is an appropriate completion of $W \otimes \mathcal{A}$.) The deformation is called infinitesimal (formal) if $\mathcal{A}$ is an infinitesimal (formal) base.

In general, the cohomology $H(D)$ of d given by the operator $D : L \to L$ with $D(\varphi) = [\varphi, d]$ may not be finite dimensional. However, L has a natural filtration, which induces a filtration H^n on the cohomology. That means if W is finite dimensional, $H(D)$ is always of finite type, that is H^n/H^{n+1} is finite. A set δ_i will be called a graded *basis of the cohomology*, if any element δ of the cohomology can be expressed uniquely as a formal sum $\delta = \delta_i a^i$.

For each δ_i, let u^i be a parameter of opposite parity. Then the infinitesimal deformation $d^1 = d + \delta_i u^i$, with base $\mathcal{A} = \mathfrak{K}[u_i]/(u_i u_j)$ is universal in the sense that if $\tilde{d}$ is any infinitesimal deformation with base $\mathcal{B}$, then there is a unique homomorphism $f : \mathcal{A} \to \mathcal{B}$, such that the morphism $f_* = 1 \otimes f : L_\mathcal{A} \to L_\mathcal{B}$ satisfies $f_*(\tilde{d}) \sim d$.

For formal deformations, there is no universal object in the sense above. A *versal deformation* is a deformation d^∞ with formal base $\mathcal{A}$ such that if $\tilde{d}$ is any formal deformation with base $\mathcal{B}$, then there is some morphism $f : \mathcal{A} \to \mathcal{B}$ such that $f_*(d^\infty) \sim \tilde{d}$. If f is unique whenever $\mathcal{B}$ is infinitesimal, then the versal deformation is called *miniversal*. In [6], we constructed a miniversal deformation for L_∞ algebras with finite type cohomology.

The method of construction is as follows. Define a coboundary operator D by $D(\varphi) = [\varphi, d]$. First, one constructs the universal infinitesimal deformation $d^1 = d + \delta_i u^i$ as before. The infinitesimal assumption that the products of parameters are equal to zero gives the property that $[d^1, d^1] = 0$. Actually, we can express

$$[d^1, d^1] = (-1)^{\delta_j(\delta_i+1)}[\delta_i, \delta_j]u^i u^j = \delta_k a^k_{ij} u^i u^j + \beta_k b^k_{ij} u^i u^j,$$

where the β_i form a basis of the coboundaries, because the bracket of d^1 with itself is a cocycle. Note that the right hand side is of degree 2 in the parameters, so it is zero up to order 1 in the parameters.

If we suppose that $D(\gamma_i) = -\frac{1}{2}\beta_i$, then by replacing d^1 with

$$d^2 = d^1 + \gamma_k b_{ij}^k u^i u^j,$$

one obtains

$$[d^2, d^2] = \delta_k a_{ij}^k u^i u^j + 2[\delta_l u^l, \gamma_k b_{ij}^k u^i u^j] + [\gamma_k b_{ij}^k u^i u^j, \gamma_l b_{ij}^l u^i u^j]$$

Thus we are able to get rid of terms of degree 2 in the coboundary terms β_i, but those which involve the cohomology terms δ_i can not be eliminated. This gives rise to a set of second order relations on the parameters. One continues this process, taking the bracket of the *n-th order deformation d^n*, adding some higher order terms to cancel coboundaries, obtaining higher order relations, which extend the n-th order relations.

Either the process continues indefinitely, in which case the miniversal deformation is expressed as a formal power series in the parameters, or after a finite number of steps, the right hand side of the bracket is zero after applying the n-th order relations. In this case, the miniversal deformation is simply the n-th order deformation. We obtain a set of relations R_i on the parameters, one for each δ_i, and the algebra $A = \mathbb{C}[[u^i]]/(R_i)$ is called the base of the miniversal deformation. For details see [5, 6].

3. Equivalence Classes of 3-Dimensional Lie Algebras

Suppose that $W = \langle f_1, f_2, f_3 \rangle$. Then $S(W)$ decomposes into three pieces.

$$
\begin{aligned}
S^1(W) &= \langle f_1, f_2, f_3 \rangle, & \dim(S^1(W)) &= 0|3 \\
S^2(W) &= \langle f_1 f_2, f_1 f_3, f_2 f_3 \rangle, & \dim(S^2(W)) &= 3|0 \\
S^3(W) &= \langle f_1 f_2 f_3 \rangle, & \dim(S^3(W)) &= 0|1.
\end{aligned}
$$

Let $L = \mathrm{Hom}(S(W), W)$ and $L_n = \mathrm{Hom}(S^n(W), W)$. Then

$$
\begin{aligned}
L_1(W) &= \{\varphi_j^I \mid I \in \{100, 010, 001\}, j = 1\ldots 3\}, & \dim(L_1) &= 9|0 \\
L_2(W) &= \{\varphi_j^I \mid I \in \{110, 101, 011\}, j = 1\ldots 3\}, & \dim(L_2) &= 0|9 \\
L_3(W) &= \{\varphi_j^{111} \mid j = 1\ldots 3\}, & \dim(L_3) &= 3|0
\end{aligned}
$$

It follows that the only candidate for an odd codifferential is of the form

$$
\begin{aligned}
d = \varphi_1^{110} a_1 + \varphi_2^{110} a_2 + \varphi_3^{110} a_3 + \\
\varphi_1^{101} a_4 + \varphi_2^{101} a_5 + \varphi_3^{101} a_6 + \\
\varphi_1^{011} a_7 + \varphi_2^{011} a_8 + \varphi_3^{011} a_9 \quad (2)
\end{aligned}
$$

Being a quadratic codifferential, we see that d gives an L_∞ structure precisely when it determines a Lie algebra structure. It is natural to consider the derived subalgebra $W' = d(S^2(W))$. Let

$$A = \begin{pmatrix} a_1 & a_2 & a_3 \\ a_4 & a_5 & a_6 \\ a_7 & a_8 & a_9 \end{pmatrix}.$$

It is easy to see that the rank of A is precisely equal to the dimension of the derived subalgebra. In particular, when $\det(A) = 0$, the derived subalgebra has dimension less than three.

The codifferential condition $[d, d] = 0$ is equivalent to a system of three quadratic equations.

If $\det A \neq 0$ then the only possibility is $a_6 = -a_2$, $a_9 = a_1$, and $a_8 = -a_4$. Thus

$$A = \begin{pmatrix} a_1 & a_2 & a_3 \\ a_4 & a_5 & -a_2 \\ a_7 & -a_4 & a_1 \end{pmatrix}, \tag{3}$$

whose determinant does not vanish in general. Consequently we have only this one pattern to consider for when the derived subalgebra has dimension three.

When the derived subalgebra has dimension one, we can choose a basis such that the codifferential d has the simple form

$$d = \varphi_1^{110} a_1 + \varphi_1^{101} a_4 + \varphi_1^{011} a_7. \tag{4}$$

Moreover, it is easy to check that any coderivation of this form is a codifferential.

Finally, suppose that the derived subalgebra has dimension two. Then we can express d in the form

$$d = \varphi_1^{101} a_4 + \varphi_2^{101} a_5 + \varphi_1^{011} a_7 + \varphi_2^{011} a_8. \tag{5}$$

Each of these three cases can be reduced to a much simpler form, up to equivalence. Let us begin with the one dimensional derived subalgebra case first. Let us suppose that g is a linear automorphism of the symmetric coalgebra of W.

Two codifferentials d' and d are equivalent precisely when there is some automorphism g such that $d' = gdg^{-1}$, in other words, $d'g = gd$. We get that all one dimensional solutions are equivalent to one of two codifferentials $d = \varphi_1^{110}$ which is the Lie algebra $\mathfrak{n}_2 \otimes \mathbb{C}$ or $d = \varphi_1^{011}$ which is the 3-dimensional nilpotent Lie algebra $\mathfrak{n}_3$.

Next, let us consider the two dimensional solutions.

We get that the nonequivalent solutions have the form $d(\lambda) = \varphi_1^{101} + \lambda\varphi_2^{011}$ parameterized by the punctured unit disc in $\mathbb{C}$, with the boundary glued together. Moreover, if we consider the case $\lambda = 0$, which is a one dimensional solution, it is then easy to see that it is equivalent to the solution $d = \varphi_1^{110}$. We therefore add this one-dimensional solution as a limit point to the family of two dimensional solutions. (For details see [8].)

Now, if the matrix associated to d is not diagonalizable, then they all arise from the single codifferential $d = \varphi_1^{101} + \varphi_2^{101} + \varphi_2^{011}$ which is the solvable 3-dimensional Lie algebra.

Finally, in the three dimensional case we obtain one equivalence class with codifferential of the form $d' = \varphi_3^{110} + \varphi_2^{101} + \varphi_1^{011}$ which gives the simple Lie algebra $sl(2, \mathbb{C})$.

Thus, up to equivalence, we obtain precisely the codifferentials

$$d_1 = \varphi_1^{011}$$
$$d(\lambda) = \varphi_1^{101} + \varphi_2^{011}\lambda$$
$$d_2 = \varphi_1^{101} + \varphi_2^{101} + \varphi_2^{011}$$
$$d_3 = \varphi_3^{110} + \varphi_2^{101} + \varphi_1^{011},$$

where $d(\lambda)$ is identified with $d(\lambda^{-1})$.

4. Versal Deformations and the Moduli Space

First, we compute the cohomology of the codifferential, use it to write the universal infinitesimal deformation, and then apply the bracket process above to determine a miniversal deformation and the relations on the parameters. Along the way, we will discover that the cohomology of the differentials reveals a lot of information about the moduli space of three dimensional Lie algebras.

4.1. *The Codifferential* $d_3 = \varphi_3^{110} + \varphi_2^{101} + \varphi_1^{011}$

Here the space H^1 is $3|0$-dimensional, and all higher cohomology vanishes.

Nevertheless, since we have three cocycles, all even, we do have a non-trivial infinitesimal deformation. Let us adopt the convention to use the Greek letter θ for an odd parameter, and the Roman letter t for an even one. For an odd parameter θ, we have $\theta^2 = 0$ as a consequence of the graded commutativity, so we do not consider this to be a relation on our parameter algebra. Thus, we have

$$d_3^1 = \varphi_3^{110} + \varphi_2^{101} + \varphi_1^{011} + (\varphi_3^{010} + \varphi_2^{001})\theta_1 + (\varphi_3^{100} - \varphi_1^{001})\theta_2 + (\varphi_2^{100} + \varphi_1^{010})\theta_3.$$

In computing $[d_3^1, d_3^1]$, we note that the brackets of the cocycles with d_3 vanish, so we only need to compute the brackets of the cocycles with each other. We obtain

$$[d_3^1, d_3^1] = (\varphi_2^{100} + \varphi_1^{010})\theta_1\theta_2 + (\varphi_3^{100} - \varphi_1^{001})\theta_1\theta_3 + (\varphi_3^{010} + \varphi_2^{001})\theta_2\theta_3.$$

Of course, these are all cocycles which are not coboundaries, so we obtain the relations $\theta_i\theta_j = 0$ for all i, j. Thus the infinitesimal deformation is miniversal, and the base of the miniversal deformation is given by $\mathbb{C}[\theta_1, \theta_2, \theta_3]/(\theta_1\theta_2, \theta_1\theta_3, \theta_2\theta_3)$.

Note that the vanishing of H^2 is consistent with the observation that any small change in the codifferential d_3 will give rise to a codifferential d' which will still have a 3-dimensional derived subalgebra. Thus any small change in d_3 gives rise to the same codifferential, and we see that d_3 does not deform into any of the other codifferentials.

4.2. The Codifferential $d_2 = \varphi_1^{101} + \varphi_2^{101} + \varphi_2^{011}$

Let us first determine its cohomology. We get three obvious 1-cocycles, φ_2^{100}, φ_1^{001} and φ_2^{001}, and in addition also $\varphi_1^{100} + \varphi_2^{010}$. The space H^2 is 1-dimensional with a basis φ_2^{011} and the higher cohomology is zero. The universal infinitesimal deformation is given by

$$d_2^1 = \varphi_1^{101} + \varphi_2^{101} + (1+t)\varphi_2^{011} + \varphi_2^{100}\theta_1 + \varphi_1^{001}\theta_2 + \varphi_2^{001}\theta_4 + (\varphi_1^{100} + \varphi_2^{010})\theta_4.$$

We have

$$[d_2^1, d_2^1] = 2\varphi_2^{101}t\theta_1 + \varphi_2^{001}(\theta_1\theta_2 - \theta_3\theta_4) - \varphi_1^{001}\theta_2\theta_4.$$

Of the three cocycles appearing on the right hand side of this equation, only the first is a coboundary. Thus we obtain two second order relations, $\theta_1\theta_2 - \theta_3\theta_4 = 0$ and $\theta_2\theta_4 = 0$, and we need to add something to d_2^1 to obtain a second order deformation. Since $D(\varphi_1^{100}) = -\varphi_2^{101}$, we can express $d_2^2 = d_2^1 + \varphi_1^{100}t\theta_1$. We compute

$$[d_2^2, d_2^2] = \varphi_2^{001}(\theta_1\theta_2 - \theta_3\theta_4) - \varphi_1^{001}(\theta_2\theta_4 + t\theta_1\theta_2).$$

Since no coboundary terms occur, we obtain that d_2^2 is a miniversal deformation, and the base is given by

$$A = \mathbb{C}[[t, \theta_1, \theta_2, \theta_3, \theta_4]]/(\theta_1\theta_2 - \theta_3\theta_4, \theta_2\theta_4 + t\theta_1\theta_2).$$

Let us study the induced topology on the moduli space of equivalence classes of codifferentials. This topology is not Hausdorff. If every neighborhood of a point a contains the point b, then we note that a is in the closure of b. In this case, we shall say that a is *infinitesimally close* to b.

Since there is a nontrivial deformation in the Lie algebra direction itself, we can explore how the deformation moves our codifferential in the moduli space. Note that d_2^1 is a codifferential which has two dimensional derived algebra, and it has eigenvalues 1 and $1 + t$, so it lies in the family $d(\lambda)$, and is near to $d(1)$. In fact, a punctured neighborhood of d_2 looks exactly like a neighborhood of $d(1)$. This phenomena shows up in the classical deformation theory as jump deformation. However, $d(1)$ is not close to d_2, in the sense that a small neighborhood of d_2 does not contain $d(1)$. When we study $d(1)$, we shall see that the opposite statement is not true.

4.3. *The Family of Codifferentials* $d(\lambda) = \varphi_1^{101} + \varphi_2^{011}\lambda$

The cohomology will depend to some extent on the value of λ.

We shall see that the only thing special about the case $\lambda = 0$ is that the dimension of the derived algebra drops to 1, but as far as deformations go, it will behave like a generic element of the family. The cases $\lambda = \pm 1$, however, are not generic in terms of their deformations. This makes sense, because in the identification of $td\lambda$ with $\tilde{d}\lambda^{-1}$, we see that every point in the unit disc has a neighborhood that is like a usual disc in $\mathbb{C}$, with the exception of ± 1, which are orbifold points. So it is not surprising to find some kind of exceptional behavior for these codifferentials.

4.3.1. *Generic case of* $d(\lambda)$

First, we treat the generic case. Clearly, $H^1 = \langle \varphi_1^{100}, \varphi_2^{010}, \varphi_1^{001}, \varphi_2^{001} \rangle$. The space H^2 is one-dimensional and we can choose φ_2^{011} as a basis. All the higher cohomology classes are zero. Thus, the universal infinitesimal deformation is given by

$$d(\lambda)^1 = \varphi_1^{101} + \varphi_2^{011}(\lambda + t) + \varphi_1^{100}\theta_1 + \varphi_2^{010}\theta_2 + \varphi_1^{001}\theta_3 + \varphi_2^{001}\theta_4.$$

It is easy to verify that

$$[d(\lambda)^1, d(\lambda)^1] = 2\varphi_1^{001}\theta_1\theta_3 + 2\varphi_2^{001}\theta_2\theta_4.$$

Thus $d(\lambda)^1$ is miniversal and the base of the miniversal deformation is

$$A = \mathbb{C}[[t, \theta_1, \theta_2, \theta_3, \theta_4]]/(\theta_1\theta_3, \theta_2\theta_4).$$

Looking at the deformation in the Lie algebra direction, we see that the deformation simply moves along the $d(\lambda)$ family.

4.3.2. *The special case $d(-1)$*

Now, let us consider the special case $\lambda = -1$. Then we have an extra cocycle φ_3^{110} in H^2, and correspondingly, an extra cocycle φ_3^{111} in H^3. Thus, the universal infinitesimal deformation becomes

$$d(-1)^1 = \varphi_1^{101} + \varphi_2^{011}(-1 + t_1) + \varphi_1^{100}\theta_1 + \varphi_2^{010}\theta_2 + \varphi_1^{001}\theta_3 + \varphi_2^{001}\theta_4$$
$$+ \varphi_3^{110}t_2 + \varphi_3^{111}\theta_5.$$

Then

$$\frac{1}{2}[d(-1)^1, d(-1)^1] = \varphi_1^{001}\theta_1\theta_3 + \varphi_2^{001}\theta_2\theta_4 + \varphi_3^{110}(t_2\theta_1 + t_2\theta_2)$$
$$+ \varphi_3^{111}(\theta_5\theta_1 + \theta_5\theta_2 - t_1t_2) - \varphi_1^{111}\theta_5\theta_3 - \varphi_2^{111}\theta_5\theta_4$$
$$+ (-\varphi_1^{110} - \varphi_3^{011})t_2\theta_3 + (-\varphi_2^{110} + \varphi_3^{101})t_2\theta_4.$$

Note that the first four terms are cocycles, so they give rise to second order relations, while the last four terms are coboundaries, so we need to add corresponding terms to obtain the second order deformation

$$d(-1)^2 = d(-1)^1 + \varphi_3^{011}\theta_3\theta_5 + \varphi_2^{110}\theta_4\theta_5 - \varphi_3^{010}t_2\theta_3 - \varphi_3^{100}t_2\theta_4.$$

Finally, let us compute the bracket of the second order deformation with itself. We obtain

$$\frac{1}{2}[d(-1)^2, d(-1)^2] = \varphi_1^{001}\theta_1\theta_3 + \varphi_2^{001}\theta_2\theta_4 + \varphi_3^{110}(t_2\theta_1 + t_2\theta_2) +$$
$$\varphi_3^{111}(\theta_5\theta_1 + \theta_5\theta_2 - t_1t_2) + \varphi_3^{011}(\theta_3\theta_5\theta_2 - t_1t_2\theta_3) +$$
$$\varphi_2^{110}(\theta_4\theta_5\theta_1 - t_1t_2\theta_4) - \varphi_3^{010}t_2\theta_3\theta_2 + \varphi_2^{010}t_2\theta_3\theta_4 -$$
$$\varphi_3^{100}t_2\theta_4\theta_1 + \varphi_1^{100}t_2\theta_4\theta_3.$$

All the terms except φ_3^{011}, φ_2^{110}, φ_3^{010}, and φ_3^{100} are cocycles, and these exceptional terms are not coboundaries. Thus, by the general theory, their coefficients must be zero.

Note that $d(-1)$ is infinitesimally close to d_3, but not the other way around. In some sense, this explains the extra infinitesimal deformation in the Lie algebra direction.

Now at first it may seem strange that this is the first case where one of our codifferentials deforms into d_3. After all, generically, we would expect the matrix of a codifferential to be invertible. But looking carefully at the codifferential with matrix (3) and the form of a solution in the family, it becomes clear that only when $\lambda = -1$ can a small change in the codifferential give a solution satisfying (3).

4.3.3. *The special case $d(1)$*

In this case, there are two additional 1-cocycles, φ_2^{100} and φ_1^{010}. Thus $\dim H^1 = 6$, $\dim H^2 = 3$ and $\dim H^3 = 0$. So we pick up two extra 2-cocycles, φ_2^{101} and φ_1^{011}. It is convenient to replace the cocycle φ_2^{011}, which we used as a basis of the cohomology in the generic case, with $\varphi_1^{101} - \varphi_2^{011}$, because it simplifies the interpretation of the bracket of the universal infinitesimal deformation with itself. Thus,

$$d(1)^1 = \varphi_1^{101} + \varphi_2^{011}\lambda + (\varphi_1^{101} - \varphi_2^{011})t_1 + \varphi_2^{101}t_2 + \varphi_1^{011}t_3$$
$$+ \varphi_1^{100}\theta_1 + \varphi_2^{010}\theta_2 + \varphi_1^{001}\theta_3 + \varphi_2^{001}\theta_4 + \varphi_2^{100}\theta_5 + \varphi_1^{010}\theta_6,$$

and we compute that

$$\frac{1}{2}[d(1)^1, d(1)^1] = (\varphi_1^{101} - \varphi_2^{011})(t_3\theta_5 - t_2\theta_6) - \varphi_2^{101}(2t_1\theta_5 + t_2(\theta_2 - \theta_1))$$
$$+ \varphi_1^{011}(2t_1\theta_6 + t_3(\theta_2 - \theta_1)) - \varphi_1^{100}\theta_5\theta_6 + \varphi_2^{010}\theta_5\theta_6$$
$$+ \varphi_1^{001}(\theta_1\theta_3 - \theta_4\theta_6) + \varphi_2^{001}(\theta_2\theta_4 - \theta_3\theta_5)$$
$$+ \varphi_2^{100}(\theta_2\theta_5 - \theta_1\theta_5) + \varphi_1^{010}(\theta_1\theta_6 - \theta_2\theta_6),$$

which is precisely the set of cocycles appearing in the universal infinitesimal deformation, multiplied by the second order relations. Thus the infinitesimal deformation is miniversal.

Now let us interpret how $d(1)$ fits into the moduli space. Note that there is a deformation along the family, given by the cocycle $\varphi_1^{101} - \varphi_2^{011}$, and two other directions of deformation, each of which corresponds to a deformation of $d(1)$ into the special codifferential d_2. In fact, if we consider the three dimensional deformation space parameterized by (t_1, t_2, t_3), we see that precisely two curves correspond to a deformation in the d_2 direction. Thus we see that $d(1)$ is infinitesimally close to d_2, although the converse is not true.

So far, we have been able to associate two of the three special codifferentials with the family in some manner.

4.4. *The Codifferential $d_1 = \varphi_1^{011}$*

This gives a nilpotent Lie algebra structure, so that we expect to find a lot of deformations.

In this case $\dim H^1 = 6$, $\dim H^2 = 5$ and $\dim H^3 = 2$. Let us analyze how this codifferential sits in the moduli space. Clearly there are a lot of directions one can deform. As $\varphi_1^{101} - \varphi_2^{011}$ and $\varphi_1^{110} + \varphi_3^{011}$ are coboundaries, there remain three ways to deform d_1 into d_3, via the cocycles φ_2^{110},

$\varphi_3^{101} + \varphi_2^{101}$ and $\varphi_3^{101} - \varphi_2^{101}$. Generically, if we add small multiples of these cocycles, we will obtain a codifferential which is equivalent to d_3. Thus d_1 is infinitesimally close to d_3.

Next, if we add an appropriate multiple of $\varphi_1^{101} + a\varphi_2^{101}$, then we are constructing an element in the family $d(\lambda)$ with a matrix B given by $B = \begin{pmatrix} t & at \\ 1 & 0 \end{pmatrix}$. (The fraction of its eigenvalues determines the element of the family, see [8].) This way we get an element of the family for any value of λ except $\lambda = \pm 1$. The reason that we do not obtain an element of the family for $\lambda = 1$ is that the resulting matrix is defective, so we obtain d_2 instead. There is a solution for $\lambda = -1$, however, given by the cocycle φ_2^{101}. Thus we see that d_1 is infinitesimally close to d_2 and to every member of the family except for $\lambda = 1$. Moreover, it is not hard to check that if we take d to be of the form

$$d = \varphi_1^{011} + \varphi_1^{110}t_1 + \varphi_1^{101}t_2 + (\varphi_2^{110} - \varphi_3^{101})t_3 + \varphi_3^{110}t_4 + \varphi_2^{101}t_5,$$

then there is no automorphism which takes it to $d(1)$.

Thus we conclude that d_1 is infinitesimally close to every codifferential except $d(1)$. From the behavior of the elements d_2 and $d(1)$, it is more natural to consider d_2 as a member of the family, because it has the same cohomology as the other members of the family except $d(1)$. Note, for example, that $\dim H^2(d_2) = 1$, the same as the generic elements of the family, while $\dim H^2(d(1)) = 3$.

For constructing a miniversal deformation for this Lie algebra, we have to do three steps which we will not discuss now in this talk.

4.5. *Deformations of the Trivial Codifferential $d_0 = 0$*

This codifferential evidently must deform into every possible type, so we know that it is infinitesimally close to every point in the moduli space. Moreover, since there are no coboundaries in the bracket, the infinitesimal deformation is obviously miniversal.

On the other hand, we do obtain some relations, and these relations carry some information about how the moduli space is put together. In addition, all second order relations can be determined from the relations on the zero codifferential, by using appropriate values for the coefficients.

References

1. Y. Agaoka, *On the variety of 3-dimensional Lie algebras*, Lobachevskii Journal of Mathematics **3** (1999), 5–17.

2. A. Fialowski, *Deformations of Lie algebras*, Mathematics of the USSR-Sbornik **55** (1986), no. 2, 467–473.

3. A. Fialowski, *An example of formal deformations of Lie algebras*, Deformation Theory of Algebras and Structures and Appl., Kluwer (1988), 375–401.

4. A. Fialowski and D. Fuchs, *Singular deformations of Lie algebras on an example*, Topics in Singularity Theory (Providence, RI) (A. Varchenko and V. Vassilie, eds.), A.M.S. Translation Series 2, **180**, Amer. Math. Soc., 1997, V. I. Arnold 60-th Anniversary Collection.

5. ——, *Construction of miniversal deformations of Lie algebras*, Journal of Functional Analysis, **161** (1), (1999), 76–110.

6. A. Fialowski and M. Penkava, *Deformation theory of infinity algebras*, Journal of Algebra **255** (2002), no. 1, 59–88, math.RT/0101097.

7. ——, *Examples of infinity and Lie algebras and their versal deformations*, Geometry and Analysis on Lie groups, Banach Center Publications, **55** (2002), pp. 27–42, math.QA/0102140.

8. ——, *Versal deformations of three dimensional Lie algebras as infinity algebras*, Preprint math.RT/0303346, submitted for publication.

9. M. Gerstenhaber, *On the Deformations of Rings and Algebras I*, Ann. Math. **79** (1964), 59–103.

10. Nathan Jacobson, *Lie algebras*, John Wiley & Sons, 1962.

11. M. Schlessinger and J. Stasheff, *The Lie algebra structure of tangent cohomology and deformation theory*, Journal of Pure and Applied Algebra **38** (1985), 313–322.

12. H. Tasaki and M. Umehara, *An invariant on 3-dimensional Lie algebras*, Proceedings of the American Mathematical Society **115** (1992), no. 2, 293–294.

QUADRATIC ALGEBRAS OF SKEW TYPE*

ERIC JESPERS

*Department of Mathematics, Vrije Universiteit Brussel,
Pleinlaan 2, 1050 Brussel, Belgium
E-mail: efjesper@vub.ac.be*

JAN OKNIŃSKI

*Institute of Mathematics, Warsaw University,
Banacha 2, 02-097 Warsaw, Poland
E-mail: okninski@mimuw.edu.pl*

We consider algebras over a field K with generators x_1, x_2, ..., x_n subject to $\binom{n}{2}$ square-free relations of the form $x_i x_j = x_k x_l$, with every product $x_p x_q$, $p \neq q$, appearing in one of the relations. Main properties of such algebras and of the underlying semigroups are presented. This is done under some natural non-degenerate conditions. Some open problems are discussed. Certain results on an important structural tool, the so called structural chains, are proved.

1. Introduction

We consider finitely generated unitary algebras $K\langle x_1, \ldots, x_n : R \rangle$ over a field K defined by a system R of relations of the form $u = v$ and $w = 0$, where u, v, w are words in the generators x_i. In other words, these are semigroup algebras $K[S]$, or contracted semigroup algebras $K_0[S]$, of the monoids given by the same presentations. Our main interest is in noetherian algebras of this type. Therefore, the methods and results presented in this paper heavily depend on background on noetherian rings, Gelfand-Kirillov dimension, graded rings, group and semigroup algebras and semigroup theory. We refer to [3, 18, 19, 20, 21, 23, 24] for the necessary aspects of these topics.

*The authors were supported in part by Onderzoeksraad of Vrije Universiteit Brussel, Fonds voor Wetenschappelijk Onderzoek (Belgium), Flemish-Polish bilateral agreement BIL 01/31, and KBN research grant 2P03A 033 25 (Poland).

Our focus is on special algebras of the above type, called quadratic algebras of skew type. Recently, certain classes of such algebras and related semigroups and groups have been investigated by several authors and for many reasons, such as homological properties of important classes of associative algebras [10, 25], Gröbner bases techniques [8, 17], structural problems concerning semigroups and semigroup algebras [12, 15] and set theoretical solutions of the quantum Yang-Baxter equation [5, 10]. In this paper our aim is to present general results on the structure of the algebra and of the underlying semigroup and to discuss certain open questions.

Let G be a group. For any non-negative integer n let $\mathcal{M}_n(G)$ denote the semigroup of all matrices in $M_n(K[G])$ with entries in the set $G \cup \{0\}$ and with at most one non-zero entry in each row and in each column. We call it the semigroup of monomial matrices over G. Then $\mathcal{M}(G, n, n; \nabla)$ denotes the subsemigroup of all matrices with at most one non-zero entry, and a typical non-zero element of this semigroup can be denoted (g, i, j) with $g \in G$ and $1 \leq i, j \leq n$. By a semigroup of matrix type over G we mean any subsemigroup T of $\mathcal{M}(G, n, n; \nabla)$ that contains some (g, i, j) for every $i, j = 1, \ldots, n$. Such T provide us with the simplest examples of uniform semigroups in the sense of [21], see Section 4 for the definition. For an ideal I of S we denote by S/I the corresponding Rees factor and we identify the contracted semigroup algebra $K_0[S/I]$ with $K[S]/K[I]$.

We start with a result obtained in [22] which gives an unexpected condition for $K[S]$ to be noetherian.

Theorem 1.1. *Assume that S is a finitely generated monoid with an ideal chain $I_1 \subseteq I_2 \subseteq \cdots \subseteq I_t = S$ such that I_1 and every factor I_j/I_{j-1} is either nilpotent or a semigroup of matrix type. If S has the ascending chain condition on right ideals and the Gelfand-Kirillov dimension of $K[S]$ is finite then $K[S]$ is right noetherian.*

The converse of this result often holds [22]. Moreover, due to a result of Anan'in [1], every finitely generated right noetherian PI K-algebra embeds into a matrix ring over a field extension F of the base field K. If $K[S]$ is a right noetherian PI algebra then S is finitely generated by Theorem 19.14 in [20]. So S is a linear semigroup, that is, S is a subsemigroup of the multiplicative semigroup $M_n(F)$ for some n and some field F. Therefore the following weak form of the general structure theorem for arbitrary semigroups of matrices [21] can be applied to recover the flavor of Theorem 1.1.

Theorem 1.2. *Let $S \subseteq M_r(F)$ be a linear semigroup over a field F. Then* *there is an ideal chain $J_1 \subseteq \cdots \subseteq J_m = S$ such that J_1 and every J_i/J_{i-1} is either nilpotent or uniform.*

We note that this result remains valid if F is a division ring and S satisfies the ascending chain condition on right ideals.

The case where the defining relations of the form $u = v$ satisfy $|u| = |v|$ (where $|u|$ denotes the length of the word u in the generators x_i) is of special interest. In this case the algebra inherits the natural $\mathbf{Z}$-gradation of the free algebra on $x_1, \ldots, x_n$. We then say that S is defined by a set of homogeneous relations. The following surprising result was proved in [11].

Theorem 1.3. *Let S be a monoid such that the algebra $K[S]$ is right noetherian and of finite Gelfand-Kirillov dimension. Then S is finitely generated. If, moreover, S has a monoid presentation of the form*

$$S = \langle x_1, \ldots, x_n \mid R \rangle$$

with a set R of homogeneous relations, then $K[S]$ satisfies a polynomial identity.

Because of Theorem 1.1, Theorem 1.2 and Theorem 1.3 it seems natural to approach various problems concerning noetherian semigroup algebras via the ideal chains of the above type.

In this paper we consider algebras over a field K, with generators x_1, x_2, ..., x_n subject to $\binom{n}{2}$ square-free relations $x_i x_j = x_k x_l$ in which every product $x_p x_q, p \neq q$, appears in one of the relations (and thus in exactly one relation). Such an algebra can be treated as the semigroup algebra $K[S]$ of the monoid

$$S = \langle x_1, x_2, \ldots, x_n \mid x_i x_j = x_k x_l \rangle$$

defined by the same relations. It is called a quadratic monoid of skew type. The assumption that there are $\binom{n}{2}$ relations is a requirement for $K[S]$ to satisfy the noetherian property. Indeed, for any $i \neq j$ consider the right ideals

$$x_i x_j S \subseteq x_i x_j S \cup x_i^2 x_j S \subseteq \cdots \subseteq \bigcup_{q=1}^{m} x_i^q x_j S \subseteq \cdots$$

If $K[S]$ is right noetherian then we must have $x_i^k x_j = x_i^m x_j s$ for some $0 < m < k$ and $s \in S$. But since, by assumption, the relations are square free, we need that $x_i x_j$ appears in one of the defining relations. Hence we have shown that each word $x_i x_j$, with $i \neq j$, appears in one of the relations. Since there are $2\binom{n}{2}$ such words, we need at least $\binom{n}{2}$ relations.

As will be explained in the next section, another motivation for studying algebras with $\binom{n}{2}$ 'rewriting relations' $x_i x_j = x_k x_l$ comes from the study of set theoretical solutions of the quantum Yang-Baxter equation.

It turns out that several properties of the algebra $K[S]$ can be investigated and characterized in terms of the properties of the underlying monoid S. In order to illustrate this, we shall restrict our attention to the case where S satisfies the so called cyclic condition, introduced and first studied in [7, 8]. This is motivated by certain important special classes of algebras in which the cyclic condition became one of the key combinatorial tools. These special classes will be recalled in the next section.

We say that S satisfies the cyclic condition if for every pair $z, y \in X = \{x_1, \ldots, x_n\}$ there exist $z_1 = z, \ldots, z_k, y' \in X$ so that

$$yz_1 = z_2 y', \quad yz_2 = z_3 y', \quad \ldots, yz_k = z_1 y'.$$

It is known ([11]) that if S satisfies the cyclic condition, then the following property holds:

(C1) for any pair $z, y \in X$, there exist two sequences: $z = z_1, z_2, \ldots, z_k$ and $y = y_1, y_2, \ldots, y_p$ in X such that

$$y_1 z_1 = z_2 y_2, \ y_1 z_2 = z_3 y_2, \ \ldots, \ y_1 z_k = z_1 y_2,$$
$$y_2 z_1 = z_2 y_3, \ y_2 z_2 = z_3 y_3, \ \ldots, \ y_2 z_k = z_1 y_3,$$
$$\vdots$$
$$y_p z_1 = z_2 y_1, \ y_p z_2 = z_3 y_1, \ \ldots, \ y_p z_k = z_1 y_1.$$

The above can be reformulated as follows. Let $Z = \{(x, x) \in X\} \subseteq X \times X$. Then there exists a disjoint union decomposition $(X \times X) \setminus Z = X_1 \times Y_1 \cup \cdots \cup X_t \times Y_t$ such that for every i we have $X_i \cap Y_i = \emptyset$ and there exist cyclic permutations $\sigma : Y_i \to Y_i, \tau : X_i \to X_i$ such that

$$xy = \sigma(y)\tau(x)$$

for all $x \in X_i, y \in Y_i$. Moreover, for every i there exists j such that $Y_j = X_i$ and $X_j = Y_i$.

2. Basic properties of algebras of skew type

We list some of the important properties of monoids of skew type obtained in [11]. First, assume that S satisfies the cyclic condition.

(C2) for some positive integer p the monoid $A = \langle x_1^p, \ldots, x_n^p \rangle$ is abelian and $S = \bigcup_{f \in F} fA$, where $F = \{z_1^{a_1} \cdots z_k^{a_k} \mid 0 \le a_i < p, z_i \in X, k \le n\}$ and $Af = fA$ for all $f \in F$.

(C3) S is right non-degenerate, that is, for every $x \in X$ there exists a permutation f_x of X so that for every $x_i \in X$ we have $xx_i = f_x(x_i)x_l$ for some $x_l \in X$. Similarly, one defines the left non-degenerate condition and it follows that S satisfies the latter as well.

If S satisfies the right non-degenerate condition (but the cyclic condition is not necessarily satisfied), then the following properties hold:

(RN1) S satisfies the ascending chain condition on right ideals and has finite Gelfand-Kirillov dimension,

(RN2) if S is also left non-degenerate, then $K[S]$ is a right and left noetherian PI algebra, whence it embeds into a matrix ring over a field.

To prove (RN2) one uses (RN1) and shows that S has a finite ideal chain $I_1 \subseteq I_2 \subseteq \cdots \subseteq I_t = S$ with properties required in Theorem 1.1. In order to give a more precise statement we recall some more notation that will also be used later. For a non-empty subset Y of X let $S_Y = \bigcap_{y \in Y} yS$ and $D_Y = \{s \in S_Y \mid \text{if } s = xt \text{ for some } x \in X \text{ and } t \in S \text{ then } x \in Y\}$. The left-right symmetric duals of these sets are denoted by S_Y' and D_Y' respectively. For $1 \le i \le n$, let $S_i = \bigcup_{Y, |Y|=i} S_Y$. Still under the assumption that S is right non-degenerate one gets

(RN3) (a) each S_Y is non-empty,
  (b) each S_i is an ideal of S,
  (c) S is a union of sets of the form $\{z_1^{a_1} \cdots z_k^{a_k} \mid \text{each } a_i \ge 0\}$ with each $z_i \in X$ and $k \le n$.

Now the desired chain $I_1 \subseteq I_2 \subseteq \cdots \subseteq I_t = S$ is obtained by a refinement of the chain $S_n \cap S_n' \subseteq S_{n-1} \cap S_{n-1}' \subseteq \cdots \subseteq S_1 \cap S_1' \subseteq S$.

In case S is left and right non-degenerate we also obtain some information on the prime ideals of $K[S]$.

(RN4) (a) S_X^m is a cancellative ideal for some positive integer m, and S_X^m has a group of quotients,
  (b) the least right cancellative congruence on S coincides with the least cancellative congruence ρ on S,
  (c) $\operatorname{ann}(S_X^m) = \operatorname{ann}_l(S_X^m) = I(\rho)$, where $I(\rho)$ is the (augmentation) ideal determined by ρ and the annihilators are considered in $K[S]$,

 (d) $I(\rho) \subseteq P$ for any prime ideal P of $K[S]$ with $P \cap S = \emptyset$, and $S_X \subseteq P$ for any prime ideal P of $K[S]$ with $P \cap S \neq \emptyset$,

 (e) if $\operatorname{char}(K) = 0$ then the prime radical $\mathcal{P}(K[S])$ of $K[S]$ is equal to $I(\rho) \cap \left(\bigcap_{P, P \cap S \neq \emptyset} P \right)$, where the intersection is taken over all minimal primes P satisfying the mentioned condition.

Among the examples of monoids $S = \langle x_1, \ldots, x_n \mid R \rangle$ of skew type that satisfy the cyclic condition are the so called binomial monoids [7, 8, 12]. Here the $\binom{n}{2}$ relations in R are of the form $x_i x_j = x_k x_l$ with $i, l > k, j$ and also each element of S has a unique presentation in the form $x_1^{a_1} \cdots x_n^{a_n}$ with each $a_i \geq 0$ (and thus $\{ x_1^{a_1} \cdots x_n^{a_n} \mid$ each $a_i \geq 0 \}$ is a K-basis for the algebra $K[S]$). The latter condition can be rephrased by saying that the ideal of the free K-algebra on $X = \{ x_1, \ldots, x_n \}$ determined by the defining relations of $K[S]$ has a Gröbner basis consisting of the polynomials $x_i x_j - x_k x_l$ determined by these relations, where one considers the degree-lexicographic order on the free monoid generated by X. It turns out that S is contained in a solvable torsion-free abelian-by-finite group (so, $K[S]$ is a domain, in particular), $S = S_{(1)} S_{(2)} \cdots S_{(r)}$, where each $S_{(i)}$ is again a binomial monoid (called a component of S) with generators a subset Y_i of X, $S_{(i)} S_{(j)} = S_{(j)} S_{(i)}$, and each element of S has a unique representation of the form $s_1 s_2 \cdots s_r$ with $s_i \in S_{(i)}$. Furthermore, the sets Y_i form a partition of X and two generators are in the same component if they are in the transitive closure of the relation $\sim$ defined as follows: $x_i \sim x_j$ if there exists a relation of the form $x_i x_k = x_l x_j$ or $x_l x_i = x_j x_k$ for some $x_l, x_k \in X$, or equivalently, $s x_i^p = x_j^p s$ for some $s \in S$ (where p is a positive integer as in (C2)). Also, S is the infinite cyclic monoid if and only if S only has one component.

Another motivation for investigating the cyclic condition comes from the study of set theoretical solutions of the quantum Yang-Baxter equation, proposed by Drinfeld in [4]. Here one considers bijective maps $r : X \times X \to X \times X, r^2 = Id$, and the associated monoids $S = \langle X \rangle$ defined by relations $x_i x_j = x_k x_l$ if $r(x_i, x_j) = (x_k, x_l)$. It can be checked that providing solutions of the quantum Yang-Baxter equation is equivalent to giving solutions of the equation $r_2 r_1 r_2 = r_1 r_2 r_1$ where $r_1, r_2 : X^3 \to X^3$ are given by $r_1(x, y, z) = (r(x, y), z)$ and $r_2(x, y, z) = (x, r(y, z))$ [5]. A combinatorially defined class of monoids, called monoids of left I-type, was introduced in [10]. In particular, it was shown that square-free monoids of this type (that is, with no nontrivial relations of the form $x_i x_i = x_k x_l$)

are exactly the left non-degenerate monoids of skew type that provide set theoretical solutions to the quantum Yang-Baxter equation and that do not have defining relations of the form $x_i x_j = x_p x_j$ with $i \neq p$. Moreover, every binomial monoid is of this type. This characterization allows to check easily whether a given monoid of skew type belongs to this class.

Surprisingly, algebras of monoids of left I-type share many properties with commutative polynomial algebras. In particular, they are noetherian domains of finite global dimension, satisfy a polynomial identity, are Koszul, Auslander-Gorenstein and Cohen-Macaulay [10]. Reasons and tools for dealing with monoids satisfying these properties came from the study of homological properties of Sklyanin algebras by Tate and Van den Bergh [25] and from the work of Gateva-Ivanova on so called skew-polynomial rings with binomial relations [8]. For proofs we refer the reader to [5, 8, 10, 11]. More results on the structure and combinatorics of monoids of I-type and of the associated groups can be found in the recent preprints [9, 16].

3. Questions and comments

We start with a list of questions on monoids of skew type $S = \langle x_1, \ldots, x_n \rangle$ satisfying the cyclic condition. These natural questions arise from a detailed study of several examples described in [15]. All of them are answered affirmatively if S is a binomial monoid.

(1) Do we have $S_i = S_i'$ for all i? In Proposition 3.1 it is shown that the answer is positive for binomial monoids.

(2) Is S periodic modulo an abelian submonoid A as in (C2)? If S is binomial, then the group generated by A is of finite index in the group of quotients G of S and we have $A = AA^{-1} \cap S$ [12], whence the answer is positive.

  Note that the example

$$S = \langle x_1, x_2, x_3, x_4 \mid x_4 x_1 = x_3 x_4, \ x_4 x_2 = x_2 x_4, \ x_4 x_3 = x_1 x_4,$$
$$x_3 x_1 = x_2 x_3, \ x_3 x_2 = x_1 x_3, \ x_2 x_1 = x_1 x_2 \rangle,$$

  with $A = \langle x_1^2, x_2^2, x_3^2, x_4^2 \rangle$, shows that in general S is not a union of disjoint cosets of A, see [15]. Indeed, in S we have $x_1 x_2 x_4^2 = x_1 x_3 x_4^2$ (because $x_1 x_2 x_4 = x_2 x_1 x_4 = x_2 x_4 x_3 = x_4 x_2 x_3 = x_4 x_3 x_1 = x_1 x_4 x_1 = x_1 x_3 x_4$), while $x_1 x_2 A \neq x_1 x_3 A$.

(3) Is the prime radical of $K[S]$ always determined by a congruence η, that is, $\mathcal{P}(K[S]) = I(\eta)$? When is $K[S]$ semiprime? Is it always the

case in characteristic zero? Notice that $K[S]$ is a domain if S is a binomial monoid [10].

(4) Is the abelian submonoid A always a separative semigroup? (If this is not true then $\mathcal{P}(K[A]) \neq 0$ by [20], Theorem 21.2, whence $\mathcal{P}(K[S]) \neq 0$.) Notice that A is free abelian if S is binomial [10, 12].

(5) Is S_n cancellative?

The following example shows that S_n does not have to be cancellative if S is non-degenerate. Let

$$S = \langle x_1, x_2, x_3, x_4 \mid x_2 x_1 = x_1 x_3,\ x_2 x_3 = x_3 x_4,\ x_2 x_4 = x_4 x_2,$$
$$x_1 x_2 = x_4 x_3,\ x_1 x_4 = x_3 x_1,\ x_3 x_2 = x_4 x_1 \rangle.$$

We first list all elements of length 3 that are equal to one of the following three distinct elements of S.

$$b = x_1^2 x_3 = x_1 x_2 x_1 = x_4 x_3 x_1 = x_4 x_1 x_4$$
$$= x_3 x_2 x_4 = x_3 x_4 x_2 = x_2 x_3 x_2 = x_2 x_4 x_1$$
$$= x_4 x_2 x_1 = x_4 x_1 x_3 = x_3 x_2 x_3$$
$$= x_3^2 x_4 \in S_4 \cap S_4'$$
$$c = x_2 x_3^2 = x_3 x_4 x_3 = x_3 x_1 x_2 = x_1 x_4 x_2 = x_1 x_2 x_4$$
$$= x_4 x_3 x_4 = x_4 x_2 x_3 = x_2 x_4 x_3 = x_2 x_1 x_2$$
$$= x_1 x_3 x_2 = x_1 x_4 x_1 = x_3 x_1^2 \in S_4 \cap S_4'$$
$$d = x_3^2 x_2 = x_3 x_4 x_1 = x_2 x_3 x_1 = x_2 x_1 x_4 = x_1 x_3 x_4$$
$$= x_1 x_2 x_3 = x_4 x_3^2 \in S_4 \cap S_4'$$

Then

$$cx_4 = (x_2 x_3 x_3) x_4 = x_2 (x_3 x_3 x_4)$$
$$= x_2 (x_1 x_2 x_1) = (x_2 x_1 x_2) x_1 = cx_1$$

and hence $cd = cx_1 x_2 x_3 = cx_4 x_2 x_3 = cc$. Since $c \neq d$, it follows that indeed S_4 is not cancellative.

(6) Does S have an ideal chain whose factors are semigroups of matrix type? More precisely, is every factor S_i/S_{i+1} a union of semigroups of matrix type intersecting different sets D_Y and different sets D_Y' (whence these semigroups of matrix type are orthogonal)? For a binomial monoid S this follows from a recent result in [16].

In characteristic zero it would follow that $K[S]$ is semiprime. To prove this, recall that if T is a cancellative semigroup so that $K[T]$ is a PI algebra then the prime radical of $K[T]$ is trivial (see for example [20], Theorem 21.8). Hence it easily follows that also the prime radical of a semigroup of matrix type over the group of quotients of T is trivial. The result now follows by induction on the length of the ideal chain of S with factors of matrix type. Notice that every S_i is then a semiprime ideal of S.

(7) If P is a minimal prime ideal of $K[S]$ with $P \cap S \neq \emptyset$, is then $K[P \cap S]$ a semiprime ideal? If the answer is positive and the characteristic is zero then by (RN4) $\mathcal{P}(K[S]) = I(\rho) \cap K[Q]$ for some ideal Q of S. Hence, $\mathcal{P}(K[S]) = I(\rho_Q)$, where ρ_Q is the restriction of ρ to Q, and thus the first part of question (3) also has a positive answer.

If P is a minimal prime of $K[S]$ with $P \cap S \neq \emptyset$ then $P = K[P \cap S]$ in case S is a binomial monoid [12]. It is worth mentioning that the origin of investigating primeness and semiprimeness of $K[P \cap S]$ comes from the following result proved in [14]. Let T be a cancellative monoid with the ascending chain condition on right ideals. If its group of quotients $G = TT^{-1}$ is torsion-free and polycyclic-by-finite then $K[Q \cap S]$ is a prime ideal if Q is a prime ideal of $K[T]$ with $Q \cap T \neq \emptyset$.

By (RN4) S_n^N is cancellative for some positive integer N and S_n^N embeds as an ideal into S/ρ. So the group of quotients $(S_n^N)(S_n^N)^{-1}$ can be identified with $(S/\rho)(S/\rho)^{-1}$ and it is abelian-by-finite (because $K[S]$ is a noetherian PI algebra, see Chapter 19 in [20]). As in the special case considered in [5], we may call this the structure group of S. (Notice that by [10], if S is of I-type, then S is cancellative, so the structure group is actually the group of quotients of S.) For example, the structure group of C is cyclic-by-finite.

(8) Are the groups $(S/\rho)(S/\rho)^{-1} = (S_n^N)(S_n^N)^{-1}$ and $S/(x_i^p = 1, 1 \leq i \leq n)$ (where p is as in (C2)) solvable? It is shown in [12] that this holds if S is a binomial monoid, and in [5, 10] it is shown to hold for the wider class of monoids of I-type.

The following example shows that a monoid S of skew type that does not satisfy the cyclic condition but is non-degenerate can have $S_i \neq S_i'$ for some i. Indeed let

$$S = \langle x_1, x_2, x_3, x_4 \mid x_4 x_1 = x_2 x_4, \; x_4 x_2 = x_3 x_1, \; x_4 x_3 = x_1 x_2,$$
$$x_3 x_2 = x_1 x_4, \; x_3 x_4 = x_2 x_3, \; x_2 x_1 = x_1 x_3 \rangle.$$

Then $x_2 x_1 x_1 = x_1 x_3 x_1 = x_1 x_4 x_2 = x_3 x_2 x_2 \in S_3 \setminus S_3'$.

On the other hand, for a binomial monoid $S = \langle x_1, \ldots, x_n \rangle$ we will show that always $S_i = S_i'$. For this we introduce some notation. If $s \in S$ then there is a unique subset Y of X such that $s \in D_Y'$. We define $t_S(s) = Y$. Similarly, $i_S(s) = \{x_i \mid s \in x_i S\}$. We shall write $t(s)$, $i(s)$ if unambiguous. So these are the sets of terminal and initial letters of s, respectively. Clearly, for $i = 1, 2, \ldots, n$, $S_i = \{s \in S \mid |i(s)| \geq i\}$ and $S_i' = \{s \in S \mid |t(s)| \geq i\}$. We know that these are ideals of S. Recall that there exists a positive integer p such that $A = \langle x_1^p, \ldots, x_n^p \rangle$ is a free abelian monoid, and that any element s of S can be written uniquely in the form $s = x_{i_1}^{n_1} \cdots x_{i_k}^{n_k}$ with $i_1 < \cdots < i_k$ and all n_j nonnegative. Furthermore $sA = As$ for every $s \in S$. Define $s^\perp = x_{i_k}^{q_k} \cdots x_{i_1}^{q_1}$ where $q_i \geq 0$ is minimal such that p divides $n_i + q_i$. Clearly $ss^\perp \in A$.

Proposition 3.1. *Assume that $S = \langle x_1, \ldots, x_n \rangle$ is a binomial monoid. We have*

(1) $S_i = S_i'$ for every i,
(2) if $s \in S$ then $|i(ss^\perp)| = |i(s)| = |t(ss^\perp)| = |t(s)|$.

Proof. We proceed by induction on n. If $n = 1$ then both assertions are clear. So suppose $n > 1$. From [12] we know that $S = S_{(1)} \cdots S_{(k)}$ where $k \geq 2$, each $S_{(i)}$ is a binomial monoid generated by a subset X_i of X and X_i, X_j are disjoint if $i \neq j$. Moreover $S_{(i)}S_{(j)} = S_{(j)}S_{(i)}$, and so the decomposition is unique up to a permutation. Also every $s \in S$ is uniquely represented in the form $s = s_1 \cdots s_k$ where $s_i \in S_{(i)}$ for each i.

Then $|i(s_i)| = |i_{S_{(i)}}(s_i)|$ and $|t_{S_{(i)}}(s_i)| = |t(s_i)|$ for every i. Hence, by the inductive hypothesis

$$|i(s_i)| = |i_{S_{(i)}}(s_i)| = |i_{S_{(i)}}(s_i s_i^\perp)| = |t_{S_{(i)}}(s_i s_i^\perp)| = |t_{S_{(i)}}(s_i)| = |t(s_i)|.$$

Let $s' = s_1 \cdots s_{k-1}$. Since s' normalizes A, it follows that $ss_k^\perp = s's_k s_k^\perp = a_k s'$ for some $a_k \in A$. Because S is right non-degenerate, this implies easily that $a_k s'$ is left divisible by at least $|i(s_k s_k^\perp)|$ elements of the set $S_{(k)} \cap X$. From the definition of the components $S_{(j)}$ and the uniqueness of the representation of elements, it then follows that a_k is left divisible by at least $|i(s_k s_k^\perp)|$ elements of $S_{(k)} \cap X$. Therefore $|i(s_k s_k^\perp)| \leq |i(a_k)|$. Similarly, since S is left non-degenerate, the equality $s's_k s_k^\perp = a_k s'$ yields $|t(a_k)| \leq |t(s_k s_k^\perp)|$. By the inductive hypothesis applied to a_k and $s_k s_k^\perp$ this yields

$$|i(s_k s_k^\perp)| = |i(a_k)| = |t(a_k)| = |t(s_k s_k^\perp)|.$$

Proceeding this way we get $ss^{\perp} = a_k \cdots a_1$ with $a_j \in A$ and

$$|i(s_j)| = |i(a_j)| = |i(s_j s_j^{\perp})| = |t(s_j s_j^{\perp})| = |t(a_j)| = |t(s_j)|.$$

Suppose $ss^{\perp} \in \bigcap_{z \in Z} Sz$ for some $Z \subseteq S_{(1)} \cap X = X_1$. As above it follows that $a_1 \in \bigcap_{z \in Z} Sz$. Therefore we get

$$|Z| \leq |t(a_1)| = |i(a_1)| = |i(s_1)|$$

which implies that $s_1 \in \bigcap_{y \in Y} yS$ for some $Y \subseteq S_{(1)}$ with $|Y| = |Z|$. It follows that s can be written with at least $|Z|$ different initial letters from X_1.

Since the a_j's commute, using a similar argument applied to all $a_j \in S_{(j)}$ we get that $|t(ss^{\perp})| \leq |i(s)|$. Since S is left non-degenerate, we also have $|t(s)| \leq |t(ss^{\perp})|$. This yields $|t(s)| \leq |i(s)|$. A symmetric argument allows us to establish equality, and assertions (1) and (2) follow. $\qquad\square$

Arbitrary non-degenerate monoids of skew type can differ a lot not only from binomial monoids but also from monoids satisfying the cyclic condition. In particular, they can be used to construct examples of semiprime algebras $K[S]$ of Gelfand-Kirillov dimension 1 and with many generators [2].

4. Monomial semigroups

In case question (6) has a positive answer, we obtain a surprising consequence on the structure of the semigroup, see Proposition 4.2 below. The proof is based on a technical lemma for which some more terminology is needed. Recall that a subsemigroup U of a completely 0-simple semigroup $Z = \mathcal{M}(G, X, Y; P)$ over a group G with a sandwich matrix P is called uniform if it intersects all non-zero $\mathcal{H}$-classes $Z_{xy} = \{(g, x, y) \mid g \in G\}$, $x \in X$, $y \in Y$, of Z. The partition of the set of non-zero elements of Z into subsets Z_{xy} determines the so called 'egg-box' pattern on Z. The non-zero $\mathcal{R}$-classes (and $\mathcal{L}$-classes respectively) of Z are of the form $\{(g, x, y) \mid g \in G, y \in Y\}$ for $x \in X$ (and $\{(g, x, y) \mid g \in G, x \in X\}$ for $y \in Y$, respectively), see [3, 21]. Moreover, the algebra $K_0[T] \subseteq K_0[\mathcal{M}(G, n, n; \nabla)] \cong M_n(K[G])$ has a natural structure of a generalized matrix ring. By the cancellative components of T we mean the subsemigroups $T \cap \mathcal{M}_{ii}$, where $\mathcal{M}_{ii} = \{(g, i, i) \mid g \in G\} \subseteq \mathcal{M}(G, n, n; \nabla)$ for $i = 1, \ldots, n$.

In order to give a general statement, recall also that if J is a completely 0-simple semigroup of quotients (in the sense of [6]) of its subsemigroup I, then for every non-zero $x, y \in I$ such that $x\mathcal{L}y$ in J, where $\mathcal{L}$ stands for the Green's relation on J, there exist $x', y' \in I$ such that $x'x = y'y \neq 0$. Similarly, for every non-zero $x, y \in I$ such that $x \mathcal{R} y$ in J there exist $x', y' \in I$ such that $xx' = yy' \neq 0$. For example, if I is an ideal of a monoid S such that $K[S]$ is noetherian or $K[S]$ satisfies a polynomial identity and I is a uniform subsemigroup of a completely 0-simple semigroup J', then I has a semigroup of quotients J. This is an easy consequence of [6], Theorem 2. If $J = \mathcal{M}(G, X, Y; P)$ and for every $a, b \in J$ the condition $ax = bx$ for all $x \in J$ implies that $a = b$, and the condition $xa = xb$ for all $x \in J$ implies that $a = b$, then we say that J is annihilator-free. This applies for example if $J = \mathcal{M}(G, n, n; \nabla)$, whence if I is a semigroup of matrix type.

Lemma 4.1. *Let I be an ideal of a semigroup S. Assume that J is a completely 0-simple semigroup such that $I \subseteq J$ and I is a uniform subsemigroup of J. If either J is a semigroup of (right and left) quotients of I or J is annihilator-free, then there is a unique semigroup structure on the disjoint union $S' = (S \setminus I) \cup J$ that extends the operation on S.*

Proof. Let $J = \mathcal{M}(G, X, Y; P)$. Choose a maximal subgroup $H = J_{xy} = \{(g, x, y) \mid g \in G\}$ of J, with $x \in X, y \in Y$. Since I is a uniform subsemigroup of J, we know that there exist $a_i \in I \cap J_{iy}$ and $b_j \in I \cap J_{xj}$ for all $i \in X, j \in Y$. Then $J = \bigcup_{i,j} a_i H b_j \cup \{0\}$ because the right side is a completely 0-simple semigroup intersecting every $\mathcal{H}$-class of J and containing H. Also every non-zero element of J has a unique presentation of the form $a_i h b_j$ where $i \in X, j \in Y, h \in H$. Let $s \in S \setminus I$ and $h \in H$. On the disjoint union $(S \setminus I) \cup J$ we define the products

$$s \cdot a_i h b_j = (sa_i)(hb_j), \quad a_i h b_j \cdot s = (a_i h)(b_j s),$$

where the products sa_i and $b_j s$ are considered in S, while the other products on the right sides of these equalities are considered in J. Also, define $s \cdot t = st$ (the latter product is considered in S) if $s, t \in S \setminus I$. Finally, let $s \cdot 0 = 0 \cdot s = 0$ for $s \in S$. First we claim that this extends the operation on S. Clearly $s \cdot t = st$ if $s, t \in I$. So we have to show that $s \cdot t = st$ and $t \cdot s = ts$ if $s \in S, t \in I$. We may assume that $t \neq 0$. Write $t = a_i h b_j$.

In order to prove that $t \cdot s = ts$, first notice that for every $u \in I$ we get

$$
\begin{aligned}
(t \cdot s)u &= [(a_i h b_j) \cdot s]u &&\text{(using the definition)} \\
&= [(a_i h)(b_j s)]u &&\text{(using associativity in } J) \\
&= (a_i h)[(b_j s)u] &&\text{(using associativity in } S) \\
&= (a_i h)[b_j(su)] &&\text{(using associativity in } J) \\
&= [(a_i h)b_j](su) = t(su) &&\text{(using associativity in } S) \\
&= (ts)u.
\end{aligned}
\tag{1}
$$

If $t \cdot s = 0$ then $(ts)u = 0$ for every $u \in I$. Since $IJ = J$, we get $tsJ = 0$, and hence $ts = 0$. The converse follows similarly. So we may assume that $t \cdot s \neq 0 \neq ts$. Choosing u such that $(ts)u \neq 0$ we see that $ts\mathcal{R}(t \cdot s)$ in J.

If J is annihilator-free then (1) implies that $ts = t \cdot s$. Thus, because of the hypothesis, we may assume that for every $0 \neq x, y \in I$ such that $x\mathcal{L}y$ in J there exist $x', y' \in I$ such that $x'x = y'y \neq 0$. Suppose that, for $s \in S$, one of the elements xs, ys is non-zero, say $xs \neq 0$. Then $xsz \neq 0$ for some $z \in I$ and we get $y'y(sz) = x'x(sz) \neq 0$. This implies that $xs\mathcal{L}ys$ in J. So S acts by right multiplication on I in such a way that it permutes the sets $L \cap I$, where L are $\mathcal{L}$-classes of J. Since $b_j\mathcal{L}t$ in J, this implies that $b_j s\mathcal{L}ts$. As we also have $b_j s\mathcal{L}(t \cdot s)$, it follows that $ts\mathcal{L}(t \cdot s)$. Therefore (1) easily implies that $t \cdot s = ts$ also in this case.

A symmetric argument shows that $s \cdot t = st$. This completes the proof of the claim.

Furthermore, if we get a semigroup structure on S', that extends those on S and on J, it must be unique. So it remains to show that the defined operation is associative. For this consider any $s, t \in S$ and $a, b \in J$. We need to check associativity only for the products $sta, tas, ast, sab, asb, abs$ (we simplify notation, writing as, sa in place of $a \cdot s$, $s \cdot a$, respectively). Let $a = a_m f b_n$ and $b = a_p f' b_q$ for some $m, p \in X, n, q \in Y, f, f' \in H$. Put $z = f b_n a_p f'$. Then $z \in H \cup \{0\}$ and we get

$$
\begin{aligned}
(ab)s &= (a_m f b_n a_p f' b_q)s = (a_m z b_q)s = (a_m z)(b_q s) \\
&= (a_m f b_n a_p f')(b_q s) = (a_m f b_n)(a_p f' b_q s) = a(bs)
\end{aligned}
\tag{2}
$$

(the third equality by the definition above and the fifth because J is a semigroup). The case sab is symmetric. For the cases ast, sta, by symmetry we may consider ast only:

$$
\begin{aligned}
(as)t &= [(a_m f b_n)s]t = [(a_m f)(b_n s)]t &&\text{(using (2))}\\
&= (a_m f)[(b_n s)t] &&\text{(using associativity in } S)\\
&= (a_m f)[b_n(st)] &&\text{(using (2))}\\
&= [(a_m f)b_n](st) = a(st).
\end{aligned}
$$
$$(3)$$

The cases asb and tas are dealt with as follows:

$$
\begin{aligned}
(as)b &= [(a_m f b_n)s]b = [(a_m f)(b_n s)]b &&\text{(using associativity in } J)\\
&= (a_m f)[(b_n s)b] &&\text{(using associativity in } S)\\
&= (a_m f)[b_n(sb)] &&\text{(using associativity in } J)\\
&= [(a_m f)b_n](sb) = a(sb),\\
t(as) &= t[(a_m f b_n)s] = t[(a_m f)(b_n s)] &&\text{(by the case symmetric to (2))}\\
&= [t(a_m f)](b_n s) &&\text{(using (2))}\\
&= [(t(a_m f))b_n]s &&\text{(by the case symmetric to (2))}\\
&= [t((a_m f)b_n)]s = (ta)s.
\end{aligned}
$$

This completes the proof. $\qquad\square$

Proposition 4.2. *Let S be a monoid that has a finite ideal chain whose factors are semigroups of matrix type over groups $G_1,\ldots,G_k$. Then S is a semigroup of monomial matrices over the group $\tilde{G} = \prod_{i=1}^{k} G_i$. Moreover $K[S]$ embeds into the product $\bigoplus_{i=1}^{k} M_{n_i}(K[G_i])$.*

Proof. Let I be an ideal of S that is a semigroup of matrix type over a group G. So I is a subsemigroup of $\overline{I} = \mathcal{M}(G,n,n,\nabla)$ and I intersects every $\mathcal{M}_{ij} = \{(g,i,j) \mid g \in G\}$, $1 \leq i,j \leq n$. Using Lemma 4.1 we form the semigroup $\overline{S} = (S \setminus I) \cup \overline{I}$, the disjoint union. Then $K_0[\overline{I}] \cong M_n(K[G])$ and $K_0[\overline{S}] \cong K_0[S/I] \oplus M_n(K[G])$. Let e be the identity in $K_0[\overline{I}]$, and thus e is a central idempotent in $K_0[\overline{S}]$. For every $x \in I$ and $s \in S$ we have

$$x(es) = xs \in I \text{ and } (es)x = sx \in I.$$

The elements xs and sx, interpreted as matrices in $M_n(K[G])$, have at most one non-zero entry (and it belongs to G). Since for every $1 \leq i,j \leq n$ there exists $g \in G$ such that $x = (g,i,j) \in I$, it follows easily that es is a monomial matrix over G. So we obtain that eS is a monomial semigroup over G. By induction on the length of the ideal chain of S, applied to S/I, it follows that S embeds into $S/I \times eS \subseteq S/I \times \mathcal{M}_{n_1}(G_1) \subseteq \cdots \subseteq \mathcal{M}_{n_1}(G_1) \times \cdots \times \mathcal{M}_{n_k}(G_k) \subseteq \mathcal{M}_{n_1+\cdots+n_k}(G_1 \times \cdots \times G_k)$ for some $n_i \geq 1$, with $n_1 = n$ and $G_1 = G$. The second assertion follows similarly. $\qquad\square$

Corollary 4.3. *Under the assumptions of Proposition 4.2, if S is finitely generated and all groups G_i are polycyclic-by-finite, then $K[S]$ is right noetherian if and only if S has the ascending chain condition on right ideals.*

Proof. This follows at once from Proposition 4.2 and [13]. However it is also an immediate consequence of one of the main results of [22] (Theorem 3.3) and the comment following it. $\square$

There is also a link between questions (6) and (7). If question (6) has a positive solution then the following property yields that for every prime P of $K[S]$ intersecting S the ideal $P \cap S$ is a union of 'blocks' that are semigroups of matrix type. In particular $S/(P \cap S)$ is the union of the remaining blocks, and hence $K[P \cap S]$ is a semiprime ideal if K is of characteristic zero.

Proposition 4.4. *Let S be a left and right non-degenerate monoid of skew type. If P is a prime ideal of $K[S]$ that intersects S then $P \cap S = \bigcup D_Y$ where the summation runs over all subsets $Y \subseteq X$ such that $P \cap D_Y \neq \emptyset$. Moreover, if $\emptyset \neq D_Y \subseteq P$ and $Y \subseteq Z$ then $D_Z \subseteq P$.*

Proof. Let P be a prime ideal of $K[S]$ so that $P \cap S \neq \emptyset$. Because of property (RN4)(d) we know that $S_X \subseteq P$.

Suppose that $P \cap D_Y \neq \emptyset$ but $D_Y \not\subseteq P$ for some Y. Choose Y with maximal possible $i = |Y|$. So $P \cap S_{i+1}$ is the union of all D_Z nontrivially intersecting P. Let $a \in P \cap D_Y$. Let $Z \subseteq X$ be such that $Y \subseteq Z$ and $D_Z \neq \emptyset$. Suppose first that $Z \neq Y$ (so $|Z| > i$). By Lemma 4.6 in [11] we have

$$D_Z^{|a|} \cap D_Z \subseteq aS \subseteq P. \tag{4}$$

If this intersection is non-empty then $D_Z \subseteq P$ by the maximality of i. If this intersection is empty then $D_Z^{|a|} \subseteq \bigcup_{|W|>|Z|, Y \subseteq W} D_W$. Repeating this argument for every W with $D_W \neq \emptyset$ (and continuing this argument a number of times) we get that $\bigcup_{Z \neq Y, Y \subseteq Z} D_Z$ is nilpotent modulo P. Since this is a right ideal of S, it must be contained in P. Now, by (4) applied to $Z = Y$ we also get

$$D_Y^{|a|} \subseteq (D_Y^{|a|} \cap D_Y) \cup \bigcup_{Z \neq Y, Y \subseteq Z} D_Z \subseteq aS \cup P \subseteq P.$$

Hence $D_Y \subseteq P$. This contradiction completes the proof of the first assertion.

The second assertion follows from the above proof. $\square$

5. Structural chains

Let S be a monoid of skew type that is left and right non-degenerate. Proposition 4.4 also can be proved via improving the ideal chain obtained in property (RN3). One such improvement is already contained in the proof of Theorem 5.2 in [11]. It is shown that S has an ideal chain

$$S_X^N = I_1 \subseteq I_2 \subseteq \cdots \subseteq I_t = S, \tag{5}$$

for some $N \geq 1$, so that each factor I_j/I_{j-1} has the following properties:

(i) it is either nilpotent or a semigroup of matrix type,
(ii) if I_j/I_{j-1} is a semigroup of matrix type, then $(I_j \setminus I_{j-1}) \subseteq (S_i \cap S_i') \setminus S_{i+1}$ for some i,
(iii) if a, b are in a cancellative component T of a semigroup of matrix type I_j/I_{j-1} then $a^{|b|} \in bS \cap Sb$.

(The last statement follows from Lemma 4.6 in [11]).

On the other hand, we know that S is a linear semigroup by (RN2). Hence S has an ideal chain as in Theorem 1.2. Of course it can be embedded in many different ways in a full matrix ring over a field. A natural question that arises here is concerned with compatibility of such chains with the chain from (5). We now show that we can embed S in a full matrix ring in such a manner that the above ideal chain (5) can be even further refined so that the structure of matrix factors of that chain corresponds with that of the the full matrix ring (as a multiplicative semigroup) and that they are at different rank levels. We start with a general observation.

Lemma 5.1. *Let S be a finitely generated monoid such that $K[S]$ is a noetherian PI algebra. Suppose that S has an ideal chain $I_1 \subset I_2 \subset \cdots \subset I_t = S$. Then there exists an embedding $K[S] \hookrightarrow M_r(F)$, for some $r \geq 1$ and a field F, such that if $2 \leq j \leq t$ then every matrix in $I_j \setminus I_{j-1}$ has rank bigger than the rank of every matrix in I_{j-1}.*

Proof. As mentioned earlier, by a theorem of Anan'in [1], $K[S]$ and every $K[S]/K[I_j]$ embed into a matrix ring over a field. Say, $f_0 : K[S] \to M_{r_0}(F_0)$ and $g_j : K[S]/K[I_j] \to M_{r_j}(F_j)$, $j = 1, \ldots, t-1$, for some fields $F_0, \ldots, F_{t-1}$ and integers $r_0, \ldots, r_{t-1}$. Define $f_j : K[S] \to M_{r_j}(F_j)$ as the composition of the natural map $K[S] \to K[S]/K[I_j]$ with g_j. Let n_1 be the minimal rank of non-zero matrices in $f_1(S)$ and let m_0 be the maximal rank of matrices in $f_0(I_1)$. Choose a non-zero integer k_1 so that $k_1 n_1 > m_0$. Then the product map $f' = f_0 \times f_1^{k_1} : K[S] \to M_{r_0}(F_0) \times M_{r_1}(F_1)^{k_1}$ is an embedding and the latter algebra embeds (block diagonally) into $M_{r_0 + k_1 r_1}(F')$, where F'

is a field containing F_0 and F_1. Clearly, the ranks of matrices in $f'(I_1)$ do not exceed m_0, whence they are smaller than the rank of every matrix in $f'(S \setminus I_1)$ (which is at least $k_1 n_1$). Proceeding this way we may find positive integers $k_1, \ldots, k_{t-1}$ and an embedding $f = f_0 \times f_1^{k_1} \times \cdots \times f_{t-1}^{k_{t-1}} : K[S] \to M_{r_0}(F) \times M_{r_1}(F_1)^{k_1} \times \cdots \times M_{r_{t-1}}(F_{t-1})^{k_{t-1}} \to M_r(F)$, where F is a field containing all F_j and $r = r_0 + \sum_{j=1}^{t-1} r_j k_j$, that satisfies the assertion of the lemma. $\qquad\square$

In the special case of monoids considered in this paper we can say more. Recall that, for any $j = 1, \ldots, n$, the set M_j of matrices of rank at most j in $M_r(F)$ is an ideal of the monoid $M_r(F)$ and the completely 0-simple semigroups M_j/M_{j-1} are called principal factors of $M_n(F)$, [21].

Proposition 5.2. *Let S be a left and right non-degenerate monoid of skew type with ideal chain (5) and a matrix embedding provided in Lemma 5.1. Then the ideal chain (5) can be refined to an ideal chain with the same properties in such a way that every factor of matrix type in the new chain consists of matrices of the same rank and has the egg-box pattern agreeing with that of the corresponding principal factor of $M_r(F)$. Furthermore, different matrix factors are of different ranks.*

Actually, if I_j/I_{j-1} is a semigroup of matrix type then in the refined chain we introduce an ideal J_j of S so that $I_{j-1} \subseteq J_j \subseteq I_j$, $I_j^{n_j} \subseteq J_j$ for some positive integer n_j and J_j/I_{j-1} is a subsemigroup of matrix type of both I_j/I_{j-1} and of a principal factor of $M_r(F)$.

Proof. Consider the chain (5). Because of Lemma 5.1 we may consider S as a submonoid of $M_r(F)$ so that each element in $I_j \setminus I_{j-1}$ is of strictly larger rank than each element of $I_k \setminus I_{k-1}$ if $j > k$. Define $I_0 = \emptyset$ and $I_1/I_0 = I_1$.

Let $T = I_j/I_{j-1}$ be a factor that is a semigroup of matrix type. Let k be a minimal positive integer so that $(I_j \setminus I_{j-1}) \cap (M_k \setminus M_{k-1}) \neq \emptyset$. Let $J = I_j \cap M_k$. Notice that $I_{j-1} \subseteq J$. Clearly J/I_{j-1} is non-empty, whence an ideal of T. Because T is a semigroup of matrix type it follows easily that J/I_{j-1} also is of matrix type. Write the set of the non-zero elements as a disjoint union $\bigcup_{1 \leq i,j \leq m} J_{ij}$ with each J_{ii} a cancellative semigroup and $J_{ij} J_{pq} \subseteq I_{j-1}$ if $j \neq p$, otherwise this product is contained in J_{iq}.

Each cancellative component J_{ii} satisfies the left an right Ore condition (because it does not have free nonabelian subsemigroups by (RN1)). Therefore, since all its elements have the same rank, it follows that J_{ii} is contained in a maximal subgroup of $M_k \setminus M_{k-1}$, see [21], Section 2.1.

By $\mathcal{H}, \mathcal{L}, \mathcal{R}$ we denote the Green's relations in $M_r(F)$. Suppose $a \in J_{pp}$

and $b \in J_{qq}$ and $a \mathcal{H} b$ in $M_r(F)$. Then the previous paragraph shows that a and b are in the same maximal subgroup of $M_k \setminus M_{k-1}$, whence $ab \notin I_{j-1}$ because elements of I_{j-1} have smaller ranks than the elements of $I_j \setminus I_{j-1}$. This implies that $p = q$. Hence, distinct cancellative components of J/I_{j-1} are contained in different $\mathcal{H}$-classes of $M_r(F)$.

Let $Z = \{x \in M_k/M_{k-1} \mid x \mathcal{R} J_{pp}$ and $x \mathcal{L} J_{qq}$ for some $1 \le p, q \le m\} \cup \{0\}$. Then Z is a subsemigroup of M_k/M_{k-1}. Because each $\mathcal{R}$- and $\mathcal{L}$-class of Z contains a group (namely the group containing one of the cancellative components J_{pp}) the semigroup Z is completely 0-simple. Because $J_{pq} J_{qp} \subseteq J_{pp}$ and $J_{qp} J_{pq} \subseteq J_{qq}$ we obtain easily that $J/I_{j-1} \subseteq Z$ and also that different components of J/I_{j-1} as a semigroup of matrix type are in different $\mathcal{H}$-classes of M_k/M_{k-1}.

Furthermore, if a and b are in distinct cancellative components of J/I_{j-1} then $ab \in I_{k-1}$ and thus $ab \in M_{k-1}$. It follows that the product in J/I_{j-1} agrees with that of the egg-box pattern for M_k/M_{k-1}.

Next we show that T/J is a nilpotent semigroup. Since S, and thus also S/J satisfies the ascending chain condition on right and left ideals by (RN2), it is sufficient to show that T/J is a nil semigroup (see Proposition 2.13 in [20]). So let $t \in T \setminus J$. Since $I_{j-1} \subseteq J$ it is clear that if t is not in a cancellative component of T, then $t^2 \in J$. On the other hand if t is in a cancellative component C then, because of the third property of the chain (5), choosing $b \in C \cap J$ and $a = t$, we get $t^{|b|} \in bS \subseteq J$. Hence indeed T/J is nil.

Thus we have shown that $I_{j-1} \subseteq I_j$ can be refined to the chain

$$I_{j-1} \subseteq J \subseteq I_j,$$

so that I_j/J is nilpotent and J/I_{j-1} is a semigroup of matrix type contained in M_k/M_{k-1} and its egg-box pattern corresponds with that of the principal factor M_k/M_{k-1} of $M_r(F)$. The result now follows. $\square$

As an application of the above structure on the ideal chain one can now reprove Proposition 4.4. This proof will be omitted. In the notation of the ideal chains we thus obtain.

Proposition 5.3. *Let S be a left and right non-degenerate monoid of skew type. Let P be a prime ideal of $K[S]$. If I_j/I_{j-1} is a semigroup of matrix type in the chain (5) and $I_j \setminus I_{j-1}$ intersects P then $I_j \setminus I_{j-1} \subseteq P$.*

References

1. Anan'in A. Z., An intriguing story about representable algebras, Ring Theory 1989, Israel Math. Conf. Proc., 31–38, Weizmann, Jerusalem, 1989.
2. Cedo F., Jespers E. and Okniński J., Semiprime quadratic algebras of Gelfand-Kirillov dimension one, J. Algebra and its Applications, to appear.
3. Clifford A. H. and Preston G. B., The Algebraic Theory of Semigroups, Vol. I, Amer. Math. Soc., Providence, 1961.
4. Drinfeld V. G., On some unsolved problems in quantum group theory, in: Quantum Groups, Lect. Notes Math. vol. 1510, pp. 1–8, Springer-Verlag, 1992.
5. Etingof P., Schedler T. and Soloviev A., Set-theoretical solutions of the quantum Yang-Baxter equation, Duke Math. J. 100 (1999), 169–209.
6. Fountain J. and Petrich M., Completely 0-simple semigroups of quotients III, Math. Proc. Camb. Phil. Soc. 105 (1989), 263–275.
7. Gateva-Ivanova T., Noetherian properties of skew polynomial rings with binomial relations, Trans. Amer. Math. Soc., 343 (1994), 203–219.
8. Gateva–Ivanova T., Skew polynomial rings with binomial relations, J. Algebra 185 (1996), 710–753.
9. Gateva–Ivanova T., A combinatorial approach to the set-theoretic solutions of the Yang-Baxter equation, preprint.
10. Gateva–Ivanova T. and Van den Bergh M. Semigroups of I-type, J. Algebra 206 (1998), 97–112.
11. Gateva–Ivanova T., Jespers E. and Okniński J., Quadratic algebras of skew type and the underlying semigroups, J. Algebra 270 (2003), 635–659.
12. Jespers E. and Okniński J., Binomial semigroups, J. Algebra 202 (1998), 250–275.
13. Jespers E. and Okniński J., On a class of noetherian algebras, Proc. Roy. Soc. Edinburgh, 129A (1999), 1185–1196.
14. Jespers E. and Okniński J., Submonoids of polycyclic-by-finite groups and their algebras, Algebras and Repres. Theory 4 (2001), 133–153.
15. Jespers E. and Okniński J., Quadratic algebras of skew type satisfying the cyclic condition, Int. J. Algebra and Computation, to appear.
16. Jespers E. and Okniński J., Monoids and groups of I-type, preprint.
17. Kramer X. H., The Noetherian property in some quadratic algebras, Trans. Amer. Math. Soc. 351 (2000), 4295–4323.
18. Krause G. R. and Lenagan T. H., Growth of Algebras and Gelfand-Kirillov Dimension, Revised edition. Graduate Studies in Mathematics 22. American Mathematical Society, Providence, RI, 2000.
19. McConnell J. C. and Robson J. C., Noncommutative Noetherian Rings, Wiley, New York, 1987.
20. Okniński J., Semigroup Algebras, Marcel Dekker, New York, 1991.
21. Okniński J., Semigroups of Matrices, World Scientific, Singapore, 1998.
22. Okniński J., In search for noetherian algebras, NATO ASI Series: Algebra – Representation Theory, pp. 235-247, Kluwer, 2001.
23. Passman D. S., The Algebraic Structure of Group Rings, Wiley, New York,

1977.
24. Passman D. S., Infinite crossed products, Academic Press, New York, 1989.
25. Tate J. and Van den Bergh M., Homological properties of Sklyanin algebras,
    Invent. Math. 124 (1996), 619–647.

REPRESENTATION TYPE OF COMMUTATIVE NOETHERIAN RINGS (INTRODUCTION)

LEE KLINGLER

Mathematics Department, Florida Atlantic University,
Boca Raton, Florida 33431–0991, USA
E-mail: klingler@fau.edu

LAWRENCE S. LEVY[*]

Mathematics Department, University of Nebraska,
Lincoln, NE 68588–0323 USA
E-mail: levy@math.wisc.edu

We introduce a series of papers [KL1, KL2, KL3] that describe the isomorphism classes of finitely generated modules *and their direct sum relations* over all commutative noetherian rings that do not have wild representation type [see Definition 2.1]. (There is a possible slight exception to our structure results, involving characteristic 2 [see Remarks 2.4].)

1. Background

Let Ω be a commutative noetherian ring whose finitely generated module category fingen(Ω) we wish to describe. For example, the case $\Omega = \mathbb{Z}$ is given by the structure theorem for finitely generated abelian groups. A well-known slight extension of the case $\Omega = \mathbb{Z}$ is given by the situation that Ω is any principal ideal domain. This was further extended to the case that Ω is any Dedekind domain by Steinitz in 1911–1912 [S]. Steinitz's motivation was to describe fingen(Ω) where Ω is the ring of integers in any algebraic number field.

In modern terminology, a *Dedekind domain* is any integral domain (necessarily noetherian) in which every ideal is a projective Ω-module. Over Dedekind domains, fingen(Ω) is not a *Krull-Schmidt category*; that is, modules in fingen(Ω) are not necessarily *unique* direct sums of indecomposables

---

[*]Levy's research was partially supported by an NSA grant. The paper is based on a talk presented at University of Lisbon, Portugal, on July 17, 2003 by L. S. Levy

(up to isomorphism). One of the most interesting ingredients in Steinitz's papers is that he was able to describe the direct-sum relations in fingen(Ω).

As far as we know, Dedekind domains remained the only class of noetherian domains for which fingen(Ω) could be described until 1985, when Levy described a very restricted class of such rings called "Dedekind-like" [L2], a proper subset of the rings called Dedekind-like in the present paper. There were, however, an assortment of results for commutative local noetherian rings Ω, some of them non-artinian (Nazarova-Roiter [NR, 1969], corrected in [NRSB, 1975], Ringel [R, 1975], Drozd [D2, 1991]).

The obstruction to getting a very general theorem is "wild representation type", which for more than 30 years has been a well-known obstruction to obtaining a structure theorem for fingen(Ω) when Ω is a finite dimensional (possibly noncommutative) algebra over an algebraically closed field. Furthermore, the vast majority of finite dimensional algebras are known to have wild representation type.

Thus the present project overlaps two research areas that do not normally have much interaction: (i) modules over commutative noetherian rings (and the associated local versus global, and direct-sum relations, which have no counterpart in finite dimensional representation theory); and (ii) tameness versus wildness (and the associated matrix problems), a familiar subject in finite dimensional representation theory, but not very familiar over commutative noetherian rings. Since the two communities—commutative noetherian ring theorists and finite dimensional representation theorists—are sometimes unfamiliar with ideas that are well-known in the other community, we ask the forbearance of readers from each community as we define terms that "everybody knows" in their community but not the other.

2. Introduction

Throughout this paper, *ring* means "commutative ring" unless otherwise specified, and *local ring* means "noetherian local ring". For a maximal ideal $\mathfrak{m}$ of a noetherian ring Ω, the notation $\Omega_\mathfrak{m}$ and $\hat{\Omega}_\mathfrak{m}$ denotes $\mathfrak{m}$-localization and $\mathfrak{m}$-adic completion, respectively.

Definitions 2.1 (tame, wild). Let Ω be a noetherian ring, not necessarily an algebra over a field. We say that Ω fingen-*tame*—or more completely, that fingen(Ω) has *tame representation type*—if we can describe all isomorphism classes of finitely generated modules *and their direct-sum relations*, and give some information about the homomorphisms in this category.

We supplement this informal definition of "tameness" by the following formal definition of wildness.

Let $k = \Omega/\mathfrak{m}$ be some residue field of our noetherian ring Ω. We say that Ω is finlen-*wild* (with respect to $\mathfrak{m}$)—or more completely, that finlen(Ω), the category of Ω-modules of finite length, has *wild representation type (with respect to $\mathfrak{m}$)*—if the following condition holds for every finite dimensional (possibly noncommutative!) k-algebra A. There exist $\mathcal{W} = \mathcal{W}_A$, a full subcategory of finlen(Ω) and an additive functor: $\Phi = \Phi_A \colon \mathcal{W} \to$ finlen(A) (say, left A-modules) such that Φ is a *representation equivalence*. In other words:

(i) Φ maps $\mathcal{W}$ onto all isomorphism classes in finlen(A);

(ii) For $M, N \in \mathcal{W}$: $M \cong N \iff \Phi(M) \cong \Phi(N)$ in finlen(A); and

(iii) Φ is a surjection on Hom groups.

$$(2.1.1)$$

It follows easily that $_\Omega M$ is indecomposable $\iff \Phi(M)$ is indecomposable.

If we say that Ω is finlen-wild without mentioning any $\mathfrak{m}$, we mean that Ω is finlen-wild (with respect to some $\mathfrak{m}$).

The definition of "wild" is not completely standard. Often one requires each Φ to be an isomorphism on Hom groups in (2.1.1)(iii). Then each functor Φ becomes a category equivalence of $\mathcal{W}$ with finlen(A); and we call Ω *strictly wild (with respect to $\mathfrak{m}$)*.

This strict wildness is not what occurs in our main results. But it does occur in our discussion of them.

Remark 2.2 (Introduction to wildness). For readers not familiar with the notion of wildness, the following comments may be helpful. Suppose that we know that Ω is finlen-wild (with respect to $\mathfrak{m}$), and that we actually know all of the associated functors Φ_A. Let $k = \Omega/\mathfrak{m}$, let A be any finite dimensional k-algebra, and choose any two modules $X, Y \in$ finlen(A). Are they isomorphic?

Since $\Phi = \Phi_A$ is onto all isomorphism classes, there exist $M, N \in$ finlen(Ω) such that $\Phi(M) \cong X$ and $\Phi(N) \cong Y$. Moreover, by property (2.1.1)(ii) of wildness we have $X \cong Y \iff M \cong N$ in mod(Ω). Thus we have transferred the original isomorphism question for A-modules—for an arbitrary finite dimensional k-algebra—to an isomorphism question about modules over the single ring Ω.

In less formal language, finlen(Ω) has such a rich system of modules of finite length that any classification of their isomorphism classes would result in a corresponding A-isomorphism classification in finlen(A) for every finite dimensional k-algebra A. The seeming hopelessness of ever describing such a classification is responsible for the terminology "wild representation type". Indeed no Ω of wild representation type has ever had its finite-length module category explicitly described. [But we know of no theorem from mathematical logic saying that such a classification is impossible.]

Our starting point, in this survey, is the following theorem.

Theorem 2.3. *Let Ω be a noetherian ring, and suppose that there is no maximal ideal $\mathfrak{m}$ of Ω such that Ω is finlen-wild (with respect to $\mathfrak{m}$). Then one of the following two situations holds.*

(i) *Ω is fingen-tame—we can explicitly describe the isomorphism classes and direct-sum relations in* fingen(Λ); *or*

(ii) *Ω is one of the "exceptional" rings described in Definition 4.3 (and we do not know whether Ω is tame, wild, or neither).*

Remarks 2.4 (on the "exception"). This exceptional class of rings is very small. For example, it contains no rings whose residue fields are all finite (e.g., rings of algebraic integers). Nor does it contain any rings that are algebras over a field of characteristic $\neq 2$, or over a perfect (e.g., algebraically closed) field of characteristic 2.

Another "loose end" is that we cannot prove that none of our tame rings is also wild, except in special cases [KL3, §37 Problem 2].

In order to make Theorem 2.3 useful, we need to say which rings are wild, which are tame, and—in the tame case—give some of the basics of the module-classification in fingen(Λ). We begin with the wild case.

3. Artinian Triads and Drozd Rings

The notation $\mu_\Omega(M)$ denotes the minimum number of elements needed to generate a module $M \in$ fingen(Ω).

Definitions 3.1 ([KL1, 2.4]). We call an artinian local ring $(\Omega, \mathfrak{m}, k)$ (maximal ideal $\mathfrak{m}$, residue field k) an *artinian triad* if $\mu_\Omega(\mathfrak{m}) = 3$ and $\mathfrak{m}^2 = 0$.

The simplest example is the k-algebra $\Omega = k[X, Y, Z]/(X, Y, Z)^2$, where k is any field and X, Y, and Z are indeterminates.

We call an artinian local ring $(\Omega, \mathfrak{m}, k)$ a *Drozd ring* if $\mu_\Omega(\mathfrak{m}) = \mu_\Omega(\mathfrak{m}^2) = 2$, $\mathfrak{m}^3 = 0$, and there is an element $x \in \mathfrak{m} - \mathfrak{m}^2$ such that $x^2 = 0$.

Remarks 3.2 (Drozd Rings). The above definition of Drozd rings has the advantage of brevity, but leaves one wondering what such rings really look like.

(i) The simplest example is the well-known 5-dimensional *Drozd algebra* over any field k. This is the 5-dimensional k-algebra with k-basis 1, x, y, xy, y^2 and all other monomials equal to zero.

(ii) Every Drozd ring $(\Omega, \mathfrak{m}, k)$ has the following easily-proved property [KL1, Lemma 4.2] that makes it look very much like the Drozd algebra. *For each element $c \in \mathfrak{m}$, there is an expression $c = u_1 x + u_2 y + u_3 xy + u_4 y^2$ with each u_i a unit or 0.* (We do not claim uniqueness of the coefficients u_i.)

(iii) An example of a Drozd ring that is not an algebra over a field (probably the simplest such example) is easily seen to be $A_p = \mathbb{Z}[X]/(X^2, p^3, p^2 X)$, where p denotes any prime number and X is an indeterminate [KL1, 6.1].

Definition 3.3 ([KL1, 2.8]). We call a (necessarily artinian) local ring $(\Omega, \mathfrak{m}, k)$ a *Klein ring* if $\mu_\Omega(\mathfrak{m}) = 2$, $\mu_\Omega(\mathfrak{m}^2) = 1$, $\mathfrak{m}^3 = 0$, and $x^2 = 0$ $(\forall x \in \mathfrak{m})$.

We will not dwell on Klein rings in this survey. They have characteristic 2 or 4 [KL1, Lemma 2.9] and are quasi-Frobenius rings. For a few more details and references, see subsection 7.7. An example is the group ring of the Klein 4-group over any field of characteristic 2. For an example that has characteristic 4 (and hence is not an algebra over a field), see [KL1, Example 5.4].

Theorem 3.4. *Artinian triads and Drozd rings are* finlen-*wild (with respect to their maximal ideal). Klein rings are* fingen-*tame (=* finlen-*tame since Klein rings are artinian).*

Artinian triads are wild by a theorem of Warfield [GLW, Lemma 3].

Remarks 3.5 ($\mathcal{S}(1\frac{1}{2})$, $\mathcal{S}(2)$, $\mathcal{S}(m)$). In order to discuss our proof of wildness of Drozd rings, we need to review some related categories that are familiar to everyone working in representations of finite dimensional algebras. Let k be a field.

The classical unsolved "wildness problem" is that of classifying ordered pairs (A, B) of square matrices over k, up to simultaneous similarity. Define the category $\mathcal{S}(2)$ to be the category whose objects are ordered pairs (A, B) of $n \times n$ matrices over k, where n ranges over all positive integers.

Morphisms are matrices τ (of appropriate sizes) such that the following diagrams commute, where $k^{(n)}$ denotes columns of length n.

$$
\mathcal{S}(2): \qquad
\begin{array}{ccccc}
k^{(n)} & \xrightarrow{\ A\ } & k^{(n)} & \xrightarrow{\ B\ } & k^{(n)} \\
\ \downarrow{\scriptstyle\tau} & & \ \downarrow{\scriptstyle\tau} & & \ \downarrow{\scriptstyle\tau} \\
k^{(n')} & \xrightarrow{\ A'\ } & k^{(n')} & \xrightarrow{\ B'\ } & k^{(n')}
\end{array}
\qquad (3.5.1)
$$

More generally, the category $\mathcal{S}(m)$ is the analogously formed category of m-tuples of matrices over k.

The connection of this with wildness, in the modern sense, is made by considering the noncommutative free algebra F on m indeterminates. Every element $(A_1, \ldots, A_m) \in \mathcal{S}(m)$, where the matrices have size (say) $n \times n$, makes the k-vector space $k^{(n)}$ into a left F-module if we define multiplication by the i-th indeterminate to be left multiplication by A_i. A straightforward computation shows that the category $\mathcal{S}(m)$ is equivalent to $\mathrm{finlen}(F)$—and therefore contains $\mathrm{finlen}(C)$ for every m-generated finite dimensional noncommutative k-algebra C. A simple, but ingenious observation of S. Brenner [B, Theorem 3] is that $\mathcal{S}(2)$ contains a full subcategory equivalent to $\mathcal{S}(m)$ for every m. Thus $\mathcal{S}(2)$ is strictly wild in the sense defined in the present paper.

Define the category $\mathcal{S}(1\frac{1}{2})$ ("one and one-half similarity") over k, to consist of ordered triples (m, n, ϕ) where m, n, are positive integers with $m \leq n$ and ϕ is an $n \times n$ matrix over k. Morphisms in $\mathcal{S}(1\frac{1}{2})$ are pairs of (σ, τ) of matrices such that the following diagram commutes (where I is an identity matrix).

$$
\mathcal{S}\!\left(1\tfrac{1}{2}\right): \qquad
\begin{array}{ccccc}
k^{(m)} & \xrightarrow{\ i=[I\ 0]\ } & k^{(n)} & \xrightarrow{\ \phi\ } & k^{(n)} \\
\ \downarrow{\scriptstyle\sigma} & & \ \downarrow{\scriptstyle\tau} & & \ \downarrow{\scriptstyle\tau} \\
k^{(m')} & \xrightarrow{\ i'=[I'\ 0]\ } & k^{(n')} & \xrightarrow{\ \phi'\ } & k^{(n')}
\end{array}
\qquad (3.5.2)
$$

A theorem of Nazarova [Nz, Lemma 1] is that $\mathcal{S}(1\frac{1}{2})$ contains a subcategory equivalent to $\mathcal{S}(2)$, and hence is strictly wild.

Wildness of Drozd rings 3.6. This wildness is proved in [KL1, §4]. The proof follows Ringel's suggested simplification [R] of Drozd's original proof [D1] for k-algebras. Let $(\Omega, \mathfrak{m}, k)$ be a Drozd ring. For each triple (m, n, ϕ) where $m \leq n$ are positive integers and ϕ is an $n \times n$ matrix over k, Ringel defines a module $M(m, n, \phi)$ over the Drozd k-algebra that makes sense over an arbitrary Drozd ring Ω. However, morphisms in fingen(Ω) are no longer k-linear maps when Ω is not a k-algebra.

Let $\mathcal{W}$ be the full subcategory of finlen(Ω) consisting of all $M(m, n, \phi)$. Our basic idea is that modules over artinian rings have projective covers. Let $f \colon M \to N$ be a homomorphism in finlen(Ω). Then f can be lifted to a homomorphism F of the projective covers of M and N respectively. Since projective modules over commutative local rings are free, F can be represented by a matrix—which we again call F—over Ω (rather than over its residue field k). Reduction of the entries of F modulo the maximal ideal $\mathfrak{m}$ of Ω yields a matrix $\bar{F}$ over k. If the given modules M, N are in $\mathcal{W}$, extraction of certain diagonal blocks of $\bar{F}$ yield a morphism $\mathcal{E}(f) \in \mathcal{S}(1\frac{1}{2})$. A rather complicated matrix computation then shows that the pair of correspondences $M(m, n, \phi) \to \phi$ and $f \to \mathcal{E}(f)$ forms the desired representation equivalence $\mathcal{W} \to \mathcal{S}(1\frac{1}{2})$ showing that Ω is finlen-wild.

An interesting aspect of the proof is that our matrices are substantially smaller than Ringel's. However, this is deceptive because an element of Ω holds more information than an element of k holds, roughly five times as much since the Drozd algebra has dimension 5. Extraction of the needed k-information from Ω-information is less convenient than one might expect, and is responsible for the complications in our proof.

Tameness of Klein rings is proved in [KL2, §11, especially Theorem.11.3]. See subsection 7.7 of the present survey for a few more details.

The next step, in our discussion of which rings are tame and which are wild, is the following bit of pure commutative algebra. Neither its statement nor its proof involves the concept of tameness or wildness. One of the types of rings that it mentions—Dedekind-like rings—will be defined in the next section.

Theorem 3.7 (Ring-theoretic Dichotomy). *Let Ω be an indecomposable noetherian ring. Then exactly one of the following two possibilities occurs.*

(i) *Ω has an artinian triad or a Drozd ring as a homomorphic image.*
(ii) *Ω is a Klein ring or a homomorphic image of a Dedekind-like ring.*

This is proved for complete local rings in [KL1, Theorem 3.1], and extended to the nonlocal case in [KL3, Theorem 14.3].

Any ring that maps onto a finlen-wild ring is obviously itself finlen-wild. Therefore our tame-wild problem is now reduced to deciding whether Dedekind-like rings and their homomorphic images are tame or wild or neither. It is here that we encounter the possible exception mentioned earlier.

Historical Remark 3.8. The idea of fingen-tame versus finlen-wild is older than one might suspect. In [NR, 1969] Nazarova and Roiter described the isomorphism classes in fingen(Λ), where Λ is what we call a strictly split local Dedekind-like ring in subsection 7.1.

Subsequently, Drozd [D1, 1972] considered complete local noetherian rings Ω with residue field k, assuming that k is algebraically closed and Ω is a finitely generated k-algebra. He proved that the only the following possibilities occur.

(i) Ω is a homomorphic image of $k[[xy]]/(xy)$.
(ii) k has characteristic 2 and Ω is the group algebra of the Klein 4-group.
(iii) Ω maps onto $k[[x, y, z]]/(x, y, z)^2$.
(iv) Ω maps onto the Drozd algebra [defined in our Remarks 3.2(ii)].

The rings in (i) were known to be fingen-tame by [NR], and Drozd showed that the ring (ii) is isomorphic to a ring known to be fingen-tame (= finlen-tame in this case). Drozd proved that the rings in (iii) and (iv) are finlen-wild.

In the subsequent explosion of work on finite-dimensional representations of noncommutative algebras, this fingen-tame/finlen-wild phenomenon seems to have been overlooked except by the present authors and one later paper of Drozd [D2, 1991], which is mainly about the noncommutative case. (New results proved there do not go beyond [D1] in the commutative case.)

4. Dedekind-like Rings

Notation 4.1. If a ring Ω is *reduced* (no nonzero nilpotent elements) it has a *normalization* (integral closure in its total quotient ring). We denote the set of maximal ideals of Ω by maxspec(Ω).

Definition 4.2. We call a noetherian ring Λ *Dedekind-like* if Λ is reduced and its normalization Γ has the following properties.

(i) Γ is a direct sum of Dedekind domains;
(ii) $(\Gamma/\Lambda)_\mathfrak{m}$ is either a simple $\Lambda_\mathfrak{m}$-module or 0 $\left(\forall \mathfrak{m} \in \text{maxspec}(\Lambda)\right)$;
(iii) $\mathfrak{m}_\mathfrak{m} = J(\Gamma_\mathfrak{m})$, the Jacobson radical $\left(\forall \mathfrak{m} \in \text{maxspec}(\Lambda)\right)$; and
(iv) (nontriviality:) $\Lambda_\mathfrak{m}$ is never a field $\left(\forall \mathfrak{m} \in \text{maxspec}(\Lambda)\right)$.

This form of the definition is taken from [KL3, Definition 10.1]. Dedekind-like rings Λ have the following property [KL3, Proposition 10.9]:

$$\text{Every ideal of } \Lambda \text{ is generated by 2 elements.} \qquad (4.2.1)$$

An immediate consequence of statement (ii) is that $\left(\forall \mathfrak{m} \in \mathrm{maxspec}(\Lambda)\right)$ the normalization $\Gamma_{\mathfrak{m}}$ of $\Lambda_{\mathfrak{m}}$ is a finitely generated $\Lambda_{\mathfrak{m}}$-module. However, $_{\Lambda}\Gamma$ *is not necessarily finitely generated.* Examples should eventually appear in [HL].

We always consider Γ and $\mathrm{fingen}(\Gamma)$ to be "approximations" to Λ and $\mathrm{fingen}(\Lambda)$ respectively, in the sense that almost every time we describe a property of Λ we do so by relating it to a property of Γ. This already occurred in the definition of "Dedekind-like" given above.

Our reason for calling this generalization of Dedekind domains "Dedekind-like" is given in the Epilog on Terminology, Section 9.

As with Drozd rings and Klein rings, our definition of Dedekind-like rings leaves one wondering what these rings really look like. We display the structure of complete local Dedekind-like rings quite explicitly in subsections 7.1–7.3. These explicit descriptions are related to the general (i.e., nonlocal) situation by the fact that a noetherian ring is Dedekind-like if and only all of its completions at maximal ideals are Dedekind-like [Proposition 5.1].

Definition 4.3 (Exceptional Dedekind-like rings). Let Λ be a Dedekind-like ring with normalization Γ. We say that Λ is *exceptional* if Λ has a maximal ideal $\mathfrak{m}$ such that the ring $\Gamma/\Gamma\mathfrak{m}$ is a 2-dimensional inseparable field-extension of the field $\Lambda/\mathfrak{m}$ (and hence both fields have characteristic 2). (The terminology "exceptional" is not used in [KL1]–[KL3]. Instead, the equivalent "Additional Hypothesis" [KL3, 10.2] and [KL2, (1.1.3)] are implicitly invoked whenever a tameness result is stated.)

For an example of an exceptional Dedekind-like ring, see subsection 7.2.

Theorem 4.4. *Every non-exceptional Dedekind-like ring is* fingen-*tame. (We do not know whether exceptional Dedekind-like rings are tame or wild or neither.)*

Remarks 4.5 (Module Classification, part 1). The proof of Theorem 4.4, in the complete local case, occupies almost all of [KL2]. The extension to the non-local situation occupies almost all of [KL3]. Tameness is more complicated than wildness for the two reasons given in items (i) and (ii) below.

(i) The form of wildness that we use is finlen-wild. Moreover, for every maximal ideal $\mathfrak{m}$ of Λ the category $\mathrm{finlen}(\hat{\Lambda}_\mathfrak{m})$—$\hat{\Lambda}_\mathfrak{m}$ the $\mathfrak{m}$-adic completion of Λ—is a full subcategory of $\mathrm{finlen}(\Lambda)$. Also, every indecomposable module in $\mathrm{finlen}(\Lambda)$ is an indecomposable module in $\mathrm{finlen}(\hat{\Lambda}_\mathfrak{m})$ for some maximal ideal $\mathfrak{m}$, and $\mathrm{finlen}(\Lambda)$ is a Krull-Schmidt category. This reduces the description of module structure in $\mathrm{finlen}(\Lambda)$ to the the description of indecomposable modules in the complete local case, described in [KL2].

In other words, finlen-wildness is determined in the simpler, complete local situation. We postpone further discussion of the complete local case to Section 7, where we also discuss the role of "1-parameter families" of modules. (These are always families of modules of finite length.)

(ii) The relationship between $\mathrm{fingen}(\hat{\Lambda}_\mathfrak{m})$ and $\mathrm{fingen}(\Lambda)$ is much more complicated, because $\mathrm{fingen}(\hat{\Lambda}_\mathfrak{m})$ is not a subcategory of $\mathrm{fingen}(\Lambda)$. Moreover, $\mathrm{fingen}(\Lambda)$ is not a Krull-Schmidt category. Therefore, unlike the situation in finitely dimensional algebras, we can no longer focus exclusively on indecomposable Λ-modules. In fact, they play a rather minor role in our module-description, even though indecomposable modules in $\mathrm{fingen}(\Lambda)$ play a major role in the complete local case, and their properties are needed in the nonlocal case.

(iii) Let $\mathrm{fingen}_\infty(\Lambda)$ denote the category of finitely generated Λ-modules with no direct summands of finite length. Since Λ is noetherian, it is easy to see that every $M \in \mathrm{fingen}(\Lambda)$ has a unique decomposition (up to isomorphism) $M = M_\infty \oplus M_0$ where $M_\infty \in \mathrm{fingen}_\infty(\Lambda)$ and $M_0 \in \mathrm{finlen}(\Lambda)$. Moreover, if $N = N_\infty \oplus N_0$, then $M_\infty \oplus N_\infty \in \mathrm{fingen}_\infty(\Lambda)$ and $M_0 \oplus N_0 \in \mathrm{finlen}(\Lambda)$. This separates the study of $\mathrm{fingen}(\Lambda)$ into two disjoint parts (except for the zero module, which belongs to both categories). We comment briefly on each part separately, in items (iv) and (v) below. For proofs and more details of this reduction, which applies to rings much more general than Dedekind-like rings, see [KL3, §7].

(iv) It therefore remains to discuss the category $\mathrm{fingen}_\infty(\Lambda)$. It turns out that a module $M \in \mathrm{fingen}(\Lambda)$ belongs to $\mathrm{fingen}_\infty(\Lambda)$ if and only if every $\mathfrak{m}$-adic completion $\hat{M}_\mathfrak{m} \in \mathrm{fingen}_\infty(\hat{\Lambda}_\mathfrak{m})$. And, as mentioned in (iii), $\mathrm{fingen}_\infty(\Lambda)$ is closed under (finite) direct sums. Thus we have reduced the study of modules and their direct sums in $\mathrm{fingen}(\Lambda)$ to the analogous problem in $\mathrm{fingen}_\infty(\Lambda)$. This is subject of Section 5.

We continue this discussion in Remarks 5.2. The following theorem summarizes what we know so far, except for a clear description of what the tameness in part (iii)(a) really means. That occupies most of the rest of this survey.

Theorem 4.6. *Let Ω be an indecomposable noetherian ring. Then exactly one of the following occurs.*

(i) Ω *maps onto an artinian triad or a Drozd ring. Here Ω is* finlen-*wild.*

(ii) Ω *is a Klein ring. Here Ω is* fingen-*tame [see subsection 7.7].*

(iii) *Some Dedekind-like ring Λ maps onto Ω.*

    (a) *If Λ is non-exceptional, then Ω is* fingen-*tame.*

    (b) *If every Λ that maps onto Ω is exceptional [Definition 4.3], then we do not know whether Ω is tame or wild or neither.*

Since every homomorphic image of a fingen-tame ring is obviously finlen-tame, we limit our discussion of the tameness in Theorem 4.6(iii)(a) to modules over Dedekind-like rings. This discussion begins in the next section.

Corollary 4.7. *Every noetherian ring Ω with Krull dimension ≥ 2 is* finlen-*wild.*

Proof. Dedekind domains have Krull dimension 1. Thus the normalization Γ of every Dedekind-like ring Λ has Krull dimension 1. It follows easily that Dedekind-like rings—exceptional or not—have Krull dimension 1; in fact, all maximal ideals of Λ have height 1 [KL3, Proposition 10.6]. Therefore no Dedekind-like ring can map onto Ω. Since rings of Krull dimension ≥ 2 are not artinian, it follows from our ring-theoretic dichotomy [Theorem 3.7] that Ω maps onto an artinian triad or Drozd ring, and hence is finlen-wild [Theorem 4.6]. $\qquad\square$

Corollary 4.8. *Let Ω be a noetherian ring with Krull dimension ≤ 1.*

(i) *If some ideal of Ω requires 3 or more generators, then Ω is* finlen-*wild.*

(ii) *If every ideal of Ω is principal, then Ω is* fingen-*tame.*

Proof. (i) As observed in (4.2.1), all ideals of Dedekind-like rings are 2-generated. Therefore no Dedekind-like ring maps onto Ω. As in the proof of the previous corollary, we conclude that Ω is finlen-wild.

(ii) The well-known fact that all principal ideal rings are fingen-tame, in the sense in which we are using the term, is a much older result than the definition of "tame". $\qquad\square$

By the previous two corollaries, "tame noetherian" is a very small class of rings of Krull dimension ≤ 1. Does the class contain any interesting non-artinian rings (other than direct sums of Dedekind domains and their homomorphic images)?

124

Examples 4.9 (Natural Examples of Dedekind-like rings).

(i) $\mathbb{Z}[\sqrt{n}\,]$ when n is squarefree.

(ii) $\mathbb{Z}G_n$ (integral group ring of a cyclic group of order n) when n is squarefree.

(iii) All subrings of squarefree index in $\mathbb{Z} \oplus \ldots \oplus \mathbb{Z}$.

(iv) (Over any algebraically closed field:) The coordinate ring of any affine curve whose singularities are all simple nodes.

(v) $k[x,y]/(xy)$ and $k[[x,y]]/(xy)$ for any field k.

(vi) $\mathbb{R} + x\mathbb{C}[x]$ and $\mathbb{R} + x\mathbb{C}[[x]]$ (polynomials and formal power series rings over the complex numbers, with real constant term).

For proofs, see the following. (i): [KL3, Example 36.3]. (ii): See [L3]. But beware of the differently-phrased definition of "Dedekind-like" in this earlier paper, a narrower class of rings than the Dedekind-like rings in the present paper. (iii): Combine Proposition 5.1 with [KL2, 12.5]. (v) and (vi): The power series rings are Dedekind-like by [KL1, 2.17]. For the polynomial version, use this and Proposition 5.1, remembering that discrete valuation rings are Dedekind-like. (iv) A purely algebraic statement of a property, from which (iv) follows, is: *Suppose that the completion of a noetherian ring Ω at each maximal ideal is either isomorphic to $k[[x,y]]/(xy)$ or $k[[x]]$, for some field k. Then Ω is Dedekind-like.* The proof is the same as that of statement (v), with the polynomial ring replaced by Ω.

We included rings (i) because rings of algebraic integers are the rings that interested Steinitz; and Dedekind-like rings of algebraic integers may be the only non-integrally closed rings of algebraic integers whose finitely generated module category has been described.

We included examples (vi) because they seem to be the simplest known fingen-tame rings that are infinite dimensional algebras over a non-algebraically closed field and have no counterpart over algebraically closed fields. We discuss its modules later [Section 8].

Example 4.10 (Super-wild ring). The ring $\Omega = \mathbb{Z}[x]/(x^2)$ is finlen-wild (and might be the most easily described nonlocal finlen-wild ring of Krull dimension 1).

This ring has an amusing property that does not occur for finite dimensional algebras, and that we call *superwild*: For each prime number p there is a maximal ideal $\mathfrak{m}$ of Λ such that $\Lambda/\mathfrak{m}$ has characteristic p and Λ is finlen-wild (with respect to $\mathfrak{m}$).

Proof. For every p, Ω maps onto the ring $\mathbb{Z}[X]/(X^2, p^3, p^2X)$, which we have observed to be a Drozd ring [Remarks 3.2]. Hence Ω is finlen-wild (with respect to the appropriate $\mathfrak{m}$) [Theorem 3.4] $\qquad\square$

5. Local-Global and Direct-Sum Relations

Throughout this section Λ denotes a non-exceptional Dedekind-like ring with normalization Γ. We focus on Dedekind-like rings for the rest of this survey, except for the short subsection 7.7 on Klein rings, because all known indecomposable fingen-tame noetherian rings except Klein rings are homomorphic images of Dedekind-like rings. Two important stability properties of the class of Dedekind-like rings are the following.

Proposition 5.1. *Let Ω be a noetherian ring.*

 (i) *Ω is Dedekind-like if and only if all localizations $\Omega_{\mathfrak{m}}$ at maximal ideals are Dedekind-like.*

 (ii) *If Ω is local with maximal ideal $\mathfrak{m}$, then it is Dedekind-like if and only its $\mathfrak{m}$-adic completion $\hat{\Omega}_{\mathfrak{m}}$ is Dedekind-like.*

When the conditions hold, the normalizations of $\Omega_{\mathfrak{m}}$ and $\hat{\Omega}_{\mathfrak{m}}$ are $\Gamma_{\mathfrak{m}}$ and $\hat{\Gamma}_{\mathfrak{m}}$ respectively, where Γ is the normalization of Ω.

Proof. (i) is an almost immediate from the definition of Dedekind-like. For details see [KL3, Corollary 10.7]. For the well-known fact that $\Gamma_{\mathfrak{m}}$ is the normalization of $\Omega_{\mathfrak{m}}$ see, for example, [KL3, Remarks 5.3].

For (ii) and its supplementary statement about the normalization in this case, see [KL3, Lemma 11.8]. $\qquad\square$

Remarks 5.2 (Module Classification, part 2). If we wish to describe the isomorphism classes and direct-sum relations in $\mathrm{fingen}(\Lambda)$, it now suffices, for the reasons given in Remarks 4.5, to work with the subcategory $\mathrm{fingen}_{\infty}(\Lambda)$. We make this assumption whenever convenient. We now outline what must be done.

Complete local case. The details of this appeared in [KL2], and we give its flavor in Sections 7 and 8.

Nonlocal case. Let $M \in \mathrm{fingen}(\Lambda)$ be given. Then we know the structure of every $\mathfrak{m}$-adic completion $\hat{M}_{\mathfrak{m}}$ by Proposition 5.1 and the complete local case. The rest of this section deals with the following three questions.

 (i) The *package deal question:* For which families $\{M(\mathfrak{m})\}$, where each $M(\mathfrak{m}) \in \mathrm{fingen}(\hat{\Lambda}_{\mathfrak{m}})$, does there exist $M \in \mathrm{fingen}(\Lambda)$ such that $\hat{M}_{\mathfrak{m}} \cong M(\mathfrak{m})$ for all $\mathfrak{m}$?

(ii) What *additional information* is needed, to determine the isomorphism class of M? Here it is usually more convenient to work in $\mathrm{fingen}_\infty(\Lambda)$.

(iii) What are the *direct-sum relations* in $\mathrm{fingen}(\Lambda)$? Again, working in $\mathrm{fingen}_\infty(\Lambda)$ is more convenient.

Question (i): package deals. The answer is straightforward. For almost all $\mathfrak{m} \in \mathrm{maxspec}(\Lambda)$, $\hat{M}_\mathfrak{m}$ must be $\hat{\Lambda}_\mathfrak{m}$-free; and the torsionfree ranks of the $M(\mathfrak{m})$ must satisfy a simple, obviously necessary consistency condition: $M(\mathfrak{m})_\mathfrak{p} \cong M(\mathfrak{m}')_\mathfrak{p}$ whenever the maximal ideals $\mathfrak{m}, \mathfrak{m}'$ contain a common minimal prime ideal $\mathfrak{p}$. We refer to these isomorphisms as "equality of torsionfree ranks" because $\hat{\Lambda}_\mathfrak{p}$ is always a field, and hence its modules are determined up to isomorphism by their vector-space dimension.

These consistency conditions remain necessary and sufficient for the existence of M in much more generality than Dedekind-like rings. They have long been well-known for torsionfree modules over arbitrary reduced noetherian rings Ω of Krull dimension 1, when the normalization is a finitely generated Ω-module. For the more general result needed here—dropping the "torsionfree" and "finite normalization" hypotheses, see [LO, §2]. In slightly more detail: Unlike the classical situation, it is not true that $\Lambda_\mathfrak{m}$ is a discrete valuation ring for all but finitely many $\mathfrak{m}$. However, for each $M \in \mathrm{fingen}(\Lambda)$ we have that $M_\mathfrak{m}$ is a free $\Lambda_\mathfrak{m}$-module for all but finitely many $\mathfrak{m}$. The difference between this and the classical situation is that the set of nontrivial ($=$ nonfree) localizations changes from module to module.

Question (ii): additional information. This additional information is given by the ideal class groups of the Dedekind-domain direct summands of Γ (discussed below) and by the units in residue fields of Λ and Γ. We omit discussion of this complicated topic, except to say that the two ingredients in it are put together in a Mayer-Vietoris sequence. See (5.5.1) for slightly more detail.

Question (iii): direct-sum relations. To answer this, we introduce the *web of genus class groups.* This consists of a system of genus class groups, together with a system of homomorphisms—called ξ-maps—between certain pairs of these groups. The rest of the present section describes, roughly, how this works.

Definitions 5.3. For $M \in \mathrm{fingen}(\Lambda)$ let

$$\mathrm{genus}(M) = \{N \in \mathrm{fingen}(\Lambda) \mid \hat{N}_\mathfrak{m} \cong \hat{M}_\mathfrak{m} \ \ \forall \mathfrak{m} \in \mathrm{maxspec}(\Lambda)\}$$
$$= \{N \in \mathrm{fingen}(\Lambda) \mid N_\mathfrak{m} \cong M_\mathfrak{m} \ \ \forall \mathfrak{m} \in \mathrm{maxspec}(\Lambda)\} \tag{5.3.1}$$

and let $[M]$ denote the isomorphism class of M. The idea conveyed by the term "genus" is that if one considers an isomorphism class to be the analog of the biological notion of "species", then the next more general class is analogous to a genus.

The next step is to define the abelian *genus class group* $\mathcal{G}(M)$. The elements of this group are the isomorphism classes of modules in genus(M). It can be shown that this collection of isomorphism classes can be made into an abelian group in such a way that $[M]$ is the zero element of this group, and:

$$\bigoplus_{i=1}^{m} M_i \cong \bigoplus_{i=1}^{n} N_i \qquad \bigl(M_i, N_i \in \text{genus}(M)\bigr)$$
$$\Longleftrightarrow$$
$$m = n \quad \text{and} \quad \sum_i [M_i] = \sum_i [N_i] \quad \text{in } \mathcal{G}(M) \tag{5.3.2}$$

In fact, this can be done in exactly one way, and the definition is implicitly contained in (5.3.2): Take $m = n = 2$ and $M_2 = M$. Since $[M] = 0$ in $\mathcal{G}(M)$, the second line of (5.3.2) becomes $[M_1] = [N_1] + [N_2]$ in $\mathcal{G}(M)$, while the first line states that this sum $[M_1]$ of $[N_1]$ and $[N_2]$ in $\mathcal{G}(M)$ must be the isomorphism class that satisfies the direct-sum condition $M_1 \oplus M \cong N_1 \oplus N_2$ (and a proof is required, to show that a unique such $[M_1]$ always exists for given N_1 and N_2).

These are well-known facts for torsionfree modules over the rings that occur in integral representation theory. In the present generality, the proof requires Serre's direct-summand theorem, Bass's cancellation theorem, and the following cancellation result of Guralnick and Levy [GL, 5.10] (extending an earlier result of Drozd). Let Ω be a reduced noetherian ring of Krull dimension 1, and suppose that $M \oplus X \cong N \oplus X$, with $M, N, X \in \text{fingen}(\Omega)$ and $X \in \text{genus}(M)$. Then $M \cong N$.

Remark on the zero element $[M]$ of $\mathcal{G}(M)$. Note that, in the above definition, $[M]$ was an arbitrarily chosen isomorphism class in its genus. We call it the *base-point* of genus(M), to emphasize this arbitrariness.

It is evident that (5.3.2) describes all direct-sum relations in genus(M), once one knows the group $\mathcal{G}(M)$. Actual computation of this group in specific instances is at least as difficult as computing ideal class groups of rings of integers in algebraic number fields. However, as in the classical situation of Dedekind domains, merely knowing that $\mathcal{G}(M)$ is an abelian group gives a lot of useful information (see below).

Remaining question: How do we adapt (5.3.2) to handle direct sums when the terms are not all one genus? Informally, we want to "add" elements in different groups. Formally, we use the web of class groups. The basic idea is simple, but since the details are complicated, we just sketch the main ideas.

Definition 5.4 (Web of genus class groups). As a set, this consists of one genus class group $\mathcal{G}(M)$—together with its arbitrarily selected base point $[M]$—for each genus in $\mathrm{fingen}_\infty(\Lambda)$. Natural maps $\xi = \xi^{M,N} : \mathcal{G}(M) \to \mathcal{G}(N)$ are defined for certain pairs M, N; and the *web of class groups* consists of these ξ-maps together with the aforementioned groups $\mathcal{G}(M)$.

Let $M, N \in \mathrm{fingen}_\infty(\Lambda)$. We say that $\xi^{M,N}$ *is defined* if, for each module $M' \in \mathrm{genus}(M)$, there exists a module $N' \in \mathrm{genus}(N)$, *unique up to isomorphism*, such that $M' \oplus N \cong M \oplus N'$. When this is the case, we set $\xi^{M,N}[M'] = [N']$. See [KL3, Definition 25.4ff] for more about this.

Instead of dwelling on the definition, we proceed to survey how these ξ-maps are used, to understand direct sums and the relation of the various genus class groups to each other.

We call these ξ-maps *loss of structure maps* because they can have nonzero kernels, indicating that information is lost when we pass from one genus to another by means of a ξ-map with nonzero kernel.

The normalization Γ is always a direct sum of Dedekind domains. We regard Γ as an approximation to Λ, and $\mathrm{fingen}(\Gamma)$ as an approximation to $\mathrm{fingen}(\Lambda)$. In particular, the (known) structure of $\mathrm{fingen}(\Gamma)$ is a special case of what we describe here. In several places we comment on this, and on the way in which it serves as an approximation. Our starting point is:

Theorem 5.5. *Let $M, N \in \mathrm{fingen}_\infty(\Lambda)$.*

(i) *If M is faithful and $_\Lambda\Gamma$ is finitely generated, then $\xi^{M,\Gamma}$ is defined and surjective. [KL3, Proposition 32.3]*

(ii) *If M, N are both faithful and $\xi^{M,N}$ is defined, then $\xi^{M,N}$ is surjective. [KL3, Lemma 25.6]*

To understand statement (i), assume that we are in the classical situation that $_\Lambda\Gamma$ is finitely generated. Then $\mathcal{G}(_\Lambda\Gamma)$ is meaningful; and it is not difficult to see that this coincides with $\mathcal{G}(_\Gamma\Gamma)$, the genus class group of Γ, considered as a Γ-module. Thus the fact that $\xi^{M,\Gamma}$ is surjective but can have nonzero kernel indicates that $\mathcal{G}(M)$ contains at least as much "global information" as $\mathcal{G}(\Gamma)$ contains. Thus $\mathcal{G}(_\Gamma\Gamma)$ is a first approximation to $\mathcal{G}(M)$ for every faithful $_\Lambda M$.

We can be more explicit about this—*whether or not $_\Lambda\Gamma$ is finitely generated.* Since Γ is a direct sum of Dedekind domains, say $\Gamma = \bigoplus_h \Gamma_h$, the group $\mathcal{G}(_\Gamma\Gamma)$ can be identified with the direct sum of the genus class groups $\mathcal{G}(_{\Gamma_h}\Gamma_h)$ of the various Γ_h, more classically known as their *ideal class groups.* Thus the direct sum of these ideal class groups forms the first approximation to every $\mathcal{G}(M)$ when $_\Lambda M$ is faithful. In slightly more detail, there is a "Mayer-Vietoris" short exact sequence for every faithful $_\Lambda M$:

$$0 \to K \to \mathcal{G}(M) \xrightarrow{\xi} \mathcal{G}(_\Gamma\Gamma) \to 0 \tag{5.5.1}$$

In the classical situation that $_\Lambda\Gamma$ is finitely generated, we can use $\xi = \xi^{M,\Gamma}$ here. In general, ξ denotes the composition of $[M] \to {}_\Gamma[\Gamma \otimes_\Lambda M]$ with the ξ-map of Γ-modules $\xi_\Gamma^{(\Gamma \otimes_\Lambda M),\Gamma}$. The kernel K is formed from subgroups of the groups of units of residue fields of Λ and Γ. Readers familiar with classical Mayer-Vietoris sequences might be interested to know that, when $_\Lambda\Gamma$ is not finitely generated—and hence there is no conductor ideal for Λ and Γ—we take advantage of the fact that local conductors always exist for Dedekind-like rings. See [KL3, Corollary 32.4 and Theorem 25.14] for details.

To help clarify the role of faithfulness in statement (ii) of Theorem 5.5, consider the trivial case $\Lambda = \Gamma$, and suppose that $\Gamma = \Gamma_1 \oplus \Gamma_2$, the direct sum of two Dedekind-domains. Then $\mathcal{G}(\Gamma) = \mathcal{G}(\Gamma_1) \oplus \mathcal{G}(\Gamma_2)$, and $\xi^{\Gamma_1,\Gamma}$ can be identified with the first coordinate projection in this direct sum. This is never a surjection unless $\mathcal{G}(\Gamma_2) = \{0\}$. In greater generality, lack of surjectivity of $\xi^{M,N}$ occurs precisely when the torsionfree part of the Γ-module $\Gamma \otimes_\Lambda N$ involves coordinate rings Γ_h of Γ that have nontrivial ideal class groups but the torsionfree part of the Γ-module $\Gamma \otimes_\Lambda M$ does not involve these coordinate rings.

The next result gives one basic situation in which the ξ-map is defined [KL3, Corollary 25.9 and Lemma 25.6].

Theorem 5.6. *Let M, N be arbitrary in* $\mathrm{fingen}_\infty(\Lambda)$, *and* $P \in \mathrm{genus}(M \oplus N)$. *Then* $\xi^{M,P}$ *and* $\xi^{N,P}$ *are defined. If* $\mathrm{ann}(M) = \mathrm{ann}(N)$, *then* $\xi^{M,P}$ *and* $\xi^{N,P}$ *are surjective. ("ann" denotes "annihilator".)*

We can finally explain how the ξ-maps describe direct-sum behavior.

Notation 5.7 (Direct sums). Let $M', N' \in \mathrm{fingen}_\infty(\Lambda)$ be given, and let M, N be the base-points in their respective genera. Also, let P be the base-point in $\mathrm{genus}(M \oplus N)$. The three boxes in diagram (5.7.1) show the three corresponding genus class groups, the given $[M']$, $[N']$, their natural images in $\mathcal{G}(P)$, and the element $[M' \oplus N']$ of $\mathcal{G}(P)$.

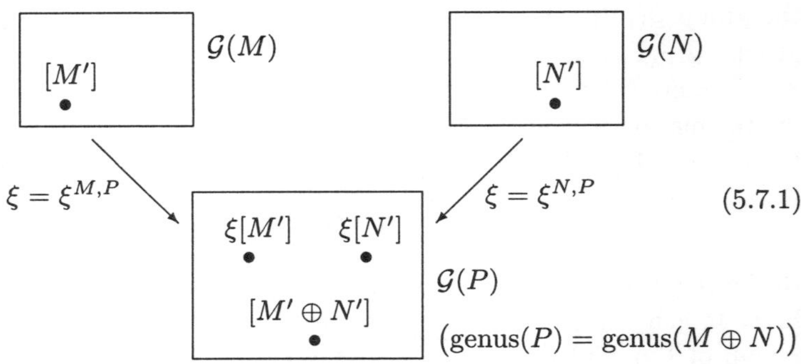

(5.7.1)

Note that the equation $\text{genus}(P) = \text{genus}(M \oplus N)$, at the bottom of the diagram, cannot be replaced by $\mathcal{G}(P) = \mathcal{G}(M \oplus N)$ because $M \oplus N$ might not be the base-point of its genus.

Theorem 5.8. *Keep the above notation. Then:*

$$[M' \oplus N'] = \xi[M'] + \xi[N'] + [P_0] \quad in \ \mathcal{G}(P) \tag{5.8.1}$$

where the "correction term" $[P_0]$ depends only on the (arbitrarily selected) base points M, N, P, and not on M' or N'. [KL3, Theorem 29.1]

Informally, the theorem says that the "sum" $[M'] + [N']$ is obtained by adding the natural images $\xi[M']$ and $\xi[N']$ in $\mathcal{G}(P)$, and then adding an (annoying) correction term whose presence is due to the completely arbitrary choice of base-points M, N, P in the three genera. This correction term seems unavoidable. For example, if we happen to have chosen $P = M \oplus N$, then the correction term becomes zero, and we get the more pleasant formula [KL3, Corollary 29.2]:

$$[M' \oplus N'] = \xi[M'] + \xi[N'] \quad in \ \mathcal{G}(P) \tag{5.8.2}$$

Nevertheless, Theorem 5.8 shows that *all non-locally-determined aspects of direct-sum behavior are determined by abelian groups and group homomorphisms.*

The reason for only saying *seems* unavoidable, above (5.8.2), is that we do not know whether it is possible to pick the set of base-points in such a way that it is closed under direct sums [KL3, §37 Problem 6].

As an application of this we get a definitive answer to the question: How can direct-sum cancellation fail? First we review a well-known result [E].

Lemma 5.9 (Evans). *The implication* $M' \oplus N \cong M'' \oplus N \implies M'' \in$ genus(M') *holds for finitely generated modules over any noetherian ring.*

Theorem 5.10. *Let* $M', N' \in$ fingen$_\infty(\Lambda)$, *let* M, N, P *be the base-points in the genera of* $M', N', (M' \oplus N')$ *respectively, and let* $M'' \in$ fingen$_\infty(\Lambda)$. *Then:*

$$M' \oplus N' \cong M'' \oplus N' \iff$$
$$M'' \in \mathrm{genus}(M) \quad and \quad [M'] - [M''] \in \ker(\xi^{M,P}) \quad in \,\, \mathcal{G}(M) \tag{5.10.1}$$

Proof. Suppose that the isomorphism in (5.10.1) holds. Then, by Evans's lemma, M' and M'' are both in the genus having base-point $[M]$. Since $[M' \oplus N'] = [M'' \oplus N']$, two applications of (5.8.1) yield:

$$\xi^{M,P}[M'] + \xi^{N,P}[N'] + [P_0] = \xi^{M,P}[M''] + \xi^{N,P}[N'] + [P_0] \qquad \text{in } \mathcal{G}(P)$$

Since $\mathcal{G}(P)$ is an abelian group and $\xi^{M,P}$ is a homomorphism, we deduce that $[M'] - [M''] \in \ker(\xi^{M,P})$.

Reversing the reasoning in the previous paragraph proves the converse part of the theorem. $\qquad\square$

Remarks 5.11. (i) *Note the uniformity of the cancellation result in Theorem 5.10:* It is independent of the choice of N' in genus(N) and independent of which M' in genus(M) one starts with.

(ii) It is very common that cancellation actually fails. For example, it can fail for rings of all of the types (i)–(iv) enumerated in Examples 4.9 [but not for types (v) and (vi)].

(iii) Failure of cancellation, when it occurs, is very different from the failure of cancellation that occurs in K-theory because of instability. The latter type of failure occurs when the stable isomorphism classes that are elements of the appropriate K_0-group are not actual isomorphism classes. However, the elements of our genus class groups are always actual isomorphism classes.

Other direct-sum behavior. Dedekind-like rings exhibit a very rich variety of describable direct-sum behavior (in addition to failure of direct-sum cancellation).

Example 5.12. Choose $n \geq 2$. Then there exist a Dedekind-like ring Λ and $M \in$ fingen(Λ) such that M is the direct sum of s indecomposables for every s in interval $2 \leq s \leq n$. (Since Λ and M are noetherian, we can never find a single M such that this decomposition property holds for all $s \geq 2$.)

In fact, Λ can be chosen to be a group ring $\mathbb{Z}G_m$ (G_m cyclic of squarefree order m) [L3, (0.1)]; and can also be chosen to be a subring of squarefree index in $\mathbb{Z} \oplus \mathbb{Z} \oplus \ldots \oplus \mathbb{Z}$ [L1].

Coherence of the web of class groups. The set of all genus class groups of faithful modules in $\mathrm{fingen}_\infty(\Lambda)$ has an inverse limit $s\mathcal{G}$, with respect to the ξ-maps. (Recall that these maps are surjections when they are defined [Theorem 5.5]).

If $_\Lambda\Gamma$ is finitely generated, then this "super genus class group" $s\mathcal{G}$ is an actual genus class group $\mathcal{G}(P)$, where P has torsionfree rank ≤ 2 [KL3, Theorem 27.12].

Suppose that Λ is an integral domain. Then there are situations in which P cannot be chosen to have torsionfree rank 1 [KL3, Example 34.1]. In such situations, the resulting P of rank 2 *cannot be torsionfree!* [KL3, Theorem 27.12] This is a counterexample to the generally held belief that torsionfree modules can hold more "global information" than any other modules.

6. Mod-Γ as Approximation to Mod-Λ

In this section Λ is again a non-exceptional Dedekind-like ring with normalization Γ. We have already discussed how the isomorphism classes in $\mathrm{fingen}_\infty(\Lambda)$ are approximated by those in $\mathrm{fingen}(\Gamma)$—via Mayer-Vietoris sequences. In the present section we discuss how homomorphisms in $\mathrm{fingen}(\Lambda)$ are approximated by those in $\mathrm{fingen}(\Gamma)$. This approximation, called a "separated cover", is the basic tool used in this series of papers to obtain the structure of Λ-modules from the known theory of Γ-modules.

We have a basic difficulty that does not occur in the study of torsionfree modules: *It is not true that every finitely generated Λ-module is contained in some Γ-module.* The following two definitions are designed to deal with this difficulty.

Definitions 6.1. We define a Γ-*separated Λ-module* to be any Λ-submodule of any Γ-module.

We define a Γ-*separated cover* $\phi\colon S \twoheadrightarrow M$ of $M \in \mathrm{fingen}(\Lambda)$ to be a Λ-module homomorphism such that:

- S is a Γ-separated Λ-module; and
- S is "as close as possible" to M, in the sense that in all (surjective) factorizations

$$\phi\colon S \xrightarrow{\ \theta\ } S' \longrightarrow M \quad \text{(with } S' \text{ Γ-separated)} \tag{6.1.1}$$

θ must be an isomorphism. In other words, in any factorization of the form (6.1.1), S' is no closer to M than S is.

Since free Λ-modules are Γ-separated, and since all of the modules we are dealing with are noetherian, it is a triviality that *every $M \in \mathrm{fingen}(\Lambda)$ has a Γ-separated cover such that the separated covering module S is again in* $\mathrm{fingen}(\Lambda)$.

We think of S as the "best approximation" to M by a Λ-submodule of a Γ-module. The reason that separated covers are useful is [KL3, Theorem 18.10]:

Theorem 6.2 (Almost functorial property). *Let $f \colon N \to M$ be a homomorphism in $\mathrm{fingen}(\Lambda)$, and let ϕ', ϕ be Γ-separated covers. Then f can be lifted to a Λ-homomorphism θ such that the following diagram commutes.*

$$\begin{array}{ccc} S' & \overset{\theta}{\dashrightarrow} & S \\ {\scriptstyle \phi'}\big\downarrow & & \big\downarrow{\scriptstyle \phi} \\ N & \xrightarrow{\ f\ } & M \end{array} \qquad (6.2.1)$$

If f is one-to-one or onto then θ has the same property.

Corollary 6.3 (Uniqueness of separated covers). *For any $M \in \mathrm{fingen}(\Lambda)$, the Γ-separated cover $\phi \colon S \twoheadrightarrow M$ of M is unique up to isomorphism over M.*

Proof. Let ϕ' be another Γ-separated cover of M, and apply Theorem 6.2 with f equal to the identity map. $\qquad\square$

See Theorem 8.5 for an explicit display of the Γ-separated covers of all indecomposable modules in $\mathrm{fingen}(\Lambda)$, for the ring $\Lambda = \mathbb{R} + x\mathbb{C}[[x]]$.

For an arbitrary pair of rings $\Upsilon \supseteq \Delta$, we define an *$\Upsilon$-separated Δ-module* and an *Υ-separated cover of a Δ-module* by replacing Γ and Λ with Υ and Δ respectively, in Definitions 6.1. This additional flexibility is needed in the next result which states, informally, that separated covers are stable with respect to localization and completion [KL3, Theorem 18.13].

Recall that if Λ is a Dedekind-like ring with normalization Γ, then $(\forall \mathfrak{m} \in \mathrm{maxspec}(\Lambda))$ $\Lambda_{\mathfrak{m}}$ and $\hat{\Lambda}_{\mathfrak{m}}$ are again Dedekind-like with normalization $\Gamma_{\mathfrak{m}}$ and $\hat{\Gamma}_{\mathfrak{m}}$, respectively [Proposition 5.1].

Corollary 6.4. *Consider a surjective Λ-homomorphism $\phi \colon S \twoheadrightarrow M$ in* $\mathrm{fingen}(\Lambda)$, *where S is Γ-separated.*

(i) ϕ *is a* Γ-*separated cover if and only if its localization* $\phi_{\mathfrak{m}} \colon S_{\mathfrak{m}} \twoheadrightarrow M_{\mathfrak{m}}$ *is a* $\Gamma_{\mathfrak{m}}$-*separated cover* $\left(\forall \mathfrak{m} \in \mathrm{maxspec}(\Lambda)\right)$.

(ii) ϕ *is a* Γ-*separated cover if and only if its* $\mathfrak{m}$-*adic completion* $\hat{\phi}_{\mathfrak{m}} \colon \hat{S}_{\mathfrak{m}} \twoheadrightarrow \hat{M}_{\mathfrak{m}}$ *is a* $\hat{\Gamma}_{\mathfrak{m}}$-*separated cover* $\left(\forall \mathfrak{m} \in \mathrm{maxspec}(\Lambda)\right)$.

7. Module Structure: Complete Local Case

We are separating this discussion from the earlier "non-local" parts of this survey for two reasons. (i) When Λ is a complete local ring, $\mathrm{fingen}(\Lambda)$ is a Krull-Schmidt category; and therefore describing the indecomposable modules becomes much more important. The full description is given in [KL2, §§2,3] and is quite long. Here we give just enough to establish its flavor and the flavor of the method used to obtain the description. (ii) All indecomposable artinian rings are complete local rings.

As our starting point we give the promised explicit description of the structure of complete local Dedekind-like rings. For a first reading, it is probably better to simply assume that all rings in this section are complete local rings. However, we invoke the completeness hypothesis only when it is needed.

Strictly Split local Dedekind-like rings 7.1. Let $(\Gamma_1, \mathfrak{m}_1, k)$ and $(\Gamma_2, \mathfrak{m}_2, k)$ be DVRs with the same residue field k, and $\rho_i \colon \Gamma_i \twoheadrightarrow k$ $(i = 1, 2)$ the natural homomorphism. Let

$$\Lambda = \{(x_1, x_2) \in \Gamma = \Gamma_1 \oplus \Gamma_2 \mid \rho_1(x_1) = \rho_2(x_2)\} \qquad (7.1.1)$$

Then $(\Lambda, \mathfrak{m}_1 \oplus \mathfrak{m}_2, k)$ *(maximal ideal* $\mathfrak{m} = \mathfrak{m}_1 \oplus \mathfrak{m}_2)$ *is a local Dedekind-like ring with normalization* Γ. [KL1, Lemma 2.15] We call such a local Dedekind-like ring *strictly split*. These are the Dedekind-like rings studied by Nazarova and Roiter in [NR]. The simplest example here (though not the example that interested Nazarova and Roiter) is $\Lambda = k[[x,y]]/(xy)$ [KL1, 2.17].

Unsplit local Dedekind-like rings 7.2. Let $(\Gamma, \mathfrak{m}, F)$ be a DVR whose residue field F is a 2-dimensional extension of a subfield k, and let $\rho \colon \Gamma \twoheadrightarrow F$ be the natural homomorphism. Let

$$\Lambda = \{x \in \Gamma \mid \rho(x) \in k\} \qquad (7.2.1)$$

Then $(\Lambda, \mathfrak{m}, k)$ *is a local Dedekind-like ring with normalization* Γ. [KL1, Lemma 2.16] We call such a local Dedekind-like ring *unsplit*. Moreover, Λ *is exceptional if and only if* F *is an inseparable extension of* k (necessarily of characteristic 2, since the dimension is 2).

The simplest example here is the power-series ring $\Lambda = k + xF[[x]]$, and is exceptional if and only if k has characteristic 2 and its dimension 2 extension F is inseparable [see Definition 4.3].

There are two other kinds of local Dedekind-like rings. One is called "nonstrictly split" [KL1, Definition 2.1] and cannot occur in the complete local case, the other is the familiar DVR.

We have [KL1, Definition 2.5, Lemmas 2.15, 2.16]:

Theorem 7.3. *Every complete local Dedekind-like ring is either strictly split, unsplit, or a DVR. (The one or two DVRs whose direct sum is the normalization of Λ are complete DVRs.)*

When Λ is complete local, Γ-separated covers [Definitions 6.1] are the basic tool used in [KL2] to find the explicit structure of isomorphism classes in fingen(Λ). Separated covers are used to convert the isomorphism question for Λ-modules to a matrix problem over the residue fields of Λ and Γ. The module-structure problem then separates naturally into three cases: strictly split, unsplit, and DVR. We ignore DVRs, since we have nothing new to say about them. We discuss the remaining two cases separately, below. An interesting curiosity is that we do not need to assume completeness, because fingen(Λ) is a Krull-Schmidt category in the strictly split and unsplit cases [KL1, Lemma 1.3].

Strictly Split Case 7.4. Here Γ is the direct sum of two DVRs, each with the same residue field as Λ. Say $\Gamma = \Gamma_1 \oplus \Gamma_2$. The solution of the associated matrix problem makes use of the "sweeping similarity" results in [KL0]. Matrix problems of this type are well-known in the finitely generated algebra community, whose methods provide an alternative solution of the matrix problem.

However, even for finite-length Λ-modules, our methods provide a description of the indecomposable Λ-modules that is slightly different from the description usually given for finite dimensional algebras. We call the two types of indecomposable modules "deleted cycle" and "block cycle" modules. These are known to the finite dimensional algebra community as "string" and "band" modules, respectively.

Our description gives a bit of new insight into the reason for their biserial nature: Consistent with our point-of-view that fingen(Γ) is an approximation to fingen(Λ), we begin with a direct sum $X = \bigoplus_i X_i$ of indecomposable modules in fingen(Γ). Each X_i is uniserial as a Γ-module, since Γ is the direct sum of two DVRs. Pairs of these modules X_i can then be glued together at the top and—when they have finite length—at the bottom. When this gluing is done appropriately, it results in the deleted cycle (string) and block cycle (band) modules.

When k is not algebraically closed, this gluing essentially amounts to another way of viewing the original Nazarova-Roiter results. We omit further details of the gluing here, since they are similar to the details in the unsplit case, which we discuss more fully below. The full details can be found in [KL2, §3].

1-parameter families 7.5. Readers familiar with representations of finite dimensional algebras are probably wondering about the role of 1-parameter families here. The block cycle (band) modules are indeed organized into 1-parameter families, as expected. But since the residue field k of Λ is not algebraically closed (and hence the Jordan canonical form is no longer available), the parameter is no longer an element of the residue field k of Λ. Instead, the parameter becomes a class of matrices that is indecomposable under similarity. Alternatively, by using the rational canonical form, the parameter can be taken to be a power of an irreducible polynomial in $k[x, x^{-1}]$.

However, all indecomposable modules of infinite length in $\mathrm{fingen}(\Lambda)$ are deleted cycle (string) modules, and therefore never have a corresponding "parameter". Thus, after separating $\mathrm{fingen}(\Lambda)$ into the two categories $\mathrm{finlen}(\Lambda)$ and $\mathrm{fingen}_\infty(\Lambda)$, and noting that $\mathrm{finlen}(\Lambda)$ is really part of the complete local case, these 1-parameter families play no further role in the nonlocal situation studied in [KL3].

Unsplit Case. 7.6. Here, Λ and Γ are both integral domains, and hence the normalization Γ of Λ is a DVR. The residue field F of Γ is a 2-dimensional separable extension of the residue field k of Λ (in the non-exceptional case). This situation, of course, does not occur when k is algebraically closed. It might be that our results are new here, even for finite dimensional Λ-modules when Λ is a k-algebra. The prototypical example is $k = \mathbb{R}$ and $F = \mathbb{C}$, and $\Lambda = \mathbb{R} + x\mathbb{C}[[x]]$. We give a detailed description of the resulting indecomposable modules in $\mathrm{fingen}(\Lambda)$ in this example in Section 8 below. The results in the general unsplit case are only slightly more complicated.

Again we obtain our indecomposable Λ-modules as certain combinations of indecomposable Γ-modules, the latter being uniserial because Γ is a DVR.

Klein Rings 7.7. To complete the complete local picture, we need to comment on Klein rings. Let $(\Omega, \mathfrak{n}, k)$ be a Klein ring. Then [KL2, Theorem 11.2]:

(i) Ω is a quasi-Frobenius ring with simple socle $\mathfrak{n}^2$ and k has characteristic 2.

(ii) $\Upsilon = \Omega/\mathfrak{n}^2$ is a homomorphic image of a strictly split (therefore non-exceptional, and hence fingen-tame) complete local Dedekind-like ring Λ.

(iii) Every Ω-module is the direct sum of a free module and an Υ-module.

Perhaps the most tantalizing fact about Klein rings is the following [KL1, Theorems 5.2, 5.1]. *A Klein ring $(\Omega, \mathfrak{n}, k)$ is a homomorphic image of some Dedekind-like ring (say) Λ if and only the residue field k is an imperfect field. When the condition holds, the Dedekind-like ring Λ must be exceptional* (in which case we do not know whether Λ is tame, wild, or neither, although Ω is necessarily tame!).

Link to Non-local Case. 7.8. The reduction to the complete local case [KL3] is obtained by intensive use of Corollary 6.4, and the resulting arguments are unfortunately quite long, as is the complete local case itself, in [KL2]. Perhaps someone with a fresh approach to the subject can find a simpler path to our results.

8. Module Structure, One Special (Complete Local) Case

Notation 8.1. In this section $(\Lambda, \mathfrak{m}, k)$ denotes the complete local unsplit Dedekind-like ring displayed in (8.1.1), together with its normalization Γ and some other details.

$$\Lambda = \mathbb{R} + x\mathbb{C}[[x]], \qquad \Gamma = \mathbb{C}[[x]]$$
$$\mathfrak{m} = x\mathbb{C}[[x]] = \Lambda x + \Lambda ix, \qquad \Lambda/\mathfrak{m} = \mathbb{R}, \quad \Gamma/\mathfrak{m} = \mathbb{C} \tag{8.1.1}$$

Since *the Krull-Schmidt theorem holds for finitely generated modules* in the complete local case, we may limit our description of module structure to the structure of indecomposable modules in fingen(Λ). We describe these modules in four steps, starting with the simple, well-known structure of Γ-modules. See [KL2, §2] for more generality (beyond this example), as well as more complete details.

8.2. *Step 1: Γ-modules*

Since Γ is a DVR, all indecomposable modules in fingen(Γ) are uniserial and their isomorphism class is determined by their length: a positive integer or

∞. In more detail, the indecomposable Γ-module of any length d is

$$\tilde{\Gamma}_d := \Gamma/\mathfrak{m}^d \tag{8.2.1}$$

where we make the convention that $\mathfrak{m}^\infty = 0$ and hence $\tilde{\Gamma}_\infty = \Gamma$. When d is finite we can conveniently visualize $\tilde{\Gamma}_d$ by setting $x^d = 0$. More formally:

$$\tilde{\Gamma}_d = \mathbb{C} + \mathbb{C}\tilde{x}_d + \mathbb{C}\tilde{x}_d^{\,2} + \cdots + \mathbb{C}\tilde{x}_d^{\,d-1} \qquad (\tilde{x}_d^{\,d} = 0,\ d < \infty) \tag{8.2.2}$$

where $\tilde{x}_d$ is the coset $x + \mathfrak{m}^d$. We refer to elements of $\tilde{\Gamma}_d$ as *truncated power series* when $d \neq \infty$.

Start with a direct sum X of indecomposable Γ-modules, for example the Γ-module X shown in (8.2.3) where each vertical bar represents one of the given indecomposable summands of X.

$$X = \tilde{\Gamma}_\infty \oplus \tilde{\Gamma}_4 \oplus \tilde{\Gamma}_5 \oplus \tilde{\Gamma}_6 : \qquad \overset{\infty}{\big|}\ \ \overset{4}{\big|}\ \ \overset{5}{\big|}\ \ \overset{6}{\big|} \tag{8.2.3}$$

We call the number over each vertical bar its *length label*.

8.3. *Step 2: gluing and reduction*

The critical fact is that the uniserial Γ-modules $\tilde{\Gamma}_d$ are never uniserial Λ-modules. We repeatedly use the facts that the Γ-top and (when $d < \infty$) Γ-bottom of $\tilde{\Gamma}_d$ have the following form.

$$\begin{array}{ll} (\text{Top } \mathbb{C} \text{ of } \tilde{\Gamma}_d:) & \Gamma/\mathfrak{m} = \mathbb{C} \\ (\text{Bottom } \mathbb{C} \text{ of } \tilde{\Gamma}_d,\ d < \infty:) & \mathfrak{m}^{d-1}/\mathfrak{m}^d = \mathbb{C}\tilde{x}_d^{\,d-1} \cong \mathbb{C} \end{array} \tag{8.3.1}$$

where the "equality" in the first line is the identification induced by mapping each power series or truncated power series to its constant term, and the isomorphism at the end of the second line is the map $\alpha \cdot \tilde{x}_d^{\,d-1} \to \alpha$ ($\alpha \in \mathbb{C}$). Thus each of these (simple) Γ-tops and Γ-bottoms has dimension 2 as an $\mathbb{R}$-vector space, and hence is the direct sum of two isomorphic simple Λ-modules.

We use these facts to build new Λ-modules in four ways, as shown in (8.3.2) and described in detail below that.

$$\begin{array}{cccc} \text{bottom-glue} & \text{top-glue} & \begin{array}{c}\text{bottom-}\\\text{reduce}\end{array} & \begin{array}{c}\text{top-}\\\text{reduce}\end{array} \\[4pt] j \qquad i & i \qquad j & i & i \\ \bigsqcup & \bigsqcap & \underline{\big|} & \overline{\big|} \end{array} \tag{8.3.2}$$

We denote the complex conjugate of an element $\alpha \in \mathbb{C}$ by $\bar{\alpha}$.

Bottom-glue: (when $i, j < \infty$) This is the Λ-module $(\tilde{\Gamma}_j \oplus \tilde{\Gamma}_i)/K$ where K is the set of ordered pairs $(\alpha \cdot \tilde{x}_j{}^{j-1}, \bar{\alpha} \cdot \tilde{x}_i{}^{i-1})$ with $\alpha \in \mathbb{C}$. Less formally, this amalgamates the bottom $\mathbb{C}$ of $\tilde{\Gamma}_j$ with that of $\tilde{\Gamma}_i$ by identifying $\alpha \cdot \tilde{x}_j{}^{j-1}$ with $-\bar{\alpha} \cdot \tilde{x}_i{}^{i-1}$. This amalgamation is $\mathbb{R}$-linear but not $\mathbb{C}$-linear, and hence is Λ-linear but not Γ-linear. Hence it defines a Λ-module but not a Γ-module.

Top-glue: This is the Λ-submodule of $\tilde{\Gamma}_i \oplus \tilde{\Gamma}_j$ consisting of all pairs $\big(f(\tilde{x}_i), g(\tilde{x}_j)\big) \in \tilde{\Gamma}_i \oplus \tilde{\Gamma}_j$ such that the constant term of $f(\tilde{x}_i)$ equals the complex conjugate of the constant term of $g(\tilde{x}_j)$. As before, this Λ-module is not a Γ-module.

Bottom-reduce: (when $i < \infty$) This diagram represents the Λ-module $\tilde{\Gamma}_i/\mathbb{R}\tilde{x}_i{}^{i-1}$, which makes sense since the Γ-socle of $\tilde{\Gamma}_i$ equals $\mathbb{C}x_i^{i-1}$, as is evident from (8.3.1). Again, this is not a Γ-module.

Top-reduce: This diagram represents the Λ-submodule of $\tilde{\Gamma}_i$ consisting of all $f(\tilde{x}_i) \in \tilde{\Gamma}_i$ whose constant term is real.

The two gluing operations look very much like the familiar string modules. The top and bottom reductions, however, are a new complication not present in the strictly split case.

8.4. *Steps 3 and 4: Combine these gluings and reductions,*

building the indecomposables described in steps 3 and 4. In each case we obtain the desired Λ-module $M(\mathcal{D})$ in the form $M(\mathcal{D}) = S/K$ from a displayed diagram $\mathcal{D}$ and Γ-module X. In step 3, the numerator S is the direct sum of Λ-submodules of X consisting of one term for each top-gluing and one for each top-reduction displayed in $\mathcal{D}$. The denominator K is the direct sum of one Λ-submodule of S for each bottom-gluing and one for each bottom-reduction displayed in $\mathcal{D}$. The same basic idea applies to step 4, but this is done in a more complicated "block form".

Before proceeding to the details of these steps, we note the following fact, which is proved at the beginning of [KL2, subsection 9.6].

Theorem 8.5. *In steps 3 and 4, the natural map $S \twoheadrightarrow S/K = M(\mathcal{D})$ is always a Γ-separated cover of $M(\mathcal{D})$.*

8.6. *Step 3: First series of combinations*

As shown in (8.6.1), begin with a direct sum $X = \tilde{\Gamma}_{i_1} \oplus \tilde{\Gamma}_{j_1} \oplus \tilde{\Gamma}_{i_2} \ldots$, and combine the terms by alternately top and bottom-gluing. Then either stop, or do a top- or bottom-reduction *at the right-hand end.*

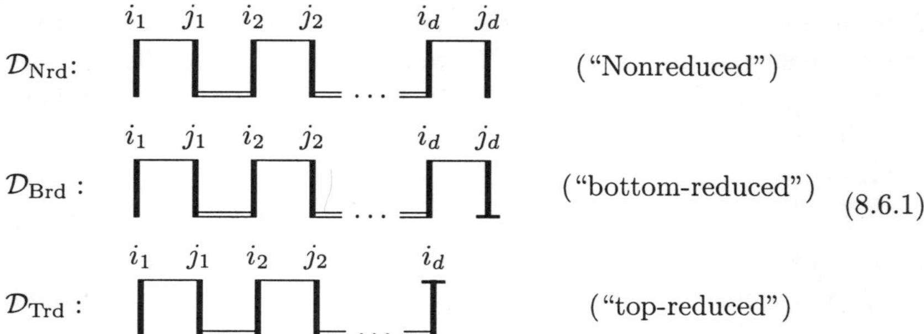

$$\mathcal{D}_{\mathrm{Nrd}}: \qquad\qquad\qquad\qquad\qquad \text{(``Nonreduced'')}$$

$$\mathcal{D}_{\mathrm{Brd}}: \qquad\qquad\qquad\qquad\qquad \text{(``bottom-reduced'')} \qquad (8.6.1)$$

$$\mathcal{D}_{\mathrm{Trd}}: \qquad\qquad\qquad\qquad\qquad \text{(``top-reduced'')}$$

Each top-gluing results in replacing a direct sum $\tilde{\Gamma}_{i_c} \oplus \tilde{\Gamma}_{j_c}$ by a Λ-submodule S_c, and each top-reduction results in replacing some $\tilde{\Gamma}_{i_c}$ by a Λ-submodule S_c. (The top-reduction occurs in only one of these diagrams.) Let $S = \bigoplus_c S_c$.

Each bottom-gluing is given by a Λ-submodule K_c of the bottom $\mathbb{C} \oplus \mathbb{C}$ of some $\tilde{\Gamma}_{j_c} \oplus \tilde{\Gamma}_{i_c}$, and each bottom-reduction is given by a Λ-submodule K_c of some $\tilde{\Gamma}_{j_c}$. (The bottom-reduction occurs in only one of these diagrams.) Let $K = \bigoplus_c K_c$.

If the conditions in the next remark are satisfied, we call $\mathcal{D}$ a *standard diagram.* We set $M(\mathcal{D}) = S/K$.

Remarks 8.7 (Restrictions in (8.6.1)). (i) In order for the above combinations to make sense: *The lengths of all uniserial Γ-modules in these diagrams must be finite except possibly for i_1 and, in the nonreduced case, j_d.* This holds because a Γ-module $\tilde{\Gamma}_e$ can be involved in bottom-gluing or bottom-reduction only if its length e is finite.

(ii) *The only situations in which we allow a uniserial Γ-module $\tilde{\Gamma}_i$ to equal the length 1 module $\tilde{\Gamma}_1$ in these diagrams is i_1 and, in the nonreduced case, j_d.* Ignoring this restriction results in construction of Λ-modules that can be constructed by not ignoring the restriction. [See (8.9.1) for a non-standard example.] Thus this restriction is included only to simplify the statement of our uniqueness Theorem 8.12.

Examples 8.8. (i) We start with the Γ-module X shown in (8.2.3), and build the Λ-module $M = M(\mathcal{D})$ determined by the nonreduced diagram in (8.8.1).

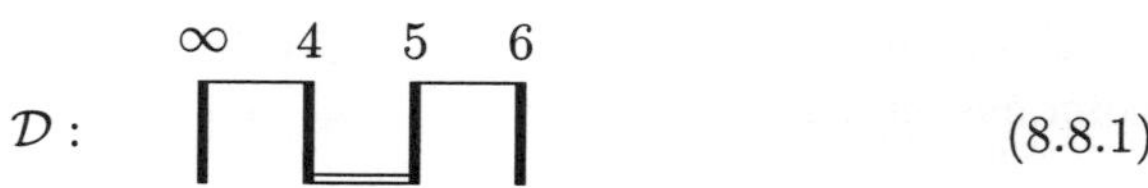

$$\mathcal{D}: \qquad\qquad\qquad\qquad\qquad\qquad (8.8.1)$$

Here, X is the set of 4-tuples:

$$(q, r, s, t) = \big(q(\tilde{x}_\infty), r(\tilde{x}_4), s(\tilde{x}_5), t(\tilde{x}_6)\big) \in \ X = \tilde{\Gamma}_\infty \oplus \tilde{\Gamma}_4 \oplus \tilde{\Gamma}_5 \oplus \tilde{\Gamma}_6$$

$$(8.8.2)$$

Note that $q = q(\tilde{x}_\infty)$ is an actual power series, and so we could have written $q = q(x)$. But r, s, t are truncated power series.

We have $S = S_1 \oplus S_2$ where S_1 is the set of elements (q, r) of $\tilde{\Gamma}_\infty \oplus \tilde{\Gamma}_4$ such that the constant term of q equals the conjugate of that of r, and S_2 is the set of elements of (s, t) of $\tilde{\Gamma}_5 \oplus \tilde{\Gamma}_6$ such that the constant term of s equals the conjugate of that of t.

$K = K_1$, a single term since $\mathcal{D}$ has only one bottom-gluing and no bottom-reduction. In fact, K_1 is the set of elements $(0, \alpha{\cdot}\tilde{x}_4{}^3, \bar{\alpha}{\cdot}\tilde{x}_5{}^4, 0)$ where α ranges through the complex numbers.

Thus we have defined $M(\mathcal{D}) = S/K$.

(ii) Suppose that, in addition to the gluing in part (i), we want a bottom-reduction at the right-hand end of the diagram. Then S is the same as in part (i). But $K = K_1 \oplus K_2$ where K_1 is as in part (i) and K_2 is the Λ-submodule $\mathbb{R}\tilde{x}_6{}^5)$ of $\tilde{\Gamma}_6$.

Examples 8.9. Before proceeding, we explicitly display the unique *standard diagrams*—that is, those of the form (8.6.1)—corresponding to the Λ-modules Λ, Γ, and $\mathbb{R} = \Lambda/\mathfrak{m}$. We also include one nonstandard diagram, in order to illustrate the kind of duplication that can occur if nonstandard diagrams are allowed.

$$\Lambda : \ \overset{\infty}{\big|} \qquad \Gamma : \ \overset{\infty}{\big|} \qquad \mathbb{R} : \ \overset{1}{\big|} \qquad \mathbb{R} \ (\text{nonstandard}) : \ \overset{1}{\big|}$$

$$(8.9.1)$$

Theorem 8.10. *Consider the Λ-modules of types (8.6.1).*

(i) *A nonreduced module is indecomposable if and only if the corresponding diagram $\mathcal{D}_{\mathrm{Nrd}}$ does not equal its left-right mirror image.*

(ii) *The reduced modules are all indecomposable.*

(iii) *Every indecomposable module of infinite length in* fingen(Λ) *is of exactly one of these types.*

This theorem is part of [KL2, Theorem 2.7], proved in [KL2, §9].

In connection with statement (iii) above, note that indecomposable modules of the types shown in (8.6.1) can also have finite length. This happens if and only if no uniserial Γ-module of length ∞ occurs in the diagram.

For examples that illustrate Theorem 8.10, note that the modules M constructed in Examples 8.8 are indecomposable, but the Λ-module determined by (8.11.1) is decomposable because it equals its left-right mirror image.

Corollary 8.11. *Every indecomposable module of infinite length in* $\mathrm{fingen}(\Lambda)$ *has torsionfree rank* ≤ 2. *If the rank equals* 2, *then the module cannot be torsionfree.*

Proof. The torsionfree rank of the module is easily seen to be the number of indices i_k and j_k that equal ∞. That this number is ≤ 2 is an immediate consequence of Theorem 8.10 and Remarks 8.7. The reason for the second (surprising!) statement of the corollary is that rank 2 can occur only in the nonreduced diagram $\mathcal{D}_{\mathrm{Nrd}}$ with $i_1 = j_d = \infty$. In this situation, if the Λ-module were torsionfree, then no $\tilde{\Gamma}_e$ could have finite length, so the diagram would have the form

$$\overset{\infty \qquad \infty}{\sqcap} \tag{8.11.1}$$

and hence equal its left-right reflection. Therefore, by Theorem 8.10, the module would be decomposable. $\qquad\qquad\square$

Theorem 8.12. *Two* Λ-*modules constructed from standard diagrams* $\mathcal{D}$ *and* $\mathcal{D}'$ *of the form* (8.6.1) *are isomorphic if and only if either:*

(i) $\mathcal{D} = \mathcal{D}'$; *or*
(ii) $\mathcal{D}$ *and* $\mathcal{D}'$ *are nonreduced and* $\mathcal{D}'$ *is the left-right mirror image of* $\mathcal{D}$.

This theorem is part of [KL2, Theorem 2.8], proved in [KL2, §9.3].

8.13. *Step 4: Second series of combinations*

Each Λ-module $M(\mathcal{D}) = S/K$ in this series is determined by one of the diagrams in (8.13.1). These diagrams display gluings, reductions, and an $m \times m$ invertible matrix U over $\mathbb{C}$, called the *blocking matrix*. We call m the *block size*. Each gluing and reduction in this series involves $2m$ or m uniserial Γ-modules, respectively; and we require every length label i_c and j_c to be both $\neq \infty$ and $\neq 1$, in order for the construction of $M(\mathcal{D})$ to make sense. In particular, the modules $M(\mathcal{D})$ in this series all have finite length.

As suggested by their labels, we call these diagrams *Bottom-bottom-reduced, Bottom-Top-reduced, Top-Top-reduced, and Cycle,* respectively.

If the restrictions described in Remarks 8.16 are satisfied, we call $\mathcal{D}$ a *standard diagram.* These somewhat technical conditions must be imposed in order to insure indecomposability. (But the construction of $M(\mathcal{D})$ makes sense without these additional restrictions.)

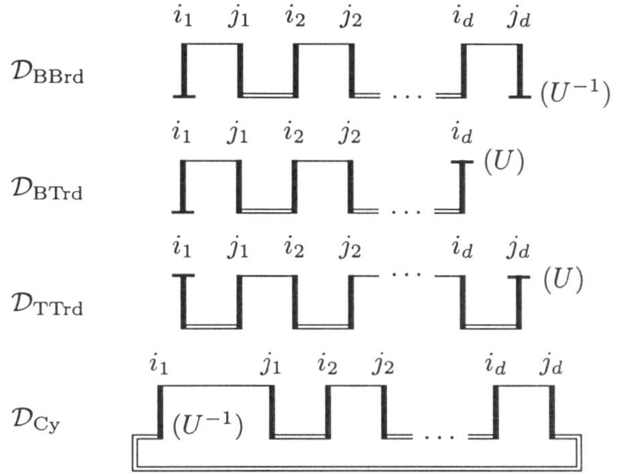

$$
\begin{array}{ll}
\mathcal{D}_{\mathrm{BBrd}} & \\
\mathcal{D}_{\mathrm{BTrd}} & \\
\mathcal{D}_{\mathrm{TTrd}} & \\
\mathcal{D}_{\mathrm{Cy}} &
\end{array}
\tag{8.13.1}
$$

Construction of $M(\mathcal{D})$. Let $\mathcal{D}$ and U (of size $m \times m$) be given. The Γ-module from which we build $M(\mathcal{D}) = S/K$ is:

$$
X = \tilde{\Gamma}_{i_1}^{(m)} \oplus \tilde{\Gamma}_{j_1}^{(m)} \oplus \ldots \oplus \tilde{\Gamma}_{i_d}^{(m)} \oplus \tilde{\Gamma}_{j_d}^{(m)}
\tag{8.13.2}
$$

Let $S = \bigoplus_c S_c$ where each S_c is (arbitrarily) associated with one of the top-gluings or top-reductions in $\mathcal{D}$, and is as defined below.

Top-glue. Suppose that a top-gluing edge connects $\tilde{\Gamma}_h$ to $\tilde{\Gamma}_k$ in $\mathcal{D}$. Then the corresponding S_c is the Λ-submodule of $\tilde{\Gamma}_h^{(m)} \oplus \tilde{\Gamma}_k^{(m)}$ consisting of all $2m$-tuples $(p_1, \ldots, p_m; q_1, \ldots, q_m)$ such that the constant term of each p_a is the complex conjugate of that of q_a.

Top-reduce. Suppose that a top-reduction occurs in $\mathcal{D}$ at some $\tilde{\Gamma}_h$ with U at its top. Then the corresponding S_c is the Λ-submodule of $\tilde{\Gamma}_h^{(m)}$ consisting of of all m-tuples $(p_1, \ldots, p_m)$ such that:

$$
\big(p_1(0), p_2(0), \ldots, p_m(0)\big) \in \mathbb{R}^{(m)} U \quad (\mathbb{R}\text{-row space of } U)
\tag{8.13.3}
$$

If, on the other hand, no matrix is attached to the top of $\tilde{\Gamma}_h$ in $\mathcal{D}$, then modify the corresponding S_c by replacing U by the identity matrix in (8.13.3) (thus getting the direct sum of m ordinary top-reductions). Let $K = \bigoplus_c K_c$

where each K_c is (arbitrarily) associated with one of the bottom-gluings or bottom-reductions in $\mathcal{D}$, as defined below.

Bottom-glue. Suppose that a bottom-gluing edge connects $\tilde{\Gamma}_h$ to $\tilde{\Gamma}_k$ in $\mathcal{D}$, and U^{-1} is attached to the bottom of $\tilde{\Gamma}_h$. Then the corresponding K_c consists of all ordered pairs (each entry of which is an m-tuple) of the form:

$$(\boldsymbol{\alpha} U^{-1}\tilde{x}_h{}^{h-1}, \bar{\boldsymbol{\alpha}}\tilde{x}_k{}^{k-1}) \in \mathbb{C}^{(m)}\tilde{x}_h{}^{h-1} \oplus \mathbb{C}^{(m)}\tilde{x}_k{}^{k-1} \subseteq \tilde{\Gamma}_h^{(m)} \oplus \tilde{\Gamma}_k^{(m)} \quad (8.13.4)$$

If, on the other hand, the bottom of $\tilde{\Gamma}_h$ is not attached to a matrix, modify the corresponding K_c by replacing U by the identity matrix in (8.13.4) (and getting the direct sum of m ordinary bottom-gluings).

Bottom-reduce. Suppose that a bottom-reduction occurs at some $\tilde{\Gamma}_h$ in $\mathcal{D}$ with U^{-1} at its bottom. Then the corresponding K_c equals the Λ-submodule $\mathbb{R}^{(m)}U^{-1}\tilde{x}_h{}^{h-1}$ of $\tilde{\Gamma}_h^{(m)}$.

If no matrix is attached to the bottom of $\tilde{\Gamma}_h$, then the corresponding K_c is the Λ-submodule $\mathbb{R}^{(m)}\tilde{x}_h{}^{h-1}$ of $\tilde{\Gamma}_h^{(m)}$ (the direct sum of m ordinary bottom-reductions).

Abuse of notation. We do not require all of the numbers i_a and j_b in (8.13.1) to be distinct. Consequently, when repetition occurs, it is necessary to refer to the diagram to identify which summand of X the notation $\tilde{\Gamma}_h$ refers to. See Examples 8.14 for an instance of this.

Examples 8.14. (i) We construct the bottom-bottom reduced Λ-module $M = M(\mathcal{D}) = S/K$ determined by diagram (8.14.1) and a 3×3 blocking matrix U.

$$\mathcal{D}_{\mathrm{BBrd}} : \quad \overset{\displaystyle 8 \quad 4 \quad 5 \quad 4}{\sqcap\sqcup\sqcap}{}_{(U^{-1})} \quad (8.14.1)$$

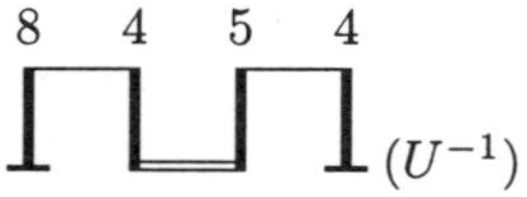

The Γ-module that we start with is:

$$X = \tilde{\Gamma}_8^{(3)} \oplus \tilde{\Gamma}_4^{(3)} \oplus \tilde{\Gamma}_5^{(3)} \oplus \tilde{\Gamma}_4^{(3)} \quad (8.14.2)$$

We have $S = S_1 \oplus S_2$ where S_1 and S_2 result from the first and second displayed top-gluings respectively. Thus, for example, S_1 is the set of 6-tuples $(p_1, p_2, p_3; q_1, q_2, q_3)$ in $\tilde{\Gamma}_8^{(3)} \oplus \tilde{\Gamma}_4^{(3)}$ such that the constant term of each p_i is the conjugate of that of q_i. (Note that one must look at the diagram to determine which of the two copies of $\tilde{\Gamma}_4^{(3)}$ we are referring to.)

We have $K = K_1 \oplus K_2 \oplus K_3$, where successive terms K_i refer to the three displayed bottom operations. In particular, K_3 is the Λ-submodule $\mathbb{R}^{(3)}U^{-1}\tilde{x}_4{}^3$ of the second $\tilde{\Gamma}_4^{(3)}$ displayed in (8.14.2).

(ii) We construct the cycle module $M = M(\mathcal{D})$ determined by diagram (8.14.3) and a 3×3 blocking matrix U.

$$\mathcal{D}_{\text{Cy}} : \qquad \overset{\displaystyle 8 \qquad\qquad 4 \quad 5 \quad 4}{\boxed{(U^{-1})}} \tag{8.14.3}$$

We have $S = S_1 \oplus S_2$, corresponding to the two top-gluings, and $K = K_1 \oplus K_2$ corresponding to the two bottom-gluings. In fact, S is the same as in part (i). We explicitly display the term K_2 corresponding to the bottom-gluing of $\tilde{\Gamma}_8^{(3)}$ and (the second displayed) $\Gamma_4^{(3)}$. By (8.13.4), K_2 is the set of elements $(\alpha U^{-1} \tilde{x}_8{}^7, \bar{\alpha} \tilde{x}_4{}^3)$ in $\tilde{\Gamma}_8^{(3)} \oplus \tilde{\Gamma}_4^{(3)}$ where α ranges through $\mathbb{C}^{(3)}$.

Remark 8.15 (Moving U). The placement of U in our standard diagrams is somewhat arbitrary. It is possible to define $M(\mathcal{D})$ with its matrix attached to the top or bottom of any vertical bar in $\mathcal{D}$. Then, for example, U can be moved from the top of any vertical bar to the bottom (or vice-versa), if we replace it by U^{-1}. This explains our use of U at the top and U^{-1} at the bottom in (8.13.1). For the complete rules about moving U, see [KL2, Proposition 2.9].

Remarks 8.16 (Restrictions in (8.13.1)). We have already stated that every length label must be $\neq \infty$ and $\neq 1$. We call $\mathcal{D}$ *standard* if it satisfies the following additional conditions.

Except for cycle diagrams we require:

(i) $U\bar{U}^{-1}$ is indecomposable under similarity; and
(ii) The sequence $\{i_1, i_2, i_3, \ldots, j_3, j_2, j_1\}$ does not equal some strictly shorter sequence repeated some number of times.

Note that condition (i), above, is stronger than indecomposability of U itself. For example, if U is any invertible matrix with entries in $\mathbb{R}$, then $U\bar{U}^{-1} = I$. For a proof that matrices of all sizes exist satisfying condition (i). see Remark 8.17.

For cycle diagrams we require all of the following three conditions.

(i) U is indecomposable under similarity.
(ii) Either:
    (a) U is not similar to $\bar{U}^{-1}$, or
    (b) The sequence $\{j_1, j_2, \ldots, j_d\}$ does not equal a cyclic shift of the sequence $\{i_d, \ldots, i_2, i_1\}$ (i.e., relocation of some subsequence from the beginning to the end).

(iii) The sequence of pairs $(i_1, j_1), \ldots, (i_d, j_d)$ does not equal some strictly shorter sequence repeated some number of times.

Remark 8.17 (Indecomposable matrices $U\bar{U}^{-1}$). *For every m there is an $m \times m$ matrix U over $\mathbb{C}$ such that $U\bar{U}^{-1}$ is indecomposable under similarity.*

Proof. Let W be the companion matrix of the polynomial $(x+1)^m$. Then W is indecomposable under similarity. With the help of the Cayley-Hamilton theorem, one can show that W is similar to $\bar{W}^{-1}$. See [KL2, (2.12.1)] for some details. A "Hilbert Theorem 90 for matrices" due to Ballantine [Ba, Lemma 8.11] states that this last similarity condition for a matrix W is equivalent to the existence of a matrix U such that $U\bar{U}^{-1} = W$. $\qquad\square$

We are grateful to Robert Guralnick for showing us this theorem of Ballantine.

The remaining theorems in this section complete our description of the indecomposable modules in fingen(Λ), for the ring $\Lambda = \mathbb{R} + x\mathbb{C}[[x]]$ in (8.1.1). These results are special cases of [KL2, Theorem 2.8], proved in [KL2, Theorem 8.18].

Theorem 8.18. *Let $\mathcal{D}$ be any of the diagrams in (8.13.1) (i.e., involving a blocking matrix). If $\mathcal{D}$ is standard [i.e., satisfies the conditions in Remarks 8.16] then $M(\mathcal{D})$ is indecomposable.*

Theorem 8.19. *Every indecomposable module in fingen(Λ) is isomorphic to $M(\mathcal{D})$ for some unique type of standard diagram (i.e., bottom-reduced, bottom-bottom reduced, ...) in one of the series (8.6.1) and (8.13.1).*

Theorem 8.20 (Isomorphism: except cycle diagrams). *Let $\mathcal{D}$ be one of the standard diagrams $\mathcal{D}_{\text{BBrd}}$, $\mathcal{D}_{\text{BTrd}}$, $\mathcal{D}_{\text{TTrd}}$, and U its blocking matrix. The isomorphism invariants of $M(\mathcal{D})$ are:*

(i) *The similarity class of $U\bar{U}^{-1}$; and*

(ii) (a) *(BTrd diagrams:) The sequences $\{i_k\}$ and $\{j_k\}$.*

   (b) *(Other two diagrams:) The sequences $\{i_k\}$ and $\{j_k\}$ modulo left-right reflection of the diagram. That is, interchanging the i-sequence with the j-sequence and then reversing each of them does not change the isomorphism class of $M(\mathcal{D})$.*

In the next theorem, $\mu(\ldots)$ ("mirror image") denotes the reversal of any finite sequence, and let $\nu(\ldots)$ denotes a cyclic-shift of a finite sequence (i.e., move some subsequence from the beginning to the end).

Theorem 8.21 (Isomorphism: cycle diagrams). *Let $\mathcal{D}, \mathcal{D}'$ be cycle diagrams with blocking matrices U, V respectively. Call the two associated length-label sequences of the first diagram I, J, and those of the second diagram I', J'.*

Then $M(\mathcal{D}) \cong M(\mathcal{D}')$ if and only if:

(i) *V is similar to U and $I' = \nu(I)$ and $J' = \nu(J)$ for some ν; or*
(ii) *V is similar to $\bar{U}^{-1}$, and $I' = \nu\mu(J)$ and $J' = \nu\mu(I)$ for some ν.*

Note that the sequence manipulations in Theorem 8.21(i) and (ii) correspond to obvious symmetries of the diagram (rotation and left-right reflection).

9. Epilog on the Concept "Dedekind-like"

We are grateful to one of the referees for asking why we use the term "Dedekind-like" to refer to rings that seem to have little to do with Dedekind domains. As he (correctly) observes: Their ideal structure, their homological properties, and their representation theory are all significantly more elaborate. Moreover, Chapter 5 of McConnell and Robson's well-known textbook *Noncommutative Noetherian Rings* [MR] is entitled "Some Dedekind-like Rings", terminology that seems to conflict with the present usage since the rings dealt with in that chapter are mostly hereditary noetherian prime rings, the most natural noncommutative noetherian generalization of commutative Dedekind domains.

We think of Dedekind-like behavior as direct-sum behavior that is controlled by two types of invariants: *counting*-invariants, and *group* invariants. In the case of Dedekind-like rings Λ, the counting-invariants come from the fact that, for each maximal ideal of Λ, the category $\operatorname{fingen}(\hat{\Lambda}_{\mathfrak{m}})$ is a Krull-Schmidt category. The group invariants come from the web of genus class groups [Definition 5.4].

These ideas come from Steinitz's original papers on modules over Dedekind domains [S], although it was not immediately evident that his invariants could be expressed in the form in which we view them. Let Λ be a Dedekind domain, a special case of Dedekind-like rings. Then every module in $\operatorname{fingen}(\Lambda)$ has a decomposition $P \oplus T$ where P is projective and T ("torsion") has finite length, and the two terms are unique up to isomorphism. Since $\operatorname{finlen}(\Lambda)$ is a Krull-Schmidt category, only counting-invariants occur in the description of T. Steinitz's really new idea involved the projective term P. There is a decomposition $P \cong \bigoplus_{i=1}^{n} M_i$ where each M_i is a nonzero ideal of Λ. Moreover, for any two finite families of nonzero ideals M_i, N_i of Λ we have

$$\bigoplus_{i=1}^{m} M_i \cong \bigoplus_{i=1}^{n} N_i$$
$$\Longleftrightarrow$$
$$m = n \quad \text{and} \quad \sum_i [M_i] = \sum_i [N_i] \quad \text{in } \mathcal{G}(\Lambda) \tag{9.0.1}$$

This becomes a special case of (5.3.2) as soon as one realizes that all nonzero ideals of Λ are in genus(Λ), because all localizations and completions of Λ at maximal ideals are principal ideal domains. Thus, in Steinitz's situation, our whole web of class groups collapses to the genus class group of the ring, more classically called the "ideal class group". More precisely, we can ignore all genus class groups except for $\mathcal{G}(\Lambda)$, because of (9.0.1).

The idea of genus class group occurs elsewhere in the literature, as mentioned earlier. One of the main new ideas in the present series of papers is that one can "add" elements of different class groups, by making use of the web of class groups—in order to describe all direct-sum relations over an arbitrary Dedekind-like ring.

Now consider (noncommutative!) HNP (hereditary noetherian prime) rings Λ. Most of the rings in McConnell-Robson's Chapter 5, "Some Dedekind-like rings" are of this type. The name refers to the fact that, as in the commutative case, every module in fingen(Λ) has a unique decomposition $P \oplus T$, up to isomorphism, where P is projective and T has finite length. Moreover, there is a decomposition $P \cong \bigoplus_{i=1}^{n} M_i$ where each M_i is a *uniform right ideal* of Λ (i.e., the intersection of any two nonzero submodules is again nonzero).

But what no-one seemed to suspect at the time is that Dedekind-like behavior—in the sense of the present paper—occurs here too, in the following form. It is possible to define a (nonunique) *normalization* Γ of Λ, a "genus class group" $\mathcal{G}(\Gamma)$, a somewhat "natural image" $[M]$ of the isomorphism class of every uniform right ideal M, in $\mathcal{G}(\Gamma)$, and *ranks* $\rho_{\mathfrak{m}}$ *at nonzero maximal ideals* $\mathfrak{m}$, in such a way that, for any two finite families, each consisting of *two or more* uniform right ideals of Λ, we have

$\bigoplus_{i=1}^{m} M_i \cong \bigoplus_{i=1}^{n} N_i$ if and only if the following conditions hold.

$$
\begin{aligned}
&\text{(i) } m = n; \\
&\text{(ii) } \sum_{i} \rho_{\mathfrak{m}}(M_i) = \sum_{i} \rho_{\mathfrak{m}}(N_i) \text{ for every } \mathfrak{m}; \text{ and} \\
&\text{(iii) } \sum_{i} [M_i] = \sum_{i} [N_i] \quad \text{in } \mathcal{G}(\Gamma)
\end{aligned}
\tag{9.0.2}
$$

Well-known counterexamples, dealing with "stable isomorphism" versus actual isomorphism show that the theorem fails unless the direct sums contain at least two terms. For details see Levy and Robson's papers [LR1, LR2]. HNP rings seem to be the only noncommutative noetherian rings whose projective modules exhibit nontrivial direct-sum behavior and possess a theorem describing that behavior.

A subsequent paper of Levy and Robson [LR3] gives a structure theorem for infinitely generated projective modules over HNP rings, again possibly the only noncommutative noetherian rings whose projective modules exhibit nontrivial direct-sum behavior and have a structure theorem describing that behavior. The main difference between this and the finitely generated case is that only conditions (i) and (ii) apply; that is, class groups disappear, and direct-sum behavior is determined by counting-invariants alone. The fact that class groups do not occur in the infinitely generated situation had been noted long ago, for commutative Dedekind domains, by Kaplansky [Ka]. He showed that, for commutative Dedekind domains, all nonfinitely generated projectives are free. Thus nontrivial direct-sum behavior does not occur until one considers noncommutative HNP rings.

There is also an interesting relationship between HNP rings and the tame-wild phenomenon. The category finlen(Λ) over an HNP ring Λ is a Krull-Schmidt category, and hence no nontrivial direct-sum behavior occurs. However, no description has been given of the indecomposable modules in finlen(Λ). Klingler and Levy [KL4] explain this by giving an example of an HNP ring Λ such that the category finlen(Λ) has wild representation type—strictly wild in this case. (Λ is the Weyl algebra $A_1(k)$ over an arbitrary field of characteristic 0.)

Thus it seems fitting to close this paper on tameness versus wildness for noetherian rings with some problems, starting with: *Is there a tame-wild theorem for modules of finite length over HNP rings?* There are probably other interesting connections between what representation theorists and other noetherian ring theorists study. For example, what happens to the main results of the present survey in the noncommutative noetherian case?

150

Klingler made a beginning in his description of fingen(Λ) where $\Lambda = \mathbb{Z}G$, the integral group ring of a nonabelian group of order pq [K]. But a full tame-wild theorem for finitely generated modules (if one exists) seems to be a significant challenge, even in the more limited context of the rings that occur in integral representation theory.

Bibliography

Ba.   C. S. Ballantine, "Products of complic cosquares and pseudo-involutory matrices", *Linear and Multilinear Algebra* **8** (1979), 73–78.

B.   S. Brenner, "Decomposition properties of some small diagrams of modules," *Symposia Mathematica* **13** (1974), 127–141.

D1.   Yu. A. Drozd, "Representations of commutative algebras" (Russian), *Funkstional'nyi Analiz i Ego Prilozheniya* **6** (1972), 41–43. English Translation in *Functional Analysis and its Applications* **6** (1972)

D2.   Yu. A. Drozd, "Finite modules over pure noetherian algebras", *Proc. Steklov Inst. Math.* **4** (1991), 97–108.

E.   E. G. Evans, Jr., "Krull-Schmidt and cancellation over local rings," *Pacific J. Math.*, **46** (1973), 115–121.

GL.   R. M. Guralnick, and L. S. Levy,"Cancellation and direct summands in dimension 1,"*J. Algebra*, **142** (1991), 310–347.

GLW.   R. M. Guralnick, L. S. Levy and R. B. Warfield, Jr., "Cancellation counterexamples in Krull Dimension 1", *Proc. Amer. Math. Soc.* **109** (1990), 323–326.

HL.   W. J. Heinzer and L. S. Levy, "Domains of dimension 1 with infinitely many singular maximal ideals," (in preparation).

Ka.   I. Kaplansky, "Modules over Dedekind rings and valuation rings", *Trans. Amer. Math. Soc.* **72** (1952), 327–340.

K.   L. Klingler, "Modules over the integral group ring of a nonabelian group of order pq", *Mem. Amer. Math. Soc.* **59** (1986).

KL0.   L. Klingler and L. S. Levy, "Sweeping-similarity of matrices", *Linear Algebra Appl.* **75** (1986), 67–104.

KL1.   L. Klingler and L. S. Levy, "Representation type of commutative noetherian rings I: local wildness," *Pacific J. Math.*, **200** (2001), 345–386.

KL2.   L. Klingler and L. S. Levy, "Representation type of commutative noetherian rings II: local tameness," *Pacific J. Math.*, **200** (2001), 387–483.

KL3.   L. Klingler and L. S. Levy, "Representation Type of Commutative Noetherian Rings III: Global Wildness and Tameness," *Mem. Amer. Math. Soc.* (to appear)

KL4.   L. Klingler and L. S. Levy, "Wild torsion modules over Weyl algebras, and general torsion modules over HNPs", *J. Algebra* **172** (1995)

L1.   L. S. Levy, "Krull-Schmidt uniqueness fails dramatically over subrings of $\mathbf{Z} \oplus \cdots \oplus \mathbf{Z}$" *Rocky Mountain J. Math.* **13** (1983), 659–678.

L2.   L. S. Levy, "Modules over Dedekind-like rings," *J. Algebra*, **93** (1985), 1–116.

L3. L. S. Levy, "$\mathbb{Z}G_n$-modules, G_n cyclic of square-free order n," *J. Algebra*, **93** (1985), 354–375.

LO. L. S. Levy and C. J. Odenthal, "Package deal theorems and splitting orders, in dimension 1," *Trans. Amer. Math. Soc.*, **348** (1996), 3457–3503.

LR1. L. S. Levy and J. C. Robson, "Hereditary noetherian prime rings 1: Integrality and simple modules", *J. Algebra* **218** (1999), 307–337.

LR2. L. S. Levy and J. C. Robson, "Hereditary noetherian prime rings 2: Finitely generated projective modules", *J. Algebra* **218** (1999), 338–372.

LR3. L. S. Levy and J. C. Robson, "Hereditary noetherian prime rings 3: Infinitely generated projective modules", *J. Algebra* **225** (2000), 275–298.

MR. J. C. McConnell and J. C. Robson, *Noncommutative Noetherian Rings*, Graduate Studies in Mathematics **30**, American Mathematical Society (1987, 2001).

NR. L. A. Nazarova and A. V. Roiter, "Finitely generated modules over a dyad of two local rings and finite groups with an abelian normal divisor of index p," *Zap. Nauch. Sem. Leningrad Odtel. Mat. Inst. Steklov (LOM1) Izv. Akad. Nauk SSSR,* Ser Mat. **33**, No. 1 (1969). English transl: *Math. USSR Izvestija* **3**, No. 1 (1969).

NRSB. L. A. Nazarova, A. V. Roiter, V. V. Sergeichuk, and V. M. Bondarenko, "Application of modules over a dyad for the classification of finite p-groups that have an abelian subgroup of index p and of pairs of mutually annihilating operators" (Russian), *Zap. Nauch. Sem. Leningrad Odtel. Mat. Inst. Steklov (LOM1)* **28** (1972), 69–92. English transl: *J. Soviet Math.* **3** (1975), 636–653.

Nz. L. A. Nazarova, "Representations of quivers of infinite type," *Izv. Akad. Nauk SSSR,* Ser Mat. **37**, No. 4 (1973). English transl: *Math. USSR Izvestija* **7**, No. 4 (1973) 749–792.

R. C. M. Ringel, "The representation type of local algebras," *Springer Lecture Notes in Mathematics* **488** (1975), 282–305.

S. E. Steinitz, "Rechteckige Systeme und Moduln in Algebraischer Zahlkörper I, II," *Math. Ann.*, **71** (1911), 328–354, **72** (1912), 297–345.

CORNER RING THEORY: A GENERALIZATION OF PEIRCE DECOMPOSITIONS, I

T. Y. LAM

Department of Mathematics, University of California,
Berkeley, Ca 94720
E-mail: lam@math.berkeley.edu

In this paper, we introduce a general theory of corner rings in noncommutative rings that generalizes the classical notion of Peirce decompositions with respect to idempotents. Two basic types of corners are the *Peirce corners eRe* ($e^2 = e$) and the *unital corners* (corners containing the identity of R). A general corner is both a unital corner of a Peirce corner, and a Peirce corner of a unital corner. The simple axioms for corners engender good functorial properties, and make possible a broader study of subrings with only *some* of the features of Peirce corners. In this setting, useful notions such as *rigid corners*, *split corners*, and *semisplit corners* also come to light. This paper develops the foundations of such a corner ring theory, with a view toward a unified treatment of various descent-type problems in ring theory in its sequel.

1. Introduction

In the fourth volume of the American Journal of Mathematics, Benjamin Peirce published a long article in 1881 with the title "Linear Associative Algebra" [Pe]. Much of Peirce's detailed study of low dimensional associative algebras in this paper is no longer read by current researchers. However, Peirce's realization of the role of nilpotent elements and idempotent elements in the study of an algebra had a lasting impact on (what is later known as) ring theory. On p. 104 of [Pe], Peirce considered an "expression" in an algebra such that, "when raised to a square or higher power, it gives itself as the result"; such an expression, he wrote, "may be called *idempotent.*"

Peirce pointed out that a (nonzero) idempotent "can be assumed as one of the independent units" (or basis elements) of the algebra. On p. 109 of [Pe], he wrote:

"The remaining units can be selected as to be separable into four distinct groups. With reference to the basis, the units of the first

> *group are idemfactors; those of the second group are idemfaciend
> and nilfacient; those of the third group are idemfacient and nilfa-
> ciend; and those of the fourth group are nilfactors."*

This quotation from [Pe] seems to be the origin of the *Peirce decomposition*
of an algebra R with respect to an idempotent element $e \in R$. Replacing the
arcane terminology with modern notations and proceeding in a basis-free
manner, we may identify Peirce's four distinct groups as eRe, eRf, fRe,
and fRf, where f denotes the "complementary idempotent" $1 - e$. The
sum of these four additive groups is a direct sum, equal to the whole ring
(or algebra) R. The first group eRe is a ring in its own right, with identity
element e: this is the *Peirce corner* of R associated to the idempotent e.

Through the last century, the study of Peirce corners has played a major
role in noncommutative ring theory. The use of rings of the type eRe has
proved to be important in the consideration of many ring-theoretic issues,
such as the decompositions and extensions of rings, continuous geometry,
Boolean algebras, projective modules, Morita equivalences and dualities,
rings of operators, and path algebras of quivers, etc. However, Peirce corners have a nontrivial presence only in rings with idempotents, so for several
important types of rings (e.g. domains or local rings), the theory of Peirce
corners cannot be expected to be of any direct impact.

In this paper, we use Peirce corner rings as a model for building a general
theory of corner rings in arbitrary rings. The notion of a Peirce corner is
generalized as follows. A subring $S \subseteq R$ is called a (general) corner of
R if $R = S \oplus C$ for a subgroup $C \subseteq R$ (called a *complement* of S) that
is closed with respect to left and right multiplications by elements of S.
Peirce corners $R_e := eRe$ are a special case, since C may be taken to be
(the "Peirce complement") $C_e := eRf \oplus fRe \oplus fRf$.

The advantage of the corner ring definition above lies in its simplicity
and flexibility. Unlike the case of Peirce corners, a general corner $S \subseteq R$
may contain the identity of R without being the whole ring: such S is
called a *unital corner* of R. Examples of unital corners are also ubiquitous
in ring theory; for instance, they show up as ring retracts, as the 0-th
components of monoid graded rings or general crossed products, and (in
many significant cases) as rings of invariants with respect to group actions
on rings. Peirce corners and unital corners play a special role in our general
corner ring theory, since any corner is a unital corner of a Peirce corner,
and also a Peirce corner of a unital corner. On the other hand, the simple
definition for corner rings in general provides a common axiomatic ground

for understanding Peirce corners and unital corners simultaneously.

A nice feature of general corner rings is their very tractable functorial behavior. This is expounded in §2, where we prove (among other things) the transitivity and descent properties of corners ((2.3) and (2.4)). We also characterize Peirce corners and unital corners, respectively, by using properties of their complements ((2.10) and (2.14)). This work, in part, brings forth a main theme of the present paper, namely, that the properties of the complements are as important as those of the corner rings themselves. For another simple illustration, if $R = S \oplus C$ as in the definition of corner rings above, then $C \cdot C = 0$ amounts to R being a "trivial extension" of S, while $C \cdot C \subseteq S$ amounts to R being a $\mathbb{Z}_2$-graded ring, with 0-component S and 1-component C. These and various other examples of corner rings are given in §3, where we draw freely from the many constructions in commutative and noncommutative ring theory.

The general axiomatic formulation of corner ring theory led to some interesting concepts that seemed to have escaped earlier notice. For instance, in studying a general corner ring S, it is natural to ask when S has a *unique* complement, or when S has a complement that is an *ideal* in the ambient ring R. These conditions define the notions of "rigid corners" and "split corners" respectively. For instance, rigid corners occur naturally in the study of crossed products (see (3.6)). In the case of Peirce corners $R_e = eRe$, (2.8) shows that we have automatic rigidity (that is, $C_e = eRf \oplus fRe \oplus fRf$ is the *only* complement for R_e). However, R_e need not split in general; we show (in (4.5) and (4.9)) that it does iff $e(RfR)e = 0$ (where $f = 1 - e$), iff any composition of R-homomorphisms $eR \to fR \to eR$ is zero. In this case, we say that the idempotent e is *split*: this seems to be a useful new notion on idempotents that is worthy of further study. For instance, 1-sided semicentral idempotents (used extensively in studying the triangular representation of rings) are always split, but split idempotents need not be 1-sided semicentral (see (4.10), (4.15)). Incidentally, Peirce corners arising from 1-sided semicentral idempotents are precisely corner rings that are 1-sided ideals, and Peirce corners arising from central idempotents are precisely corner rings that are 2-sided ideals ((2.11) and (2.12)). All of these results serve to show how nicely the classical Peirce corners fit into the general theory of corner rings.

The last section of the paper (§5) is devoted to the aforementioned theme that any corner can be represented as a unital corner of a Peirce corner, and also as a Peirce corner of a unital corner. Along with this work, we prove several results giving a one-one correspondence between the complements of certain pairs of corner rings; see (5.1) and (5.10). The former shows, for instance, that a corner S is rigid in R iff it is rigid in some (or in all) Peirce corner(s) of R containing S.

The conference lecture given in Lisbon (on which this paper is based) also reported on some of the applications of the corner ring theory. This part of our work will appear later in a sequel to this paper, [La$_4$], in which we shall study the multiplicative structure of corner rings and various descent problems of ring-theoretic properties. Some further applications of the viewpoint of corner rings are presented in [LD].

Throughout this note, R denotes a ring with an identity element $1 = 1_R$, and by the word "subring", we shall always mean a subgroup $S \subseteq R$ that is closed under multiplication (hence a ring in its own right), but with an identity element *possibly different from* 1_R. If 1_R happens to be in S (so it is also the identity of S), we say that S is a *unital subring* of R. Other general ring-theoretic notations and conventions in this paper follow closely those used in [La$_1$] and [La$_3$].

2. Different Types of Corner Rings

We introduce the following general definition of a "corner" in a ring R.

Definition 2.0. A ring $S \subseteq R$ (with the same multiplication as R, but not assumed to have an identity initially) is said to be a *corner ring* (or simply a corner) of R if there exists an additive subgroup $C \subseteq R$ such that

$$R = S \oplus C, \quad S \cdot C \subseteq C, \quad \text{and} \quad C \cdot S \subseteq C. \tag{2.0}'$$

In this case, we write $S \prec R$, and we call any subgroup C satisfying $(2.0)'$ a *complement* of the corner ring S in R.

Of course, in general, such a complement C is far from being unique. For instance, if R contains $\mathbb{Z}$ as a unital subring, then any additive subgroup $C \subseteq R$ such that $\mathbb{Z} \oplus C = R$ is a complement of $\mathbb{Z}$ in R in the sense of (2.0). If a corner S of a ring R happens to have a unique complement, we shall call S a *rigid corner of R, and write $S \prec_r R$.*

Proposition 2.1. *A ring $S \subseteq R$ is a corner of R iff the inclusion map $S \hookrightarrow R$ has a splitting $\tau : R \to S$ that is both left and right S-linear.*[a]

---

[a]We could have called τ a splitting in the category of (S, S)-bimodules if, by the word "bimodule", we mean a bimodule that is not necessarily unital on either side.

Proof. If $\tau : R \to S$ exists, one checks easily that $C := \ker(\tau)$ satisfies $(2.0)'$, so $S \prec R$. Conversely, if $S \prec R$, with a complement C, the inclusion map $S \hookrightarrow R$ splits by the map $\tau : R \to S$ given by $\tau(s + c) = s$ for $s \in S$ and $c \in C$. For $s_0 \in S$, we have

$$\tau(s_0(s + c)) = \tau(s_0 s + s_0 c) = s_0 s = s_0 \tau(s + c)$$

since $s_0 s \in S$ and $s_0 c \in C$. Thus, τ is left S-linear, and a similar check shows that τ is also right S-linear. $\qquad\square$

The following easy proposition shows that a corner ring of any ring R must have an identity (although this may not be the identity of R).

Proposition 2.2. *Let $S \prec R$, with a complement C, and let $1 = e + f$, where $e \in S$ and $f \in C$. Then e is an identity of the ring S. In particular, the decomposition $1 = e + f$ is independent of the choice of the complement C, and e, f are complementary idempotents in R.*

Proof. For any $s \in S$, $s = s \cdot 1 = se + sf$. Since $s, se \in S$ and $sf \in C$, we have $s = se$. Similarly, $s = es$, so e is an identity for S. Since the identity element of S is unique, the remaining statements in the Proposition follow immediately. $\qquad\square$

The next two propositions serve to show the robustness of our chosen definition of corners in rings.

Proposition 2.3 (Descent). *Let $S \prec R$, with a complement C.*

(1) *If S' is any subring of R containing S, then $S \prec S'$ (with complement $C \cap S'$).*
(2) *If R' is any subring of R containing C, then $S \cap R' \prec R'$ (with complement C).*

Proof. (1) Let $C_0 = S' \cap C$. Then $S' = S \oplus C_0$, and

$$SC_0 \subseteq S' \cap SC \subseteq S' \cap C = C_0.$$

By symmetry, we have also $C_0 S \subseteq C_0$. This shows that $S \prec S'$, with a complement C_0. (2) is proved similarly. $\qquad\square$

Proposition 2.4 (Transitivity). *Suppose $S \prec S'$ and $S' \prec R$. Then $S \prec R$. If $S \prec_r R$, then $S \prec_r S'$.*

Proof. Let C_0 be a complement of S in S' and C' be a complement of S' in R. Then, for $C := C_0 + C'$, we have $S \oplus C = R$. Moreover,

$$SC \subseteq SC_0 + SC' \subseteq C_0 + S'C' \subseteq C_0 + C' = C,$$

and similarly $CS \subseteq C$. Thus, $S \prec R$, with a complement C. Now assume $S \prec_r R$. If C_1 is another complement of S in S', then $C_1 \oplus C'$ is also a complement of S in R. Therefore, $C_0 \oplus C' = C_1 \oplus C'$. Contracting these to S', we see that $C_0 = C_1$, so $S \prec_r S'$. $\qquad\square$

Remark. *If $S \prec R$, a subring $S' \subseteq R$ containing S need not be a corner of R. For instance, let $S = \mathbb{Z}$, and $R = \mathbb{Z}[x]$ with the relation $x^2 = 0$. Certainly, $S \prec R$, but the subring $S' = \mathbb{Z} \oplus 2\mathbb{Z}x$ is not a corner of R. (If S' has a complement C' in R, take a nonzero element $f = a + bx \in C'$. Then $2f = 2a + 2bx \in S' \cap C' = 0$ implies that $f = 0$, a contradiction.) For some cases in which we can infer that $S' \supseteq S$ is a corner in R, see (5.9).*

An easy consequence of (2.3) and (2.4) is the following.

Corollary 2.5. *If $S \subseteq S'$ are both corners of a ring R, then every complement C' for S' in R can be enlarged to one for S.*

Proof. By (2.3)(1), we know that $S \prec S'$, so we can fix some complement C_0 of S in S'. By the proof of (2.4), $C_0 \oplus C'$ is a complement of S: this is the complement we seek, since it contains C'. $\qquad\square$

In view of (2.5), it is natural to ask the following

Question 2.6. *If $S \subseteq S'$ are both corners of a ring R, does every complement C for S in R contain some complement for S'?*

In general, the answer is "no". We construct a counterexample as follows. For any field k, let R be the commutative local ring $k[x]$ with the relation $x^4 = 0$, and let $S = k$, $S' = k[x^2]$. Then R has k-basis $\{e_i\}$, where

$$e_1 = 1, \quad e_2 = x^2, \quad e_3 = 1 + x, \quad \text{and} \quad e_4 = 1 + x^3.$$

Thus, the span C of e_2, e_3, e_4 is a complement to S in R. Assume, for the moment, that C contains a complement C' to S'. Since $R = S' \oplus xS'$, we have $C' \cong R/S' \cong S'$ as S'-modules, so there exists an element $y \in C'$ such that C' has a k-basis $\{y, x^2y\}$. Write $y = ae_2 + be_3 + ce_4$, where $a, b, c \in k$.

Then

$$x^2 y = x^2[ax^2 + b(1+x) + c(1+x^3)]$$
$$= (b+c)x^2 + bx^3$$
$$= -be_1 + (b+c)e_2 + be_4$$

implies that $b = 0$ since $x^2 y \in C' \subseteq C$. But then $x^2 y = ce_2 \in S'$, which is a contradiction. This shows that C does not contain any complement of S' in R.

In spite of examples such as the above, it turns out that Question (2.6) has an affirmative answer if one of the corners S, S' in question is a Peirce corner. Before we come to the proof of this (in (2.9)), let us first prove a key lemma on complements of general corner rings.

Lemma 2.7. *Let $S \prec R$, with identity e and complement C, and let $r \in R$. Then $r \in C$ iff $ere \in C$. In particular, $ere = 0 \implies r \in C$.*

Proof. The "only if" part follows from $(2.0)'$. For the "if" part, assume that $ere \in C$, and write $r = s + c$, where $s \in S$ and $c \in C$. Then

$$s = ese = e(r - c)e = ere - ece \in C$$

implies that $s = 0$. Thus, $r = c \in C$. $\qquad\square$

We now come to the following basic result on Peirce corners.

Theorem and Definition 2.8 (Peirce Corners). *Let e, f be complementary idempotents in a ring R. Then:*

(1) *$R_e := eRe \prec R$; it is the largest subring (resp. corner) of R having e as identity element.*

(2) *$R_e \prec_r R$ (that is, R_e is rigid in R), with a unique complement*

$$C_e := fRe \oplus eRf \oplus fRf = \{r \in R : ere = 0\}. \qquad (2.8)'$$

We shall call R_e the Peirce corner of R (arising from the idempotent e), and call C_e its Peirce complement. The notations R_e and C_e will be fixed in the sequel of this paper, and we shall use the notation $S \prec_P R$ to refer to the fact that S is a Peirce corner of R.

(3) *$R_e \cap R' \prec_P R'$ for any subring $R' \subseteq R$ containing e.*

(4) *(Transitivity of Peirce Corners) For any subring $S \subseteq R_e$, $S \prec_P R_e$ iff $S \prec_P R$.*

Proof. (1) By Peirce's theory,[b] the sum in (2.8)$'$ is direct, and we have $R = R_e \oplus C_e$. An easy calculation shows that

$$R_e C_e = eRf \subseteq C_e, \quad \text{and} \quad C_e R_e = fRe \subseteq C_e, \tag{2.8}$''$$

so $R_e \prec R$, with a complement C_e. If S is any subring of R having e as its identity, then for any $s \in S$, we have $s = ese \in R_e$, so $S \subseteq R_e$.

(2) Consider *any* complement C to R_e. Let $K = \{r \in R : ere = 0\}$. Clearly, $C_e \subseteq K$, and by (2.7), $K \subseteq C$. Since C_e and C are *both* complements of R_e, the inclusions $C_e \subseteq K \subseteq C$ must all be equalities! In particular, $R_e \prec_r R$.

(3) Since $e \in R'$, $eR'e \subseteq R' \cap eRe$. On the other hand, for any $r' \in R'$ of the form ere (for some $r \in R$), we have $r' = er'e \in eR'e$. Therefore, $R_e \cap R' = R'_e \prec_P R'$.

(4) The "if" part follows from (3). For the "only if" part, assume that $S \prec_P R_e$. Then $S = e'R_e e'$ where e' is an idempotent in R_e. But then $e'e = e' = ee'$, so $e'R_e e' = e'(eRe)e' = e'Re' \prec_P R$. $\qquad\square$

Corollary 2.9. *Let $S \subseteq S'$ be both corners of R, with complements C and C' respectively. If one of S, S' is a Peirce corner of R, then $C' \subseteq C$. (In particular, (2.6) has an affirmative answer if one of S, S' is a Peirce corner of R.)*

Proof. First assume that $S = eRe$, for some idempotent $e \in R$. Since R_e is rigid, we have here $C = C_e$. By (2.5), C' can be enlarged to a complement of $S = R_e$, which must be C_e. Thus, $C' \subseteq C_e = C$. Next, assume instead, that $S' = e'Re'$, for some idempotent $e' \in R$. Here, $C' = C_{e'}$. Let e be the identity of S. Since $e \in S \subseteq S'$, we have $ee' = e'e = e$. For any $r \in C' = C_{e'}$, we have $e're' = 0$, and hence $ere = ee're'e = 0$. By (2.7), this implies that $r \in C$. Thus, $C' \subseteq C$, as desired. The last conclusion of (2.9) is now obvious. $\qquad\square$

For the corollary above, the following remark is relevant. In the first part of the result, the argument would have worked as long as the corner S is rigid. But for the second part of the result, we do require that the corner S' be Peirce. If S' is only rigid, the desired conclusion $C' \subseteq C$ need not hold. For instance, if $R = \mathbb{R}\{1, i, j, k\}$ is the division ring of the real quaternions, then $S = \mathbb{R} \prec R$ is contained in $S' = \mathbb{R}\{1, i\} \prec R$. Here, $C' := \mathbb{R}\{j, k\}$ is a complement of S', $C = \mathbb{R}\{i, j, 1 + k\}$ is a complement of

---

[b]For an exposition, see [La$_1$:p. 308].

S, but $C' \not\subseteq C$. (It is easy to see that S' here is indeed rigid in R. For a more general fact, see (3.6) below.)

We can give some easy characterizations of Peirce corners, as follows.

Proposition 2.10. *For $S \prec R$ with identity e, the following are equivalent:*

(1) $S \prec_P R$;

(2) S *has a complement* C *such that* $eCe = 0$;

(3) *the subring* $S \subseteq R$ *is "hereditary", in the sense that* $sRs' \subseteq S$ *for all* $s, s' \in S$.

Proof. $(1) \Rightarrow (2)$ follows by taking C to be a Peirce complement (in case $S \prec_P R$).

$(2) \Rightarrow (1)$. If C is as in (2), then

$$eRe = e(S + C)e = eSe + eCe = eSe = S \implies S = R_e \prec_P R.$$

$(1) \Rightarrow (3)$. If $S = R_e$, we have $s = es$ and $s' = s'e$ for any $s, s' \in S$. Therefore, $sRs' = esRs'e \subseteq eRe = S$.

$(3) \Rightarrow (1)$. It suffices to assume only $sRs \subseteq S$ for all $s \in S$. For, if so, then $e \in S$ implies that $eRe \subseteq S$, and we must have equality here since $S = eSe \subseteq eRe$. $\qquad\square$

After my lecture at the Lisbon Conference, Professors J. Okniński and F. Perera both pointed out to me the importance of the notion of hereditary subalgebras in algebra and in analysis. This prompted me to include the result (3) above, according to which the hereditary corner rings (in our general sense) are precisely the classical Peirce corner rings.

Next, we shall characterize corners in R that are one-sided ideals. These turn out to be necessarily Peirce corners, but they are Peirce corners of a special kind. To see this, let us first recall some standard definitions in the theory of idempotents. In [Bi], [BH], and [HT], an idempotent $e \in R$ with complementary idempotent $f = 1 - e$ is said to be *left semicentral* if $fRe = 0$, and *right semicentral* if $eRf = 0$. Take, for instance, the former: it is easy to show that

$$\begin{aligned}
fRe = 0 &\iff f(ReR)e = 0 \\
&\iff ere = re \quad (\forall r \in R) \iff eRe = Re \qquad (*)\\
&\iff eR \text{ is an ideal} \qquad \iff Rf \text{ is an ideal},
\end{aligned}$$

so each of these conditions is a characterization for e to be a left semicentral idempotent. A similar remark applies to right semicentral idempotents.

Proposition 2.11. *For a corner $S \prec R$ with identity e and complement C, the following are equivalent:*

(1) *S is a left ideal in R;*
(2) *$C \cdot S = 0$;*
(3) *e is a left semicentral idempotent in R and $S = R_e$.*

Proof. (1) $\Leftrightarrow$ (2). If S is a left ideal, then $C \cdot S \subseteq S \cap C = 0$. Conversely, if $C \cdot S = 0$, then $R \cdot S = (S + C) \cdot S = S \cdot S = S$, so S is a left ideal.

(3) $\Rightarrow$ (1). If $S = R_e$ with e left semicentral, then by (*) above, $S = eRe = Re$, so it is a left ideal.

(1) $\Rightarrow$ (3). If S is a left ideal, then $e \in S$ yields $Re \subseteq S \subseteq eSe \subseteq eRe$. Therefore, equality holds throughout, so $S = eRe$, and $Re = eRe$ implies that e is left semicentral by (*). $\qquad\square$

If the idempotent e is *both* left and right semicentral, then $re = ere = er$ for all $r \in R$, so e is in fact central. Then $R_e = eR$ and $C_e = R_f = fR$ are both ideals of R, so R is a ring direct product $R_e \times R_f$. In this case, we call $R_e = eR$ a *direct Peirce corner* of R. We can thus conclude:

Corollary 2.12. *A corner $S \prec R$ is a direct Peirce corner iff S is an ideal of R.*

We move on now to consider another important type of corner rings.

Definition 2.13 (Unital Corners). Let S be a corner of R, with identity e. We say that S is a *unital corner* (and write $S \prec_u R$) if $e = 1$ (that is, if S is a unital subring of R).

It is easy to see that $S \prec_P R$ and $S \prec_u R$ iff $S = R$. In parallel to (2.10), the following is a characterization of unital corners.

Proposition 2.14. *For $S \prec R$ with an identity e, the following are equivalent:*

(1) *$S \prec_u R$;*
(2) *S has a complement C with $eCe = C$;*
(3) *every complement C of S satisfies $eCe = C$.*

Proof. (1) $\Rightarrow$ (3) is clear, since $e = 1$ under (1). (3) $\Rightarrow$ (2) is also clear, since a complement of S always exists. To prove (2) $\Rightarrow$ (1), suppose S has a complement C with $eCe = C$. Then, e acts as the identity map by left and by right multiplication on C, as well as on S. This clearly implies that $e = 1$; that is, $S \prec_u R$. $\qquad\square$

We shall now introduce two more kinds of corner rings.

Definition 2.15 (Split and Rigid-Split Corners). A corner S in R is called a *split corner* (written $S \prec_s R$) if it has a complement C that is an ideal in R. Note that, in view of $(2.0)'$, this is equivalent to S having a complement C that is closed under multiplication. In this case, we have a unital ring isomorphism $S \cong R/C$, although, as a subring of R, S may still be not unital. If $S \prec_s R$ happens to have a unique ideal complement, we shall call it a *rigid-split corner*, and write $S \prec_{rs} R$.

A word of caution is necessary on this piece of terminology. If a corner S in R is both rigid and split, then clearly $S \prec_{rs} R$. However, if $S \prec_{rs} R$, then S is split, *but it may not be rigid,* as there may be other complements to S besides the guaranteed unique ideal complement. For an example illustrating this situation, see (3.9) below.

For split corners, it is easy to verify the following analogues of (2.3) and (2.4) (for descent and transitivity), and of (2.5).

Proposition 2.16. (1) *If $S \prec_s R$, then $S \prec_s S'$ for any subring S' of R containing S. And, for any subring R' containing an ideal complement of S, $S \cap R' \prec_s R'$.*

(2) *$S \prec_s S'$ and $S' \prec_s R$ imply $S \prec_s R$. If $S \prec_{rs} R$, then $S \prec_{rs} S'$ for any subring $S' \supseteq S$.*

(3) *If $S \subseteq S'$ are both split corners of R, then any ideal complement of S' can be enlarged into one for S.*

Proof. (1) is obvious as an ideal complement C of S in R contracts to an ideal complement of S in S', and C remains an ideal complement to $S \cap R'$ in R'. To prove (2), take C_0 to be an ideal complement of S in S', and C' to be an ideal complement of S' in R. Then, for the complement $C := C_0 + C'$ of S in R, we have

$$RC = R(C_0 + C') \subseteq (S' + C')C_0 + C' \subseteq C_0 + C' = C,$$

and similarly, $CR \subseteq C$. Thus, C is an ideal in R, so $S \prec_s R$. The second part of (2) can be proved in the same way as the second part of (2.4): in the argument there, we simply replace complements by ideal complements. The proof for (3) follows by a similar modification of that for (2.5). $\qquad \square$

Split unital corners in a ring R are very familiar objects in ring theory; they are called "retracts" of R. On the other hand, split Peirce corners did not seem to have received much attention; we shall study them in more detail in §4. Here, let us give an example of a split corner that is neither unital nor Peirce.

Example 2.17. Take a split unital corner S_0 in some ring A, with an ideal complement $C_0 \neq 0$, and let $R = \left(\begin{smallmatrix} A & A \\ 0 & A \end{smallmatrix}\right)$. Then, for $S := \left(\begin{smallmatrix} S_0 & 0 \\ 0 & 0 \end{smallmatrix}\right)$ and $C := \left(\begin{smallmatrix} C_0 & A \\ 0 & A \end{smallmatrix}\right)$, we have $R = S \oplus C$, and a direct calculation shows that C is a complement of S in R, with $C \cdot C \subseteq C$. Thus, $S \prec_s R$, with the identity element $e := \left(\begin{smallmatrix} 1 & 0 \\ 0 & 0 \end{smallmatrix}\right)$. This is not the identity of R, so S is not a unital corner. It is also *not* a Peirce corner, since $R_e = \left(\begin{smallmatrix} A & 0 \\ 0 & 0 \end{smallmatrix}\right)$ properly contains $S = \left(\begin{smallmatrix} S_0 & 0 \\ 0 & 0 \end{smallmatrix}\right)$.

In ring theory, there is a very useful construction of "trivial extensions", whereby, for any ring S and a unital (S, S)-bimodule C, a ring $R := S \oplus C$ is produced in which S is a unital subring, $C \cdot C = 0$, and the left/right multiplications of S on C are given by the (S, S)-bimodule structure. Such a ring R is called a *trivial extension* of S by $C = {}_S C_S$; see [La$_3$: p. 37]. Clearly, S is a retract of R, and C is a complement of S with a trivial multiplication. Note that R may also be viewed as the unital subring of the triangular ring $\left(\begin{smallmatrix} S & C \\ 0 & S \end{smallmatrix}\right)$ consisting of matrices of the form $\left(\begin{smallmatrix} s & c \\ 0 & s \end{smallmatrix}\right)$, with $s \in S$ and $c \in C$.

To relate the construction of trivial extensions to corner rings, we make the following:

Definition 2.18 (Trivial Corners). A corner S of a ring R is said to be a *trivial corner* if it has a complement C with $C \cdot C = 0$.

Proposition 2.19. *A trivial corner S of R is a retract of R, and R is a trivial extension of S.*

Proof. Say C is a complement of S with $C \cdot C = 0$. Then C is an ideal of R, so S is a split corner. Let $1 = e + f$, where $e \in S$ and $f \in C$. Then $f = f \cdot f \in C \cdot C = 0$ leads to $e = 1$. Thus, S is a unital corner, and hence a retract of R. Here, under the ring structure on R, C is a unital (S, S)-bimodule, and, for $s, s' \in S$ and $c, c' \in C$:

$$(s + c)(s' + c') = ss' + sc' + cs',$$

so R is precisely the trivial extension of S by ${}_S C_S$. $\qquad\square$

In conclusion, let us also point out that standard constructions in ring theory can be used to give various examples of new corners from old ones. We note for instance the following three types of constructions, starting from any $S \prec R$ with a complement C.

Construction 2.20. For any integer $n \geq 1$, it is routine to check that $\mathbb{M}_n(C)$ is a complement to $\mathbb{M}_n(S)$ in $A := \mathbb{M}_n(R)$. Thus, $\mathbb{M}_n(S) \prec A$. If $S \prec_P R$, say $S = R_e$ in R, then $\mathbb{M}_n(S) \prec_P A$. In fact, an easy calculation shows that $\mathbb{M}_n(S) = A_e$, where e is viewed as usual as the (idempotent) matrix $eI_n \in A$. (The same applies to rings of upper triangular matrices.)

Construction 2.21. $S[(x_i)_{i \in I}] \prec R[(x_i)_{i \in I}]$; a complement is given by $C[(x_i)_{i \in I}]$. (The same applies to power series constructions.)

Construction 2.22. For any (multiplicative) monoid G, the monoid ring $S[G] \prec R[G]$; a complement is given by $C[G]$.

3. Examples of Unital Corners (and Their Complements)

Peirce corners are easy to find since they are as ubiquitous as idempotents in rings. However, a Peirce corner cannot be unital unless it is the whole ring. In (3.1) below, we shall collect some examples of unital corners (and their complements). Note that, *if R is a ring with only trivial idempotents $\{0, 1\}$, then any nonzero corner ring S is necessarily unital.* (The identity e of S is a nonzero idempotent, and so $e = 1$.) One major difference between Peirce corners and unital corners is the following: by (2.8)(2), Peirce corners are rigid, but, as we'll see from the examples below, unital corners need not be rigid, and split unital corners need not be rigid either.

Examples 3.1. (A) Let S be a *central unital* subring of R. If S is a self-injective ring or $_SR$ is a semisimple S-module, then $S \prec_u R$. In fact, in either case, we have $R = S \oplus C$ for a suitable submodule C of $_SR$. Since S is central, we have $CS = SC \subseteq C$, so $S \prec_u R$.

(B) Let H be a subgroup of a group G. Then, for any ring k, $kH \prec_u kG$. In fact, a complement for kH can be taken to be $C = \bigoplus_{g \notin H} kg$.

(C) Let S be a unital subring of a nonzero ring R. If R is free as a left S-module with a basis G containing 1 such that $gS \subseteq Sg$ for every $g \in G$, then $S \prec_u R$, with a complement $\bigoplus_{g \neq 1} Sg$. For instance, for any base ring k, take R to be the polynomial ring $k[x]$. Then, for any $n \geq 1$, $S = k[x^n] \prec_u R$ since R is a free (left, right) S-module on the basis $\{1, x, \ldots, x^{n-1}\}$.

(D) For $R = \mathbb{M}_n(k)$ where k is any ring, let S be the subring of diagonal matrices in R. Then $S \prec_u R$, with a complement C given by the group of all matrices with a zero diagonal. (It is easy to check that $SC \subseteq C$ and $CS \subseteq C$.) It can be seen, actually, that this is a special case of (C): for instance, for $n = 3$, we can take G in (C) to be the 3-element set consisting of

$$I_3 = e_{11} + e_{22} + e_{33}, \quad e_{12} + e_{23} + e_{31}, \quad \text{and} \quad e_{13} + e_{21} + e_{32}.$$

(E) Another interesting special case of (C) above is a monoid ring $R = S[G]$, where G is any multiplicative monoid (and S is some ring). If G has an invertible element $g \neq 1$, then the complement $C := \bigoplus_{g \neq 1} Sg$ (given in (C) above) is *not* an ideal, since $C \cdot g^{-1} \nsubseteq C$ (if $S \neq 0$). Nevertheless, S is a split corner, since we can choose, as another complement, the augmentation ideal $\sum_{g \neq 1} S(g-1)$. This shows that *not every complement of a split unital corner need to be an ideal.* More generally, if R is a ring graded by a monoid G, say $R = \bigoplus_{g \in G} R_g$ (with $R_g R_h \subseteq R_{gh}$ for all $g, h \in G$), then $R_1 \prec_u R$, with complement $C := \bigoplus_{g \neq 1} R_g$. If $gh = 1 \in G$ can happen only if $g = h = 1$, then C is an ideal, so in fact $R_1 \prec_s R$. On the other hand, if we consider the case $G \cong \mathbb{Z}_2$, a $\mathbb{Z}_2$-*graded ring corresponds exactly to a ring R with a given unital corner $S \subseteq R$ having a complement C such that $C \cdot C \subseteq S$.*

(F) Various twisted versions of the monoid ring example in (E) (of the "crossed product" variety) also give examples of unital corners. We shall only mention (here and in (G) below) two of the most basic types of examples. Let S be a ring with an endomorphism $\sigma : S \to S$, and let R be the skew polynomial ring $S[x; \sigma]$, whose elements have the form $\sum_i a_i x^i$ ($a_i \in S$), and are multiplied by the rule $xa = \sigma(a)x$. Here, we have $x^i S \subseteq \sigma^i(S)x^i$ for every $i \geq 0$, so $S \prec_u R$, with a complement

$$C := \left\{ \sum_{i \geq 1} a_i x^i : a_i \in S \right\} = \bigoplus_{i=1}^{\infty} Sx^i.$$

This complement is an ideal, so $S \prec_s R$. This example serves to show the existence of nonrigid split corners. For instance, taking $\sigma = \mathrm{Id}_S$, we can choose a new variable $y = x + a$ (with a in the center of S) and write $R = S[y]$. With respect to this expression, $C' = \bigoplus_{i=1}^{\infty} Sy^i$ is another ideal complement of S. In case S has an infinite center, this gives infinitely many ideal complements to S.

(G) Yet another interesting special case of (C) is the following example from the theory of central simple algebras. Let R be a central simple algebra of degree d over a field F containing a subfield K that is Galois over F of dimension d. It is well-known that R can be written as a crossed product algebra $\bigoplus_{g \in G} Ku_g$, where $G = \mathrm{Gal}(K/F)$, with $u_1 = 1$, $u_g k = g(k)u_g$, and $u_g u_h \in K^* u_{gh}$ (for $g, h \in G$ and $k \in K$). In particular, we have $u_g K = Ku_g$

for all $g \in G$. Thus, $K \prec_u R$, with a complement $C := \bigoplus_{g \in G \setminus \{1\}} K u_g$. More general crossed products (in the sense of [Pa]) can be treated similarly.

(H) If R is a *commutative* ring, there is an interesting criterion for a unital subring $S \subseteq R$ to be a corner; namely,

$$S \prec_u R \implies \mathrm{tr}(_S R) = S. \qquad (3.2)$$

Here, the "trace ideal" $\mathrm{tr}(_S R)$ is the sum of the images of all S-linear functionals on the S-module R. This criterion is due to G. Azumaya and B. Müller; a proof of it can be found in [La$_3$: (2.49)]. From this, it can be shown that, if R is a finitely generated projective module over S, or if $_S R$ is projective and S is a noetherian ring, then $S \prec_u R$ (see [La$_3$: (2.50)]). Thus, for instance, if S is a Dedekind domain, then S is a corner in any domain $R \supseteq S$ that is finitely generated as an S-module.

(I) Let G be a finite group acting on a ring R, and let $S = R^G$ be the subring of G-invariant elements of R. *If $|G|^{-1}$ exists in R, then $S \prec_u R$.* To see this, let $\tau : R \to S$ be (the "averaging map") defined by

$$\tau(r) = |G|^{-1} \cdot \sum_{g \in G} r^g \in S. \qquad (3.3)$$

This map clearly splits the inclusion $S \hookrightarrow R$, and for $s \in S$, we have $\tau(sr) = s\tau(r)$ (since $(sr)^g = s^g r^g = s \cdot r^g$), and similarly $\tau(rs) = \tau(r)s$. Thus, $S \prec_u R$ by (2.1).[c] Furthermore, S can be thought of, *in two different ways*, as a corner in the skew group ring $A := R * G$. Here, $R * G$ consists of (finite) formal sums $\sum_{g \in G} r_g g$ ($r_g \in R$), which are multiplied by using the rule $gr = r^{g^{-1}} g$ for $r \in R$ and $g \in G$. Upon identifying $r \in R$ with $r \cdot 1 \in A$, we have $R \prec_u A$ (with complement $\sum_{g \neq 1} Rg$), so the transitivity property (2.4) implies that $S \prec_u A$. Secondly, let e be the idempotent $|G|^{-1} \sum_{g \in G} g$ in A, and let

$$\varphi : S \to eAe \quad \text{be defined by} \quad \varphi(s) = es = se = ese \quad (\forall s \in S). \qquad (3.4)$$

It is easy to check that φ is a ring isomorphism: see [Mo: Lemma 2.1], or [Al]. Thus, $S = R^G$ *is also isomorphic to the Peirce corner eAe in A.* These examples of corner rings in R and in $R * G$ set the stage for some useful applications of corner ring theory to the study of rings of the type

---

[c]Analogues of this also exist for various actions of *infinite* groups G on R. In the case where R is commutative, for instance, it is often possible to replace the averaging map τ in (3.3) by a suitable "Reynolds operator" (an R^G-linear retraction from R to R^G). Cayley's "Ω process" and Weyl's "unitarian trick" are among the best known examples of this in classical invariant theory.

R^G; such applications will be more fully explored in Part II of this paper
([La$_4$]).

(J) If R is a finite von Neumann algebra and S is the center of R, then
the center-valued trace $\Delta : R \to S$ is S-linear and is the identity on S (see
[KR: (8.4.3)]). Therefore, by (2.1), S is a (unital) corner of R.

Let us now give some examples of *rigid* unital corners, partly drawing
from the list of examples above. Specifically, consider the unital corners
arising in the manner of (3.1)(C). If R is a commutative ring, then in the
notations there, we can produce other complements for S by changing the
given S-basis $G = \{1, g_1, g_2, \dots\}$ to, say, $\{1, s_1 + g_1, s_2 + g_2, \dots\}$ (where
$s_i \in S$). Therefore, we do not expect the unital corner S to be rigid in
this case. However, if R is noncommutative, our odds are better, as the
following three examples show.

Example 3.5. *The corner* $S \prec_u R = \mathbb{M}_n(k)$ *in* (3.1)(D) *is rigid.* To see
this, consider any complement C_0 for S, and use the notations in (3.1)(D).
Let $a \in k$, and let i, j be two distinct indices in $\{1, \dots, n\}$. Then $ae_{ij} =$
$\text{diag}(b_1, \dots, b_n) + M$ for some $b_1, \dots, b_n \in k$ and $M \in C_0$. Thus, $M =$
$ae_{ij} - \sum_\ell b_\ell e_{\ell\ell}$. Since $e_{ii}, e_{jj} \in S$, C_0 contains the matrix

$$e_{ii} M e_{jj} = ae_{ii}e_{ij}a_{jj} - \sum_\ell b_\ell e_{ii}e_{\ell\ell}e_{jj} = ae_{ij}.$$

This shows that C_0 contains the group C of *all* matrices with a zero diagonal, and hence $C_0 = C$, proving that S is rigid in $R = \mathbb{M}_n(k)$.

Example 3.6. *The corner* $K \prec_u R$ *in the crossed product example in*
(3.1)(G) *also turns out to be rigid.* To see this, consider any complement
C_0 for K, and keep the notations in (3.1)(G). To show that $C_0 = C$, we
exploit the same ideas used in the usual proof for the simplicity of the
crossed product algebra R. Given any $g \neq 1$ in G, decompose the element
u_g into $a + c$, where $a \in K$ and $c \in C_0$. Take any element $k \in K$ such that
$g(k) \neq k$. Then C_0 contains the element

$$kc - ck = (ku_g - ka) - (u_g k - ak) = ku_g - g(k)u_g = (k - g(k))u_g.$$

Since $k - g(k) \in K^*$, left multiplication of the above element by $(k - g(k))^{-1}$
yields $u_g \in C_0$. Thus, C_0 contains $C = \bigoplus_{g \in G \setminus \{1\}} Ku_g$, and hence $C_0 = C$,
proving that K is rigid in R. (Again, the case of more general crossed
products can be treated similarly.)

Example 3.7. Let k be any ring, and $R = k\langle x, y \rangle$ with the relations $yx = y^2 = 0$. Then $R = S \oplus Sy$, where $S = k[x]$. Here, $C := Sy$ is an ideal with square zero, so S is a trivial unital corner in R, with complement C. We claim that S is rigid. Indeed, if D is another complement, let $y = s + d$, where $s \in S$ and $d \in D$. Then $0 = yx = sx + dx$ implies that $sx = 0$, so $s = 0$, and $y = d \in D$. It follows that $C = Sy \subseteq SD \subseteq D$, so $C = D$, proving that $S \prec_r R$.

Of course, there are good examples of *rigid-split* unital corners too. Some of the most natural examples are given by the following result.

Proposition 3.8. *Let S be a (not necessarily unital) corner of a ring R, with a complement C. If $C = \mathrm{rad}(R)$ (the Jacobson radical), $\mathrm{Nil}_*(R)$ (the lower nilradical), or $\mathrm{Nil}^*(R)$ (the upper nilradical), then $S \prec_{rs} R$.*

Proof. Suppose C' is another ideal complement of S. We would like to prove that $C \subseteq C'$, for then $C' = C$, and we'll have $S \prec_{rs} R$. First assume that $C = \mathrm{rad}(R)$. Let $\pi : R \to R/C'$ be the projection map modulo C'. Then $R/C' \cong S \cong R/C$ implies that $\mathrm{rad}(R/C') = 0$. Since the surjection π takes $\mathrm{rad}(R)$ into $\mathrm{rad}(R/C')$, it follows that $C = \mathrm{rad}(R) \subseteq \ker(\pi) = C'$, as desired. The two cases $C = \mathrm{Nil}_*(R)$ or $\mathrm{Nil}^*(R)$ can be handled similarly. $\square$

Corollary 3.9. (1) *If a local ring $(R, \mathfrak{m})$ has a subring S that maps isomorphically onto the residue division ring $R/\mathfrak{m}$, then $S \prec_{rs} R$, with a unique ideal complement $\mathfrak{m}$.* (2) *If a semiprime ring $S \prec R$ has a complement that is a nilpotent ideal in R, then $S \prec_{rs} R$. (In particular, a semiprime ring S is always a rigid-split corner in any trivial extension of S.)*

Example 3.10. (3.9)(1) above gives a natural source for examples of rigid-split corners that are not "rigid and split". For instance, let $R = \mathbb{Q}[x]$, with the relation $x^2 = 0$. Then R is a local ring with maximal ideal $\mathbb{Q}x$. By (3.9)(1), $\mathbb{Q} \prec_{rs} R$, with a unique ideal complement $\mathbb{Q} \cdot x$. But $\mathbb{Q}$ is *not* a rigid corner; in fact, it has infinitely many complements $\mathbb{Q} \cdot (x - a)$ for a ranging over $\mathbb{Q}$.

Example 3.11. The conclusion in (3.9)(2) is in general not true if S is not semiprime. For instance, if $R = \mathbb{Z}_4[x]$ with the relation $x^2 = 0$, then R is a trivial extension of $S = \mathbb{Z}_4$ by its ideal complement $C = \mathbb{Z}_4 \cdot x$. Here, $C' = \mathbb{Z}_4 \cdot (\bar{2} + x)$ is easily checked to be another ideal complement to S (also with square zero, since $(\bar{2} + x)^2 = \bar{4} + \bar{4}x + x^2 = 0$). Thus, S is not a rigid-split corner in R.

Example 3.12. We close by mentioning that some examples of rigid-split (unital) corners can also be gotten from group ring constructions in (E) above. For any commutative ring k, *the subring k is a rigid-split (unital) corner of a group ring kG iff there is no nontrivial homomorphism from G to the group of units $\mathrm{U}(k)$.* In this case, the augmentation ideal in kG is the unique ideal complement for k. (The proof of this is left as an easy exercise.) Thus, for instance, if G is a group with no subgroup of index 2, then $\mathbb{Z}$ is a rigid-split unital corner in $\mathbb{Z}G$.

4. Split Peirce Corners

Once the notion of corners is formulated, we have the associated notion of split corners, and in particular, split Peirce corners. Prior to this, however, split Peirce corners did not seem to have been fully scrutinized. In this section, we shall prove a few basic facts about split Peirce corners, some of which will be generalized later to arbitrary split corners.

Theorem 4.1. *Given a Peirce corner R_e $(e = e^2)$, let $\langle C_e \rangle$ be the ideal of R generated by C_e. Then, for $f = 1 - e$, we have the equations*

$$\langle C_e \rangle = RfR = e(RfR)e \oplus C_e, \tag{4.2}$$

and a ring isomorphism $R/RfR \cong eRe/e(RfR)e$.

Proof. Since $f \in C_e$, clearly $RfR \subseteq \langle C_e \rangle$. On the other hand, $C_e \subseteq RfR$, so $\langle C_e \rangle \subseteq RfR$. This proves the first equality in (4.2). As for the second equality, the inclusion "$\supseteq$" is clear, and "$\subseteq$" will follow if we can show that $e(RfR)e \oplus C_e$ is an ideal of R. This is a routine check that we can safely leave to the reader. Finally, $e(RfR)e \cap C_e \subseteq eRe \cap C_e = 0$, so the sum on the RHS of (4.2) is direct. The last conclusion of the Proposition follows from the Noether Isomorphism Theorem, as $eRe + RfR = R$, and $eRe \cap RfR = e(RfR)e$. $\qquad\square$

We record below a couple of natural consequences of (4.1).

Corollary 4.3. *Recall that an idempotent $f \in R$ is said to be full if $RfR = R$. This is the case iff $e(RfR)e = eRe$ (where $e = 1 - f$).*

Proof. This follows from the last conclusion of (4.1). $\qquad\square$

Corollary 4.4. (1) $ReR \cap RfR = e(RfR)e \oplus eRf \oplus fRe \oplus f(ReR)f$. In particular, the *RHS is an ideal in R.*

(2) *We have a ring isomorphism:*

$$R/[e(RfR)e \oplus eRf \oplus fRe \oplus f(ReR)f] \cong (R/ReR) \times (R/RfR).$$

Proof. (1) The inclusion "$\supseteq$" is clear. To prove "$\subseteq$", consider any element $r \in ReR \cap RfR$. Write $r = a + b + c + d$ where $a \in eRe$, $b \in eRf$, $c \in fRe$ and $d \in fRf$. After modifying r by $b + c \in eRf \oplus fRe \subseteq ReR \cap RfR$, we are reduced to handling the case $r = a + d$. Now $a = r - d \in eRe \cap RfR = e(RfR)e$, and $d = r - a \in fRf \cap ReR = f(ReR)f$, so $r \in e(RfR)e \oplus f(ReR)f$, which completes the proof of (1).

(2) $ReR + RfR$ is the unit ideal, since it contains $e + f = 1$. Thus, (2) follows from (1) and the Chinese Remainder Theorem. $\qquad\square$

Theorem 4.5 (Split Idempotent Criteria). *For any idempotent $e \in R$, the following conditions are equivalent:*

(1) $R_e \prec_s R$;
(2) $R_e \prec_{rs} R$;
(3) C_e *is an ideal of R;*
(4) $e(RfR)e = 0$;
(5) $exeye = exye$ *for all $x, y \in R$;*
(6) *the map $\varphi : R \to R_e$ defined by $\varphi(x) = exe$ (for any $x \in R$) is a (unital) ring homomorphism.*

If any of these conditions holds, we say that e is a split idempotent of R.

Proof. (1) $\Leftrightarrow$ (2) $\Leftrightarrow$ (3). By (2.8)(2), R_e has a unique complement C_e. Thus, R_e has an ideal complement iff C_e is an ideal of R, in which case R_e is automatically rigid-split.

(3) $\Leftrightarrow$ (4). Note that (2) holds iff $\langle C_e \rangle = C_e$. By (4.2), this holds iff $e(RfR)e = 0$.

(4) $\Leftrightarrow$ (5). This is clear since (4) amounts to $ex(1 - e)ye = 0$ for all $x, y \in R$.

(5) $\Leftrightarrow$ (6) is also clear, since φ is always additive (and unital), and (5) amounts to the fact that φ is multiplicative. $\qquad\square$

Corollary 4.6. *Let e be a split idempotent, and let $f = 1 - e$. Then (1) f is not full unless $e = 0$, and (2) e is not full unless $e = 1$.*

Proof. (1) If $RfR = R$, then (4.5)(4) implies that $eRe = 0$, and so $e = 0$. For (2), assume that $ReR = R$. Then

$$fRe \subseteq (ReR)fRe \subseteq R \cdot e(RfR)e = 0$$

by (4.5)(4), and similarly, $eRf = 0$. Thus, e is a *central* idempotent. But then $R = ReR$ implies that $e = 1$. $\qquad\qquad\square$

As a quick example, in a matrix ring $R = \mathbb{M}_n(k)$ with $n \geq 2$ and $k \neq 0$, the matrix unit $e := e_{11}$ is a *nonsplit* idempotent, and its complementary idempotent $f = e_{22} + \cdots + e_{nn}$ is full. Note that, in some cases, we may have a partial converse to (4.6)(1); for instance, for simple Peirce corners, we have the following.

Corollary 4.7. *If eRe is a simple ring, then $e \in R$ splits iff $f = 1 - e$ is not a full idempotent.*

Proof. The "only if" part follows from (4.6) since $e \neq 0$. Conversely, if f is not full, then by (4.3), $e(RfR)e \neq eRe$. Since $e(RfR)e$ is an ideal of the simple ring eRe, we must have $e(RfR)e = 0$, and so e splits by (4.5). $\qquad\square$

It turns out that the condition $e(RfR)e = 0$ in (4.5)(4) has another nice interpretation in terms of R-module homomorphisms. To formulate the ideas more broadly, we take the viewpoint that any ring R is the endomorphism ring of some (say right) module over some other ring (e.g. the right module R_R).

Proposition 4.8. *Let $R = \mathrm{End}(M_A)$, where M_A is a right module over some ring A. Let $M = P \oplus Q$ be a direct sum decomposition of M_A, and let $e, f \in R$ be, respectively, the projections of M onto P and Q with respect to this decomposition. Then $e(RfR)e = 0$ iff the composition*

$$P \xrightarrow{\alpha} Q \xrightarrow{\beta} P$$

is zero for any A-homomorphisms $\alpha : P \to Q$ and $\beta : Q \to P$.

Proof. Define a map $\varphi : fRe \to \mathrm{Hom}_A(P, Q)$ by $\varphi(g) = g|P$. This is an additive group isomorphism, since it has an inverse φ' given by $\varphi'(h) = h'$ where $h' \in R$ denotes the extension of $h : P \to Q \subseteq M$ with $h'(Q) = 0$. (It is easy to see that $h' \in fRe$.)[d] Similarly, we have an additive group isomorphism $\psi : eRf \to \mathrm{Hom}_A(Q, P)$ defined by restriction to Q. If we think of the isomorphisms φ and ψ as "identifications", the condition $0 = e(RfR)e = (eRf)(fRe)$ translates into the statement that any composition of A-homomorphisms $P \to Q \to P$ is zero. $\qquad\square$

---

[d]This implies, incidentally, that the idempotent e is left semicentral in R iff $\mathrm{Hom}_A(P, Q) = 0$.

Corollary 4.9. *An idempotent $e \in R$ with complementary idempotent f is split iff any composition of R-homomorphisms $eR \to fR \to eR$ is zero.*

Proof. This follows by applying the Proposition to the case $R = \mathrm{End}(R_R)$ and taking $B = eR$, $C = fR$. $\qquad\square$

Corollary 4.10. *If $e = e^2$ is left semicentral, then e and $f = 1 - e$ are split idempotents. Moreover, $eRf = ReR \cap RfR$ is an ideal, and we have a ring isomorphism $R/eRf \cong eRe \times fRf$. (Thus, as long as $eRf \neq 0$, e, f are non-central and non-full idempotents.)*

Proof. From $fRe = 0$, we have of course $e(RfR)e = f(ReR)f = 0$. Thus, e and f are both split according to (4.5).[e] Furthermore, $(4.4)(1)$ simplifies to $ReR \cap RfR = eRf$, so eRf is an ideal of R. The isomorphism $R/eRf \cong eRe \times fRf$ follows from the Peirce decomposition (and is, in fact, a special case of the isomorphism in $(4.4)(2)$). $\qquad\square$

Of course, the second part of this Corollary also follows easily from the usual representation of R as a formal triangular ring $\left(\begin{smallmatrix} eRe & eRf \\ 0 & fRf \end{smallmatrix} \right)$, where eRf is viewed as an (eRe, fRf)-bimodule in the obvious way (by multiplication in R). In general, if S, T are rings and $M = {}_S M_T$ is an (S, T)-bimodule, then the triangular ring $R := \left(\begin{smallmatrix} S & M \\ 0 & T \end{smallmatrix} \right)$ has the property $fRe = 0$ for the complementary idempotents $e = \left(\begin{smallmatrix} 1_S & 0 \\ 0 & 0 \end{smallmatrix} \right)$ and $f = \left(\begin{smallmatrix} 0 & 0 \\ 0 & 1_T \end{smallmatrix} \right)$ in R. Here, $eRf = \left(\begin{smallmatrix} 0 & M \\ 0 & 0 \end{smallmatrix} \right) \neq 0$ if $M \neq 0$.

The following examples show that, for two complementary idempotents $e, f \in R$, the splittings of e and f are, in general, *independent* conditions. The same examples also show that it is possible for e and/or f to be split without being left or right semicentral.

Example 4.11. Take a ring A with a pair of complementary idempotents $\varepsilon, \varepsilon'$ such that $\varepsilon' A \varepsilon = 0 \neq \varepsilon A \varepsilon'$ (that is, ε is left semicentral but not right semicentral), and let $R = \left(\begin{smallmatrix} A & A \\ 0 & A \end{smallmatrix} \right)$. Consider in R the complementary idempotents $e = \left(\begin{smallmatrix} \varepsilon' & 0 \\ 0 & 0 \end{smallmatrix} \right)$ and $f = \left(\begin{smallmatrix} \varepsilon & 0 \\ 0 & 1 \end{smallmatrix} \right)$. These are *not* one-sided semicentral in R, since

$$ fRe = \begin{pmatrix} \varepsilon A \varepsilon' & 0 \\ 0 & 0 \end{pmatrix} \neq 0, \quad \text{and} \quad eRf = \begin{pmatrix} \varepsilon' A \varepsilon & \varepsilon' A \\ 0 & 0 \end{pmatrix} = \begin{pmatrix} 0 & \varepsilon' A \\ 0 & 0 \end{pmatrix} \neq 0. $$

Taking the products of these, we see that $e(RfR)e = 0$, but $f(ReR)f \neq 0$ (using $\varepsilon A \varepsilon' \neq 0$). Thus, e is a split idempotent in R, while its comple-

---

[e]Alternatively, $C_e = eRf \oplus fRf = Rf$ and $C_f = eRe \oplus eRf = eR$ are both ideals by the display $(*)$ prior to (2.11), which gives the same conclusions.

mentary idempotent f is *not* split. More explicitly, we can check that the Peirce corner

$$eRe = \begin{pmatrix} \varepsilon' A \varepsilon' & 0 \\ 0 & 0 \end{pmatrix} \quad \text{has the complement} \quad C_e = \begin{pmatrix} \varepsilon A & A \\ 0 & A \end{pmatrix},$$

which is an ideal since εA is an ideal in A. On the other hand, the Peirce corner

$$fRf = \begin{pmatrix} \varepsilon A \varepsilon & \varepsilon A \\ 0 & A \end{pmatrix} \quad \text{has the complement} \quad C_f = \begin{pmatrix} A\varepsilon' & \varepsilon' A \\ 0 & 0 \end{pmatrix},$$

which is not an ideal since $\varepsilon' A$ is not an ideal in A.

For a more concrete construction, let $A = \mathbb{T}_2(k)$ be the ring of 2×2 upper triangular matrices over a nonzero ring k, and let $\varepsilon = \left(\begin{smallmatrix} 1 & 0 \\ 0 & 0 \end{smallmatrix}\right)$, $\varepsilon' = \left(\begin{smallmatrix} 0 & 0 \\ 0 & 1 \end{smallmatrix}\right)$ in A. Here, indeed, $\varepsilon' A \varepsilon = 0 \neq \varepsilon A \varepsilon'$. The construction above yields the ring

$$R = \{(a_{ij}) \in \mathbb{T}_4(k) : a_{23} = 0\} \subseteq \mathbb{T}_4(k), \tag{4.12}$$

with the complementary idempotents $e = e_{22}$ and $f = e_{11} + e_{33} + e_{44}$, where $\{e_{ij}\}$ are the matrix units. Here, the split corner eRe is just $k \cdot e_{22}$, with the ideal complement

$$C_e = \{(a_{ij}) \in \mathbb{T}_4(k) : a_{22} = a_{23} = 0\} \subseteq \mathbb{T}_4(k). \tag{4.13}$$

On the other hand, the nonsplit corner fRf is

$$\begin{pmatrix} k & 0 & k & k \\ 0 & 0 & 0 & 0 \\ 0 & 0 & k & k \\ 0 & 0 & 0 & k \end{pmatrix}, \quad \text{with the (non-ideal) complement} \quad C_f = \begin{pmatrix} 0 & k & 0 & 0 \\ 0 & k & 0 & k \\ 0 & 0 & 0 & 0 \\ 0 & 0 & 0 & 0 \end{pmatrix}. \tag{4.14}$$

Example 4.15. A suitable modification of the construction above can be used to produce complementary idempotents e, f (in a new ring R) that are *both* split, but not 1-sided semicentral. For $k \neq 0$ as above, let

$$R = \{(a_{ij}) \in \mathbb{T}_4(k) : a_{12} = a_{23} = a_{34} = 0\} \subseteq \mathbb{T}_4(k), \tag{4.16}$$

and take $e = e_{11} + e_{44}$, $f = e_{22} + e_{33}$. Then $eRf = ke_{13} \neq 0$, and $fRe = ke_{24} \neq 0$, so e, f are not one-sided semicentral. But here,

$$e(RfR)e = k \cdot e_{13}e_{24} = 0, \quad \text{and} \quad f(ReR)f = k \cdot e_{24}e_{13} = 0,$$

so e, f are both split. The Peirce corners

$$eRe = ke_{11} + ke_{14} + ke_{44}, \quad \text{and} \quad fRf = ke_{22} + ke_{33}$$

have, respectively, the *ideal* complements

$$C_e = ke_{13} + ke_{22} + ke_{24} + ke_{33}, \quad \text{and} \quad C_f = ke_{11} + ke_{13} + ke_{14} + ke_{24} + ke_{44}.$$

5. Reduction of Corners, and Correspondence of Complements

In this section, we shall prove a number of results that will clarify the special roles played by Peirce corners and unital corners in the general theory of corner rings. Specifically, we shall see that *any corner of a ring R is a unital corner of a Peirce corner of R, and is also a Peirce corner of a unital corner of R.* The significance of this is that, in many cases, the consideration of corners can be reduced to the two cases of Peirce corners and unital corners. We recall from (2.17), however, that there are examples of corners that are neither Peirce corners nor unital corners.

We start with the theme that any corner is a unital corner of a Peirce corner. This is quite easy to see: if $S \prec R$, say with identity e_0, then by (2.3)(1), $S \prec_u R_{e_o} \prec_P R$. Indeed, $R_{e_o} = e_0 R e_0$ is the smallest Peirce corner of R containing S; we shall call it the *associated Peirce corner of S*. In general, by considering *any* Peirce corner R_e containing S, we have the following reduction result relating the complements of S in R and in R_e, which follows easily from (2.9).[f]

Theorem 5.1. *Given $R_e \supseteq S \prec R$, let $\mathcal{C}$ be the set of complements for S in R, and $\mathcal{C}'$ be the set of complements for S in R_e. Then there is a natural one-one correspondence between $\mathcal{C}$ and $\mathcal{C}'$, defined by $C \mapsto C \cap R_e$, and $C' \mapsto C' \oplus C_e$ (where C_e denotes the Peirce complement of R_e, as defined in (2.8)'). In particular, $S \prec_r R$ iff $S \prec_r R_e$.*

The next step is to try to develop some criteria for the corner S above to split in R. Again, we try to make a reduction to the splitting of S in any given Peirce corner R_e containing S (for instance, the associated Peirce corner R_{e_o} of S). For this, we shall use (4.1) and (4.5) as our blueprints, and try to extend them from the case R_e to the case $S \prec R_e$. [In particular, if we choose $e = e_0$ (the identity of S), we'll be reduced to the case of the splitting of a *unital* corner.] We start by taking any complement C of S in R (that is, $C \in \mathcal{C}$) and computing the ideal $\langle C \rangle$ generated by C in R. Throughout, we let $f = 1 - e$; recall that $f \in C$ by (2.2).

---

[f]In particular, (5.1) can be applied to the associated Peirce corner R_{e_o} of S. In this case, there is a slight simplification: for $C \in \mathcal{C}$ in (5.1), the contraction $C \cap R_{e_o}$ can also be expressed as $e_0 C e_0$.

Proposition 5.2. *Let $C' = C \cap R_e$ so that $C = C' \oplus C_e$ (as in (5.1)).
Then*

$$\langle C \rangle = \langle C' \rangle_e + RfR, \qquad (5.3)$$

*where $\langle C' \rangle_e$ denotes the ideal generated by C' in the ring R_e. In particular,
C is an ideal of R iff C' is an ideal of R_e containing $e(RfR)e$.*

Proof. Since $f \in C$ and $C' \subseteq C$, the inclusion "$\supseteq$" is clear in (5.3). For
the reverse inclusion, let $I := \langle C' \rangle_e + RfR$, which contains $C' \oplus C_e = C$.
Thus, equality holds in (5.3) if we can show that I is an *ideal* in R. By
left/right symmetry, it suffices to show that $R \cdot I \subseteq I$. Since $R = S \oplus C$,
this amounts to showing $S \cdot I \subseteq I$ and $C \cdot I \subseteq I$. The former is immediate,
since RfR is an ideal in R, and $\langle C' \rangle_e$ is an ideal in $R_e \supseteq S$. For the latter,
we need only show that $C \cdot \langle C' \rangle_e \subseteq I$. But this is clear since

$$C = C' \oplus C_e \subseteq R_e + RfR,$$

and we have $R_e \cdot \langle C' \rangle_e \subseteq \langle C' \rangle_e$, and $RfR \cdot \langle C' \rangle_e \subseteq RfR \subseteq I$.

If C is an ideal in R, then $C' = C \cap R_e$ is an ideal in R_e, and $f \in C$
implies that $RfR \subseteq C$, whence $e(RfR)e \subseteq C'$. Conversely, assume that
C' is an ideal of R_e containing $e(RfR)e$. By (5.3) and (4.1), we have

$$\langle C \rangle = \langle C' \rangle_e + RfR = C' + e(RfR)e + C_e = C' + C_e = C,$$

so C is an ideal in R. $\qquad\square$

We are now ready to prove the following result on the characterization
of split corners.

Theorem 5.4 (Split Corner Criteria). *Let $R_e \supseteq S \prec R$ as before,
and let $f = 1 - e$. Then S splits in R iff it has an ideal complement in
R_e containing $e(RfR)e$, iff $S \cap e(RfR)e = 0$ and the image of S under
the map $S \to R_e/e(RfR)e$ is a split corner of the ring $R_e/e(RfR)e$. In
particular, $S \prec_{rs} R$ iff $S \cap e(RfR)e = 0$ and the image of S under the map
$S \to R_e/e(RfR)e$ is a rigid-split corner in $R_e/e(RfR)e$.*

Proof. It is sufficient to prove the first "iff" statement. If $S \prec_s R$, choose
for it an *ideal* complement C. By (5.2), $e(RfR)e \subseteq C' := C \cap R_e$, and so

$$S \cap e(RfR)e \subseteq S \cap C' \subseteq S \cap C = 0.$$

Conversely, assume S has an *ideal* complement J in R_e containing $e(RfR)e$.
Let $C := J \oplus C_e$, which is a complement for S in R. Since $C' := C \cap R_e =
J \supseteq e(RfR)e$, (5.2) implies that C is an ideal in R, so we have $S \prec_s R$, as
desired. $\qquad\square$

Corollary 5.5. *If a Peirce corner R_e contains a nonzero split corner S of R, then $f = 1 - e$ is not a full idempotent in R.*

Proof. By (5.4), $S \cap e(RfR)e = 0$. If f was a full idempotent, this would give $0 = S \cap eRe = S$. $\qquad\square$

Remark. If $S \prec_s R_e$ and $R_e \prec_s R$, then (2.16)(2) implies that $S \prec_s R$. However, the converse is not true; namely, $S \prec_s R$ implies only $S \prec_s R_e$, but in general does not imply that $R_e \prec_s R$. We shall demonstrate this with an example below.

Example 5.6. Let k is a field of characteristic $\neq 2$, and let σ be the k-algebra automorphism on $A = k[x]$ defined by $\sigma(x) = -x$. Let $R = A \oplus Ay$, which is made into a ring using the rules $y^2 = 1$, and $y\alpha = \sigma(\alpha)y$ for every $\alpha \in A$. (In other words, R is the quotient obtained from the skew polynomial ring $A[y; \sigma]$ by factoring out the ideal generated by $y^2 - 1$.) Let e, f be the complementary idempotents $(1 + y)/2$ and $(1 - y)/2$ in R. In the factor ring R/RfR, y is identified with 1, so every $\alpha \in A$ is identified with $\sigma(\alpha)$. From this, it is easy to see that $R/RfR \cong k[x]/(x) \cong k$. Thus, $e \notin RfR$ (in particular f is non-full), and the subring $S := k \cdot e \prec_s R$, with an ideal complement RfR. However, by an easy computation,

$$exfxe = \frac{1}{8}(1 + y)x(1 - y)x(1 + y) = x^2(1 + y)/2 = x^2 e \neq 0,$$

so $e(RfR)e \neq 0$. Thus, the associated Peirce corner R_e of S fails to split, although S itself splits. A similar calculation shows that the other Peirce corner R_f is also non-split.

We now introduce the last type of corners in this paper, which is a certain weakening of split corners.

Definition 5.7. We say that a corner $S \prec R$ (with identity e) is *semisplit* in R (written $S \prec_{ss} R$) if S is split in its associated Peirce corner $R_e = eRe$. A split corner is always semisplit, though not conversely. For instance, any Peirce corner is always semisplit, but not necessarily split. (For unital corners, of course, "split" and "semisplit" are synonymous.)

Note that, in this definition, the identity of the corner $S \prec R$ is denoted by e; in other words, the earlier notation e_0 is now replaced simply by e. This will be more convenient since, in the following, we shall only work with the associated Peirce complement R_e of S (instead of any Peirce complement containing S). The following result offers a couple of easy criteria for semisplit corners.

Theorem 5.8 (Semisplit Corner Criteria). *For a corner $S \prec R$ with identity element e, the following are equivalent:*

(1) $S \prec_{ss} R$;
(2) *S has a complement C in R such that $R_e C \subseteq C$ and $CR_e \subseteq C$;*
(3) *S has a complement C in R such that $(Re)C \subseteq C$ and $C(eR) \subseteq C$.*

Proof. $(1) \Rightarrow (2)$. Take an ideal complement I for S in R_e. Then $C := I \oplus C_e$ is a complement for S in R. We have

$$R_e C \subseteq R_e I + R_e C_e \subseteq I + C_e = C,$$

and similarly, $CR_e \subseteq C$, as desired.

$(2) \Rightarrow (3)$. Suppose C exists as in (2). By (2.9), we have $C \supseteq C_e$. Therefore,

$$(Re)C = (R_e + fRe)C \subseteq R_e C + fR \subseteq C + C_e = C.$$

Similarly, we can check that $C(eR) \subseteq C$.

$(3) \Rightarrow (1)$. Suppose C exists as in (3). Its contraction $C \cap R_e = eCe$ is a complement of S in R. The hypotheses on C imply that $R_e(eCe) \subseteq e(ReC)e \subseteq eCe$, and similarly, $(eCe)R_e \subseteq eCe$. Thus, eCe is an *ideal* in R_e, showing that $S \prec_{ss} R$. $\qquad\square$

Next, we take up the second theme of this section, which is that of realizing an arbitrary corner ring as a Peirce corner of a unital corner. This requires some nontrivial work. We begin more generally with the following observation.

Lemma 5.9. *Let S and T be subrings of R with identities e, f such that $ef = fe = 0$. Then*

(1) *$S + T \prec R$ iff $S \prec R$ and $T \prec R$;*
(2) *$S + T \prec_r R$ iff $S \prec_r R$ and $T \prec_r R$.*

Proof. First note that $ST = (Se)(fT) = 0$, and $TS = (Tf)(eS) = 0$. Second, $S \cap T = 0$, for, if $r \in S \cap T$, then $r = er = e(fr) = 0$. Therefore, $S' := S + T \subseteq R$ is a subring with identity $e + f$, and S and T are direct Peirce corners of S'. For convenience, we may identity S' with the ring direct product $S \times T$.

(1) The "only if" part follows from the transitivity of corners. For the "if" part, assume that $S, T \prec R$, say with complements C, D respectively.

For any $t \in T$, we have $ete = e(ft)e = 0$, so (2.7) implies that $t \in C$. Thus, $T \subseteq C$, and hence $C = C' \oplus T$, where $C' := C \cap D$. Now we have

$$R = S \oplus C = S \oplus (T \oplus C') = S' \oplus C',$$

so C' will be a complement to S' if we can show that $S'C' \subseteq C'$ and $C'S' \subseteq C'$. By symmetry, it suffices to show the former, which can be reduced to showing that $SC' \subseteq C'$ and $TC' \subseteq C'$. These in turn will follow if we can show the following four inclusions:

$$SC \subseteq C, \quad SD \subseteq D, \quad TC \subseteq C, \quad \text{and} \quad TD \subseteq D.$$

The first and the fourth are given. For the third, let $t \in T$ and $c \in C$, Then $e(tc)e = e(ft)ce = 0 \Rightarrow tc \in C$ by (2.7). Similarly, for $s \in S$ and $d \in D$, $f(sd)f = f(es)df = 0$ implies $sd \in D$, again by (2.7). This checks that $S' \prec R$, as desired.

(2) For the "only if" part, assume that $S' \prec_r R$. Let C, D, and $C' := C \cap D$ be as in (1) above. Since C' is a complement of S', it is uniquely determined. But then the earlier equation $C = C' \oplus T$ shows that C is also uniquely determined. Therefore, $S \prec_r R$, and similarly, $T \prec_r R$. The "if" part will follow from (5.10)(1) below. $\qquad\square$

The last main result of this section is the following theorem, which is in some sense parallel to (5.1) and (5.4).

Theorem 5.10. *Let S, T be corners of R with identities e, f that are orthogonal idempotents. By (5.9), we have $S' := S \times T \prec R$. If T is rigid in R, with a (unique) complement D, the following conclusions hold:*

(1) *there is a one-one correspondence between $\mathcal{C}$, the set of complements for S in R, and $\mathcal{C}'$, the set of complements for S' in R, given by*

$$\alpha(C) \mapsto C \cap D \quad \text{for} \quad C \in \mathcal{C}, \quad \text{and} \quad \beta(C') = T \oplus C' \quad \text{for} \quad C' \in \mathcal{C}'.$$

In particular, $S \prec_r R$ (and $T \prec_r R$) $\implies S' \prec_r R$.

(2) *$S' \prec_s R$ iff $S \prec_s R$ and $T \prec_s R$. In this case, under the one-one correspondence in (1), $C \in \mathcal{C}$ is an ideal in R iff the corresponding $C' \in \mathcal{C}'$ is an ideal in R; in particular, $S \prec_{rs} R$ iff $S' \prec_{rs} R$.*

Proof. (1) For $C \in \mathcal{C}$, we have shown in the proof of (5.9) that $C \cap D \in \mathcal{C}'$, so $\alpha(C) \in \mathcal{C}'$. On the other hand, for $C' \in \mathcal{C}'$, we have

$$R = S' \oplus C' = S \oplus (T \oplus C').$$

Also, $S \cdot (T \oplus C') = ST + SC' = SC' \subseteq S'C' \subseteq C'$, and similarly, $(T \oplus C')S \subseteq C'$. Thus, $\beta(C') := T \oplus C'$ is a complement to S; that is, $\beta(C') \in \mathcal{C}$. Now, for $C \in \mathcal{C}$, the proof of (5.9) gives

$$\beta(\alpha(C)) = \beta(C \cap D) = (C \cap D) \oplus T = C,$$

so $\beta \circ \alpha$ is the identity on $\mathcal{C}$. Finally, consider any $C' \in \mathcal{C}'$. Since $T \subseteq S'$ are both corners of R, C' can be "enlarged" into a complement for T (by (2.5)). Thus, the rigidity assumption on T forces $C' \subseteq D$. Therefore,

$$\alpha(\beta(C')) = \alpha(T \oplus C') = (T \oplus C') \cap D \supseteq C'.$$

Since C' and $\alpha(\beta(C'))$ are *both* complements of S', this implies that $\alpha(\beta(C')) = C'$, so $\alpha \circ \beta$ is also the identity on $\mathcal{C}'$. We have thus shown that α and β are mutually inverse one-one correspondences between $\mathcal{C}$ and $\mathcal{C}'$. This, of course, implies the last statement in (1).

(2) Assume that $S' \prec_s R$. Since $S' = S \times T$, (2.16)(2) implies that $S \prec_s R$ and $T \prec_s R$. In fact, the proof of (2.16)(2) shows that, for any *ideal* complement $C' \in \mathcal{C}'$, $\beta(C') = C' \oplus T$ is an ideal complement to S. Conversely, assume that $S \prec_s R$ and $T \prec_s R$, and consider any *ideal* complement $C \in \mathcal{C}$ (for S in R). Since T is rigid, $T \prec_s R$ implies that D is an ideal of R. Then $\alpha(C) = C \cap D$ is the intersection of two ideals, and is thus also an ideal. (In particular, $S' \prec_s R$.) This proves the one-one correspondence between the *ideal* complements in $\mathcal{C}$ and those in $\mathcal{C}'$, which, of course, also gives the last conclusion of (2). $\qquad\square$

Corollary 5.11. *If $e_1, \ldots, e_n$ are mutually orthogonal idempotents in R, then $R_{e_1} \times \cdots \times R_{e_n}$ is a rigid corner in R. Its unique complement is $C_e \oplus \bigoplus_{i \neq j} e_i R e_j$, where $e := e_1 + \cdots + e_n$.*

Proof. Since each R_{e_i} is rigid, the rigidity of $R_{e_1} \times \cdots \times R_{e_n}$ follows from the last conclusion of (5.10)(1), plus induction on n. The computation of the (unique) complement of $R_{e_1} \times \cdots \times R_{e_n}$ is left to the reader. (Note that (3.5) is a special case of the present result.) $\qquad\square$

To see how Theorem (5.10) applies to our "second theme" (of realizing a corner as a Peirce corner of a unital corner), let us start with *any* corner $S \prec R$, with identity e. For the complementary idempotent $f := 1 - e$, we have $ef = fe = 0$, so the (rigid) Peirce corner $T := R_f = fRf$ satisfies the hypotheses of (5.9) and (5.10). Since $e + f = 1$, (5.9) implies that $S' := S + T = S \times R_f$ is a *unital* corner of R, and so S is a (direct) Peirce corner of this unital corner. The map $S \mapsto S \times R_f$ is a canonical "suspension process" that produces a unital corner from an arbitrary corner. As we shall see from the sequel of this paper [La$_4$], this suspension process is very useful in analyzing the behavior of the arbitrary corner S. To summarize, let us restate our main conclusions (from (5.10)) about $S \times R_f$, with the appropriate amendments in the present special case.

Corollary 5.12. *For any corner $S \prec R$ with identity e, the "suspension" $S' := S \times R_f$ (for $f = 1 - e$) has the following properties:*

(0) *S' is a unital corner of R, containing S as a direct Peirce corner;*

(1) *the complements of S and those of S' are in one-one correspondence, with*

$$C(\text{complement of } S) \mapsto C \cap C_f = eCe \oplus eRf \oplus fRe, \qquad (5.13)$$

and $C'(\text{complement of } S') \mapsto C' \oplus R_f$. In particular, $S \prec_r R$ iff $S' \prec_r R$; and

(2) *$S' \prec_s R$ iff $S \prec_s R$ and f is a split idempotent. In this case, under the one-one correspondence in (1), ideal complements of S correspond to ideal complements of S'; in particular, $S \prec_{rs} R$ iff $S' \prec_{rs} R$.*

Proof. In (2) of (5.10), the condition that $T = R_f$ be split translates here into the splitting of the idempotent f in (2) above. Besides this, the only other point that requires an explanation is (5.13). Here, C_f is the Peirce complement of the idempotent f; that is, $C_f = eRf \oplus fRe \oplus fRf$. By (5.1), $C = (C \cap R_e) \oplus C_e = eCe \oplus C_e$. Therefore,

$$C \cap C_f = (eCe \oplus C_e) \cap C_f = eCe \oplus eRf \oplus fRe,$$

as claimed in (5.13). $\qquad\qquad\qquad\qquad\qquad\qquad\qquad\qquad\qquad\square$

Note Added in Proof

After the writing of this article, I received an interesting email communication from Professor C. M. Ringel. In this communication, Professor Ringel pointed out that the notion of "split corners" discussed in this article occurred very naturally, and have in fact been used, in the representation theory of finite-dimensional algebras. More specifically, in dealing with "controlled embeddings" in the consideration of the representation types of algebras, one encounters split corners in certain endomorphism algebras. While I am not able to expound on this interesting connection in this paper, I was pleased to learn that representation theory provides another interesting source of examples of split corners.

References

Al. J. Alev: *Quelques propriétés de l'anneau R^G*, II, Comm. Algebra **10** (1982), 203–216.

Bi. G. F. Birkenmeier: *Idempotents and completely semiprime ideals*, Comm. Algebra **11** (1983), 567–580.

BH. G. F. Birkenmeier, H. E. Heatherly, J. Y. Kim, and J. K. Park: *Semicentral idempotents and triangular representations*, Proc. 31st Symp. on Ring Theory and Representation Theory (Osaka, 1998), 1–5, Shinshu Univ., Matsumoto, 1999.

HT. H. E. Heatherly and R. P. Tucci: *Central and semicentral idempotents*, Kyungpook Mth. J. **40** (2000), 255–258.

KR. R. V. Kadison and J. R. Ringrose: *Fundamentals of the Theory of Operator Algebras*, Vol.II, Academic Press, New York, 1983. Cambridge, MA, 1983.

La$_1$. T. Y. Lam: *A First Course in Noncommutative Rings*, Second Edition, Graduate Texts in Math., Vol. **131**, Springer-Verlag, Berlin-Heidelberg-New York, 2001.

La$_2$. T. Y. Lam: *Exercises in Classical Ring Theory*, Second Edition, Problem Books in Mathematics, Springer-Verlag, Berlin-Heidelberg-New York, 2003.

La$_3$. T. Y. Lam: *Lectures on Modules and Rings*, Graduate Texts in Math., Vol. **189**, Springer-Verlag, Berlin-Heidelberg-New York, 1998.

La$_4$. T. Y. Lam: *Corner ring theory: a generalization of Peirce decompositions*, II, in preparation.

LD. T. Y. Lam and A. S. Dugas: *Quasi-duo rings and stable range descent*, J. Pure Appl. Algebra **195** (2005), 243–259.

Mo. M. S. Montgomery: *Fixed Rings of Finite Automorphism Groups of Associative Rings*, Lecture Notes in Math., Vol. **818**, Springer-Verlag, Berlin-Heidelberg-New York, 1980.

Pa. D. S. Passman: *Infinite Crossed Products*, Pure and Applied Mathematics, Vol. **135**, Academic Press, Inc., Boston, Mass., 1989.

Pe. B. Peirce: *Linear associative algebra. With notes and addenda, by C. S. Peirce, Son of the Author*, Amer. J. Math. **4** (1881), 97–229.

UNITS, PRINCIPAL PRIMES, AND GENERALIZATIONS OF FACTORIALITY

PETER MALCOLMSON and FRANK OKOH

Department of Mathematics, Wayne State University,
Detroit, Michigan (USA)
E-mail: petem@math.wayne.edu
E-mail: okoh@math.wayne.edu

A reduced commutative ring R is said to have the *Roquette-Samuel* property if the group of units of any reduced finitely generated extension S of R is finitely generated over the unit group of R. If only finitely many principal prime ideals of R can become the unit ideal in any such S, then we say that R is *robust*. In this paper we extend these notions to the non-commutative setting and to commutative rings with zero-divisors. For integral domains it is known that robustness is equivalent to a generalization of factoriality called PPF (principal primes finite). A domain is PPF if every non-zero element of R is contained in only a finite number of principal prime ideals of R. We show how robustness and PPF transfer between R and $R/\operatorname{Nil}(R)$. We contrast the situation where the PPF condition is required for only non-zero divisors; in this case robustness implies this weaker condition. We compare the generalizations of factoriality due to Fletcher and Bouvier and Galovich with the various versions of PPF. We position Noetherian rings amongst these generalizations of factoriality. The paper ends with some problems suggested by the results therein.

1. PPF for the non-commutative case

A commutative integral domain R is said to have the *R-S property* if for any reduced finitely generated commutative extension S of R, the group of units of S is finitely generated over the group of units of R. Both Roquette and Samuel ([Roq], [S]) gave proofs that the ring of rational integers has this property. This result is used in papers in various areas: manifolds ([Lue]), Diophantine equations ([Lau]), and infinite group theory ([R] and [W]). In several of these applications, it is enough that only a finite number of prime elements of R become units in such an extension S, see for instance p. 112 of [W]. Indeed, there are not many classes of rings with the R-S property. More generally we call a commutative ring R *robust* if only a finite number of principal prime ideals of R become unit ideals in such an extension S.

184

We use a more restricted version of this concept in Section 2.2.

In [MO1] it is shown that robustness is equivalent to a generalization of factoriality called PPF. A domain is PPF if every non-zero element of R is contained in only a finite number of principal prime ideals of R. As remarked in [MO1] (where PPF is called GD(1)), a Dedekind domain is PPF. Apart from the motivation mentioned above for studying robustness, another motivation is the specialization of skew fields to finite characteristic. A finitely generated (over the integers) subring of a skew field should have a non-zero reduction to finite characteristic for all but a finite number of primes. Thus only a finite number of integral primes should be units in the subring. Also the reduction should not depend on the choice of subring (except for finitely many primes). These ideas are due to L. Makar-Limanov and have been mentioned in Section 6 of [J].

As a simple example of this sort of idea, consider the fraction field $\mathbf{Q}(i)$ of the ring of Gaussian integers. Reducing the Gaussian integers $\mathbf{Z}[i]$ mod p results in three possibilities:

> the direct product of two copies of $\mathrm{GF}(p)$ if $p \equiv 1 \mod 4$;
>
> the field $\mathrm{GF}(p^2)$ if $p \equiv 3 \mod 4$; and
>
> the four-element non-reduced ring if $p = 2$.

If instead we reduce the subring $\mathbf{Z}[\frac{i}{15}]$, the results will be the same except for the primes 3 and 5, for which the reduction will be zero. These answers serve as a kind of invariant of the original skew field.

A more complicated situation is the skew field of rational quaternions, for which the corresponding invariants are the 2-by-2 matrix rings over $\mathrm{GF}(p)$. Thus there is some interest in non-commutative versions of the R-S property and robustness. We can try to generalize the R-S property by saying that an integral domain R has the R-S property if for any finitely generated (not necessarily commutative) extension domain S of R, the group of units of S is finitely generated over the group of units of R. Or we may define R *robust* if for any finitely generated (not necessarily commutative) extension domain S, only a finite number of principal prime ideals of R become unit ideals in S. The reason we have restricted to domains is the following example.

Example 1.1. The ring of integers $\mathbf{Z}$ does not have the R-S property in the non-commutative setting. Consider the extension S of the integers $\mathbf{Z}$ generated by $\frac{1}{2}$ and two non-commuting indeterminates X and Y with the relation $XY = 1$. Then the unit group of S includes $u_n = 1 + Y^n X^n$ which has inverse $1 - \frac{1}{2}Y^n X^n$, each of infinite order for every positive integer n. Hence the unit group of S is infinitely generated. This S has no nilpotents, as well.

The generality of the non-commutative definition leads us to expect that some factor ring of the free algebra over $\mathbf{Z}$ on two generators has infinitely many integral primes as units. However this is still unresolved.

The following lemma is an analog, in some non-commutative cases, of the basic lemma of [MO1]. First we define the *skew polynomial ring* $R[X; \phi]$, where R is a ring with identity. Here ϕ is a (surjective) ring automorphism of R. The skew polynomial ring $R[X; \phi] = \{\Sigma_{i=0}^{n} a_i X^i : n \geq 0, a_i \in R\}$ with the usual addition of polynomials, and multiplication determined by $Xr = \phi(r)X$ for all $r \in R$. We say that X *normalizes* R.

Lemma 1.2. *Let J be an ideal (two-sided) of the skew polynomial ring $R[X; \phi]$ with $J \cap R = 0$. Then there is a polynomial $f(X)$ in J and a non-zero element a in R such that for every $g(X)$ in J there is a positive integer m and a polynomial $q(X)$ in $R[X]$ with $a^m g(X) = f(X)q(X)$.*

Proof. Assume $J \neq 0$. Let $f(X)$ be a non-zero polynomial in J of least possible degree $n > 0$ and leading term aX^n.

Suppose the set $S = \{g(X) \in J : \text{no such } m \text{ and no such } q(X) \text{ exist}\}$ is not empty. The zero polynomial is not in S. Choose an element $g(X)$ in S of minimum degree k and leading term bX^k. By the choice of $f(X)$, $k \geq n$.

The leading term of $f(X)\phi^{-n}(b)X^{k-n}$ is abX^k, which is also the leading term of ag. The polynomial $r(X) = ag(X) - f(X)\phi^{-n}(b)X^{k-n} \in J$ and has lower degree than g, so $r(X) \notin S$. Hence $a^m r(X) = f(X)h(X)$ for some positive integer m and some polynomial h in $R[X; \phi]$. Therefore, $a^{m+1}g(X) = a^m f(X)\phi^{-n}(b)X^{k-n} + f(X)h(X)$. Now, $af(X) - f(X)\phi^{-n}(a) \in J$ has lower degree than $f(X)$. Hence $af(X) = f(X)\phi^{-n}(a)$. Applying this m times gives $a^m f(X) = f(X)\phi^{-n}(a^m)$, so that $a^{m+1}g(X) = f(X)q(X)$ for the polynomial $q(X) = \phi^{-n}(a^m b)X^{k-n} + h(X) \in R[X; \phi]$. Thus $g(X) \notin S$. This contradicts the choice of g as an element in S. Therefore, S is an empty set. $\square$

Remark. The lemma is also valid if we let $rX = X\phi(r) + \delta(r)$, where δ is a ϕ-derivation of R. A ϕ-derivation is a map $\delta : R \to R$ satisfying $\delta(rs) = \phi(r)\delta(s) + \delta(r)s$.

2. PPF for commutative rings with zero-divisors

In this section the rings R are commutative with identity and may have zero-divisors. An element in R will be called *regular* if it is not a zero-divisor. To introduce a class of rings between the class of commutative rings and the class of integral domains, we define an element r in R to be *semi-regular* if $\mathrm{Ann}(r) \subseteq \mathrm{Nil}(R)$. The ring R is *semi-regular* if every non-zero element is semi-regular.

We have the following implications: r is regular $\Rightarrow$ $\mathrm{Ann}(r) \subseteq \mathrm{Nil}(R)$ $\Rightarrow$ $r \neq 0$, where $\mathrm{Ann}(r)$ and $\mathrm{Nil}(R)$ stand respectively for annihilator of r and nil radical of R.

2.1. *New versions of PPF*

We introduce the following versions of PPF: PPF(NZ), PPF(SR), and PPF(R) for the cases where every non-zero (respectively, semi-regular and regular) element of R is contained in only finitely many principal prime ideals of R. We shall obtain some relationships amongst them and some relationships between them and robustness. The following is immediate from the definitions.

Proposition 2.1. *PPF(NZ) $\Rightarrow$ PPF(SR) $\Rightarrow$ PPF(R).*

In this section, we say that R is *robust* if for every constant-free ideal I in $R[X]$, only finitely many principal prime ideals become the unit ideal in $R[X]/I$. It is shown in Theorem 1.14 of [MO1] that, at least for domains, this definition is equivalent to that in Section 1. The proof of the next proposition is merely a rearrangement of the proof of Proposition 1.3 of [MO1].

Proposition 2.2. *PPF(NZ) $\Rightarrow$ Robust $\Rightarrow$ PPF(R).*

Proof. Suppose R is PPF(NZ) and let I be a constant-free ideal in $R[X]$. Let $f(X) \in R[X]$ and let $a \in R$ be the leading coefficient of f as in Lemma 1.2. (Here ϕ is the identity automorphism.) Now suppose $p \in R$ is a prime element which is a unit in $R[X]/I$. Then $pg(X) - 1 \in I$ for some polynomial $g(X) \in R[X]$. From Lemma 1.2, there exist $q(X) \in R[X]$ and an integer $m > 0$ such that

$$a^m(pg(X) - 1) = f(X)q(X). \tag{1}$$

Considering (1) $\mod p$ gives

$$-a^m \equiv f(X)q(X) \mod P, \tag{2}$$

where P is the prime ideal in R generated by p. We may consider (2) as an equation in $(R/P)[X]$. If $q(X) = 0$ in $(R/P)[X]$, then $a^m \in P$. Hence $a \in P$ and p divides a. On the other hand, if $q(X) \neq 0$ in $(R/P)[X]$, then $f(X)$ is a constant in $(R/P)[X]$. But $\deg f(X) \geq 1$, so $a \in P$. Since R is PPF(NZ), this implies that only finitely many prime elements become units in $R[X]/I$, i.e., R is robust.

Suppose R is not PPF(R). Then there is a regular element a in R that is in infinitely many principal prime ideals P of R. Let $\langle aX - 1 \rangle$ be the ideal generated by $aX - 1$. Since a is regular, this ideal is constant-free. If $a \in P = \langle p \rangle$, then p divides a. Hence infinitely many primes become units in $R[X]/\langle aX - 1 \rangle$. Thus R is not robust. $\qquad\square$

The rest of the paper will study to what extent the implications in Propositions 2.1 and 2.2 can be reversed. In the process we construct interesting examples of commutative rings. First we look at the transfer of properties between R and $\overline{R}$, where $\overline{R} = R/\operatorname{Nil}(R)$. We refer to the latter as the *reduced quotient of R*.

Lemma 2.3. *If r is regular in R, then $\overline{r} = r + \operatorname{Nil}(R)$ is regular in $R/\operatorname{Nil}(R)$.*

Proof. Suppose $\overline{rx} = 0 \in \overline{R}$. Then $(rx)^k = 0$ in R. Since r is regular, r^k is also regular. Thus $x^k = 0$. Hence $\overline{x} = 0$. $\qquad\square$

Lemma 2.4. *Let R be a commutative ring.*

(a) *Let p_1 and p_2 be principal prime ideals in R. Then p_1 and p_2 are associates in R if and only if $\overline{p_1}$ and $\overline{p_2}$ are associates in the reduced quotient $\overline{R}$ of R.*

(b) *p prime in R implies that $\overline{p}$ is prime in $\overline{R}$.*

(c) *The overline map is injective on the set of principal prime ideals.*

Proof. (a) $\langle \overline{p_1} \rangle = \langle \overline{p_2} \rangle$ implies that $(p_1 - xp_2)^k = 0$ for some positive integer k and some element $x \in R$. Expanding this leads to $p_1^k \in \langle p_2 \rangle$. Thus $\langle p_1 \rangle \subseteq \langle p_2 \rangle$. Symmetry implies that $\langle p_1 \rangle = \langle p_2 \rangle$. Thus p_1 and p_2 are associates.

(b) Assume that p is prime, and also assume that $\overline{yz} = \overline{px}$. Then $(px - yz)^k = 0$ leads to $y^k z^k \in \langle p \rangle$. Since $\langle p \rangle$ is prime, either y or z is in $\langle p \rangle$. Thus $\langle \overline{p} \rangle$ is a prime ideal.

(c) This follows from (a) and (b). $\qquad\square$

The following example shows that the respective converses of Lemmas 2.3 and 2.4(b) are false.

Example 2.5. Let $R = \mathbf{Z}[x, y]$ with $y^2 = 0$, $xy = 0$. Then $\overline{R} = \mathbf{Z}[\overline{x}]$. While x is a zero-divisor, $\overline{x}$ is not a zero-divisor.

Since $\overline{R}$ in Example 2.5 is an integral domain, while R is not an integral domain, the converse of Lemma 2.4(b) is false: $\overline{r}$ prime in $\overline{R}$ does not imply r prime in R. However it is primary:

Lemma 2.6. *Let P be an ideal in R. Then $\overline{P}$ prime in $\overline{R}$ implies that P is a primary ideal in R.*

Proof. Let $xy \in P$. Then $\overline{xy} \in \overline{P}$ implies that $\overline{x}$ (say) is in $\overline{P}$. Therefore, for some $p \in P$, $x - p$ is nilpotent. This implies that some power of x is in P. Thus, P is primary. $\qquad\square$

2.2. *Transfer from the reduced quotient*

We now show that the reduced quotient of R transfers only PPF(R) and robustness to R. Later we will show that R transfers none of these properties to its reduced quotient.

Proposition 2.7. $\overline{R}$ *PPF(R) implies R PPF(R).*

Proof. Suppose R is not PPF(R), then some regular element a is in infinitely many principal prime ideals. Then $\overline{a}$ is not zero in $\overline{R}$. We deduce from Lemma 2.4 that $\overline{R}$ is not PPF(R). $\qquad\square$

Proposition 2.8. $\overline{R}$ *robust implies R robust.*

Proof. Let S be a finitely generated extension of R. We need to show that only finitely many principal prime ideals in R become units in S.

Let i_R, $i_{\overline{R}}$ be the respective embeddings of R and $\overline{R}$ into S and $\overline{S}$, while π_R and π_S are the respective projections of R and S onto $\overline{R}$ and $\overline{S}$. Since i_R is an embedding, $(i_R(x))^n = 0$ implies that $x \in \mathrm{Nil}(R)$. Hence the map $i_{\overline{R}}$ defined by $r + \mathrm{Nil}(R) \mapsto i_R(r) + \mathrm{Nil}(S)$ is a well-defined embedding. If G is a finite set of generators of S over R, then $\pi_S(G)$ is a set of generators of $\overline{S}$ over $\overline{R}$.

Now, let $P = \{\langle p_1 \rangle, \ldots, \langle p_n \rangle, \ldots\}$ be an infinite set of principal prime ideals in R. Then Lemma 2.4 implies that $\pi_R(P)$ is an infinite set of principal prime ideals of $\overline{R}$. By hypothesis, only a finite subset of $\pi_R(P)$

become units in $\overline{S}$. Since $i_{\overline{R}} \circ \pi_R = \pi_S \circ i_R$, we deduce that only a finite subset of P is a set of units in S. $\qquad\qquad\square$

We shall show that $\overline{R}$ transfers no other PPF-property except those in Propositions 2.7 and 2.8. We will be using various quotients of the rings in the following example.

Example 2.9. (Template Example 1.) Let $R_1 = \mathbf{Z}[X, Y_1, Y_2, \ldots, \frac{X}{\mathbf{Y}}]$, where $\mathbf{Y}$ is the set of monomials in the variables $\{Y_1, Y_2, \ldots\}$ and $\frac{X}{\mathbf{Y}} = \{\frac{X}{Y} : Y \in \mathbf{Y}\}$. Template Example 1 is the localization of R_1 at the complement of $\cup_{i=1}^{\infty} \langle Y_i \rangle$. We will denote this ring by T_1.

(Template Example 2.) Let $R_2 = \mathbf{Z}[X, Y_1, Y_2, \ldots, \frac{X}{\mathrm{rad}\,\mathbf{Y}}]$, where $\mathrm{rad}\,\mathbf{Y}$ is the set of monomials in $\mathbf{Y}$ with no square factors and $\frac{X}{\mathrm{rad}\,\mathbf{Y}} = \{\frac{X}{Y} : Y \in \mathrm{rad}\,\mathbf{Y}\}$. Template Example 2 is the localization of R_2 at the complement of $\cup_{i=1}^{\infty} \langle Y_i \rangle$. We will denote this ring by T_2.

Let us note that in T_2, $\frac{X}{Y_i} \notin \langle Y_i \rangle$. To see this, fix i and define $deg(X) = deg(Y_i) = -1$ and deg of other variables and non-zero constants to be 0. As usual $deg(0) = -\infty$ and degree of a product is the sum of the degrees of the factors. Here degree of $\frac{X}{Y}$ is $deg(X) - deg(Y)$ where $Y \in \mathrm{rad}\,\mathbf{Y}$. The degree of a sum is the maximum of the degrees of its terms. In R_2 every element has non-positive degree. Every element in the complement of $\cup_{n=1}^{\infty} \langle Y_n \rangle$ has degree 0. Therefore, the degree function extends to T_2 and is still non-positive. Comparison of degrees shows $\frac{X}{Y_i} \notin \langle Y_i \rangle$.

The Template Examples appear in Sections 2 and 3 of [MO2].

Lemma 2.10 (Lemmas 2.2.1 and 3.3.1 of [MO2]). *For each n, the principal prime ideal $\langle Y_n \rangle$ in T_1 (respectively, T_2) contains the ideal $\langle X \rangle$. Every other prime ideal of T_1 is contained in some $\langle Y_n \rangle$. Moreover, the only principal prime ideals in T_1 (respectively T_2) are $\langle Y_n \rangle$, $n = 1, 2, \ldots$.*

Proof. All but the last statement for T_2 are proved in [MO2]. So let $\langle f \rangle$ be a principal prime ideal in T_2. Since f is not a unit, $f \in \langle Y_i \rangle$ for some i. Since T_2 is a domain, this implies that $\langle f \rangle = \langle Y_i \rangle$. $\qquad\square$

Lemma 2.11. *Let R be T_1 (respectively, T_2). Let $n_1, n_2, \ldots$, be an infinite set of positive integers. Then $\cap_{i=1}^{\infty} \langle Y_{n_i} \rangle \subseteq \langle \frac{X}{\mathbf{Y}} \rangle$ (respectively, $\langle \frac{X}{\mathrm{rad}\,\mathbf{Y}} \rangle$). Furthermore $\langle \frac{X}{\mathbf{Y}} \rangle$ is a prime ideal of T_1 and $\langle \frac{X}{\mathrm{rad}\,\mathbf{Y}} \rangle$ is a maximal ideal of T_2 that does not contain any Y_i.*

Proof. When R is the Template Example T_1 (respectively, T_2), we observe that $R/\langle \frac{X}{\mathbf{Y}} \rangle$ (respectively, $R/\langle \frac{X}{\mathrm{rad}\,\mathbf{Y}} \rangle$) is isomorphic to a localization of a factorial domain. So $\langle \frac{X}{\mathbf{Y}} \rangle$ is a prime ideal of T_1. Also, no non-zero element in $R/\langle \frac{X}{\mathbf{Y}} \rangle$ (respectively, $R/\langle \frac{X}{\mathrm{rad}\,\mathbf{Y}} \rangle$) can have infinitely many prime divisors. But each nonunit irreducible is a Y_i. Therefore the first assertion of the lemma follows.

For the maximality of $\langle \frac{X}{\mathrm{rad}\,\mathbf{Y}} \rangle$, we need only show that the Y_i have inverses in $T_2/\langle \frac{X}{\mathrm{rad}\,\mathbf{Y}} \rangle$. But recall that $Y_i + \frac{X}{Y_i}$ is not in $\cup_{i=1}^{\infty} \langle Y_i \rangle$, using the degree argument. Thus it has an inverse in T_2 and this supplies an inverse to Y_i in $T_2/\langle \frac{X}{\mathrm{rad}\,\mathbf{Y}} \rangle$. This implies that this factor ring is a field. $\qquad\square$

Example 2.12. This is an example of a ring that is not PPF(SR) but its reduced quotient is PPF(NZ). Consider the following quotient of T_1: $R = T_1/I$, where $I = \langle \frac{X}{\mathbf{Y}} \rangle^2$. It follows from Lemma 2.11 that $\mathrm{Nil}(R) = \langle \frac{X}{\mathbf{Y}} + I \rangle$.

To show that R is not PPF(SR), we first show that $X + I$ is semi-regular. We use a variation of the degree function before Lemma 2.10. Here the degree of X is -1 but the degrees of nonzero constants and every other variable is zero. Then the degree extends to T_1 in a similar fashion. If $(X + I)(f + I) = 0$ in R, then $Xf \in I$. It follows that $deg(f)$ is negative, so that $f \in \langle \frac{X}{\mathbf{Y}} \rangle$.

So $X + I$ is a semi-regular element which is in infinitely many principal prime ideals $\langle Y_n + I \rangle$. Thus, R is not PPF(SR).

In contrast, $R/\mathrm{Nil}(R)$ is a localization of a factorial domain. Thus the reduced quotient of R has robustness and all the PPF-properties by Proposition 1.3 of [MO1]. In particular, it is PPF(NZ).

Example 2.13. This is an example of a ring that is not PPF(SR) but its reduced quotient is robust and PPF(SR). Consider the following quotient of the Template Example T_2: $R = T_2/I$, where $I = \langle X \rangle^2$.

Since $X^2 \in \langle \frac{X}{\mathrm{rad}\,\mathbf{Y}} \rangle$, Lemma 2.11 implies $\mathrm{Nil}(R) \subseteq \langle \frac{X}{\mathrm{rad}\,\mathbf{Y}} + I \rangle$.

Suppose that g is any element in $\langle \frac{X}{\mathrm{rad}\,\mathbf{Y}} \rangle$. Then $g = \frac{X}{Y}h$ for some monomial $Y \in \mathrm{rad}\,\mathbf{Y}$, where h has no factor in common with Y. If for such g, $g + I$ is nilpotent, then $g^k \in \langle X^2 \rangle$ for some k. If Y is not a constant, choose some factor Y_i of Y. Use this i to define the degree function as before Lemma 2.10. The degree argument there shows that the degree of g is negative, so either h is divisible by X or Y is a constant. In either case $g + I \in \langle X + I \rangle$. Hence, $\mathrm{Nil}(R) = \langle X + I \rangle = \mathrm{Ann}(X + I)$.

Therefore, the element $X + I \in R$ is semi-regular. Since $\langle X + I \rangle$ is contained in $\langle Y_n + I \rangle$, $n = 1, 2, \ldots$ and the latter are principal prime ideals of R, R is not PPF(SR). In $\overline{R}$, $(Y_n + \langle X \rangle)(\frac{X}{Y_n} + \langle X \rangle) = 0$. Since every non-unit element of $\overline{R}$, is in some $\langle Y_n \rangle + \langle X \rangle$ and $\mathrm{Nil}(\overline{R}) = 0$, we deduce that $\overline{R}$ has no nonunit semi-regular elements. Hence it is vacuously PPF(SR).

We show that $\overline{R}$ is robust by showing that every prime element in $\overline{R}$ is a zero-divisor. Since $I \subseteq \langle X \rangle$, $\overline{R}$ is isomorphic to $T_2/\langle X \rangle$.

Let J be $\langle X \rangle \subseteq T_2$ and suppose $f + J$ is a prime element in $\overline{R}$. Since it is not a unit, $f = Y_i q$ for some q in T_2 and some positive integer i. Since $f + J = (Y_i + J)(q + J)$, it divides $Y_i + J$ or $q + J$. In the former case $f + J$ is a zero-divisor because, as seen above, $Y_i + J$ is a zero-divisor in $\overline{R}$. In the latter case, we get that $f + J = Y_i f r + J$ for some $r \in R$. Thus $(f + J)((1 - Y_i r) + J) = 0$. To complete the argument we show that $1 - Y_i r \notin J$. But if $1 - Y_i r = X h$ for some h in T_2, then $1 = Y_i(r + \frac{X}{Y_i}h)$. This contradicts Y_i is not a unit in T_2.

Since all prime elements are zero-divisors, they cannot be units in any extension of $\overline{R}$. Thus $\overline{R}$ is robust.

The next proposition summarizes Examples 2.5, 2.9, 2.12, and 2.13.

Proposition 2.14. *The reduced quotient of an arbitrary ring transfers only robustness and PPF(R) to the ring.*

2.3. *Transfer to the reduced quotient*

Our next goal is to show that an arbitrary ring R does not transfer robustness nor any of the PPF's to its reduced quotient. The ring in the example will be PPF(NZ) while its reduced quotient, $\overline{R}$, is a domain that is not PPF(R). Hence by Propositions 2.1 and 2.2, $\overline{R}$ has no PPF-property nor is it robust.

Example 2.15. As in the Template Examples, $\frac{X}{Y}$ stands for the set $\{\frac{X}{Y} : Y$ a monomial in Y_i, $i = 1, 2, \ldots\}$. Let $R = \mathbf{Q}[X, Z, Y_i, i = 1, 2, \ldots, \frac{X}{Y}]/I$, where $\mathbf{Q}$ is the field of rational numbers and $I = \langle Z^2, Y_i Z, i = 1, 2, \ldots \rangle$.

Each element F of R has a unique representation in the form $F_1(X, Y) + cZ$, where $c \in \mathbf{Q}$ and $F_1(X, Y)$ is a polynomial in X and the variables Y_i, $i = 1, 2, \ldots$. (This is because $XZ + I = (\frac{X}{Y_1} + I)(Y_1 Z + I) = 0$.)

Under the above representation of elements of R, let M be the multiplicative set of elements of R with non-zero constant term as a term in F_1. The required ring is the localization of R at M, but we still denote it by R.

This R is a quasi-local ring with maximal ideal consisting of the elements of R with zero constant term.

Lemma 2.16. $\mathrm{Nil}(R) = \langle Z + I \rangle$. *Moreover, the reduced quotient of R is an integral domain.*

Proof. When we set $Z = 0$ in R, the ring R becomes an integral domain. Thus $\langle Z + I \rangle$ is a prime ideal. The fact that $\langle Z + I \rangle$ is also nilpotent implies that it is $\mathrm{Nil}(R)$. $\qquad\square$

Lemma 2.17. *The ideal $\langle Z + I \rangle$ is the only principal prime ideal of R.*

Proof. Since $Z + I$ is nilpotent, every prime ideal contains $Z + I$. Suppose $\langle f + I \rangle$ is a principal prime ideal of R. (As R is not a domain, $f \neq 0$.) Then $Z + I = (f + I)(g + I)$. Write $f = F_1(X, Y) + cZ$, $g = G_1(X, Y) + dZ$.

Letting $Z = 0$ in the above equation leads to $\overline{f}\,\overline{g} = 0$ in the domain $\overline{R}$. If $\overline{f} = 0$ in $\overline{R}$, then $f = cZ$ in R. We have $c \neq 0$, since $f \neq 0$. Since $c \in \mathbf{Q}$, $\langle f + I \rangle = \langle Z + I \rangle$. If $\overline{g} = 0$ in $\overline{R}$, then $g = dZ$, d a non-zero constant. Then $Z = dZ(F_1(X, Y))$ in the domain $\mathbf{Q}[X, Z, Y_i, i = 1, 2, \ldots]$. Therefore the constant term of f is a unit. Thus $f \in M$ and so f is a unit, contradicting the assumption that f is a principal prime. This proves the lemma. $\qquad\square$

It follows immediately from Lemma 2.17 that R is PPF(NZ). Hence by Propositions 2.1 and 2.2, R is PPF(SR) and PPF(R), and also robust.

Lemma 2.18. *The reduced quotient of the ring R in Example 2.15 is not PPF(R).*

Proof. As already remarked, the reduced quotient of R is an integral domain. In particular, X is a regular element in $\overline{R}$. Each $\langle Y_i \rangle + \mathrm{Nil}(R)$ is a prime ideal in $\overline{R}$; indeed, when we set $Y_i = 0$, then $\frac{X}{Y} = Y_i\frac{X}{Y} = 0$. Therefore, $\overline{R}/\langle Y_i \rangle$ is a localization of a polynomial ring in the other Y's with coefficients in $\mathbf{Q}$.

Since $X \in \langle Y_i \rangle$ for each $i = 1, 2, \ldots$, $\overline{R}$ is not PPF(R). Hence by Propositions 2.1 and 2.2, $\overline{R}$ is not robust, not PPF(NZ), nor PPF(SR). $\qquad\square$

We summarize the above results in the next proposition.

Proposition 2.19. *An arbitrary ring R may not transfer robustness nor any of the PPFs to its reduced quotient.*

3. New generalizations of factoriality

In this section we shall show that none of the implications from Propositions 2.1 and 2.2 can be reversed. Our main tool will be the Template Examples in 2.9. We show that our results are new generalizations of factoriality to commutative rings with zero-divisors. The variety of examples in this section shows that the PPF-properties are less restrictive than other generalizations of factoriality to commutative rings with zero-divisors.

3.1. *Reversing Implications*

We recall the following implications from Propositions 2.1 and 2.2.

$$\text{PPF(NZ)} \Rightarrow \text{PPF(SR)} \Rightarrow \text{PPF(R)}$$
$$\text{PPF(NZ)} \Rightarrow \text{Robust} \Rightarrow \text{PPF(R)}.$$

Proposition 3.1. *PPF(R) $\not\Rightarrow$ PPF(SR). Hence PPF(R) $\not\Rightarrow$ PPF(NZ). Also Robust $\not\Rightarrow$ PPF(SR) and Robust $\not\Rightarrow$ PPF(NZ).*

Proof. Our starting point is the Template Example T_2. Let $R = T_2/\langle X \rangle^2$. This is Example 2.13 from a different point of view.

Since $Y_n \frac{X^2}{Y_n} = X^2$, we deduce that $Y_n + \langle X \rangle^2$ is a zero-divisor for each positive integer n. Every non-unit is a multiple of some $Y_n + \langle X \rangle^2$ by the definition of T_2. Therefore, every non-unit in R is a zero-divisor. Since zero-divisors cannot be units, R is vacuously robust. Hence R is PPF(R).

As proved in Example 2.13, $\text{Nil}(R) = \langle X + \langle X \rangle^2 \rangle = \text{Ann}(X + \langle X \rangle^2)$. Therefore, $X + \langle X \rangle^2$ is semi-regular. Since it is contained in the principal prime ideals $\langle Y_n + \langle X \rangle^2 \rangle$, $(n = 1, 2, \ldots)$, R is not PPF(SR). $\qquad\square$

If we had used T_1 (i.e., $\mathbf{Y}$ in place of $\text{rad}\,\mathbf{Y}$) in Proposition 3.1, there would have been more nilpotent elements to deal with, e.g., $(\frac{X}{Y_i})^3 = \frac{X^3}{Y_i^3} = X^2(\frac{X}{Y_i^3})$. Thus $\frac{X}{Y_i}$ would be nilpotent.

For the ring R of Proposition 3.1 every non-unit is a zero-divisor. The following more familiar examples of such rings will be relevant at the end of the paper.

1. A direct product of at least two copies of the two-element field.
2. Start with a quasi-local ring (R, M). Then every non-unit element in R/M^2 is a zero-divisor.

Proposition 3.2. *There is a robust ring that also has the PPF(SR)-property but does not have the PPF(NZ)-property.*

Proof. Again we use the Template Example T_2. The required ring is $R = T_2/I$, where $I = \langle XY_1 \rangle$. In this ring $X + I$ is nilpotent since $X^2 = XY_1 \frac{X}{Y_1}$. Therefore, $\langle X + I \rangle \subseteq \text{Nil}(R)$. Since $XY_1 \in \langle \frac{X}{\text{rad}\,\mathbf{Y}} \rangle$ Lemma 2.11 implies $\text{Nil}(R) \subseteq \langle \frac{X}{\text{rad}\,\mathbf{Y}} + I \rangle$.

Suppose that g is any element in $\langle \frac{X}{\text{rad}\,\mathbf{Y}} \rangle$. Then $g = \frac{X}{Y} h$ for some monomial $Y \in \text{rad}\,\mathbf{Y}$, where h has no factor in common with Y. If for such g, $g + I$ is nilpotent, then $g^k \in \langle XY_1 \rangle$ for some k. If Y is not a constant, choose some factor Y_i. (We repeat a prior argument.) Use this i to define

194

the degree function as before Lemma 2.10. The degree argument there shows that the degree of g is negative, so either h is divisible by X or Y is a constant. In either case $g + I \in \langle X + I \rangle$. Hence, $\mathrm{Nil}(R) = \langle X + I \rangle$.

Now suppose $g + I$ is semi-regular in R and that $g + I$ is in a principal prime ideal, $\langle f + I \rangle$. Since f is not a unit, $f \in \langle Y_i \rangle$ for some i. Thus $f + I = (Y_i + I)(q + I)$, for some $q \in T_2$.

Suppose $f + I$ divides $q + I$. This leads to $f + I = (Y_i + I)(f + I)(r + I)$ for some r in T_2. Thus $(f + I)(1 - Y_i r) + I) = 0$. Since $g + I$ is in $\langle f + I \rangle$, $(g+I)(1-Y_i r)+I) = 0$. Since $g+I$ is semi-regular, $(1-Y_i r)+I \in \langle X+I \rangle = \mathrm{Nil}(R)$. Since $I \subseteq \langle X \rangle$, this implies that $1 - Y_i r = Xh$ for some $h \in T_2$. Thus $1 = Y_i(r + \frac{X}{Y_i}h)$ in T_2. But Y_i is not a unit in T_2.

Therefore, we must have that $f + I$ divides $Y_i + I$. Hence $\langle f + I \rangle = \langle Y_i + I \rangle$. Since $\langle Y_i \rangle \supseteq I$, if $g + I$ is in infinitely many $\langle Y_n \rangle + I$, then g is in infinitely many $\langle Y_n \rangle$. Then Lemma 2.11 implies $g \in \langle \frac{X}{\mathrm{rad}\,\mathbf{Y}} \rangle$.

We now show that no element of $\langle \frac{X}{\mathrm{rad}\,\mathbf{Y}} \rangle + I$ is semi-regular. Suppose $Y \in \mathrm{rad}\,\mathbf{Y}$ and $\frac{X}{Y} + I$, is semi-regular. Since $(Y_1 Y + I)(\frac{X}{Y} + I) = 0 + I$, we get that $Y_1 Y + I \in \langle X + I \rangle = \mathrm{Nil}(R)$. Therefore $Y_1 Y - Xg \in I$ for some $g \in T_2$. This leads to $Y_1 Y - Xg = XY_i h$ for some $h \in T_2$. Thus $Y = \frac{X}{Y_1}g + Xh$. Hence $Y \in \langle \frac{X}{Y_1} \rangle \subseteq T_2$. This contradicts the last part of Lemma 2.11.

We have proved that R is PPF(SR). Since the non-zero element $X + I$ is in infinitely many principal prime ideals of R, namely $\langle Y_n + I \rangle, n = 1, 2, \ldots$, R is not PPF(NZ).

Since $\overline{R} \cong T_2/\langle X \rangle$, the same ring in the last paragraph of Example 2.13, $\overline{R}$ is robust. Hence R is robust by Proposition 2.14. $\qquad\square$

The Template Example T_1 could not have been used, because there would have been more nilpotent elements. For example, $\frac{X^2}{Y_1^2} = (XY_1)\frac{X}{Y_1^3}$.

Remark. Here is another illustration of the difference between the Template Examples 1 and 2. Let $S_1 = T_1/I_1$ where $I_1 = \langle \frac{X}{\mathbf{Y}} \rangle^2$ and let $S_2 = T_2/I_2$ where $I_1 = \langle \frac{X}{\mathbf{Y}} \rangle^2$ and $I_2 = \langle \frac{X}{\mathrm{rad}\,\mathbf{Y}} \rangle^2$. The ideals $\langle Y_n + I_1 \rangle$ are principal prime ideals in S_1, while the ideals, $\langle Y_n + I_2 \rangle$ are not prime ideals in S_2 because $(\frac{X}{Y_n})^2 \in I_1 \subseteq \langle Y_n \rangle$ and $\frac{X}{Y_n} \notin \langle Y_n \rangle$, by the degree argument before Lemma 2.10. The ring S_1 is robust but not PPF(SR), as shown in Example 2.12. The ring S_2 is a quasi-local ring with only one prime ideal $\langle \frac{X}{\mathrm{rad}\,\mathbf{Y}} \rangle$, as implied by Lemma 2.11. Hence it is PPF(NZ).

The fact that PPF(NZ) implies robustness raises the following problem.

Problem 1. Does PPF(R) or PPF(SR) imply robustness?

As noted in [MO1] there are domains that are not PPF(R). The product of infinitely many copies of $\mathbf{Z}$ is not PPF(R) because for any prime integer p the regular element $(p, p, \ldots)$ is in all of the principal prime ideals $\mathbf{Z} \times \mathbf{Z} \times \ldots \times \mathbf{Z} \times p\mathbf{Z} \times \mathbf{Z} \times \ldots$. This ring is not *connected* in the sense that there are nontrivial idempotents.

Proposition 3.3. *There is a connected commutative ring (with zero-divisors) that is not PPF(R).*

Proof. Let $R_1 \subseteq \mathbf{Q}[X, Y]$ be defined by

$$R_1 = \{f(X, Y) \in \mathbf{Q}[X, Y] : f(0, 0) \in \mathbf{Z}\} = \mathbf{Z} + X\mathbf{Q}[X, Y] + Y\mathbf{Q}[X, Y].$$

The variables X and Y are in R_1. Since $X = p(\frac{1}{p}X)$, $X \in \langle p \rangle$ for all rational primes p. So R_1 is not PPF(NZ); since R_1 is a domain it is not PPF(R) either. To get the required example, let $R = R_1[Z]/I$, where $I = \langle XZ, (Y-1)Z \rangle$. In this ring, $X + I$ and $Z + I$ are zero-divisors. Later we shall see that $Y + I$ is not a zero-divisor.

The composition of the natural maps $R_1 \to R \to R/\langle Z \rangle \to R_1$ is the identity map, so R_1 embeds in R. Now we show that rational primes p remain primes in R. As already noted, $\langle p \rangle$ contains X and Y, hence also Z since $YZ = Z$. Since $R_1/\langle X, Y \rangle \cong \mathbf{Z}$ it follows that $R/\langle p \rangle \cong \mathbf{Z}/\langle p \rangle$. Thus p remains prime in R.

To see that the element $Y + I$, where $I = \langle XZ, (Y-1)Z \rangle$, is not a zero-divisor in R, we note that every element in R has a unique representative of the form $g(X, Y) + h(Z) + m$, where m is an integer and $g(X, Y)$ is a polynomial in X and Y with coefficients in $\mathbf{Q}$ with $g(0, 0) = 0$ and $h(Z)$ is a polynomial in Z with coefficients in $\mathbf{Q}$ with $h(0) = 0$.

Suppose $(Y + I)(f + I) = 0$ in R. Using the above representation of elements of R we get that $Yf = Yg(X, Y) + Yh(Z) + mY = XZG_1 + (Y-1)ZG_2$ in $R_1[Z]$, for some $G_1, G_2 \in R_1[Z]$. Letting $Z = 0$ leads to $g(X, Y) + m = 0$. Therefore, $f = h(Z)$. Hence, $Yh(Z) = XZG_1 + (Y-1)ZG_2$ in $R_1[Z]$. Letting $X = 0$ and $Y = 1$ in the last equation leads to $f = 0$. Thus $Y + I$ is not a zero-divisor.

Since $Y + I$ is in $\langle p \rangle$ for each rational prime p, R is not PPF(R).

Expressing an alleged idempotent $f = f^2$ in terms of its unique representative $g(X, Y) + h(Z) + m$, we must have $g(X, Y) = g(X, Y)^2 + 2mg(X, Y)$. Such a g must be a constant by degree arguments. A similar argument eliminates h. Hence idempotents are trivial and thus R is connected. $\square$

On the positive side, the next proposition gives many examples that are PPF(R).

Proposition 3.4. *A Noetherian ring is PPF(SR), hence PPF(R).*

Proof. We first note some oft-used facts.

(a) Any factor of a semi-regular element is semi-regular.
(b) If $r \in R$ is semi-regular and $rs = 0$ for some s in R, then s is nilpotent and $1 - s$ is a unit.

The proof of the proposition begins in a standard way. Let $S = \{\langle x \rangle : x$ is semi-regular and x has infinitely many prime divisors$\}$. If S is not empty, let $\langle x \rangle = M$ be a maximal element in S. Let $x = p_1 y$, where p_1 is prime. So $\langle x \rangle \subseteq \langle y \rangle$. If $\langle x \rangle = \langle y \rangle$, we would get $x(1 - p_1 q_1) = 0$ for some element q_1 in R. By (b), $1 - p_1 q_1$ is nilpotent. However this would lead to the contradiction that p_1 is a unit. Therefore $\langle x \rangle$ is strictly contained in $\langle y \rangle$. By (a), y is semi-regular.

Therefore by the choice of x, y has only finitely many non-associate prime divisors, $q_1, \ldots, q_r$ (say). We claim that the set of non-associate prime divisors of x is $F = \{p_1, q_1, \ldots, q_r\}$. Suppose

q is prime, $q \mid x$ and q is not an associate of any element in F.

Since $x = p_1 y$, we get that $q \mid p_1$ or $q \mid y$. If the latter holds then q is associated to an element of F, a contradiction. So $q \mid p_1$. Then $qr = p_1$ for some $r \in R$. Therefore $p_1 \mid q$ or $p_1 \mid r$.

Suppose $p_1 \mid q$. Then $q = p_1 s$ for some $s \in R$; thus $qr = p_1$ leads to $q(1 - rs) = 0$. Since $q \mid x$ and x is semi-regular, so is q by (a). Therefore, by (b), $1 - rs$ is nilpotent. This implies that rs, hence r and s are units. Since $q = p_1 s$, this implies that q and p_1 are associate. This contradicts the assumption on q.

If instead $p_1 \mid r$, then $qr = p_1$ leads to $p_1(1 - tq) = 0$ for some $t \in R$. Since x is semi-regular and $q \mid x$, (a) and (b) imply that tq is a unit, contradicting the primeness of q.

Therefore, there is no such q. We have now proved that the finite set F is the set of non-associate prime divisors of x . This contradicts the choice of x, so S is empty. $\qquad\square$

3.2. *Factorial rings in the literature*

Bouvier, Galovich, and Fletcher have investigated factoriality in commutative rings with zero-divisors, see [B], [G], and [F]. (There is an alternative treatment of Fletcher's results in [AM]. As pointed out there, Bouvier anticipated Galovich.) To state their generalizations, we recall some definitions.

A non-zero non-unit element $r \in R$ is *irreducible* if its only divisors in R are units and associates of r. Two elements a and b are *associates* if $aR = bR$. Bouvier and Galovich define R to be a UFR if for every non-zero non-unit element a in R, $a = p_1 \cdots p_n$, where the p_i's are irreducible and if also $a = q_1 \cdots q_m$, where the q_j's are irreducibles, then $m = n$ and, after a suitable renumbering, p_i and q_i are associates for $i = 1, \ldots, n$.

Fletcher's definitions are somewhat different. A *refinement* of a factorization $r = a_1 \cdots a_n$ is obtained by factoring one or more of the a_i's. The *U-class* of $r \in R$ is defined by $U(r) = \{b \in R : bcr = r$ for some $c \in R\}$. (Note that for r semi-regular, $U(r)$ is just the group of units of R.) In [F] a non-zero non-unit element $r \in R$ is said to be *irreducible* if each factorization of r has a refinement with r as a factor. A *U-decomposition* of $r \in R$ is a factorization $r = (p_1' \cdots p_k')(p_1 \cdots p_n)$ such that p_i', p_j are irreducible, $p_i' \in U(p_1 \cdots p_n)$ and $p_j \notin U(p_1 \cdots \hat{p_j} \cdots p_n)$ for $i = 1, \ldots k$ and $j = 1, \ldots, n$, where $\hat{p_j}$ means that p_j is omitted.

Then R is *factorial in the sense of Fletcher* if every non-zero, non-unit element r in R has a U-decomposition and $r = (p_1' \cdots p_k')(p_1 \cdots p_n) = (q_1' \cdots q_l')(q_1 \cdots q_m)$ implies that $m = n$ and p_i and q_i are associate (after a suitable renumbering) for $i = 1, \ldots, n$.

The next two theorems are amalgams of Theorems 3.1 and 3.2 in [A]. To state them we recall that a *special principal ideal ring* (SPIR) is a principal ideal ring with a unique prime ideal that is nilpotent. An SPIR is trivially PPF(NZ).

Theorem 3.5 ([B] and [G]). *A commutative ring R is a UFR (in the sense of Bouvier-Galovich) if and only if R is either a factorial domain, SPIR, or a quasi-local ring (R, M) with $M^2 = 0$.*

Theorem 3.6 ([F]). *A commutative ring is a factorial ring in the sense of Fletcher if and only if it is a finite product of factorial domains and SPIR's.*

Corollary 3.7. *A commutative ring that is a UFR is PPF(NZ).*

Proof. A factorial domain or an SPIR is PPF(NZ). Suppose (R, M) is quasi-local with $M^2 = 0$. Then M is the only prime ideal of R because every prime ideal contains all nilpotent elements. So R is clearly PPF(NZ). $\square$

Corollary 3.8. *A commutative ring R that is factorial in the sense of Fletcher is PPF(SR) but not necessarily PPF(NZ).*

Proof. The ring $\mathbf{Z} \times \mathbf{Z}$ is a Fletcher factorial ring. The non-zero element $(0,1)$ is contained in infinitely many principal prime ideals: $\langle(\langle p\rangle, \mathbf{Z})\rangle$ as p ranges over the prime ideals of $\mathbf{Z}$. Hence $\mathbf{Z} \times \mathbf{Z}$ is not PPF(NZ).

Let $R = R_1 \times R_2 \times \cdots \times R_n$ where each R_i is either a factorial domain or an SPIR. Note the following:

(a) A prime ideal of R is of the form $R_1 \times R_2 \times \cdots \times P_i \times R_{i+1} \times \cdots \times R_n$, where P_i is a prime ideal in R_i;
(b) The product of those components of R that are factorial is reduced;
(c) The product of the SPIR's, while not reduced, has only finitely many prime ideals.

Suppose $r = (r_1, \ldots, r_n)$ is a semi-regular element in R. Then each component r_i is non-zero and therefore is contained in only finitely many principal prime ideals of R_i because each R_i is PPF(NZ). That R is PPF(SR) now follows from (a), (b), and (c). $\qquad\square$

As seen in Proposition 3.4, a Noetherian ring is PPF(SR), hence PPF(R). We now show that a familiar non-Noetherian ring is also PPF(SR). The direct product R of infinitely many copies of the two-element field is reduced and any nonzero nonunit element is a zero-divisor. Thus R is vacuously PPF(SR). On the other hand, the element $(1, 0, 0, \ldots)$ is in infinitely many principal prime ideals, so R is not PPF(NZ). But by Theorems 3.5 and 3.6, R is not factorial in the sense of Bouvier-Galovich nor in the sense of Fletcher.

The following supplementary comments about Noetherian rings follow from Theorems 3.5 and 3.6. The Noetherian domain $\mathbf{R} + X\mathbf{C}[X]$ is shown in Proposition 2.7 of [MO1] to be PPF(NZ), but it is not factorial in the sense of Bouvier-Galovich nor in the sense of Fletcher. By contrast the Noetherian domain $\mathbf{Z} \times \mathbf{Z}$ is factorial in the sense of Fletcher but not factorial in the sense of Bouvier-Galovich. As seen in the proof of Corollary 3.8, $\mathbf{Z} \times \mathbf{Z}$ is not PPF(NZ).

In light of Corollaries 3.7 and 3.8, PPF(NZ), PPF(SR), PPF(R), and robustness are new generalizations of factoriality to commutative rings with zero-divisors. Theorems 3.5 and 3.6 suggest the following problem.

Problem 2. Do any of the conditions PPF or robustness have a complete characterization analogous to Theorems 3.5 and 3.6?

For domains, where all PPF's and robustness coincide, it is shown in Proposition 1.4 of [MO1] that $R[X]$ is PPF(NZ) if and only if R is PPF(NZ).

Theorem 2.7 of [AM] states, in part, that $R[X]$ is factorial in the sense of Fletcher if and only if R is a finite direct product of factorial domains. These results motivate the next problem.

Problem 3. In the presence of zero-divisors, which of the PPF-classes and robustness are closed under polynomial ring extensions?

In [MO1] a commutative ring R is defined to be 1-*robust* if only finitely many principal prime ideals of R become unit ideals in every $R[X]/I$ with I a constant-free ideal (this is *robust* as it appears in Section 2 of this paper), while R is said to be $\aleph_0$-*robust* if for every positive integer n and every constant-free ideal I in $R[X_1, \ldots, X_n]$, only finitely many principal prime ideals of R become unit ideals in $R[X_1, \ldots, X_n]/I$ (this is *robust* as it appears in Section 1 of this paper). In Theorem 1.14 of [MO1] it was shown that an integral domain is 1-robust if and only if it is $\aleph_0$-robust. The fact that for a field k the polynomial ring $k[X]$ is 1-robust, hence $\aleph_0$-robust, gives an alternative approach to Zariski's version of the Nullstellensatz in Corollary 1.15 of [MO1].

For any commutative ring R, $\aleph_0$-*robust* clearly implies n-robust for every positive integer n.

Problem 4. Let R be a commutative ring with zero-divisors. Does 1-robust imply $\aleph_0$-robust? Or is there a bound on n for which 1-robust implies k-robust for $2 \le k \le n$?

References

A.   D. D. Anderson, "Extensions of unique factorization: a survey", in Advances in commutative ring theory (Fez), *Lecture Notes in Pure and Appl. Math.*, vol. 205, Dekker, New York, 1999, 31–53.

AM.   D. D. Anderson and R. Markanda, Unique factorization rings with zero divisors, *Houston J. Math* **11** (1985), 15–30, corrigendum 423–426.

B.   A. Bouvier, Structure des anneaux à factorisation unique, *Publ. Dép. Math. (Lyon)* **11** (1974), 39–49.

F.   C. R. Fletcher, The structure of unique factorization rings, *Math. Proc. Cambridge Philos. Soc.* **67** (1970), 535–540.

G.   S. Galovich, Unique factorization rings with zero divisors, *Math. Mag.* **51** (1978), 276–283.

J.   A. Joseph, "Some ring theoretic techniques and open problems in enveloping algebras" in Noncommutative rings, edited by S. Montgomery and L. Small, *MSRI Publications* vol. 24, Springer Verlag, 1992.

Lau.   M. Laurent, Exponential differential equations, *Invent. Math.* **78**(2) (1984), 299–327.

Lue. J. Luecke, Finite covers of 3-manifolds containing essential tori, *Trans. Amer. Math. Soc.* **310** (1988), 381–391.

MO1. P. Malcolmson and F. Okoh, Expansions of prime ideals, To appear in *Rocky Mountain J. Math.*.

MO2. P. Malcolmson and F. Okoh, "Minimal prime ideals and generalizations of factorial domains", in Rings, modules, algebras, and abelian groups, edited by A. Facchini, E. Houston, and L. Salce, *Lecture Notes in Pure and Appl. Math.*, vol. 236, Dekker, New York, 2004, 401–410.

R. D. J. S. Robinson, Decision problems for soluble groups of finite rank, *Illinois J. Math* **30**(2) (1986), 197–213.

Roq. P. Roquette, Einheiten und Divisorklassen in endlich erzeugbaren Körpern, *Jahresber. Deutsch. Math.-Verein.* **10** (1958), 1–17.

S. P. Samuel, A propos du théorème des unités, *Bull. Soc. Math.* (2) **90** (1966), 89–96.

W. B. A. F. Wehrfritz, Groups of automorphisms of solvable groups, *Proc. London Math. Soc.* (3) **20** (1970), 101–122.

ALGEBRA ASSOCIATED WITH THE PRINCIPAL BLOCK OF CATEGORY $\mathcal{O}$ FOR $sl_3(\mathbb{C})$

FRANTIŠEK MARKO

Pennsylvania State University
76 University Drive, Hazleton, PA 18202, USA
and
Mathematical Institute, Slovak Academy of Sciences
Štefánikova 49, 814 38 Bratislava, Slovakia
E-mail: fxm13@psu.edu

In this paper we describe explicitly the algebra associated with the principal block of category $\mathcal{O}$ for complex Lie algebra $sl_3(\mathbb{C})$ by a quiver and relations and represent it as an endomorphism algebra of a semilocal module over a commutative local selfinjective algebra.

Introduction

This paper was motivated by a desire to gain a better understanding of explicit examples of quasi-hereditary algebras $A(\mathcal{G})$ associated to integral regular blocks of category $\mathcal{O}$ for semisimple complex Lie algebras $\mathcal{G}$.

Many important properties of the algebras $A(\mathcal{G})$ were established by [2], [12], [3] and [1]. In particular, the algebras $A(\mathcal{G})$ were presented as endomorphism algebras of modules over a commutative selfinjective algebra in a fundamental work [12].

Motivated by [12], the quasi-heredity of endomorphism algebras of semilocal modules over commutative local selfinjective algebras were investigated in [7]. Subsequently, the extensive study [4] on stratifications of endomorphism algebras originated as an effort to understand the connection between the results of [7] and those of [12]. The quasi-hereditary algebras constructed in [7] are believed to cover algebras associated with integral regular blocks with simple multiplicities, see [4], (5.1.6).

For the principal block of $\mathcal{G} = sl_2(\mathbb{C})$ the algebra $A(\mathcal{G})$ is a 5-dimensional algebra given by the quiver $\Gamma : 1 \leftrightarrows 2$ and the relation $121 = 0$. In the next case of $G = sl_3(\mathbb{C})$ the algebra $A(\mathcal{G})$ was investigated earlier by König [10] who gave its Loewy filtration. Its structure is presented in a brief survey by Dlab [6].

We describe the algebra $A(sl_3(\mathbb{C}))$ explicitly by a quiver and relations and reconfirm that this algebra is obtained by construction of [7] and represent it in the simple form.

After this note was written we have learned about the work [13]. That work was started after the results of this note were presented at the Colloquium of the Quebec Mathematical Society in October 1997. Apparently each author was unaware of the work of the other. Results of [13], where the algebras $A(\mathcal{G})$ were computed for Lie algebras of rank 1 and 2 as well as for $\mathcal{G} = sl_4(\mathbb{C})$, are based on the explicit presentation of $A(\mathcal{G})$ as an endomorphism algebra given in [12]. The methods used in our work are different and describe the projective modules and maps explicitly. Besides that, the relations in [13] involve fractional coefficients whereas our relations involve only integral coefficients (plus or minus 1). However, both presentations of $A(sl_3(\mathbb{C}))$, in terms of a quiver and relations, are isomorphic (C. Stroppel, personal communication). Also, these presentations are isomorphic to the description of the algebra $A(sl_3(\mathbb{C}))$ given in [6] (V. Dlab, personal communication).

1. Basic properties of category $\mathcal{O}$

For basic description, additional explanations and properties of category $\mathcal{O}$ consult [2], [8] and [9].

Let $\mathcal{G}$ be a finite-dimensional semisimple complex Lie algebra, $\mathcal{H}$ be its Cartan subalgebra, $\mathcal{H}^*$ be a dual space of $\mathcal{H}$, $\mathcal{W}$ be the Weyl group of $\mathcal{G}$, R be the root system of $\mathcal{G}$ with fixed system of positive roots $R^+ \subset R$ and $\mathcal{G} = \mathcal{N}_- \oplus \mathcal{H} \oplus \mathcal{N}_+$ be a Cartan decomposition. We have the Borel subalgebras $\mathcal{B}_- = \mathcal{N}_- \oplus \mathcal{H}$ and $\mathcal{B}_+ = \mathcal{N}_+ \oplus \mathcal{H}$ of $\mathcal{G}$ and the universal enveloping algebras $\mathfrak{U}(\mathcal{G})$, $\mathfrak{U}(\mathcal{B}_+)$ and $\mathfrak{U}(\mathcal{N}_-)$, respectively of $\mathcal{G}$, $\mathcal{B}_+$ and $\mathcal{N}_-$, respectively.

The objects of the *category* $\mathcal{O}$ are finitely generated left $\mathcal{G}$-modules M that are

— $\mathcal{H}$-diagonalizable (meaning that $M = \oplus_{\mu \in \mathcal{H}^*} M_\mu$, where $M_\mu = \{m \in M : h.m = \mu(h)m$ for all $h \in \mathcal{H}\}$ is the μ-weight space),
— $\mathcal{N}_+$-finite (that is the vector space $\mathfrak{U}(\mathcal{N}_+)m$ is finite dimensional for all $m \in M$). The morphisms in the category $\mathcal{O}$ are $\mathcal{G}$-module morphisms.

The category $\mathcal{O}$ is an abelian category with finite-dimensional homomorphism spaces that contains all Verma modules. The Verma module $\Delta(\lambda)$ associated to a weight $\lambda \in \mathcal{H}^*$ is given as

$$\Delta(\lambda) = \mathfrak{U}(\mathcal{G}) \otimes_{\mathfrak{U}(\mathcal{B}_+)} \mathbb{C}v,$$

where the one-dimensional vector space $\mathbb{C}v$ is made into a $\mathcal{B}_+$-module by defining $\mathcal{N}_+.v = 0, h.v = \lambda(h)v$ for $h \in \mathcal{H}$.

It follows from the Poincare-Birkhoff-Witt theorem that each $\Delta(\lambda)$ is a free $\mathfrak{U}(\mathcal{N}_-)$-module of rank 1. Every Verma module $\Delta(\lambda)$ has a unique maximal submodule. The corresponding factor module $L(\lambda)$ is an irreducible $\mathcal{G}$-module.

Let $\mathcal{Z}(\mathcal{G})$ be the center of $\mathfrak{U}(\mathcal{G})$ and Θ be the set of characters of $\mathcal{Z}(\mathcal{G})$. The category $\mathcal{O}$ decomposes into a direct sum of blocks $\mathcal{O} = \oplus_{\theta \in \Theta} \mathcal{O}_\theta$, where the subcategory $\mathcal{O}_\theta$ consists of modules M from $\mathcal{O}$ annihilated by some power of $(z - \theta(z))$ for every $z \in \mathcal{Z}(\mathcal{G})$. Let us fix a character θ and the corresponding block $\mathcal{O}_\theta$. By the Harish-Chandra theorem, the set $\Lambda = \Lambda_\theta$ of weights λ, for which the irreducible module $L(\lambda)$ belongs to the block $\mathcal{O}_\theta$, is an orbit of the dot action of $\mathcal{W}$ on $\mathcal{H}^*$ and is therefore finite. Here the dot action of $w \in \mathcal{W}$ on weight λ is defined as $w.\lambda = w(\lambda + \rho) - \rho$, where $\rho = \frac{1}{2} \sum_{\alpha \in R^+} \alpha$. Every Verma module $\Delta(\lambda)$ admits a finite filtration by irreducible modules. By $[\Delta(\lambda) : L(\mu)]$ we will denote the multiplicity of $L(\mu)$ in the composition series of $\Delta(\lambda)$.

A theorem of Bernstein-Gelfand-Gelfand (see Theorem 2.10 of [8]) describes a partial order $\prec$ on Λ that determines morphisms between Verma modules, namely, $\lambda \prec \mu$ if and only if there is a nonzero $\mathcal{G}$-morphism from $\Delta(\lambda)$ to $\Delta(\mu)$. If such a morphism exists then it is unique up to a scalar multiple and it is injective. The partial order $\prec$ is described with the help of the Bruhat order on $\mathcal{W}$ and the partial order on $\mathcal{H}^*$ given by positive roots of $\mathcal{G}$. Let us note that there is a unique maximal element (which is a dominant weight) of Λ with respect to the partial order $\prec$.

For each $\lambda \in \Lambda$, there is an indecomposable projective cover $P(\lambda) \in \mathcal{O}_\theta$ of the irreducible module $L(\lambda)$. Every projective module $P(\lambda)$ has a filtration by Verma modules (called Verma composition series). We will denote by $[P(\lambda) : \Delta(\mu)]$ the multiplicity of $\Delta(\mu)$ in the Verma composition series of $P(\lambda)$. The Bernstein-Gelfand-Gelfand reciprocity law states that

$$[P(\lambda) : \Delta(\mu)] = [\Delta(\mu) : L(\lambda)].$$

The category $\mathcal{O}_\theta$ is a highest weight category that is equivalent to its dual category $\mathcal{O}_\theta^o$. This equivalence is given by a duality functor $F : \mathcal{O}_\theta \to \mathcal{O}_\theta^o$ induced by an anti-involution σ on $\mathcal{G}$ fixing $\mathcal{H}$ pointwise and sending $\mathcal{N}_+$

to $\mathcal{N}_-$. Moreover, F fixes irreducible objects and sends projective objects $P(\lambda)$ to injective objects $I(\lambda)$. The image $F(\Delta(\lambda))$ of a Verma module $\Delta(\lambda)$ is denoted by $\nabla(\lambda)$ and is called a dual Verma module.

The category $\mathcal{O}_\theta$ is equivalent to the category of right modules over the (associative) algebra $A = A_\theta = End_{\mathcal{G}}(P)$, where $P = \oplus_{\lambda \in \Lambda} P(\lambda)$. The images of the projective modules $P(\lambda)$, irreducible modules $L(\lambda)$, Verma modules $\Delta(\lambda)$ and dual Verma modules $\nabla(\lambda)$, respectively, under this equivalence are the right projective A-modules $P_A(\lambda)$, simple modules $S_A(\lambda)$, standard modules $\Delta_A(\lambda)$ and costandard modules $\nabla_A(\lambda)$, respectively. The algebra A is a quasi-hereditary algebra with respect to the partial order $\prec$ on Λ and the duality $F : \mathcal{O}_\theta \to \mathcal{O}_\theta^o$ leads to an antiisomorphism i of A such that $i^2 = 1$.

2. Indecomposable projective modules in $\mathcal{O}_0$ for $sl_3(\mathbb{C})$

From now on, we will assume that $\mathcal{G} = sl_3(\mathbb{C})$ and the block $\mathcal{O}_\theta$ is the integral regular block $\mathcal{O}_0$ given by the character θ corresponding to the weight 0 (see Section 3.4 of [9] for details).

This special case describes all integral blocks that are regular because by translation principle (see Section 2.1 of [12] and Section 4.12 of [9]) every integral regular block of $sl_3(\mathbb{C})$ is Morita equivalent to $\mathcal{O}_0$.

Denote by α and β simple roots of $\mathcal{G}$. Then the root system Φ of $\mathcal{G}$ has roots $\Phi = \{\alpha, \beta, \alpha + \beta, -\alpha, -\beta, -\alpha - \beta\}$. We will work with the representation of $sl_3(\mathbb{C})$ given by generators $e_\alpha, e_\beta, e_{\alpha+\beta}, f_\alpha, f_\beta, f_{\alpha+\beta}, h_\alpha, h_\beta$ over $\mathbb{C}$ and relations

$$[h_\alpha, h_\beta] = 0, [h_\alpha, e_\alpha] = 2e_\alpha, [h_\alpha, e_\beta] = -e_\beta, [h_\beta, e_\alpha] = -e_\alpha,$$
$$[h_\beta, e_\beta] = 2e_\beta, [e_\alpha, f_\alpha] = h_\alpha, [e_\beta, f_\beta] = h_\beta, [e_\alpha, e_\beta] = e_{\alpha+\beta},$$
$$[f_\alpha, f_\beta] = -f_{\alpha+\beta}, [e_\alpha, e_{\alpha+\beta}] = [e_\beta, e_{\alpha+\beta}] = 0, [f_\alpha, f_{\alpha+\beta}] = [f_\beta, f_{\alpha+\beta}] = 0.$$

For the block $\mathcal{O}_0$, the set Λ has six weights $\Lambda = \{0, -\alpha, -\beta, -2\alpha - \beta, -\alpha - 2\beta, -2\alpha - 2\beta\}$. Indecomposable projective modules $P(\lambda)$ are direct summands of bigger projective modules defined by $\Pi(\lambda) = \mathfrak{U}(sl_3(\mathbb{C}))/I(\lambda)$, where $I(\lambda)$ is generated by relations $h.1 = \lambda(h)$ for $h \in \mathcal{H}$ and weight spaces $\mathfrak{U}(sl_3(\mathbb{C}))_\mu$ for $\lambda + \mu \not\leq 0$, see Proposition 15 of Section (2.10) in [11].

Denote the image of the unit element 1 in $\Pi(\lambda)$ by v_λ. Then $\Pi(\lambda)$ is a free $\mathfrak{U}(\mathcal{N}_-)$-module by Lemma 6 of Section (2.10) in [11]. It is generated by

v_λ for $\lambda = 0$;

$v_\lambda, e_\alpha v_\lambda$ for $\lambda = -\alpha$;

$v_\lambda, e_\beta v_\lambda$ for $\lambda = -\beta$;

$v_\lambda, e_\alpha v_\lambda, e_\beta v_\lambda, e_\alpha e_\beta v_\lambda, e_{\alpha+\beta} v_\lambda, e_\alpha^2 v_\lambda, e_\alpha^2 e_\beta v_\lambda, e_\alpha e_{\alpha+\beta} v_\lambda$ for $\lambda = -2\alpha - \beta$;

$v_\lambda, e_\alpha v_\lambda, e_\beta v_\lambda, e_\alpha e_\beta v_\lambda, e_{\alpha+\beta} v_\lambda, e_\beta^2 v_\lambda, e_\alpha e_\beta^2 v_\lambda, e_\beta e_{\alpha+\beta} v_\lambda$ for $\lambda = -\alpha - 2\beta$

and

$$v_\lambda, e_\alpha v_\lambda, e_\beta v_\lambda, e_\alpha^2 v_\lambda, e_\alpha e_\beta v_\lambda, e_{\alpha+\beta} v_\lambda, e_\beta^2 v_\lambda, e_\alpha^2 e_\beta v_\lambda, e_\alpha e_{\alpha+\beta} v_\lambda,$$
$$e_\alpha e_\beta^2 v_\lambda, e_\beta e_{\alpha+\beta} v_\lambda, e_\alpha^2 e_\beta^2 v_\lambda, e_\alpha e_\beta e_{\alpha+\beta} v_\lambda, e_{\alpha+\beta}^2 v_\lambda \text{ for } \lambda = -2\alpha - 2\beta.$$

The generators described above are highest vectors of the Verma modules they generate. Therefore it is easy to determine a filtration of each projective module $\Pi(\lambda)$ by these Verma modules and their multiplicities. Using a theorem of Bernstein-Gelfand-Gelfand (see Theorem 2.10 of [8]) mentioned earlier we determine that Verma modules generated by elements $e_\alpha v_\lambda, e_\beta v_\lambda$ for $\lambda = -2\alpha - \beta$ and $\lambda = -\alpha - 2\beta$, and $e_\alpha^2 v_\lambda$, $e_\alpha e_\beta v_\lambda$, $e_{\alpha+\beta} v_\lambda$, $e_\beta^2 v_\lambda$ for $\lambda = -2\alpha - 2\beta$ are filtration factors of projective modules that do not belong to the block $\mathcal{O}_0$.

Clearly, $[\Pi(\lambda) : \Delta(\lambda)] = 1$. Since $[\Pi(-\alpha) : \Delta(0)] = 1$ and $[\Pi(-\beta) : \Delta(0)] = 1$, we have

$$\Pi(0) = P(0), \Pi(-\alpha) = P(-\alpha) \quad \text{and} \quad \Pi(-\beta) = P(-\beta).$$

Denote by $\Pi(\lambda)_0$ the direct sum of indecomposable projective summands of $\Pi(\lambda)$ belonging to the block $\mathcal{O}_0$.

Since $[\Pi(-2\alpha - \beta) : \Delta(-\alpha)] = 2$, $[\Pi(-2\alpha - \beta) : \Delta(-\beta)] = 1$ and $[\Pi(-2\alpha - \beta) : \Delta(0)] = 2$, the projective module $\Pi(-2\alpha - \beta)_0$ decomposes as

$$\Pi(-2\alpha - \beta)_0 = P(-2\alpha - \beta) \oplus P(-\alpha).$$

In the same manner we obtain the decomposition

$$\Pi(-\alpha - 2\beta)_0 = P(-\alpha - 2\beta) \oplus P(-\beta).$$

Finally, $[\Pi(-2\alpha - 2\beta) : \Delta(-2\alpha - \beta)] = [\Pi(-2\alpha - 2\beta) : \Delta(-\alpha - 2\beta)] = 1$, $[\Pi(-2\alpha - 2\beta) : \Delta(-\alpha)] = [\Pi(-2\alpha - 2\beta) : \Delta(-\beta)] = 2$ and $[\Pi(-2\alpha - 2\beta) : \Delta(0)] = 3$ implies that

$$\Pi(-2\alpha - 2\beta)_0 = P(-2\alpha - 2\beta) \oplus P(-\alpha) \oplus P(-\beta).$$

Recall that an element v of a $\mathcal{G}$-module M is called a primitive vector if it is a weight vector relative to $\mathcal{H}$, and there exists a submodule N of M not containing v such that $\mathcal{B}_+ v \subset N$. To find the indecomposable projective modules, we describe a Verma composition series of $P(\lambda)$ by computing a set U_λ of primitive vectors u_λ^μ, for $\lambda \prec \mu$, which generate $P(\lambda)$ as $\mathfrak{U}(\mathcal{N}_-)$-module and satisfy the property that $\mathfrak{U}(\mathcal{N}_+)u_\lambda^\mu$ is included in the $\mathfrak{U}(\mathcal{N}_-)$-module generated by u_λ^ν for $\mu \npreceq \nu$.

Lemma 2.1. *The sets of primitive vectors for $P(0), P(-\alpha)$ and $P(-\beta)$ are*

$$U_0 = \{u_0^0 = v_0\},$$
$$U_{-\alpha} = \{u_{-\alpha}^{-\alpha} = v_{-\alpha}, u_{-\alpha}^0 = e_\alpha v_{-\alpha}\} \text{ and}$$
$$U_{-\beta} = \{u_{-\beta}^{-\beta} = v_{-\beta}, u_{-\beta}^0 = e_\beta v_{-\beta}\}.$$

The set $U_{-\alpha-2\beta}$ of primitive vectors in $P(-\alpha - 2\beta)$ is

$$u_{-\alpha-2\beta}^0 = e_\alpha e_\beta^2 v_{-\alpha-2\beta},$$
$$u_{-\alpha-2\beta}^{-\alpha} = (e_\beta^2 + af_\alpha e_\beta e_{\alpha+\beta})v_{-\alpha-2\beta},$$
$$u_{-\alpha-2\beta}^{-\beta} = \left(e_\alpha e_\beta + e_{\alpha+\beta} + (-a+1)f_\beta e_\beta e_{\alpha+\beta}\right)v_{-\alpha-2\beta},$$
$$u_{-\alpha-2\beta}^{-\alpha-2\beta} = \Big(2 - f_\alpha e_\alpha + 2f_\beta e_\beta + (a+1)f_\alpha f_\beta e_{\alpha+\beta} + 2af_{\alpha+\beta}e_{\alpha+\beta}$$
$$+ (2a-2)f_\beta f_{\alpha+\beta}e_\beta e_{\alpha+\beta}\Big)v_{-\alpha-2\beta},$$

where a is an arbitrary complex number.

The set $U_{-2\alpha-\beta}$ of primitive vectors for $P(-2\alpha - \beta)$ is given as

$$u_{-2\alpha-\beta}^0 = e_\beta e_\alpha^2 v_{-2\alpha-\beta},$$
$$u_{-2\alpha-\beta}^{-\alpha} = (e_\alpha e_\beta - 2e_{\alpha+\beta} + bf_\alpha e_\alpha^2 e_\beta)v_{-2\alpha-\beta},$$
$$u_{-2\alpha-\beta}^{-\beta} = \left(e_\alpha^2 - (b + \frac{1}{2})f_\beta e_\alpha^2 e_\beta\right)v_{-2\alpha-\beta},$$
$$u_{-2\alpha-\beta}^{-2\alpha-\beta} = \Big(2 + 2f_\alpha e_\alpha - f_\beta e_\beta - (2b+2)f_\alpha f_\beta e_{\alpha+\beta} + 2bf_{\alpha+\beta}e_{\alpha+\beta}$$
$$+ 4bf_\alpha f_{\alpha+\beta}e_\alpha e_{\alpha+\beta}\Big)v_{-2\alpha-\beta},$$

where b is any complex number.

Finally, the set $U_{-2\alpha-2\beta}$ of primitive vectors of $P(-2\alpha-2\beta)$ is described *by*

$$u^0_{-2\alpha-2\beta} = (e_\alpha^2 e_\beta^2 - 2e_\alpha e_\beta e_{\alpha+\beta})v_{-2\alpha-2\beta},$$

$$u^{-\alpha}_{-2\alpha-\beta} = (e_\alpha e_\beta^2 - 2e_\beta e_{\alpha+\beta} + f_\alpha e_\alpha e_\beta e_{\alpha+\beta} - f_\alpha e_{\alpha+\beta}^2)v_{-2\alpha-2\beta},$$

$$u^{-\beta}_{-2\alpha-2\beta} = (e_\alpha^2 e_\beta - f_\beta e_\alpha e_\beta e_{\alpha+\beta})v_{-2\alpha-2\beta},$$

$$\begin{aligned}
u^{-2\alpha-\beta}_{-2\alpha-2\beta} = (&+2e_\beta - f_\beta e_\beta^2 + 2f_\alpha e_\alpha e_\beta - 2f_\alpha f_\beta e_\beta e_{\alpha+\beta} + f_\alpha^2 e_\alpha e_{\alpha+\beta} \\
&- cf_\alpha^2 f_\beta e_\alpha e_\beta e_{\alpha+\beta} - \frac{3}{2}f_\alpha^2 f_\beta e_{\alpha+\beta}^2 + f_\alpha f_{\alpha+\beta} e_\alpha e_\beta e_{\alpha+\beta} \\
&- f_\alpha f_{\alpha+\beta} e_{\alpha+\beta}^2)v_{-2\alpha-2\beta},
\end{aligned}$$

$$\begin{aligned}
u^{-\alpha-2\beta}_{-2\alpha-2\beta} = \Big(&2e_\alpha - f_\alpha e_\alpha^2 + 2f_\beta e_\alpha e_\beta - 2f_\beta e_{\alpha+\beta} + 2f_\alpha f_\beta e_\alpha e_{\alpha+\beta} \\
&+ 2f_{\alpha+\beta} e_\alpha e_{\alpha+\beta} - f_\beta^2 e_\beta e_{\alpha+\beta} + df_\alpha f_\beta^2 e_\alpha e_\beta e_{\alpha+\beta} \\
&- (d+\frac{3}{2})f_\alpha f_\beta^2 e_{\alpha+\beta}^2 + (2d+1)f_\beta f_{\alpha+\beta} e_\alpha e_\beta e_{\alpha+\beta} \\
&- (2d+3)f_\beta f_{\alpha+\beta} e_{\alpha+\beta}^2\Big)v_{-2\alpha-2\beta},
\end{aligned}$$

$$\begin{aligned}
u^{-2\alpha-2\beta}_{-2\alpha-2\beta} = \Big(&-2 + f_\alpha e_\alpha + f_\beta e_\beta + 6f_\alpha f_\beta e_\alpha e_\beta - 3f_\alpha f_\beta e_{\alpha+\beta} \\
&+ 3f_{\alpha+\beta} e_\alpha e_\beta - 2f_{\alpha+\beta} e_{\alpha+\beta} + (3-d)f_\alpha^2 f_\beta e_\alpha e_{\alpha+\beta} \\
&+ (5-2d)f_\alpha f_{\alpha+\beta} e_\alpha e_{\alpha+\beta} + (c-3)f_\alpha f_\beta^2 e_\beta e_{\alpha+\beta} \\
&- f_\beta f_{\alpha+\beta} e_\beta e_{\alpha+\beta} + \Big(\frac{d}{2}+\frac{3}{4}\Big)f_\alpha^2 f_\beta^2 e_{\alpha+\beta}^2 \\
&+ (c+d+3)f_\alpha f_\beta f_{\alpha+\beta} e_\alpha e_\beta e_{\alpha+\beta} \\
&+ \Big(\frac{3}{2}+d\Big)f_{\alpha+\beta}^2 e_\alpha e_\beta e_{\alpha+\beta} \\
&- (d+\frac{3}{2})f_{\alpha+\beta}^2 e_{\alpha+\beta}^2\Big)v_{-2\alpha-2\beta},
\end{aligned}$$

where c, d are complex numbers.

Proof. The defining properties of primitive vectors are expressed by systems of linear equations and the primitive vectors are computed as solutions of such linear systems. $\qquad\square$

3. Morphisms between projective modules in $\mathcal{O}_0$ for $sl_3(\mathbb{C})$

An important step towards the understanding of morphisms in the category $\mathcal{O}_0$ is a description of irreducible morphisms between indecomposable projective modules $P(\lambda)$.

A morphism $f : M \to N$ is called irreducible if neither one of the identity morphisms 1_M and 1_N factors through f, and if $f = hg$ for some

$g : M \to X$ and $h : X \to N$, then either 1_M factors through g or 1_N factors through h. Informally it means that an irreducible morphism cannot be factored nontrivially.

First, we will describe irreducible morphisms $P(\lambda) \to P(\mu)$ in the case when $\mu \not\leqq \lambda$.

Lemma 3.1. *The irreducible morphism $P(\lambda) \to P(\mu)$ for $\mu \not\leqq \lambda$ are the following:*

$\iota_{0,-\alpha} : P(0) \to P(-\alpha)$ *is given by* $\iota_{0,-\alpha}(u_0^0) = u_{-\alpha}^0,$

$\iota_{0,-\beta} : P(0) \to P(-\beta)$ *is given by* $\iota_{0,-\beta}(u_0^0) = u_{-\beta}^0,$

$\iota_{-\alpha,-\alpha-2\beta} : P(-\alpha) \to P(-\alpha - 2\beta)$ *is given by* $\iota_{-\alpha,-\alpha-2\beta}(u_{-\alpha}^{-\alpha}) = u_{-\alpha-2\beta}^{-\alpha},$

$\iota_{-\alpha,-2\alpha-\beta} : P(-\alpha) \to P(-2\alpha - \beta)$ *is given by* $\iota_{-\alpha,-2\alpha-\beta}(u_{-\alpha}^{-\alpha}) = u_{-2\alpha-\beta}^{-\alpha},$

$\iota_{-\beta,-\alpha-2\beta} : P(-\beta) \to P(-\alpha - 2\beta)$ *is given by* $\iota_{-\beta,-\alpha-2\beta}(u_{-\beta}^{-\beta}) = u_{-\alpha-2\beta}^{-\beta},$

$\iota_{-\beta,-2\alpha-\beta} : P(-\beta) \to P(-2\alpha - \beta)$ *is given by* $\iota_{-\beta,-2\alpha-\beta}(u_{-\beta}^{-\beta}) = u_{-2\alpha-\beta}^{-\beta},$

$\iota_{-2\alpha-\beta,-2\alpha-2\beta} : P(-2\alpha - \beta) \to P(-2\alpha - 2\beta)$ *is given by*

$$\iota_{-2\alpha-\beta,-2\alpha-2\beta}(u_{-2\alpha-\beta}^{-2\alpha-\beta}) = u_{-2\alpha-2\beta}^{-2\alpha-\beta},$$

$\iota_{-\alpha-2\beta,-2\alpha-2\beta} : P(-\alpha - 2\beta) \to P(-2\alpha - 2\beta)$ *is given by*

$$\iota_{-\alpha-2\beta,-2\alpha-2\beta}(u_{-\alpha-2\beta}^{-\alpha-2\beta}) = u_{-2\alpha-2\beta}^{-\alpha-2\beta}.$$

Proof. It is easy to check that the above maps are morphisms using the values of these maps at other primitive vectors. They are:

$$\iota_{-\alpha,-\alpha-2\beta}(u_{-\alpha}^0) = u_{-\alpha-2\beta}^0, \quad \iota_{-\alpha,-2\alpha-\beta}(u_{-\alpha}^0) = u_{-2\alpha-\beta}^0,$$
$$\iota_{-\beta,-\alpha-2\beta}(u_{-\beta}^0) = u_{-\alpha-2\beta}^0, \quad \iota_{-\beta,-2\alpha-\beta}(u_{-\beta}^0) = u_{-2\alpha-\beta}^0,$$
$$\iota_{-2\alpha-\beta,-2\alpha-2\beta}(u_{-2\alpha-\beta}^\lambda) = u_{-2\alpha-2\beta}^\lambda, \quad \iota_{-\alpha-2\beta,-2\alpha-2\beta}(u_{-\alpha-2\beta}^\lambda) = u_{-2\alpha-2\beta}^\lambda$$

for $\lambda = -\alpha, -\beta, 0$, showing that the choice of primitive vectors was correct. It is clear that the above morphisms are irreducible. $\square$

Lemma 3.2. *The irreducible morphisms $P(\lambda) \to P(\mu)$ in the case $\lambda \not\leqq \mu$ are given as follows:*

$\varphi_{-\alpha,0} : P(-\alpha) \to P(0)$ *is determined by* $\varphi_{-\alpha,0}(u_{-\alpha}^{-\alpha}) = f_\alpha u_0^0,$

$\varphi_{-\beta,0} : P(-\beta) \to P(0)$ *is determined by* $\varphi_{-\beta,0}(u_{-\beta}^{-\beta}) = -f_\beta u_0^0,$

$\varphi_{-2\alpha-\beta,-\alpha} : P(-2\alpha - \beta) \to P(-\alpha)$ *is determined by*

$$\varphi_{-2\alpha-\beta,-\alpha}(u_{-2\alpha-\beta}^{-2\alpha-\beta}) = (-f_\alpha f_\beta + f_{\alpha+\beta})u_{-\alpha}^{-\alpha} + (2f_\alpha f_{\alpha+\beta} + xf_\alpha^2 f_\beta)u_{-\alpha}^0,$$

$\varphi_{-2\alpha-\beta,-\beta} : P(-2\alpha - \beta) \to P(-\beta)$ *is determined by*

$$\varphi_{-2\alpha-\beta,-\beta}(u_{-2\alpha-\beta}^{-2\alpha-\beta}) = f_\alpha^2 u_{-\beta}^{-\beta} + (f_\alpha f_{\alpha+\beta} + y f_\alpha^2 f_\beta) u_{-\beta}^0,$$

$\varphi_{-\alpha-2\beta,-\alpha} : P(-\alpha - 2\beta) \to P(-\alpha)$ *is determined by*

$$\varphi_{-\alpha-2\beta,-\alpha}(u_{-\alpha-2\beta}^{-\alpha-2\beta}) = -f_\beta^2 u_{-\alpha}^{-\alpha} + \left((1 + 2z) f_\beta f_{\alpha+\beta} + z f_\alpha f_\beta^2 \right) u_{-\alpha}^0,$$

$\varphi_{-\alpha-2\beta,-\beta} : P(-\alpha - 2\beta) \to P(-\beta)$ *is determined by*

$$\varphi_{-\alpha-2\beta,-\beta}(u_{-\alpha-2\beta}^{-\alpha-2\beta}) = (f_\alpha f_\beta + 2 f_{\alpha+\beta}) u_{-\beta}^{-\beta} + \left((2 + 2t) f_\beta f_{\alpha+\beta} + t f_\alpha f_\beta^2 \right) u_{-\beta}^0,$$

$\varphi_{-2\alpha-2\beta,-\alpha-2\beta} : P(-2\alpha - 2\beta) \to P(-\alpha - 2\beta)$ *is determined by*

$$\varphi_{-2\alpha-2\beta,-\alpha-2\beta}(u_{-2\alpha-2\beta}^{-2\alpha-2\beta}) = -f_\alpha u_{-\alpha-2\beta}^{-\alpha-2\beta} + \left(f_\beta f_{\alpha+\beta} + (-a_1 + 3) f_\alpha f_\beta^2 \right) u_{-\alpha-2\beta}^{-\alpha}$$
$$+ \left((3 - b_1) f_\alpha^2 f_\beta + (4 - 2b_1) f_\alpha f_{\alpha+\beta} \right) u_{-\alpha-2\beta}^{-\beta}$$
$$+ \left(c_1 f_\alpha^2 f_\beta^2 + (-a_1 + b_1 + 2c_1 - 3) f_\alpha f_\beta f_{\alpha+\beta} + (b_1 - \frac{3}{2}) f_{\alpha+\beta}^2 \right) u_{-\alpha-2\beta}^0,$$

and

$\varphi_{-2\alpha-2\beta,-2\alpha-\beta} : P(-2\alpha - 2\beta) \to P(-2\alpha - \beta)$ *is determined by*

$$\varphi_{-2\alpha-2\beta,-2\alpha-\beta}(u_{-2\alpha-2\beta}^{-2\alpha-2\beta}) = -f_\beta u_{-2\alpha-\beta}^{-2\alpha-\beta}$$
$$+ \left(2 f_\beta f_{\alpha+\beta} + (\frac{7}{2} + b - a_2) f_\alpha f_\beta^2 \right) u_{-2\alpha-\beta}^{-\alpha}$$
$$+ \left((\frac{7}{2} - b_2) f_\alpha^2 f_\beta + (6 - 2b_2) f_\alpha f_{\alpha+\beta} \right) u_{-2\alpha-\beta}^{-\beta}$$
$$+ \left(c_2 f_\alpha^2 f_\beta^2 + (\frac{5}{2} - 2b - a_2 + b_2 + 2c_2) f_\alpha f_\beta f_{\alpha+\beta} + (1 - 2b + b_2) f_{\alpha+\beta}^2 \right) u_{-2\alpha-\beta}^0,$$

where x, y, t, z, a_1, b_1, c_1, a_2, b_2, c_2 *are arbitrary complex numbers.*

Proof. The fact that these maps are irreducible morphisms can be determined using the values of these morphisms at other primitive vectors. They are:

$$\varphi_{\lambda,\mu}(u_\lambda^0) = 0 \text{ for } \lambda, \mu \text{ as above,}$$

$$\varphi_{-2\alpha-\beta,-\alpha}(u_{-2\alpha-\beta}^{-\alpha}) = f_\alpha u_{-\alpha}^0, \qquad \varphi_{-2\alpha-\beta,-\alpha}(u_{-2\alpha-\beta}^{-\beta}) = -f_\beta u_{-\alpha}^0,$$

$$\varphi_{-2\alpha-\beta,-\beta}(u_{-2\alpha-\beta}^{-\alpha}) = f_\alpha u_{-\beta}^0, \qquad \varphi_{-2\alpha-\beta,-\alpha}(u_{-2\alpha-\beta}^{-\beta}) = 0,$$

$$\varphi_{-\alpha-2\beta,-\alpha}(u_{-\alpha-2\beta}^{-\alpha}) = 0, \qquad \varphi_{-\alpha-2\beta,-\alpha}(u_{-\alpha-2\beta}^{-\beta}) = -f_\beta u_{-\alpha}^0,$$

$$\varphi_{-\alpha-2\beta,-\beta}(u_{-\alpha-2\beta}^{-\alpha}) = f_\alpha u_{-\beta}^0, \qquad \varphi_{-\alpha-2\beta,-\beta}(u_{-\alpha-2\beta}^{-\beta}) = -f_\beta u_{-\beta}^0,$$

$$\varphi_{-2\alpha-2\beta,-\alpha-2\beta}(u_{-2\alpha-2\beta}^{-\alpha}) = 0, \qquad \varphi_{-2\alpha-2\beta,-\alpha-2\beta}(u_{-2\alpha-2\beta}^{-\beta}) = f_\beta u_{-\alpha-2\beta}^0,$$

210

$$\varphi_{-2\alpha-2\beta,-\alpha-2\beta}(u_{-2\alpha-2\beta}^{-2\alpha-\beta}) =$$

$$(f_\alpha f_\beta - f_{\alpha+\beta})u_{-\alpha-2\beta}^{-\alpha} + f_\alpha^2 u_{-\alpha-2\beta}^{-\beta} + (a_1 f_\alpha^2 f_\beta - f_\alpha f_{\alpha+\beta})u_{-\alpha-2\beta}^{0},$$

$$\varphi_{-2\alpha-2\beta,-\alpha-2\beta}(u_{-2\alpha-2\beta}^{-\alpha-2\beta}) = f_\beta^2 u_{-\alpha-2\beta}^{-\alpha} + \left((2b_1-1)f_\beta f_{\alpha+\beta} + b_1 f_\alpha f_\beta^2\right)u_{-\alpha-2\beta}^{0}$$

$$\varphi_{-2\alpha-2\beta,-2\alpha-\beta}(u_{-2\alpha-2\beta}^{-\alpha}) = f_\alpha u_{-2\alpha-\beta}^{0},$$

$$\varphi_{-2\alpha-2\beta,-2\alpha-\beta}(u_{-2\alpha-2\beta}^{-\beta}) = 0,$$

$$\varphi_{-2\alpha-2\beta,-2\alpha-\beta}(u_{-2\alpha-2\beta}^{-2\alpha-\beta}) = f_\alpha^2 u_{-2\alpha-\beta}^{-\beta} + (f_\alpha f_{\alpha+\beta} + a_2 f_\alpha^2 f_\beta)u_{-2\alpha-\beta}^{0},$$

$$\varphi_{-2\alpha-2\beta,-2\alpha-\beta}(u_{-2\alpha-2\beta}^{-\alpha-2\beta}) =$$

$$f_\beta^2 u_{-2\alpha-\beta}^{-\alpha} + (f_\alpha f_\beta + 2f_{\alpha+\beta})u_{-2\alpha-\beta}^{-\beta} + \left(b_2 f_\alpha f_\beta^2 + (2b_2+1)f_\beta f_{\alpha+\beta}\right)u_{-2\alpha-\beta}^{0}$$

$$\square$$

4. Algebra associated with the principal block $\mathcal{O}_0$ for $sl_3(\mathbb{C})$

To compute the algebra associated with the principal block of category $\mathcal{O}$ for the Lie algebra $sl_3(\mathbb{C})$, we have to find the compositions of various morphisms $\iota_{\kappa,\lambda}$ and $\varphi_{\mu,\nu}$.

Clearly,

$$\iota_{-\alpha,-2\alpha-\beta}\iota_{0,-\alpha} = \iota_{-\beta,-2\alpha-\beta}\iota_{0,-\beta},$$

$$\iota_{-\alpha,-\alpha-2\beta}\iota_{0,-\alpha} = \iota_{-\beta,-\alpha-2\beta}\iota_{0,-\beta},$$

$$\iota_{-2\alpha-\beta,-2\alpha-2\beta}\iota_{-\alpha,-2\alpha-\beta} = \iota_{-\alpha-2\beta,-2\alpha-2\beta}\iota_{-\alpha,-\alpha-2\beta} \text{ and}$$

$$\iota_{-\alpha-2\beta,-2\alpha-2\beta}\iota_{-\beta,-\alpha-2\beta} = \iota_{-2\alpha-\beta,-2\alpha-2\beta}\iota_{-\beta,-2\alpha-\beta}.$$

Also,

$$\varphi_{-\alpha,0}\varphi_{-2\alpha-\beta,-\alpha} = \varphi_{-\beta,0}\varphi_{-2\alpha-\beta,-\beta} \text{ and}$$

$$\varphi_{-\beta,0}\varphi_{-\alpha-2\beta,-\beta} = \varphi_{-\alpha,0}\varphi_{-\alpha-2\beta,-\alpha}.$$

Moreover,

$$\varphi_{-\alpha-2\beta,-\alpha}\varphi_{-2\alpha-2\beta,-\alpha-2\beta} = \varphi_{-2\alpha-\beta,-\alpha}\varphi_{-2\alpha-2\beta,-2\alpha-\beta}$$

$$\text{if and only if} \quad -a_2 + b + b_2 - x = -3 - b_1 - z$$

and

$$\varphi_{-\alpha-2\beta,-\beta}\varphi_{-2\alpha-2\beta,-\alpha-2\beta} = \varphi_{-2\alpha-\beta,-\beta}\varphi_{-2\alpha-2\beta,-2\alpha-\beta}$$

$$\text{if and only if} \quad \frac{7}{2} - a_2 + b - y = -a_1 + b_1 - t.$$

Among the mixed compositions, the relations

$$\varphi_{-\alpha,0}\iota_{0,-\alpha} = \varphi_{-\beta,0}\iota_{0,-\beta} = 0,$$

$$\varphi_{-2\alpha-\beta,-\alpha}\iota_{-\alpha,-2\alpha-\beta} = \iota_{0,-\alpha}\varphi_{-\alpha,0},$$

$$\varphi_{-\alpha-2\beta,-\alpha}\iota_{-\alpha,-\alpha-2\beta} = 0,$$

$$\iota_{0,-\beta}\varphi_{-\alpha,0} = \varphi_{-\alpha-2\beta,-\beta}\iota_{-\alpha,-\alpha-2\beta} = \varphi_{-2\alpha-\beta,-\beta}\iota_{-\alpha,-2\alpha-\beta},$$

$$\varphi_{-\alpha-2\beta,-\beta}\iota_{-\beta,-\alpha-2\beta} = \iota_{0,-\beta}\varphi_{-\beta,0},$$

$$\varphi_{-2\alpha-\beta,-\beta}\iota_{-\beta,-2\alpha-\beta} = 0,$$

$$\iota_{0,-\alpha}\varphi_{-\beta,0} = \varphi_{-2\alpha-\beta,-\alpha}\iota_{-\beta,-2\alpha-\beta} = \varphi_{-\alpha-2\beta,-\alpha}\iota_{-\beta,-\alpha-2\beta}$$

are immediate.

The remaining compositions

$$\varphi_{-2\alpha-2\beta,-2\alpha-\beta}\iota_{-2\alpha-\beta,-2\alpha-2\beta}, \quad \varphi_{-2\alpha-2\beta,-\alpha-2\beta}\iota_{-2\alpha-\beta,-2\alpha-2\beta},$$

$$\varphi_{-2\alpha-2\beta,-2\alpha-\beta}\iota_{-\alpha-2\beta,-2\alpha-2\beta} \quad \text{and} \quad \varphi_{-2\alpha-2\beta,-\alpha-2\beta}\iota_{-\alpha-2\beta,-2\alpha-2\beta}$$

are identified as follows:

$$\varphi_{-2\alpha-2\beta,-2\alpha-\beta}\iota_{-2\alpha-\beta,-2\alpha-2\beta} = \iota_{-\beta,-2\alpha-\beta}\varphi_{-2\alpha-\beta,-\beta}$$
$$\text{if and only if} \quad -a_2 + y = 0,$$

$$\varphi_{-2\alpha-2\beta,-\alpha-2\beta}\iota_{-2\alpha-\beta,-2\alpha-2\beta} = \iota_{-\beta,-\alpha-2\beta}\varphi_{-2\alpha-\beta,-\beta} - \iota_{-\alpha,-\alpha-2\beta}\varphi_{-2\alpha-\beta,-\alpha}$$
$$\text{if and only if} \quad a_1 + x - y = 0,$$

$$\varphi_{-2\alpha-2\beta,-2\alpha-\beta}\iota_{-\alpha-2\beta,-2\alpha-2\beta} = \iota_{-\beta,-2\alpha-\beta}\varphi_{-\alpha-2\beta,-\beta} - \iota_{-\alpha,-2\alpha-\beta}\varphi_{-\alpha-2\beta,-\alpha}$$
$$\text{if and only if} \quad b_2 - t + z = 0$$

and

$$\varphi_{-2\alpha-2\beta,-\alpha-2\beta}\iota_{-\alpha-2\beta,-2\alpha-2\beta} = -\iota_{-\alpha,-\alpha-2\beta}\varphi_{-\alpha-2\beta,-\alpha}$$
$$\text{if and only if} \quad b_1 - z = 0.$$

We choose a solution of the linear system

$$-a_2 + b + b_2 - x = -3 - b_1 - z$$
$$\frac{7}{2} - a_2 + b - y = -a_1 + b_1 - t$$
$$-a_2 + y = 0$$
$$a_1 + x - y = 0$$
$$b_2 - t + z = 0$$
$$b_1 - z = 0$$

to make the presentation of $A = A(sl_3(\mathbb{C}))$ in term of a quiver and relations as simple as possible.

We will denote the simple A-modules by $S(1)$ through $S(6)$, where labels 1, 2, 3, 4, 5, 6 corresponds to weights $-2\alpha-2\beta, -\alpha-2\beta, -2\alpha-\beta, -\beta, -\alpha, 0$, respectively. Then the vertices $\{1,2,3,4,5,6\}$ of the quiver correspond to idempotents e_λ of A given by identities on $P(\lambda)$ and the arrows of the quiver correspond to the previously defined maps $\iota_{\kappa,\lambda}$ and $\varphi_{\mu,\nu}$.

The results of previous computations can be summarized in the following theorem.

Theorem 4.1. *The algebra A associated to the principal block of category $\mathcal{O}$ for the Lie algebra $sl_3(\mathbb{C})$ is given by the quiver*

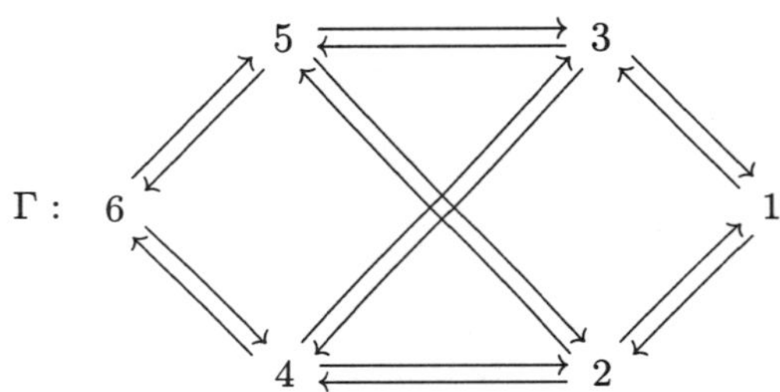

and relations

$$124 = 134, \quad 125 = 135, \quad 212 = -252, \quad 213 = 243 - 253, \quad 246 = 256,$$
$$312 = 342 - 352, \quad 313 = 343, \quad 346 = 356, \quad 421 = 431, \quad 424 = 464,$$
$$434 = 0, \quad 425 = 435 = 465, \quad 521 = 531, \quad 524 = 534 = 564, \quad 525 = 0,$$
$$535 = 565, \quad 642 = 652, \quad 643 = 653, \quad 646 = 0, \quad 656 = 0.$$

The arrows are given explicitly as follows:

$$6 \to 5 \text{ stands for } \iota_{0,-\alpha}, \quad 6 \to 4 \text{ for } \iota_{0,-\beta},$$
$$5 \to 3 \text{ for } \iota_{-\alpha,-2\alpha-\beta}, \quad 5 \to 2 \text{ for } \iota_{-\alpha,-\alpha-2\beta},$$
$$4 \to 3 \text{ for } \iota_{-\beta,-2\alpha-\beta}, \quad 4 \to 2 \text{ for } \iota_{-\beta,-\alpha-2\beta},$$
$$3 \to 1 \text{ for } \iota_{-2\alpha-\beta,-2\alpha-2\beta}, \quad 2 \to 1 \text{ for } \iota_{-\alpha-2\beta,-2\alpha-2\beta}$$

and

$$5 \to 6 \text{ stands for } \varphi_{-\alpha,0}, \quad 4 \to 6 \text{ for } \varphi_{-\beta,0},$$
$$3 \to 5 \text{ for } \varphi_{-2\alpha-\beta,-\alpha}, \quad 3 \to 4 \text{ for } \varphi_{-2\alpha-\beta,-\beta},$$
$$2 \to 5 \text{ for } \varphi_{-\alpha-2\beta,-\alpha}, \quad 2 \to 4 \text{ for } \varphi_{-\alpha-2\beta,-\beta},$$
$$1 \to 3 \text{ for } \varphi_{-2\alpha-2\beta,-2\alpha-\beta} \quad \text{and} \quad 1 \to 2 \text{ for } \varphi_{-2\alpha-2\beta,-\alpha-2\beta}.$$

The compositions of the arrows is done from left to right, that is 124 is the composition of the arrow 12 followed by the arrow 24 and so on.

Algebra A can be realized by construction of [7] as an endomorphism algebra of a semilocal module M over a commutative selfinjective algebra R.

Corollary 4.2. *The algebra A associated to the principal block of category $\mathcal{O}$ for the Lie algebra $sl_3(\mathbb{C})$ is isomorphic to algebra $End_R(M)$, where $R = \mathbb{C}[x_1, x_2]/\langle x_1 x_2^2 - x_1^2 x_2, x_1^2 - x_1 x_2 + x_2^2 \rangle$ and $M = \oplus R \oplus \langle \overline{x_1} \rangle \oplus \langle \overline{x_2} \rangle \oplus \langle \overline{x_1}^2 \rangle \oplus \langle \overline{x_2}^2 \rangle \oplus \langle \overline{x_1}^2 \overline{x_2} = \overline{x_1 x_2}^2 \rangle.$*

Proof. The isomorphism is straightforward if we use the previous description of A in terms of generators and relations. $\square$

Note that $R \cong e_1 A e_1$ if we identify x_1 with 121 and x_2 with 131 in the previous description of A. By [12], R must be isomorphic to the ring of coinvariants of $sl_3(\mathbb{C})$ given as $\mathbb{C}[u, v]/\langle u^2 + uv + v^2, 2u^3 + 3u^2v - 3uv^2 - 2v^3 \rangle$. The isomorphism is given explicitly by substitutions $x_1 = u - v, x_2 = 2u + v$.

Acknowledgment

The author would like to thank the referee for valuable suggestions that helped to improve the paper.

References

1. Beilinson, A., Ginzburg, V. and Soergel, W., *Koszul duality patterns in representation theory*, J. Amer. Math. Soc. **9** (1996), 473–527.
2. Bernstein, I. N., Gelfand I. M. and Gelfand S. I., *A category of $\mathcal{G}$-modules*, Funct. Anal. Appl. **10** (1976), 87–92.
3. Cline, E., Parshall, B. and Scott, L., *Finite dimensional algebras and highest weight categories*, J. Reine Angew. Math. **391** (1988), 85–99.
4. Cline, E., Parshall, B. and Scott, L., *Stratifying endomorphism algebras*, Memoirs AMS **591** (1996)
5. Dixmier, J., *Enveloping algebras*, Graduate Studies in Mathematics, 11. American Mathematical Society, Providence, RI, 1996.
6. Dlab, V., *Quasi-hereditary algebras revisited*, An. Ştiinţ. Univ. Ovidius Constanţa Ser. Mat., **4** (1996), 43–54.
7. Dlab, V., Heath, P., Marko, F., *Quasi-hereditary endomorphism algebras*, Can. Math. Bull. **38** (1995), 421–428.
8. Jantzen, J. C., *Moduln mit einem höchsten Gewicht*, Lecture Notes in Mathematics, **750** Springer, Berlin, 1979.
9. Jantzen, J. C., *Einhüllende Algebren halbeinfacher Lie-Algebren*, Ergebnisse der Mathematik und ihrer Grenzgebiete, **3** Springer-Verlag, Berlin, 1983.

10. König, S., *Strong exact Borel subalgebras and global dimensions of quasi-hereditary algebras*, Representation theory of algebras (Cocoyoc, 1994), CMS Conf. Proc., **18** 399–417.

11. Moody, R. V. and Pianzola, A., *Lie algebras with triangular decompositions*, J. Wiley, New York 1995.

12. Soergel, W., *Kategorie $\mathcal{O}$, perverse Garben und Moduln über den Koinvarianten zur Weylgruppe*, J. Amer. Math. Soc. **3** (1990), 421–445.

13. Stroppel, C., *Category $\mathcal{O}$: Quivers and endomorphism rings of projectives*, Represent. Theory **7** (2003), 322–345.

NONCOMMUTATIVE PROJECTIVE SCHEMES AND POINT SCHEMES

IZURU MORI

Department of Mathematics, Syracuse University,
Syracuse, NY 13244
E-mail: imori@syr.edu

Let k be a field and A be a noetherian graded algebra finitely generated in degree 1 over k. There are two basic ingredients to study A in noncommutative algebraic geometry, namely, the point scheme of A introduced by Artin, Tate and Van den Bergh, and the noncommutative projective scheme $X = \operatorname{Proj} A$ associated to A introduced by Artin and Zhang. In this paper, we will study the relationship between the point scheme of A and the noncommutative projective scheme $\operatorname{Proj} A$. In particular, we will prove that if X is a Gorenstein noncommutative projective scheme which can be embedded into a quantum projective space and whose dualizing sheaf is ample or anti-ample, then the point variety is independent of the choice of a Gorenstein homogeneous coordinate ring A for X. As an application, we will classify quantum projective planes whose point schemes are generic and singular in terms of geometric data.

1. Introduction

Throughout, let k be a field. Let A be a noetherian graded algebra finitely generated in degree 1 over k. There are two basic ingredients to study A in noncommutative algebraic geometry, namely, the point scheme of A introduced by Artin, Tate and Van den Bergh, and the noncommutative projective scheme $X = \operatorname{Proj} A$ associated to A introduced by Artin and Zhang. In this paper, we will study the relationship between the point scheme of A and the noncommutative projective scheme $\operatorname{Proj} A$.

First, we recall the definition of a noncommutative projective scheme from [5]. Let A be a graded k-algebra. The category of graded right A-modules and graded right A-module homomorphisms of degree 0 is denoted by $\operatorname{GrMod} A$. We define the quotient category $\operatorname{Tails} A = \operatorname{GrMod} A / \operatorname{Fdim} A$ where $\operatorname{Fdim} A$ is the full subcategory of $\operatorname{GrMod} A$ consisting of direct limits of finite dimensional A-modules over k. We write $\mathcal{M}$ for the image of $M \in \operatorname{GrMod} A$ in $\operatorname{Tails} A$.

Definition 1.1. We call the pair $\operatorname{Proj} A = (\operatorname{Tails} A, \mathcal{A})$ the noncommutative projective scheme associated to A. A map $F\colon \operatorname{Proj} A \to \operatorname{Proj} A'$ between noncommutative projective schemes is a functor $F\colon \operatorname{Tails} A \to \operatorname{Tails} A'$ together with an isomorphism $F(\mathcal{A}) \cong \mathcal{A}'$ in $\operatorname{Tails} A'$.

We write $X = \operatorname{Proj} A$ to pretend that there is an imaginary (noncommutative) geometric object X whose homogeneous coordinate ring is A.

Given two graded algebras A and A', the following are natural questions to ask:

(1) $A \cong A'$ as graded k-algebras?
(2) $\operatorname{GrMod} A \cong \operatorname{GrMod} A'$?
(3) $\operatorname{Proj} A \cong \operatorname{Proj} A'$?

In general, it is difficult to answer these questions in terms of generators and relations for algebras A and A'. In fact, one of the starting points of noncommutative algebraic geometry was an attempt to classify regular algebras defined by Artin and Schelter, which we call Artin-Schelter regular algebras. In their paper [2], Artin and Schelter classified 3-dimensional Artin-Schelter regular algebras in terms of generators and relations, but they used computer to complete their classification. Later, Artin, Tate and Van den Bergh [3] classified them using a geometric approach, introducing a notion of a point scheme.

Now we recall the definition of a point scheme from [3]. Let V be a finite dimensional k-vector space, and I be a k-vector subspace of $V^{\otimes r}$. Since an element of I defines a multilinear function on $(V^*)^{\times r}$, we can define a projective scheme associated to I by

$$\mathcal{V}(I) = \{(p_0, \cdots, p_{r-1}) \in \mathbb{P}(V^*)^{\times r} \mid f(p_0, \cdots, p_{r-1}) = 0 \text{ for all } f \in I\}.$$

Now, let $A = T(V)/I$ be a graded algebra finitely generated in degree 1 over k, where $T(V)$ is the tensor algebra over V. For $i \geq 1$, we define $\Gamma_i = \mathcal{V}(I_i) \subset \mathbb{P}(V^*)^{\times i}$. For $j \geq i$, let $pr_i^j\colon \Gamma_j \to \Gamma_i$ be the restriction of the projection $\mathbb{P}(V^*)^{\times j} \to \mathbb{P}(V^*)^{\times i}$ from the first i factors. Then $\{\Gamma_i, pr_i^j\}$ becomes an inverse system of schemes.

Definition 1.2. The point scheme of A is defined to be the inverse limit

$$\Gamma := \varprojlim \Gamma_i.$$

If we only consider the reduced structure of the point scheme, then we call it the point variety.

In this paper, we will always assume that the inverse system $\{\Gamma_i, pr_i^j\}$ is eventually constant when we define the point scheme Γ, so that Γ is a scheme in the usual sense.

The purpose of this paper is to use point schemes to answer the above three questions. Let A, A' be graded algebras finitely generated in degree 1 over k, and Γ, Γ' be the point schemes of A, A'. The following implications are well known (cf. [3], [20]):

$$A \cong A' \text{ as graded } k\text{-algebras}$$
$$\Downarrow$$
$$\operatorname{GrMod} A \cong \operatorname{GrMod} A' \quad \Rightarrow \Gamma \cong \Gamma'$$
$$\Downarrow$$
$$\operatorname{Proj} A \cong \operatorname{Proj} A'.$$

In order for point schemes to be useful in classifying noncommutative projective schemes, we would like to find implications between the conditions $\operatorname{Proj} A \cong \operatorname{Proj} A'$ and $\Gamma \cong \Gamma'$. It is easy to see that the implication $\Gamma \cong \Gamma' \implies \operatorname{Proj} A \cong \operatorname{Proj} A'$ is false. (If it were true, then every noncommutative projective scheme would have been isomorphic to a commutative projective scheme.) So our question is:

Question 1.3. If $\operatorname{Proj} A \cong \operatorname{Proj} A'$, then $\Gamma \cong \Gamma'$?

Unfortunately, Question 1.3 is also false in general by [6] (see section 3). In this paper, we will show that if A, A' are homologically "almost commutative" Gorenstein algebras (typically, if they are graded quotient algebras of quantum polynomial rings), then Question 1.3 is true.

The paper is organized as follows. In section 2, we study some homological conditions on an algebra, which are needed to answer Question 1.3. Then, in section 3, we will prove that if $\operatorname{Proj} A$ and $\operatorname{Proj} A'$ are Gorenstein which can be embedded into quantum projective spaces and whose dualizing sheaves are ample or anti-ample, then $\operatorname{Proj} A \cong \operatorname{Proj} A'$ implies $\Gamma \cong \Gamma'$ as varieties. At this time, we were not able to conclude that $\Gamma \cong \Gamma'$ as schemes in general, which would be a much nicer result.

In section 4, we introduce a notion of a geometric algebra following [3], which is an algebra closely tied with a geometric data (E, σ) where E is a point scheme and σ is a k-automorphism of E. Let A, A' be geometric algebras with the associated geometric data (E, σ), (E', σ'). Using the results of [20], we will show that $\operatorname{GrMod} A \cong \operatorname{GrMod} A'$ if and only if there is a sequence of isomorphisms $\tau_n \colon E \to E'$, $n \in \mathbb{Z}$ which can be extended to automorphisms of the ambient projective space such that

$\sigma' \tau_n = \tau_{n+1}\sigma \colon E \to E'$ for all $n \in \mathbb{Z}$. In section 5, we will apply the result of section 4 to some quantum polynomial rings. In particular, we will classify quantum projective planes whose point schemes are generic and singular in terms of geometric data.

2. Cohen-Macaulay Algebras

In this section, we will study some homological conditions on an algebra, which are needed to answer Question 1.3. In particular, we will find a nice class of algebras for which point modules defined below have nice characterization.

Let A be a graded algebra and M, N be graded right A-modules. The set of graded right A-module homomorphisms $M \to N$ of degree 0 is denoted by $\mathrm{Hom}_A(M, N)$, which has a natural k-vector space structure. For each integer n, the shift of M, denoted by $M(n)$, is a graded right A-module such that $M(n)_i = M_{n+i}$. We define

$$\underline{\mathrm{Ext}}^i_A(M, N) = \oplus_{n=-\infty}^{\infty} \mathrm{Ext}^i_A(M, N(n)),$$

which has a natural graded k-vector space structure for each i. For any graded k-vector space V such that $\dim_k V_i < \infty$ for all i, we define the Hilbert series of V by

$$H_V(t) = \sum_{i=-\infty}^{\infty} (\dim_k V_i) t^i \in \mathbb{Z}[[t, t^{-1}]].$$

Definition 2.1. Let $A = k \oplus A_1 \oplus A_2 \oplus \cdots$ be a connected algebra, $\mathrm{m} = A_1 \oplus A_2 \oplus \cdots$ be the augmentation ideal of A, and M be a graded right A-module. We define the i-th local cohomology of M by

$$\mathrm{H}^i_{\mathrm{m}}(M) = \lim_{n \to \infty} \underline{\mathrm{Ext}}^i_A(A/A_{\geq n}, M),$$

which is a graded right A-module. We say that M is rational if

- $H_M(t)$ and $\sum_{i=0}^{\infty}(-1)^i H_{\mathrm{H}^i_{\mathrm{m}}(M)}(t)$ are both well-defined rational functions over $\mathbb{C}$, and
- $H_M(t) = \sum_{i=0}^{\infty}(-1)^i H_{\mathrm{H}^i_{\mathrm{m}}(M)}(t)$ as rational functions over $\mathbb{C}$.

We define

- $j(M) = \inf\{i \mid \underline{\mathrm{Ext}}^i_A(M, A) \neq 0\}$,
- $\mathrm{depth}\, M = \inf\{i \mid \underline{\mathrm{Ext}}^i_A(k, M) \neq 0\}$, and
- $\mathrm{ldim}\, M = \sup\{i \mid \mathrm{H}^i_{\mathrm{m}}(M) \neq 0\}$.

We say that M is Cohen-Macaulay if depth $M = \operatorname{ldim} M < \infty$. We say that A is Cohen-Macaulay if A is Cohen-Macaulay as a graded right module over itself. If A is Cohen-Macaulay of depth $A = \operatorname{ldim} A = d$, then the canonical module is a graded A-A bimodule $\omega_A = \mathrm{H}_{\mathfrak{m}}^d(A)^*$ where $*$ denotes graded vector space dual.

We refer to [14, Chapter 11] and [13] for basic properties of a Cohen-Macaulay algebra. Let A be a Cohen-Macaulay algebra of depth $A = d$ and ω_A be the canonical module. One of the most fundamental properties of A is that A satisfies a local duality theorem: for every graded right A-module M, there are functorial isomorphisms

$$\underline{\operatorname{Ext}}_A^i(M, \omega_A) \cong \mathrm{H}_{\mathfrak{m}}^{d-i}(M)^*$$

as graded left A-modules for all i. For a graded right (resp. left) A-module M of depth m, we define

$$M^\dagger = \underline{\operatorname{Ext}}_A^{d-m}(M, \omega_A),$$

which is a graded left (resp. right) A-module. By [14, Chapter 11, Lemma 4.1], depth $M = \inf\{i \mid \mathrm{H}_{\mathfrak{m}}^i(M) \neq 0\}$, so M is Cohen-Macaulay of depth $M = m$ if and only if $\underline{\operatorname{Ext}}_A^{d-i}(M, \omega_A) \cong \mathrm{H}_{\mathfrak{m}}^i(M)^* = 0$ for $i \neq m$ and $\underline{\operatorname{Ext}}_A^{d-m}(M, \omega_A) \cong \mathrm{H}_{\mathfrak{m}}^m(M)^* \neq 0$.

Definition 2.2. A noetherian connected algebra A is called a d-dimensional Artin-Schelter regular (resp. Gorenstein) algebra if

- $\operatorname{gldim} A = d < \infty$ (resp. $\operatorname{id}(A) = d < \infty$) on both sides,
- $\operatorname{GKdim} A < \infty$, and
- (Gorenstein condition)

$$\underline{\operatorname{Ext}}_A^i(k, A) \cong \begin{cases} 0 & \text{if } i \neq d, \\ k(r) & \text{if } i = d \end{cases}$$

for some $r \in \mathbb{Z}$.

Definition 2.3. A d-dimensional Artin-Schelter regular algebra A is called a quantum polynomial ring if

- A is generated in degree 1 over k,
- $H_A(t) = (1 - t)^{-d}$, and
- (Cohen-Macaulay property with respect to GKdimension)

$$j(M) + \operatorname{GKdim} M = \operatorname{GKdim} A$$

for every finitely generated graded right A-module M.

Remark 2.4. Let A be a d-dimensional Artin-Schelter Gorenstein algebra as defined above. Then A is Cohen-Macaulay of depth $A = \operatorname{ldim} A = \operatorname{id}(A) = d$ and the canonical module ω_A is isomorphic to $A(-r)$ as a graded right and left A-module (but not necessarily as a graded A-A bimodule) by [8, Theorem 1.2]. So for any finitely generated graded right A-module M,

$$
\begin{aligned}
j(M) &= \inf\{i \mid \underline{\operatorname{Ext}}_A^i(M, A) \neq 0\} \\
&= \inf\{i \mid \underline{\operatorname{Ext}}_A^i(M, \omega_A) \neq 0\} \\
&= \inf\{i \mid \operatorname{H}_{\mathfrak{m}}^{d-i}(M) \neq 0\} \\
&= d - \sup\{i \mid \operatorname{H}_{\mathfrak{m}}^i(M) \neq 0\} \\
&= \operatorname{ldim} A - \operatorname{ldim} M.
\end{aligned}
$$

Suppose that $\operatorname{GKdim} A = d$. Then A satisfies Cohen-Macaulay property with respect to GKdimension if and only if $\operatorname{ldim} M = \operatorname{GKdim} M$ for every finitely generated graded right A-module M.

Remark 2.5. Let A be a d-dimensional Artin-Schelter Gorenstein algebra. If A satisfies SSC condition defined in [21], such as FBN algebras (including noetherian PI algebras) or 3-dimensional Artin-Schelter regular algebras, then A satisfies Cohen-Macaulay property with respect to GKdimension, and $\operatorname{GKdim} A = d$ by [21, Theorem 3.1].

Remark 2.6. If A is an Artin-Schelter regular algebra, then for every finitely generated graded right A-module M,

$$
\operatorname{pd}(M) = \operatorname{depth} A - \operatorname{depth} M = \operatorname{ldim} A - \operatorname{depth} M
$$

by Auslander-Buchsbaum formula [7, Theorem 3.2], so M is Cohen-Macaulay if and only if $\operatorname{pd}(M) = j(M)$, which is another notion of Cohen-Macaulay property used in the literature ([4], [11], etc.).

The point modules introduced in [3] are essential in the study of non-commutative algebraic geometry.

Definition 2.7. Let A be a graded algebra finitely generated in degree 1 over k. A graded right A-module M is called a point module if M is cyclic and $H_M(t) = (1 - t)^{-1}$.

If A is a commutative graded algebra, then there is a bijection between the set of closed points of $\operatorname{Proj} A$ in the classical sense and the set of isomorphism classes of point modules over A via $p \mapsto A/I(p)$ where $I(p)$ is an ideal generated by $\{f \in A_1 \mid f(p) = 0\}$. So if A is noncommutative, we can think of isomorphism classes of point modules of A as closed points of $\operatorname{Proj} A$.

For the rest of this section, we assume the condition

"A is a graded quotient algebra of a quantum polynomial ring S". (*)

Geometrically speaking, this condition (*) guarantees that there is an embedding of $\operatorname{Proj} A$ into a "quantum projective space" $\operatorname{Proj} S$. In commutative algebraic geometry, we can always embed any projective scheme into a projective space, so (*) is a natural condition in noncommutative algebraic geometry. In fact, under the assumption (*), point modules over a Cohen-Macaulay algebra have nice characterization (see Lemma 2.9 (4)).

Let M be a finitely generated graded right A-module. The main advantage of assuming the condition (*) is that the Hilbert series of M behaves well, that is,

$$H_M(t) = \frac{f(t)}{(1-t)^d}$$

for some $d \in \mathbb{N}$ and some $f(t) \in \mathbb{Z}[t, t^{-1}]$. In particular, GKdimension of M is the order of the pole of $H_M(t)$ at $t = 1$ (which is always an integer) and the multiplicity of M is

$$e(M) = \lim_{t \to 1}(1-t)^{\operatorname{GKdim} M} H_M(t)$$

(which is always an integer). The following Lemma now follows easily (cf. [4, Proposition 6.4], [11, Proposition 2.6]).

Lemma 2.8. *Let A be a graded quotient algebra of a quantum polynomial ring, and M be a finitely generated graded right A-module. If $\operatorname{GKdim} M = 1$ and $e(M) = e$ then*

$$H_M(t) = \frac{e}{1-t} + f(t)$$

for some $f(t) \in \mathbb{Z}[t, t^{-1}]$.

Let A be a connected algebra and M be a graded right A-module. We define the torsion submodule of M by $\tau(M) = \mathrm{H}^0_{\mathfrak{m}}(M) \subset M$. We say that M is torsion if $\tau(M) = M$ and torsion-free if $\tau(M) = 0$. If A is right noetherian and M is finitely generated, then M is torsion if and only if $\operatorname{GKdim} M = 0$.

Lemma 2.9. *Let A be a Cohen-Macaulay algebra of $\operatorname{depth} A = d$, which is a graded quotient algebra of a quantum polynomial ring, and M be a finitely generated graded right A-module.*

222

(1) If $M \neq 0$ is torsion, then M is Cohen-Macaulay.

(2) Suppose that $\mathrm{GKdim}\,M = 1$. Then M is Cohen-Macaulay if and only if M is torsion-free.

(3) The duality $M \mapsto M^{\dagger}$ gives a bijection between finitely generated Cohen-Macaulay graded right and left A-modules of depth m. In particular, $M^{\dagger\dagger} \cong M$, $H_{M^{\dagger}}(t) = (-1)^m H_M(t^{-1})$, and $e(M^{\dagger}) = e(M)$.

(4) M is a shift of a point module if and only if M is Cohen-Macaulay of GKdimension 1 and multiplicity 1.

(5) If $\mathrm{GKdim}\,M = 1$ and $e(M) = e$, then there exists a Cohen-Macaulay graded right A-module M' such that $H_{M'}(t) = e/(1 - t)$ and there is an exact sequence of graded right A-modules

$$0 \to T \to M_{\geq 0} \to M' \to T' \to 0$$

where T, T' are finite dimensional over k.

Proof. (1) and (2): Let A be a graded quotient algebra of a quantum polynomial ring S. Since $\mathrm{GKdim}\,S = \mathrm{ldim}\,S$ and S satisfies Cohen-Macaulay property with respect to GKdimension, $\mathrm{ldim}\,M = \mathrm{GKdim}\,M$ for every finitely generated graded right A-module M by Remark 2.4 because $\mathrm{ldim}\,M$ and $\mathrm{GKdim}\,M$ are the same whether computed over A or S by [14, Chapter 11, Proposition 3.5]. It follows that M is Cohen-Macaulay if and only if $\mathrm{depth}\,M = \mathrm{GKdim}\,M$. The results now follow easily.

(3): Since A is a Cohen-Macaulay algebra which is a graded quotient algebra of a quantum polynomial ring, the canonical module ω_A is a dualizing complex in the sense of [18, Definition 3.3] by [9, Theorem 1.6], so $M \mapsto M^{\dagger}$ gives the above duality by [18, Proposition 3.5] (cf. [14, Chapter 11, Proposition 7.1], [13, Lemma 4.6]).

Since A is a graded quotient algebra of a quantum polynomial ring, every finitely generated graded right A-module M is rational by [10, Proposition 5.5]. So if M is Cohen-Macaulay of depth $M = m$, then

$$H_{M^{\dagger}}(t) = H_{\underline{\mathrm{Ext}}_A^{d-m}(M,\omega_A)}(t) = H_{\mathrm{H}_{\mathfrak{m}}^m(M)^*}(t) = (-1)^m H_M(t^{-1}).$$

(4): Using (2) and (3), the proof is similar to that of [11, Proposition 2.7].

(5): Using (1), (2) and (3), the proof is similar to that of [4, Proposition 6.6]. $\qquad\square$

3. Gorenstein Algebras

In this section, we will prove that if X is a Gorenstein noncommutative projective scheme which can be embedded into a quantum projective space and whose dualizing sheaf is ample or anti-ample, then the point variety is independent of the choice of a Gorenstein homogeneous coordinate ring A for X.

Let A be a graded algebra. For each $r \in \mathbb{N}$, the r-th Veronese subalgebra of A, denoted by $A^{(r)}$, is a subalgebra of A such that $A_i^{(r)} = A_{ri}$. If M is a graded right A-module, then $M^{(r)}$ is a graded right $A^{(r)}$-module such that $M_i^{(r)} = M_{ri}$. Define a functor $F \colon \operatorname{GrMod} A \to \operatorname{GrMod} A^{(r)}$ by $F(M) = M^{(r)}$, and a functor $G \colon \operatorname{GrMod} A^{(r)} \to \operatorname{GrMod} A$ by $G(N) = N \otimes_{A^{(r)}} A$ with the grading $\deg(n \otimes a) = r \deg n + \deg a$. It is easy to see that $F \circ G \cong \operatorname{Id}$, but in general, $G \circ F \not\cong \operatorname{Id}$, so $\operatorname{GrMod} A \not\cong \operatorname{GrMod} A^{(r)}$. However, if A is right noetherian and generated in degree 1 over k, then these functors induce isomorphisms $\operatorname{Proj} A \cong \operatorname{Proj} A^{(r)}$. So, in order to answer Question 1.3, a natural question to ask is whether or not the point scheme of A is isomorphic to the point scheme of $A^{(r)}$.

Recall that the point scheme parameterizes point modules.

Lemma 3.1 ([3, Proposition 3.9, Corollary 3.13]). *Let A be a graded algebra finitely generated in degree 1 over k. Then there is a bijection between the set of isomorphism classes of point modules over A and the set of closed points of the point scheme Γ.*

If $A = T(V)/I$ is a graded algebra finitely generated in degree 1 over k, then $A^{(r)} \cong T(V^{\otimes r})/I^{(r)}$. Let Γ, $\Gamma^{(r)}$ be the point schemes of $A, A^{(r)}$ for $r \in \mathbb{N}$. By definition,

$$\Gamma_{ri} = \mathcal{V}(I_{ri}) \subset \mathbb{P}(V^*)^{\times ri}$$

and

$$\Gamma_i^{(r)} = \mathcal{V}(I_i^{(r)}) \subset \mathbb{P}(V^{* \otimes r})^{\times i}.$$

The product of the Segre embedding

$$\mathbb{P}(V^*)^{\times ri} = (\mathbb{P}(V^*)^{\times r})^{\times i} \to \mathbb{P}(V^{* \otimes r})^{\times i}$$

restricts to an embedding $\Gamma_{ri} \to \Gamma_i^{(r)}$, which is compatible with projections, that is, for $j \geq i$, the diagram

$$\begin{array}{ccc} \Gamma_{rj} & \longrightarrow & \Gamma_j^{(r)} \\ \downarrow{\scriptstyle pr_{ri}^{rj}} & & \downarrow{\scriptstyle (pr^{(r)})_i^j} \\ \Gamma_{ri} & \longrightarrow & \Gamma_i^{(r)} \end{array}$$

commutes. Hence, if the inverse system $\{\Gamma_i, pr_i^j\}$ is eventually constant (a basic assumption throughout the paper), then there is a natural embedding $\Gamma \to \Gamma^{(r)}$.

In order for Question 1.3 to be true, the embedding $\Gamma \to \Gamma^{(r)}$ must be an isomorphism for all $r \in \mathbb{N}$. Unfortunately, there is an example of a 3-dimensional Artin-Schelter regular algebra A on two generators such that the point scheme of $A^{(2)}$ is isomorphic to a disjoint union of the point scheme of A and a singleton, so that $\Gamma \not\cong \Gamma^{(2)}$ even as varieties (see [6]). We avoid this counterexample by assuming the condition (*) "A is a graded quotient algebra of a quantum polynomial ring", defined in the previous section.

Proposition 3.2. *If A is a Cohen-Macaulay algebra which is a graded quotient algebra of a quantum polynomial ring, then the point variety of A and the point variety of $A^{(r)}$ are isomorphic for any $r \in \mathbb{N}$.*

Proof. Let Γ, $\Gamma^{(r)}$ be point varieties of A, $A^{(r)}$. If M is a point module over A, then clearly $M^{(r)}$ is a point module over $A^{(r)}$, and the map $M \to M^{(r)}$ on the sets of isomorphism classes of point modules coincides with the embedding $\Gamma \to \Gamma^{(r)}$ on the sets of closed points via Lemma 3.1. We will show that the embedding is in fact surjective. Let N be a point module over $A^{(r)}$. Since $N \cong (N \otimes_{A^{(r)}} A)^{(r)}$, it follows that $\mathrm{GKdim}(N \otimes_{A^{(r)}} A) = 1$ and $e(N \otimes_{A^{(r)}} A) = 1$ by Lemma 2.8. By Lemma 2.9 (4), (5), there exists a point module M over A such that there is an exact sequence of graded right A-modules

$$0 \to T \to N \otimes_{A^{(r)}} A \to M \to T' \to 0$$

where T, T' are finite dimensional over k. It induces an exact sequence of graded right $A^{(r)}$-modules

$$0 \to T^{(r)} \to (N \otimes_{A^{(r)}} A)^{(r)} \cong N \xrightarrow{\varphi} M^{(r)} \to T'^{(r)} \to 0,$$

where $T^{(r)}, T'^{(r)}$ are finite dimensional over k. Since N and $M^{(r)}$ are cyclic and generated in degree zero over $A^{(r)}$, φ is either zero or surjective. If φ is zero, then $T^{(r)} \cong N$ which is a contradiction, so φ is surjective. Since

N and $M^{(r)}$ have the same Hilbert series, $N \cong M^{(r)}$. It follows that the embedding $\Gamma \to \Gamma^{(r)}$ is surjective, so $\Gamma \cong \Gamma^{(r)}$ as varieties. $\qquad\square$

Surprisingly, Proposition 3.2 is enough to answer Question 1.3 for Artin-Schelter Gorenstein algebras. I thank Zhang for his showing me how to prove the following lemma.

Lemma 3.3. *Let A, A' be Artin-Schelter Gorenstein algebras of* $\mathrm{id}(A) = d + 1$, $\mathrm{id}(A') = d' + 1$ *with the Gorenstein conditions*

$$\underline{\mathrm{Ext}}_A^{d+1}(k, A) \cong k(r), \quad \text{and} \quad \underline{\mathrm{Ext}}_{A'}^{d'+1}(k, A') \cong k(r')$$

for some $r, r' \in \mathbb{Z}$. If $\mathrm{Proj}\, A \cong \mathrm{Proj}\, A'$ and $d \geq 1, r \neq 0$, then $\mathrm{GrMod}\, A^{(|r|)} \cong \mathrm{GrMod}\, A'^{(|r'|)}$.

Proof. (Zhang) By [19, Corollary 4.3], $d = d'$, and for any noetherian object $\mathcal{M} \in \mathrm{Tails}\, A$, there are functorial isomorphisms

$$\mathrm{Ext}^i_{\mathrm{Tails}\, A}(\mathcal{M}, \mathcal{A}(-nr - r)) \cong \mathrm{Ext}^i_{\mathrm{Tails}\, A}(\mathcal{M}(nr), \mathcal{A}(-r))$$
$$\cong \mathrm{Ext}^{d-i}_{\mathrm{Tails}\, A}(\mathcal{A}, \mathcal{M}(nr))^*$$
$$\cong \mathrm{Ext}^{d-i}_{\mathrm{Tails}\, A}(\mathcal{A}(-nr), \mathcal{M})^*$$

as vector spaces for all i, n. Similarly, for any noetherian object $\mathcal{M}' \in \mathrm{Tails}\, A'$, there are functorial isomorphisms

$$\mathrm{Ext}^i_{\mathrm{Tails}\, A'}(\mathcal{M}', \mathcal{A}'(-nr' - r')) \cong \mathrm{Ext}^{d-i}_{\mathrm{Tails}\, A'}(\mathcal{A}'(-nr'), \mathcal{M}')^*$$

as vector spaces for all i, n.

Let $F\colon \mathrm{Proj}\, A \to \mathrm{Proj}\, A'$ be an isomorphism. Since $F(\mathcal{A}) \cong \mathcal{A}'$,

$$\mathrm{Hom}_{\mathrm{Tails}\, A'}(\mathcal{M}', F(\mathcal{A}(-r))) \cong \mathrm{Hom}_{\mathrm{Tails}\, A}(F^{-1}(\mathcal{M}'), \mathcal{A}(-r))$$
$$\cong \mathrm{Ext}^d_{\mathrm{Tails}\, A}(\mathcal{A}, F^{-1}(\mathcal{M}'))^*$$
$$\cong \mathrm{Ext}^d_{\mathrm{Tails}\, A'}(F(\mathcal{A}), \mathcal{M}')^*$$
$$\cong \mathrm{Ext}^d_{\mathrm{Tails}\, A'}(\mathcal{A}', \mathcal{M}')^*$$
$$\cong \mathrm{Hom}_{\mathrm{Tails}\, A'}(\mathcal{M}', \mathcal{A}'(-r'))$$

for all noetherian objects $\mathcal{M}' \in \mathrm{Tails}\, A'$, so $F(\mathcal{A}(-r)) \cong \mathcal{A}'(-r')$ by Yoneda's lemma. Inductively, we can show that $F(\mathcal{A}(-nr)) \cong \mathcal{A}'(-nr')$ for all $n \geq 0$. Since $\mathrm{Ext}^i_{\mathrm{Tails}\, A}(\mathcal{M}, \mathcal{N})$, $\mathrm{Ext}^i_{\mathrm{Tails}\, A'}(\mathcal{M}', \mathcal{N}')$ are finite dimensional over k for all i and all noetherian objects $\mathcal{M}, \mathcal{N} \in \mathrm{Tails}\, A, \mathcal{M}', \mathcal{N}' \in \mathrm{Tails}\, A'$

by [5, Corollary 7.3 (3)],

$$\mathrm{Hom}_{\mathrm{Tails}\,A'}(F(\mathcal{A}(r)), \mathcal{M}') \cong \mathrm{Hom}_{\mathrm{Tails}\,A}(\mathcal{A}(r), F^{-1}(\mathcal{M}'))$$

$$\cong \mathrm{Ext}^d_{\mathrm{Tails}\,A}(F^{-1}(\mathcal{M}'), \mathcal{A})^*$$

$$\cong \mathrm{Ext}^d_{\mathrm{Tails}\,A'}(\mathcal{M}', F(\mathcal{A}))^*$$

$$\cong \mathrm{Ext}^d_{\mathrm{Tails}\,A'}(\mathcal{M}', \mathcal{A}')^*$$

$$\cong \mathrm{Hom}_{\mathrm{Tails}\,A'}(\mathcal{A}'(r'), \mathcal{M}'),$$

for all noetherian objects $\mathcal{M}' \in \mathrm{Tails}\,A'$, so $F(\mathcal{A}(r)) \cong \mathcal{A}'(r')$ by Yoneda's lemma. Inductively, we can show that $F(\mathcal{A}(nr)) \cong \mathcal{A}'(nr')$ for all $n \geq 0$.

If $\bar{F}\colon \mathrm{Tails}\,A^{(|r|)} \to \mathrm{Tails}\,A'^{(|r'|)}$ is an equivalence functor defined by the following commutative diagram

$$
\begin{array}{ccc}
\mathrm{Tails}\,A & \xrightarrow[\cong]{F} & \mathrm{Tails}\,A' \\
{\scriptstyle(|r|)}\Big\downarrow & & \Big\downarrow{\scriptstyle(|r'|)} \\
\mathrm{Tails}\,A^{(|r|)} & \xrightarrow{\bar{F}} & \mathrm{Tails}\,A'^{(|r'|)},
\end{array}
$$

then

$$\bar{F}(\mathcal{A}^{(|r|)}(n)) \cong \bar{F}(\mathcal{A}(n|r|)^{(|r|)})$$

$$\cong F(\mathcal{A}(n|r|))^{(|r'|)}$$

$$\cong \mathcal{A}'(\pm n|r'|)^{(|r'|)}$$

$$\cong \mathcal{A}'^{(|r'|)}(\pm n)$$

for all $n \in \mathbb{Z}$. Since A, A' are Artin-Schelter Gorenstein algebras of depth $A = \mathrm{depth}\,A' = d+1 \geq 2$, we can show that $\bar{F}(\mathcal{A}^{(|r|)}(n)) \cong \mathcal{A}'^{(|r'|)}(n)$ for all $n \in \mathbb{Z}$. By [20, Theorem 3.5, Theorem 3.7], $\mathrm{GrMod}\,A^{(|r|)} \cong \mathrm{GrMod}\,A'^{(|r'|)}$. $\qquad\square$

Note that the condition $r \neq 0$ is saying that the dualizing sheaf $\mathcal{A}(-r)$ for $X = \mathrm{Proj}\,A$ is either "ample" or "anti-ample".

Now the following theorem follows easily.

Theorem 3.4. *Let A, A' be Artin-Schelter Gorenstein algebras of* $\mathrm{id}(A) = d + 1$, $\mathrm{id}(A') = d' + 1$ *with the Gorenstein conditions*

$$\underline{\mathrm{Ext}}^{d+1}_A(k, A) \cong k(r), \quad and \quad \underline{\mathrm{Ext}}^{d'+1}_{A'}(k, A') \cong k(r')$$

for some $r, r' \in \mathbb{Z}$. Suppose that A, A' are graded quotient algebras of quantum polynomial rings. If $\mathrm{Proj}\,A \cong \mathrm{Proj}\,A'$ and $d \geq 1$, $r \neq 0$, then the point variety of A and the point variety of A' are isomorphic.

Proof. Let Γ, Γ' be point varieties of A, A'. By Proposition 3.2 and Lemma 3.3,

$$\Gamma \cong \Gamma^{(|r|)} \cong \Gamma'^{(|r'|)} \cong \Gamma'. \qquad \square$$

It would be a much nicer result if we were able to prove that $\Gamma \cong \Gamma'$ as schemes in Theorem 3.4. If we know that Γ is a closed subscheme of $\mathrm{Proj}\, A$ (which is often the case), then presumably the fact that $\Gamma \cong \Gamma^{(r)}$ as varieties implies that $\Gamma \cong \Gamma^{(r)}$ as schemes. For example, if A is a 3-dimensional Artin-Schelter regular algebra, then the point scheme of A is a closed subscheme of $\mathrm{Proj}\, A$ by [3].

Corollary 3.5. *Let A, A' be 3-dimensional Artin-Schelter regular algebras generated in degree 1 over k. If $\mathrm{Proj}\, A \cong \mathrm{Proj}\, A'$, then the point scheme of A and the point scheme of A' are isomorphic.*

Proof. Let Γ, Γ' be the point schemes of A, A'.

If A, A' have 3 generators in degree 1, then A, A' are 3-dimensional quantum polynomial rings, so

$$\Gamma \cong \Gamma^{(3)} \cong \Gamma'^{(3)} \cong \Gamma'$$

as varieties by Theorem 3.4. Since Γ, Γ' are isomorphic to $\mathbb{P}^2$ or divisors of degree 3 in $\mathbb{P}^2$ by [3], they are isomorphic as schemes.

If A, A' have 2 generators in degree 1, then $\Gamma^{(2)} \cong \Gamma \cup \Gamma_{sp}$ and $\Gamma'^{(2)} \cong \Gamma' \cup \Gamma'_{sp}$ where $\Gamma_{sp}, \Gamma'_{sp}$ are either singleton or empty by [6, Lemma 5.1.3]. By [6, Proposition 3.1, Remark 3.3], there are 4-dimensional quantum polynomial rings S, S' and normalizing elements $x \in S_2$, $x' \in S'_2$ such that $A^{(2)} \cong S/(x)$, $A'^{(2)} \cong S'/(x')$. Since $A^{(2)}$, $A'^{(2)}$ are Artin-Schelter Gorenstein by [10, Theorem 3.6],

$$\Gamma^{(2)} \cong \Gamma^{(4)} \cong \Gamma'^{(4)} \cong \Gamma'^{(2)}$$

as varieties by Theorem 3.4. It follows that Γ and Γ' are isomorphic as varieties up to at most two isolated points. Since Γ, Γ' are isomorphic to $\mathbb{P}^1 \times \mathbb{P}^1$ or divisors of bidegree $(2,2)$ in $\mathbb{P}^1 \times \mathbb{P}^1$ by [3], they are isomorphic as schemes. $\qquad \square$

4. Geometric Algebras

Let A, A' be graded algebras finitely generated in degree 1 over k, and Γ, Γ' be the point schemes of A, A'. In the previous section, we showed that if A, A' are "almost commutative" and $\mathrm{Proj}\, A \cong \mathrm{Proj}\, A'$, then $\Gamma \cong \Gamma'$ (as varieties). This shows that, to some extent, the point varieties are useful in classifying such noncommutative projective schemes. However, the point scheme Γ itself does not recover $\operatorname{Proj} A$, so it is preferable to have another data to classify them. In this section, we will use another geometric data, a k-automorphism σ of the point scheme E for a geometric algebra A defined below following [3], and find a necessary and sufficient geometric condition for $\operatorname{GrMod} A \cong \operatorname{GrMod} A'$ when A, A' are geometric algebras so that such a geometric pair (E, σ) recovers $\operatorname{Proj} A$.

First, we recall that Zhang [20] have found a necessary and sufficient algebraic condition for $\operatorname{GrMod} A \cong \operatorname{GrMod} A'$, introducing a notion of a twisting system.

Definition 4.1 ([20]). Let A be a graded k-algebra. A set of graded k-linear automorphisms $\theta = \{\theta_i\}$ of A is called a twisting system on A if

$$\theta_n(a\theta_l(b)) = \theta_n(a)\theta_{n+l}(b)$$

for all l, m, n and $a \in A_l$, $b \in A_m$. The twist of A by θ, denoted by A^θ, is a graded k-algebra A with a new multiplication $*$ defined by

$$a * b = a\theta_l(b)$$

for all $a \in A_l$, $b \in A_m$. If M is a graded right A-module, then the twist of M by θ, denoted by M^θ, is a graded right A^θ-module M with a new action $*$ defined by

$$m * a = m\theta_l(a)$$

for all $m \in M_l$, $a \in A_m$.

If θ is a twisting system on A, then θ defines a functor $\theta \colon \operatorname{GrMod} A \to \operatorname{GrMod} A^\theta$ given by $M \mapsto M^\theta$. This is in fact an equivalence of categories with an inverse defined by the twisting system $\theta^{-1} = \{\theta_i^{-1}\}$ on A^θ. The converse is also true.

Theorem 4.2 ([20, Theorem 3.5]). *Let A, A' be graded algebras generated in degree 1 over k. Then $\operatorname{GrMod} A \cong \operatorname{GrMod} A'$ if and only if A' is isomorphic to a twist of A by a twisting system.*

Although Theorem 4.2 is an elegant and complete answer when $\operatorname{GrMod} A \cong \operatorname{GrMod} A'$, it is often difficult to construct a twisting system on A if A is given by generators and relations. So we will interpret the above theorem geometrically for a geometric algebra, which is an algebra closely tied with a geometric data (E, σ) where E is a point scheme and σ is a

k-automorphism of E. Let A, A' be geometric algebras with the associated geometric data (E, σ), (E', σ'). Then a suitable interpretation of Theorem 4.2 will show that $\mathrm{GrMod}\, A \cong \mathrm{GrMod}\, A'$ if and only if there is a sequence of isomorphisms $\tau_n \colon E \to E', n \in \mathbb{Z}$ which can be extended to automorphisms of the ambient projective space such that $\sigma' \tau_n = \tau_{n+1} \sigma \colon E \to E'$ for all $n \in \mathbb{Z}$. We will see later by examples that this geometric condition is often easy to check.

Let V be a finite dimensional k-vector space, $E \subset \mathbb{P}(V^*)$ be a closed k-subscheme, σ be a k-automorphism of E, and $\mathcal{L} = j^* \mathcal{O}_{\mathbb{P}(V^*)}(1)$ where $j \colon E \to \mathbb{P}(V^*)$ is the embedding. We define a map

$$\mu \colon \mathrm{H}^0(E, \mathcal{L}) \otimes \mathrm{H}^0(E, \mathcal{L}) \to \mathrm{H}^0(E, \mathcal{L}) \otimes \mathrm{H}^0(E, \mathcal{L}^\sigma) \to \mathrm{H}^0(E, \mathcal{L} \otimes_{\mathcal{O}_E} \mathcal{L}^\sigma)$$

of k-vector spaces by $v \otimes w \mapsto v \otimes w^\sigma$ where $\mathcal{L}^\sigma = \sigma^* \mathcal{L}$ and $w^\sigma = w \circ \sigma$.

Definition 4.3. A quadratic algebra $A = T(V)/(R)$ is called geometric if there is a pair (E, σ) where $E \subset \mathbb{P}(V^*)$ is a closed k-subscheme, and σ is a k-automorphism of E such that

- (G1): $\Gamma_2 = \mathcal{V}(R) \subset \mathbb{P}(V^*) \times \mathbb{P}(V^*)$ is the graph of E under σ, and
- (G2): $R = \ker \mu$ with the identification

$$\mathrm{H}^0(E, \mathcal{L}) = \mathrm{H}^0(\mathbb{P}(V^*), \mathcal{O}_{\mathbb{P}(V^*)}(1)) = V$$

  as k-vector spaces.

If A is geometric as above, then we write $A = \mathcal{A}(E, \sigma)$.

Let $A = T(V)/(R)$ be a quadratic algebra. If A satisfies the condition (G1), then the point scheme Γ of A is isomorphic to E by [3]. Moreover, if A satisfies the condition (G1), then the condition (G2) becomes equivalent to the condition (G2'): $R = \{ f \in V \otimes V \mid f|_{\Gamma_2} = 0 \}$. A quadratic algebra $A = T(V)/(R)$ satisfying the condition (G2') was called the algebra determined by the geometric data (E, σ) in [17, Definition 1.12].

Example 4.4. 1. Every 3-dimensional quantum polynomial ring is geometric by [3].

2. Every 4-dimensional quantum polynomial ring satisfying the Auslander condition satisfies the condition (G1) by [15, Theorem 1.4]. Moreover, if $\dim \Gamma_2 = 0$, then A is geometric by [16, Theorem 4.1].

Geometric data can be used to determine when geometric algebras have equivalent graded module categories. We will use the following two lemmas. The first lemma is standard and the proof is left to the reader.

Lemma 4.5. *Let $j\colon E \to \mathbb{P}^n$, $j'\colon E' \to \mathbb{P}^n$ be embeddings of closed k-subschemes, and $\mathcal{L} = j^*\mathcal{O}_{\mathbb{P}^n}(1)$, $\mathcal{L}' = j'^*\mathcal{O}_{\mathbb{P}^n}(1)$. Then an isomorphism of k-schemes $\tau\colon E \to E'$ can be extended to an automorphism of $\mathbb{P}^n$ if and only if $\tau^*\mathcal{L}' \cong \mathcal{L}$.*

Lemma 4.6 ([20, Proposition 2.8]). *A graded algebra A' is isomorphic to a twist of a graded algebra A if and only if there are graded k-linear isomorphisms $\phi_n\colon A' \to A$ for $n \in \mathbb{Z}$, such that*

$$\phi_n(ab) = \phi_n(a)\phi_{n+l}(b)$$

for all $l, m, n \in \mathbb{Z}$ and all $a \in A'_l, b \in A'_m$. If this is the case, then $\theta_n = \phi_n \circ \phi_0^{-1}$ is a twisting system on A and $\phi_0\colon A' \to A^\theta$ is an isomorphism of graded k-algebras.

Theorem 4.7. *Let $A = T(V)/I$ and $A' = T(V)/I'$ be graded algebras finitely generated in degree 1 over k.*

(1) If $A = \mathcal{A}(E, \sigma)$ is geometric and $\operatorname{GrMod} A \cong \operatorname{GrMod} A'$, then $A' = \mathcal{A}(E', \sigma')$ is also geometric and there is a sequence of automorphisms $\{\tau_n\}$ of $\mathbb{P}(V^*)$ for $n \in \mathbb{Z}$, each of which sends E isomorphically onto E', such that the diagram

$$
\begin{array}{ccc}
E & \xrightarrow{\ \tau_n\ } & E' \\
\sigma \downarrow & & \downarrow \sigma' \\
E & \xrightarrow{\ \tau_{n+1}\ } & E'
\end{array}
$$

commutes for every $n \in \mathbb{Z}$.

(2) Conversely, if $A = \mathcal{A}(E, \sigma)$ and $A' = \mathcal{A}(E', \sigma')$ are geometric and there is a sequence of automorphisms $\{\tau_n\}$ of $\mathbb{P}(V^*)$ for $n \in \mathbb{Z}$, each of which sends E isomorphically onto E', such that the diagram

$$
\begin{array}{ccc}
E & \xrightarrow{\ \tau_n\ } & E' \\
\sigma \downarrow & & \downarrow \sigma' \\
E & \xrightarrow{\ \tau_{n+1}\ } & E'
\end{array}
$$

commutes for every $n \in \mathbb{Z}$, then $\operatorname{GrMod} A \cong \operatorname{GrMod} A'$.

Proof. (1): Suppose that $A = \mathcal{A}(E, \sigma)$ is geometric and $\operatorname{GrMod} A \cong \operatorname{GrMod} A'$. Then $A' \cong A^\theta$ for some twisting system θ on A by Theorem 4.2. If $F_\bullet$ is a minimal free resolution of k as a graded right A-module, then $F_\bullet^\theta$ is a minimal free resolution of k as a graded right A^θ-module, so

$\mathrm{Tor}_i^A(k,k) \cong \mathrm{Tor}_i^{A^\theta}(k,k)$ as graded k-vector spaces for all $i \geq 0$. Since A is quadratic, $\mathrm{Tor}_2^{A^\theta}(k,k)_i \cong \mathrm{Tor}_2^A(k,k)_i = 0$ for all $i \neq 2$, so A^θ is also quadratic.

Let $A = T(V)/(R)$, $A' = T(V)/(R')$, and Γ, Γ' be the point schemes of A, A'. Since A' is isomorphic to a twist of A, there are graded k-linear isomorphisms $\phi_n\colon A' \to A$ as in Lemma 4.6. Let $\tau_n\colon \mathbb{P}(V^*) \to \mathbb{P}(V^*)$ be automorphisms induced by the duals of $\phi_n|_V\colon V \to V$. Then $\sum_i v_i \otimes w_i \in R'$ if and only if, for each $n \in \mathbb{Z}$, $\sum_i \phi_n|_V(v_i) \otimes \phi_{n+1}|_V(w_i) = \sum_i v_i^{\tau_n} \otimes w_i^{\tau_{n+1}} \in R$, so there is a commutative diagram

$$
\begin{array}{ccc}
\mathbb{P}(V^*) \times \mathbb{P}(V^*) & \xrightarrow{\ \tau_n \times \tau_{n+1}\ } & \mathbb{P}(V^*) \times \mathbb{P}(V^*) \\
\cup & & \cup \\
\Gamma_2 & \xrightarrow[\ \cong\ ]{} & \Gamma_2'.
\end{array}
$$

It follows that Γ_2' is the graph of the closed k-subscheme $E' = \tau_0(E) \subset \mathbb{P}(V^*)$ under the k-automorphism $\sigma' = \tau_1 \sigma \tau_0^{-1}$, and the diagram of isomorphisms

$$
\begin{array}{ccc}
E & \xrightarrow{\ \tau_n\ } & E' \\
\sigma \downarrow & & \downarrow \sigma' \\
E & \xrightarrow{\ \tau_{n+1}\ } & E'
\end{array}
$$

commutes for every $n \in \mathbb{Z}$.

Let $j\colon E \to \mathbb{P}(V^*)$, $j'\colon E' \to \mathbb{P}(V^*)$ be the embeddings, and $\mathcal{L} = j^*\mathcal{O}_{\mathbb{P}(V^*)}(1)$, $\mathcal{L}' = j'^*\mathcal{O}_{\mathbb{P}(V^*)}(1)$. For every $n \in \mathbb{Z}$, $\tau_n^*\mathcal{L}' \cong \mathcal{L}$ by Lemma 4.5, so the above commutative diagram induces the commutative diagram of isomorphisms

$$
\begin{array}{ccccccc}
\mathrm{H}^0(E,\sigma^*\mathcal{L}) & \xleftarrow{\ \cong\ } & \mathrm{H}^0(E,\sigma^*\tau_{n+1}^*\mathcal{L}') & \cong \mathrm{H}^0(E,\tau_n^*\sigma'^*\mathcal{L}') & \xleftarrow{\ \tau_n^*\ } & \mathrm{H}^0(E',\sigma'^*\mathcal{L}') \\
\sigma^* \uparrow & & \sigma^* \uparrow & & & \sigma'^* \uparrow \\
\mathrm{H}^0(E,\mathcal{L}) & \xleftarrow{\ \cong\ } & \mathrm{H}^0(E,\tau_{n+1}^*\mathcal{L}') & & \xleftarrow{\ \tau_{n+1}^*\ } & \mathrm{H}^0(E',\mathcal{L}').
\end{array}
$$

Hence there is a commutative diagram

$$
\begin{array}{ccccc}
\mathrm{H}^0(E,\mathcal{L}) \otimes \mathrm{H}^0(E,\mathcal{L}) & \xrightarrow{\ 1\otimes\sigma^*\ } & \mathrm{H}^0(E,\mathcal{L}) \otimes \mathrm{H}^0(E,\mathcal{L}^\sigma) & \longrightarrow & \mathrm{H}^0(E,\mathcal{L}\otimes\mathcal{L}^\sigma) \\
{\scriptstyle \phi_n\otimes\phi_{n+1}}\uparrow{\scriptstyle =\tau_n^*\otimes\tau_{n+1}^*} & & {\scriptstyle \phi_n\otimes\phi_n}\uparrow{\scriptstyle =\tau_n^*\otimes\tau_n^*} & & {\scriptstyle \phi_n}\uparrow{\scriptstyle =\tau_n^*} \\
\mathrm{H}^0(E',\mathcal{L}') \otimes \mathrm{H}^0(E',\mathcal{L}') & \xrightarrow{\ 1\otimes\sigma'^*\ } & \mathrm{H}^0(E',\mathcal{L}') \otimes \mathrm{H}^0(E',\mathcal{L}'^{\sigma'}) & \longrightarrow & \mathrm{H}^0(E',\mathcal{L}\otimes\mathcal{L}'^{\sigma'})
\end{array}
$$

where all vertical maps are isomorphisms. It follows that

$$R' = (\phi_0 \otimes \phi_1)^{-1}(R) = (\phi_0 \otimes \phi_1)^{-1}(\ker \mu) = \ker \mu',$$

and $A' = \mathcal{A}(E', \sigma')$ is a geometric algebra.

(2): Suppose that $A = T(V)/(R) = \mathcal{A}(E, \sigma), A' = T(V)/(R') = \mathcal{A}(E', \sigma')$ are geometric and that there is a sequence of automorphisms $\{\tau_n\}$ of $\mathbb{P}(V^*)$ as above. Let $\phi_n|_V \colon V \to V$ be isomorphisms induced by the duals of $\tau_n \colon \mathbb{P}(V^*) \to \mathbb{P}(V^*)$, which are determined up to nonzero scalar multiples. Define graded k-linear isomorphisms $\phi_n \colon T(V) \to T(V)$ by

$$\phi_n|_{V^{\otimes l}}(v_1 \otimes v_2 \otimes \cdots \otimes v_l) = \phi_n|_V(v_1) \otimes \phi_{n+1}|_V(v_2) \otimes \cdots \otimes \phi_{n+l-1}|_V(v_l)$$

for $v_i \in V$. By definition, clearly

$$\phi_n(v \otimes w) = \phi_n(v) \otimes \phi_{n+l}(w)$$

for all $l, m, n \in \mathbb{Z}$ and all $v \in V^{\otimes l}, w \in V^{\otimes m}$.

Let $j \colon E \to \mathbb{P}(V^*)$, $j' \colon E' \to \mathbb{P}(V^*)$ be the embeddings, and $\mathcal{L} = j^* \mathcal{O}_{\mathbb{P}(V^*)}(1)$, $\mathcal{L}' = j'^* \mathcal{O}_{\mathbb{P}(V^*)}(1)$. As before, there is a commutative diagram

$$
\begin{array}{ccccc}
\mathrm{H}^0(E, \mathcal{L}) \otimes \mathrm{H}^0(E, \mathcal{L}) & \xrightarrow{1 \otimes \sigma^*} & \mathrm{H}^0(E, \mathcal{L}) \otimes \mathrm{H}^0(E, \mathcal{L}^\sigma) & \longrightarrow & \mathrm{H}^0(E, \mathcal{L} \otimes \mathcal{L}^\sigma) \\
\phi_n|_V \otimes \phi_{n+1}|_V \uparrow {\scriptstyle = \tau_n^* \otimes \tau_{n+1}^*} & & \phi_n|_V \otimes \phi_n|_V \uparrow {\scriptstyle = \tau_n^* \otimes \tau_n^*} & & \uparrow {\scriptstyle \tau_n^*} \\
\mathrm{H}^0(E', \mathcal{L}') \otimes \mathrm{H}^0(E', \mathcal{L}') & \xrightarrow{1 \otimes \sigma'^*} & \mathrm{H}^0(E', \mathcal{L}' \otimes \mathcal{L}'^{\sigma'}) & \longrightarrow & \mathrm{H}^0(E', \mathcal{L}' \otimes \mathcal{L}'^{\sigma'})
\end{array}
$$

where all vertical maps are isomorphisms, so

$$\phi_n|_{V \otimes V}(R') = (\phi_n|_V \otimes \phi_{n+1}|_V)(\ker \mu') = \ker \mu = R$$

for every $n \in \mathbb{Z}$. Let $v \otimes r \otimes w \in V^{\otimes s} \otimes V^{\otimes 2} \otimes V^{\otimes t}$ for $s, t \geq 0$. Then $v \otimes r \otimes w \in V^{\otimes s} \otimes R' \otimes V^{\otimes t}$ if and only if

$$\phi_n(v \otimes r \otimes w) = \phi_n(v) \otimes \phi_{n+s}(r) \otimes \phi_{n+s+2}(w) \in V^{\otimes s} \otimes R \otimes V^{\otimes t}.$$

Since

$$A_i = \frac{V^{\otimes i}}{\sum_{s+t+2=i} V^{\otimes s} \otimes R \otimes V^{\otimes t}}, \qquad A'_i = \frac{V^{\otimes i}}{\sum_{s+t+2=i} V^{\otimes s} \otimes R' \otimes V^{\otimes t}},$$

as k-vector spaces for all $i \geq 2$, ϕ_n induce well-defined graded k-linear isomorphisms $\phi_n \colon A' \to A$ for all $n \in \mathbb{Z}$. Clearly

$$\phi_n(ab) = \phi_n(a)\phi_{n+l}(b)$$

for all $l, m, n \in \mathbb{Z}$ and all $a \in A'_l, b \in A'_m$ by construction, so A' is isomorphic to a twist of A by Lemma 4.6. Hence $\mathrm{GrMod}\, A \cong \mathrm{GrMod}\, A'$ by Theorem 4.2. $\qquad \square$

Remark 4.8. Let $A = \mathcal{A}(E, \sigma)$ and $A' = \mathcal{A}(E', \sigma')$ be geometric algebras. Theorem 4.7 says that $\operatorname{GrMod} A \cong \operatorname{GrMod} A'$ if and only if there is an automorphism τ_0 of $\mathbb{P}(V^*)$, which sends E isomorphically onto E', such that $\tau_n := \sigma'^n \tau_0 \sigma^{-n}$ can be extended to an automorphism of $\mathbb{P}(V^*)$ for every $n \in \mathbb{Z}$. If this is the case, then A' is isomorphic to a twist of A by a twisting system $\{\theta_n\}$ given by

$$\theta_n|_V = \tau_n^* \tau_0^{*-1} = (\tau_0^{-1} \tau_n)^* = (\tau_0^{-1} \sigma'^n \tau_0 \sigma^{-n})^*.$$

Remark 4.9. Let $A = \mathcal{A}(E, \sigma)$ and $A' = \mathcal{A}(E', \sigma')$ be geometric algebras. As a corollary to Theorem 4.7, we can easily show that A is isomorphic to A' as graded k-algebras if and only if $A_1 = A_1' =: V$ and there is an automorphism τ of $\mathbb{P}(V^*)$, which sends E isomorphically onto E', such that the diagram

$$
\begin{array}{ccc}
E & \xrightarrow{\ \tau\ } & E' \\
\sigma \downarrow & & \downarrow \sigma' \\
E\tau & \longrightarrow & E'
\end{array}
$$

commutes. This result is well known.

Example 4.10. If $A = k\langle x, y, z \rangle$ is a graded algebra with the defining relations

$$zy = \alpha yz, \qquad xz = \beta zx, \qquad yx = \gamma xy,$$

where $\alpha, \beta, \gamma \in k$ such that $\alpha\beta\gamma \neq 0, 1$, then $A = \mathcal{A}(E, \sigma)$ is geometric where $E = \ell_1 \cup \ell_2 \cup \ell_3 \subset \mathbb{P}^2$, $\ell_1 = \mathcal{V}(x), \ell_2 = \mathcal{V}(y), \ell_3 = \mathcal{V}(z)$, and $\sigma \in \operatorname{Aut} E$ is given by

$$\sigma|_{\ell_1}(0, b, c) = (0, \alpha b, c)$$
$$\sigma|_{\ell_2}(a, 0, c) = (a, 0, \beta c)$$
$$\sigma|_{\ell_3}(a, b, 0) = (\gamma a, b, 0).$$

(A is a 3-dimensional quantum polynomial ring of Type S_1. See the next section.) Let $A' = k\langle x, y, z \rangle$ be another such graded algebra with the defining relations

$$zy = \alpha' yz, \qquad xz = \beta' zx, \qquad yx = \gamma' xy.$$

Applying Theorem 4.7, we can show that

(1) $A \cong A'$ as graded k-algebras if and only if

$$(\alpha', \beta', \gamma') = (\alpha, \beta, \gamma), (\beta, \gamma, \alpha), (\gamma, \alpha, \beta),$$
$$(\alpha^{-1}, \gamma^{-1}, \beta^{-1}), (\beta^{-1}, \alpha^{-1}, \gamma^{-1}), \text{ or } (\gamma^{-1}, \beta^{-1}, \alpha^{-1}).$$

(2) $\mathrm{GrMod}\, A \cong \mathrm{GrMod}\, A'$ if and only if $\alpha'\beta'\gamma' = (\alpha\beta\gamma)^{\pm 1}$

Since the algebras A, A' above are rather simple, presumably it is possible to conclude the above results by linear-algebra calculations, but a geometric approach is often simpler (see the next section).

5. Three-dimensional Quantum Polynomial Rings

In this last section, we apply Theorem 4.7 to some 3-dimensional quantum polynomial rings. In particular, we classify quantum projective planes whose point schemes are generic and singular in terms of geometric data. In this section, we assume that k is an algebraically closed field.

If A is a 3-dimensional quantum polynomial ring, then $A = \mathcal{A}(E, \sigma)$ is geometric where E is either $\mathbb{P}^2$ or a divisor of degree 3 in $\mathbb{P}^2$. Artin, Tate and Van den Bergh [3] classified "generic" 3-dimensional quantum polynomial rings A in terms of their geometric data (E, σ). In their classification, if E is singular, then (E, σ) is one of the following:

- Type S_1: E is a triangle, and σ stabilizes each of the three components.
- Type S_2: E is a triangle, and σ interchanges two of its sides.
- Type S_1': E is a union of a line and a conic meeting the line in two points, and σ stabilizes each of the components and each of the intersection points.
- Type S_2': E is a union of a line and a conic meeting the line in two points, and σ stabilizes each of the components and interchanges the intersection points.

If E is a smooth curve, then E must be a smooth elliptic curve in $\mathbb{P}^2$. In addition, if σ is given by a translation by a point of E, then we call A a 3-dimensional Sklyanin algebra. Let $A = \mathcal{A}(E, \sigma)$, $A' = \mathcal{A}(E', \sigma')$ be 3-dimensional quantum polynomial rings. If $\mathrm{Proj}\, A \cong \mathrm{Proj}\, A'$, then $E \cong E'$ by Corollary 3.5.

For a noetherian domain R, we define the fraction ring of R by

$$\mathrm{Fract}(R) = \{ab^{-1} \mid a, b \in R \text{ and } b \neq 0\}.$$

For a noetherian graded domain, we define the graded fraction ring of A by

$$\mathrm{Fract}_{\mathrm{Gr}}(A) = \{ab^{-1} \mid a, b \in A \text{ are homogeneous elements and } b \neq 0\}.$$

If A is a noetherian graded domain over k and $X = \mathrm{Proj}\, A$, then we define the function field of X by $k(X) = (\mathrm{Fract}_{\mathrm{Gr}}(A))_0$. (Despite the terminology, $k(X)$ is a division algebra over k.)

We need the following lemma on division algebras.

Lemma 5.1 ([1, Proposition 3.2, Corollary 3.11 (c)]).

(1) $\mathrm{Fract}(k\langle u, v\rangle/(vu - quv))$ *is finite over its center if and only if q is a root of unity.*

(2) Suppose that $0 \neq q, q' \in k$ are not roots of unity. Then

$$\mathrm{Fract}(k\langle u, v\rangle/(vu - quv)) \cong \mathrm{Fract}(k\langle u, v\rangle/(vu - q'uv))$$

as k-algebras if and only if $q' = q^{\pm 1}$.

(3) If $0, 1 \neq q \in k$, then

$$\mathrm{Fract}(k\langle u, v\rangle/(vu - quv - r)) \cong \mathrm{Fract}(k\langle u, v\rangle/(vu - quv))$$

as k-algebras for any $r \in k$.

Theorem 5.2. *Let $A = \mathcal{A}(E, \sigma)$, $A' = \mathcal{A}(E', \sigma')$ be 3-dimensional quantum polynomial rings, and $X = \mathrm{Proj}\, A$, $X' = \mathrm{Proj}\, A'$. Suppose that $k(X)$ and $k(X')$ are not finite over their centers. If (E, σ) and (E', σ') are of the same generic singular Type (namely, S_1, S_2, S_1', or S_2'), then the following are equivalent:*

(1) $\mathrm{GrMod}\, A \cong \mathrm{GrMod}\, A'$.

(2) $X \cong X'$.

(3) $k(X) \cong k(X')$ *as k-algebras (birationally equivalent).*

(4) There is a sequence of automorphisms $\{\tau_n\}$ of $\mathbb{P}^2$ for $n \in \mathbb{Z}$, each of which sends E isomorphically onto E', such that the diagram

$$
\begin{array}{ccc}
E & \xrightarrow{\ \tau_n\ } & E' \\
\sigma \downarrow & & \downarrow \sigma' \\
E\tau_{n+1} & \longrightarrow & E'
\end{array}
$$

commutes for every $n \in \mathbb{Z}$.

It is well known that $(1) \Rightarrow (2) \Rightarrow (3)$ for any noetherian graded domains A, A' over k having finite GKdimension. By Theorem 4.7, $(1) \Leftrightarrow (4)$ for any geometric algebras A, A', so it remains to show that $(3) \Rightarrow (1)$ within each Type. We will give a proof for Type S_2'. The proofs for the other types are analogous and left to the reader.

Proof. Assume that both (E, σ) and (E', σ') are of Type S_2'. By [12, Lemma 5.4], we may assume that $A = k\langle x, y, z\rangle$ with the defining relations

$$z^2 = \alpha y^2 + x^2, \qquad xy = \beta zx, \qquad yx = \beta xz$$

and $A' = k\langle x, y, z\rangle$ with the defining relations

$$z^2 = \alpha' y^2 + x^2, \qquad xy = \beta' zx, \qquad yx = \beta' xz$$

where $\alpha\beta^2, \alpha'\beta'^2 \neq 0, 1$.

Clearly x is a regular normal element of degree 1 in both A and A'. It is easy to calculate

$$A[x^{-1}]_0 \cong k\langle u, v\rangle/(vu - \alpha\beta^2 uv - \beta), \qquad A'[x^{-1}]_0 \cong k\langle u, v\rangle/(vu - \alpha'\beta'^2 uv - \beta'),$$

where $u = zx^{-1}$, $v = yx^{-1}$. If $k(X)$ and $k(X')$ are not finite over their centers, then

$$k(X) \cong \mathrm{Fract}(A[x^{-1}]_0) \cong \mathrm{Fract}(A'[x^{-1}]_0) \cong k(X')$$

as k-algebras if and only if $\alpha'\beta'^2 = (\alpha\beta^2)^{\pm 1}$ by Lemma 5.1.

Note that $E = \ell \cup C$ where $\ell = \mathcal{V}(x), C = \mathcal{V}(x^2 + \gamma yz), \gamma = \alpha\beta - \beta^{-1}$, and

$$\sigma|_\ell(0, b, c) = (0, c, \alpha b)$$
$$\sigma|_C(a, b, c) = (a, \beta c, \beta^{-1} b).$$

It follows that

$$\sigma^{2n}|_\ell(0, b, c) = (0, \alpha^n b, \alpha^n c) = (0, b, c)$$
$$\sigma^{2n}|_C(a, b, c) = (a, b, c),$$

and

$$\sigma^{2n+1}|_\ell(0, b, c) = (0, c, \alpha b)$$
$$\sigma^{2n+1}|_C(a, b, c) = (a, \beta c, \beta^{-1} b).$$

Similarly, $E' = \ell' \cup C'$ where $\ell' = \mathcal{V}(x), C' = \mathcal{V}(x^2 + \gamma' yz), \gamma' = \alpha'\beta' - \beta'^{-1}$, and $\sigma', (\sigma')^{2n}$ and $(\sigma')^{2n+1}$ satisfy equations analogous to those above for σ, σ^{2n} and σ^{2n+1}. Since k is algebraically closed, there is an automorphism τ_0 of $\mathbb{P}^2$,

$$\tau_0(a, b, c) = (\sqrt{\gamma^{-1}\gamma'} a, b, c),$$

which sends E isomorphically onto E'. Since

$$\sigma'^{2n}\tau_0\sigma^{-2n}|_\ell(0,b,c) = (0,b,c)$$
$$\sigma'^{2n}\tau_0\sigma^{-2n}|_C(a,b,c) = (\sqrt{\gamma^{-1}\gamma'}a,b,c),$$

it follows that $\sigma'^{2n}\tau_0\sigma^{-2n}$ can be extended to an automorphism τ_{2n} of $\mathbb{P}^2$ for all $n \in \mathbb{Z}$.

Since

$$\sigma'^{2n+1}\tau_0\sigma^{-2n-1}|_\ell(0,b,c) = (0,b,\alpha'\alpha^{-1}c)$$
$$\sigma'^{2n+1}\tau_0\sigma^{-2n-1}|_C(a,b,c) = (\sqrt{\gamma^{-1}\gamma'}a,\beta'\beta^{-1}b,\beta'^{-1}\beta c),$$

it follows that $\sigma'^{2n+1}\tau_0\sigma^{-2n-1}$ can be extended to an automorphism τ_{2n+1} of $\mathbb{P}^2$ if and only if $\alpha\beta^2 = \alpha'\beta'^2$. By Remark 4.8, if $\alpha\beta^2 = \alpha'\beta'^2$, then $\operatorname{GrMod} A \cong \operatorname{GrMod} A'$.

Suppose that $\alpha'\beta'^2 = (\alpha\beta^2)^{-1}$. If we switch y and z, and replace x by $\sqrt{-\alpha}x$ in the relations of A, then we get the new graded algebra $A'' = k\langle x,y,z\rangle$ with the defining relations

$$z^2 = \alpha^{-1}y^2 + x^2, \qquad xy = \beta^{-1}zx, \qquad yx = \beta^{-1}xz.$$

Since $A \cong A''$ as graded k-algebras and $\operatorname{GrMod} A' \cong \operatorname{GrMod} A''$ by the previous argument, it follows that $\operatorname{GrMod} A \cong \operatorname{GrMod} A'$. $\qquad\square$

Finally, we apply Theorem 4.7 to 3-dimensional Sklyanin algebras.

Lemma 5.3. *Let $j\colon E \to \mathbb{P}^2$ be an embedding of a smooth elliptic curve, and $\sigma \in \operatorname{Aut} E$ be a translation by a point of E. Then σ can be extended to an automorphism of $\mathbb{P}^2$ if and only if $\sigma^3 = \operatorname{Id}$.*

Proof. Fix an identity $p_0 \in E$ in a group structure such that $\mathcal{O}_E(3p_0) \cong j^*\mathcal{O}_{\mathbb{P}^2}(1)$. Let $\sigma \in \operatorname{Aut} E$ be a translation by a point $p \in E$. By Lemma 4.5, σ can be extended to an automorphism of $\mathbb{P}^2$ if and only if

$$\mathcal{O}_E(3p_0) \cong j^*\mathcal{O}_{\mathbb{P}^2}(1) \cong \sigma^{*-1}j^*\mathcal{O}_{\mathbb{P}^2}(1) \cong \sigma^{*-1}\mathcal{O}_E(3p_0) \cong \mathcal{O}_E(3p).$$

But this is equivalent to p being 3-torsion, that is, $\sigma^3 = \operatorname{Id}$. $\qquad\square$

Theorem 5.4. *Let $A = \mathcal{A}(E,\sigma)$, $A' = \mathcal{A}(E',\sigma')$ be 3-dimensional Sklyanin algebras. Then the following are equivalent:*

(1) $\operatorname{GrMod} A \cong \operatorname{GrMod} A'$.

(2) There is an automorphism τ of $\mathbb{P}^2$, which sends E isomorphically onto E' such that the diagram

$$
\begin{array}{ccc}
E & \xrightarrow{\ \tau\ } & E' \\
\sigma^3 \downarrow & & \downarrow \sigma'^3 \\
E\tau & \longrightarrow & E'
\end{array}
$$

commutes.

(3) $\mathcal{A}(E,\sigma^3) \cong \mathcal{A}(E',\sigma'^3)$ as graded k-algebras (if $\sigma^9, \sigma'^9 \neq \mathrm{Id}$).

Proof. Let τ be any automorphism of $\mathbb{P}^2$, which sends E isomorphically onto E'. Since $\sigma'' := \tau^{-1}\sigma'\tau \in \mathrm{Aut}\, E$ is also a translation by a point of E, by Lemma 5.3, $\sigma''^n \sigma^{-n} = \tau^{-1}\sigma'^n \tau \sigma^{-n}$ can be extended to automorphisms of $\mathbb{P}^2$ for all $n \in \mathbb{Z}$ if and only if $(\sigma''^3 \sigma^{-3})^n = (\sigma''^n \sigma^{-n})^3 = \mathrm{Id}$ for all $n \in \mathbb{Z}$, or equivalently, $\sigma^3 = \sigma''^3 = \tau^{-1}\sigma'^3\tau$, that is, the diagram

$$
\begin{array}{ccc}
E & \xrightarrow{\ \tau\ } & E' \\
\sigma^3 \downarrow & & \downarrow \sigma'^3 \\
E\tau & \longrightarrow & E'
\end{array}
$$

commutes. By Remark 4.8, (1) $\Leftrightarrow$ (2).

Moreover, if $\sigma^9, \sigma'^9 \neq \mathrm{Id}$, then σ^3, σ'^3 cannot be extended to automorphisms of $\mathbb{P}^2$ by Lemma 5.3, so $\mathcal{A}(E,\sigma^3), \mathcal{A}(E',\sigma'^3)$ are again 3-dimensional Sklyanin algebras, in particular, they are geometric. By Remark 4.9, (2) $\Leftrightarrow$ (3). $\qquad\square$

Acknowledgments

I would like to thank Michaela Vancliff and James Zhang for many useful conversations to improve the paper.

References

1. J. Alev and F. Dumas, Sur les Corps de Fractions de Certaines Algèbres de Weyl Quantiques, *J. Algebra* **170** (1994), 229–265.
2. M. Artin and W. Schelter, Graded Algebras of Global Dimension 3, *Adv. Math.* **66** (1987), 171–216.
3. M. Artin, J. Tate and M. Van den Bergh, Some Algebras Associated to Automorphisms of Elliptic Curves, *The Grothendieck Festschrift* Vol. **1** Birkhäuser, (1990), 33–85.
4. M. Artin, J. Tate and M. Van den Bergh, Modules over Regular Algebras of Dimension 3, *Invent. Math.* **106** (1991), 335–388.

5. M. Artin and J. J. Zhang, Noncommutative Projective Schemes. *Adv. Math.* **109** (1994), 228–287.

6. K. Bauwens and M. Van den Bergh, Normalizing Extensions of the Veronese of a Three Dimensional Artin-Schelter Regular Algebra on Two Generators, *J. Algebra* **205** (1998), 368–390.

7. P. Jorgensen, Noncommutative Graded Homological Identities, *J. Lond. Math. Soc.* **57** (1998), 336–350.

8. P. Jorgensen, Local Cohomology of Non-commutative Graded Algebras, *Comm. Algebra* **25** (1997), 575–591.

9. P. Jorgensen, Properties of AS-Cohen-Macaulay Algebras, *J. Pure Appl. Algebra* **138** (1999), 239–249.

10. P. Jorgensen and J. J. Zhang, Gourmet's Guide to Gorensteinness, *Adv. Math.* **151** (2000), 313–345.

11. T. Levasseur and S. P. Smith, Modules over the 4-dimensional Sklyanin Algebra, *Bull. Soc. Math. France* **121** (1993), 35–90.

12. I. Mori, The Center of Some Quantum Projective Planes, *J. Algebra* **204** (1998), 15–31.

13. I. Mori, Homological Properties of Balanced Cohen-Macaulay Algebras, *Trans. Amer. Math. Soc.* **355** (2003), 1025–1042.

14. S. P. Smith, *Non-commutative Algebraic Geometry*, lecture notes, University of Washington, (1994).

15. B. Shelton and M. Vancliff, Embedding a Quantum Rank Three Quadric in a Quantum $\mathbb{P}^3$, *Comm. Algebra* **27** (1999), 2877–2904.

16. B. Shelton and M. Vancliff, Schemes of Line Modules I, *J. Lond. Math. Soc.* **65** (2002), 575–590.

17. M. Vancliff and K. Van Rompay, Embedding a Quantum Nonsingular Quadric in a Quantum $\mathbb{P}^3$, *J. Algebra* **195** (1997), 93–129.

18. A. Yekutieli, Dualizing Complexes over Noncommutative Graded Algebras, *J. Algebra* **153** (1992), 41–84.

19. A. Yekutieli and J. J. Zhang, Serre Duality for Non-commutative Projective Schemes, *Proc. Amer. Math. Soc.* **125** (1997), 697–707.

20. J. J. Zhang, Twisted Graded Algebras and Equivalences of Graded Categories, *Proc. Lond. Math. Soc.* **72** (1996), 281–311.

21. J. J. Zhang, Connected Graded Gorenstein Algebras with Enough Normal Elements, *J. Algebra* **189** (1997), 390–405.

QUASIDETERMINANTS AND RIGHT ROOTS OF POLYNOMIALS OVER DIVISION RINGS

BARBARA L. OSOFSKY

Department of Mathematics, Rutgers University
110 Frelinghuysen Road, Piscataway, NJ 08854-8019
E-mail: osofsky@math.rutgers.edu

1. Introduction

1.1. *Basic purpose*

In this note I will present an introduction to the Gelfand, Retakh, et al. theory of quasideterminants of square matrices, especially over division rings, and indicate its application to right roots of polynomials. The theory of quasideterminants was introduced in the 1990's by Israel Gelfand and Vladimir Retakh. For example [5], [7], [9] and [11] give their initial attempts to develop the area as something that might be useful in place of determinants over noncommutative rings. In spite of the name, quasideterminants are not the analog of determinants for commutative rings. A quasideterminant for an $n \times n$ matrix $\mathbf{A}$ is the last pivot found when attempting to invert $\mathbf{A}$ provided the $(n-1) \times (n-1)$ submatrix of $\mathbf{A}$ obtained by omitting the row and column of this last pivot is invertible. This $(n - 1) \times (n - 1)$ submatrix is inverted, and then elementary row (or column) operations are performed to clear the rest of the row (or column) and generate this last pivot. A given matrix $\mathbf{A}$ may have up to n^2 different quasideterminants, indexed by the row and column where this last pivot is found. Over a division ring, one can find $\mathbf{A}^{-1}$, if it exists, by Gaussian elimination, so $\mathbf{A}$ is invertible if and only if it has at least one nonzero quasideterminant. A standard way of computing determinants of $n \times n$ matrices over commutative fields is as a product of the pivots in Gaussian elimination. Thus in the commutative case, one usually computes a quasideterminant when computing the determinant of a matrix except in very low dimensions. The determinant of an invertible matrix over a field is a product of quasideterminants of a nested set of square submatrices.

The major area where quasideterminants have generated a very rich body of new knowledge is in the study of right roots of polynomials in a noncommuting indeterminant over a division ring. In this context, since in general we cannot pull the coefficients past the indeterminant, we consider mainly those polynomials which are linear combinations of powers of some variable with coefficients in a division ring K, and a right root will not correspond to a linear factor on the right.

The bibliography is in chronological order, to give the reader some idea of the development of the ideas presented here. There is a summary of much of the work of the Gelfand Retakh group and its connections with many areas of noncommutative algebra available online in reference [20]. Work on quasideterminants is ongoing, and new connections with other areas of noncommutative algebra, such as in [22] are being found.

The reader is presumed to be very familiar with introductory linear algebra, including Gaussian elimination, which I loosely use to refer to elementary row and column operations and the corresponding elementary matrices.

1.2. *Preliminary remarks on right roots of polynomials*

We first look at right roots of polynomials as a way of showing how quasideterminants can arise. Start with a monic polynomial

$$P(x) = a_0 + a_1 z + a_2 z^2 + \cdots + a_{n-1} z^{n-1} + z^n$$

with coefficients in some division ring K, that is, a linear combination of the left vector space spanned by the powers of z. In the commutative case, we know that the coefficients a_i are nice symmetric functions of the roots of $P(x)$ in the algebraic closure of K. Let us look at a way to get that when the roots are all distinct.

Assume that the roots $\{\rho_i : 1 \leq i \leq n\}$ in some overfield of K are all distinct. Consider

$$\mathbf{V} = \begin{bmatrix} 1 & 1 & \ldots & 1 \\ \rho_1 & \rho_2 & \cdots & \rho_n \\ \vdots & \vdots & \vdots & \vdots \\ \rho_1^{n-1} & \rho_2^{n-1} & \ldots & \rho_n^{n-1} \end{bmatrix},$$

the van der Monde matrix[a] and the row vector

$$\mathbf{a} = \begin{bmatrix} a_0 \ a_1 \ \ldots \ a_{n-1} \end{bmatrix}.$$

---

[a]Actually, Gelfand et al. write this in the opposite order to what is done here, with the

Since the ρ_i are roots of p, we have the (partitioned) matrix equation

$$\left[\,\mathbf{a}\,\middle|\,1\,\right]\left[\begin{array}{c} V \\ \hline \rho_1^n \ \cdots \ \rho_n^n \end{array}\right] = \mathbf{0}.$$

Since $\mathbf{V}$ is a van der Monde matrix with distinct columns, it is invertible, so we can post multiply the above by $\mathbf{V}^{-1}$ to get

$$\mathbf{0} = \left[\,\mathbf{a}\,\middle|\,1\,\right]\left[\begin{array}{c} V \\ \hline \rho_1^n \ \cdots \ \rho_n^n \end{array}\right]\mathbf{V}^{-1}$$

$$= \left[\,\mathbf{a}\,\middle|\,1\,\right]\left[\begin{array}{c} I \\ \hline [\rho_1^n \ \cdots \ \rho_n^n]\mathbf{V}^{-1} \end{array}\right]$$

$$= \mathbf{a} + [\rho_1^n \ \cdots \ \rho_n^n]\mathbf{V}^{-1}.$$

Since the entries of $\mathbf{V}^{-1}$ and of $[\rho_1^n \ \cdots \ \rho_n^n]$ are all symmetric rational functions of $\{\rho_i\}_{i=1}^n$ this expresses each of the coefficients of P as a rational function of the roots.

For example, if $n = 2$, we have

$$[a_0 \ a_1 \ 1]\begin{bmatrix} 1 & 1 \\ \rho & \sigma \\ \rho^2 & \sigma^2 \end{bmatrix}\begin{bmatrix} \sigma & -1 \\ -\rho & 1 \end{bmatrix}\frac{1}{\sigma - \rho}$$

$$= [a_0 \ a_1 \ 1]\begin{bmatrix} 1 & 0 \\ 0 & 1 \\ \frac{1}{\sigma-\rho}(\rho^2\sigma - \sigma^2\rho) & \frac{1}{\sigma-\rho}(-\rho^2 + \sigma^2) \end{bmatrix}$$

$$= \left[a_0 + \tfrac{1}{\sigma-\rho}(\rho^2\sigma - \sigma^2\rho) \ \ a_1 + \tfrac{1}{\sigma-\rho}(-\rho^2 + \sigma^2)\right] = 0$$

so $a_0 = -\frac{1}{\sigma-\rho}(\rho^2\sigma - \sigma^2\rho) = \rho\sigma$ and $a_1 = -\frac{1}{\sigma-\rho}(-\rho^2 + \sigma^2) = -\sigma - \rho$, exactly as expected.

Where did we use the fact that anything commuted with anything else under multiplication? One place was in inverting that van der Monde matrix (we used $\rho\sigma - \sigma\rho = 0$ to get the $2,1$ entry of the inverse equal to 0). Another was the final dividing out by $\sigma - \rho$. For everything else, factors that started on the right of other factors stayed on the right. If a van der Monde matrix $\mathbf{V}$ over a not necessarily commutative division ring is

---

powers decreasing from top to bottom. That is, the top row is $[\rho_1^{n-1} \ \cdots \ \rho_n^{n-1}]$ and the bottom row all 1's in their papers. I believe this presentation is a little more typical. In any case, if the first row is to be row 1 as opposed to row 0, the row indices will not agree with the powers of the ρ_j.

invertible then you can compute its inverse by elementary row operations, and $\mathbf{V}^{-1}$ has entries rational functions in the entries of $\mathbf{V}$. Thus you can do exactly the same thing over a division ring with a noncommuting indeterminant z written on the right as long as you never try to simplify expressions by changing the order of the factors in any product. The ρ_i are called right roots of the polynomial P, and standard linear algebra shows that the coefficients of P are rational functions of the roots.

There is an obvious way to start inverting a matrix over a division ring, namely Gaussian elimination on columns (which corresponds to postmultiplying by elementary matrices on the right). In the 2×2 case we have

$$\begin{bmatrix} 1 & 1 \\ \rho & \sigma \\ 1 & 0 \\ 0 & 1 \end{bmatrix} \rightsquigarrow \begin{bmatrix} 1 & 0 \\ \rho & \sigma - \rho \\ 1 & -1 \\ 0 & 1 \end{bmatrix} \rightsquigarrow \begin{bmatrix} 1 & 0 \\ \rho & 1 \\ 1 & -(\sigma - \rho)^{-1} \\ 0 & (\sigma - \rho)^{-1} \end{bmatrix}$$

$$\rightsquigarrow \begin{bmatrix} 1 & 0 \\ 0 & 1 \\ 1 + (\sigma - \rho)^{-1}\rho & -(\sigma - \rho)^{-1} \\ -(\sigma - \rho)^{-1}\rho & 1(\sigma - \rho)^{-1} \end{bmatrix}$$

so the coefficients of $P(x) = x^2 + a_1 x + a_0$ look like

$$\begin{bmatrix} a_0 & a_1 \end{bmatrix} = - \begin{bmatrix} \rho^2 & \sigma^2 \end{bmatrix} \begin{bmatrix} 1 + (\sigma - \rho)^{-1}\rho & -(\sigma - \rho)^{-1} \\ -(\sigma - \rho)^{-1}\rho & (\sigma - \rho)^{-1} \end{bmatrix}$$

$$= - \begin{bmatrix} \rho^2 + \rho^2(\sigma - \rho)^{-1}\rho - \sigma^2(\sigma - \rho)^{-1}\rho & -\rho^2(\sigma - \rho)^{-1} + \sigma^2(\sigma - \rho)^{-1} \end{bmatrix}$$

Then $a_1 = \rho^2(\sigma - \rho)^{-1} - \sigma^2(\sigma - \rho)^{-1} = (\rho^2 - \sigma^2)(\sigma - \rho)^{-1}$ is clearly symmetric (invariant under the permutation interchanging ρ and σ). It is not as obvious that $a_0 = -\rho^2 - \rho^2(\sigma - \rho)^{-1}\rho + \sigma^2(\sigma - \rho)^{-1}\rho$ is symmetric. We give a computation to show that it is.

$$-(\rho^2 + \rho^2(\sigma - \rho)^{-1}\rho - \sigma^2(\sigma - \rho)^{-1}\rho) + (\sigma^2 + \sigma^2(\rho - \sigma)^{-1}\sigma - \rho^2(\rho - \sigma)^{-1}\sigma)$$

$$= -\rho^2 + \sigma^2 - \rho^2((\sigma - \rho)^{-1}\rho + (\rho - \sigma)^{-1}\sigma) + \sigma^2((\sigma - \rho)^{-1}\rho - (\rho - \sigma)^{-1}\sigma)$$

$$= -\rho^2 + \sigma^2 - \rho^2(\sigma - \rho)^{-1}(\rho - \sigma) + \sigma^2(\sigma - \rho)^{-1}(\rho - \sigma)$$

$$= -\rho^2 + \sigma^2 + \rho^2 - \sigma^2 = 0.$$

The element $(\sigma - \rho)$ that you have to divide by here is a quasideterminant of the matrix $\begin{bmatrix} 1 & 1 \\ \rho & \sigma \end{bmatrix}$. You will get similar rational expressions by Gaussian elimination for higher order polynomials as long as you never divide by 0. The last pivot is a quasideterminant of the coefficient matrix.[b]

In the commutative case, right roots of polynomials correspond to right linear factors of the polynomial. In the case of twisted polynomial rings $K[z; \sigma, \delta]$ where σ is an endomorphism of K and δ is a σ-derivation of K, and $z\alpha = \sigma(\alpha)z + \delta(\alpha)$, Lam, in [3], defines evaluation of the right polynomial $P(z)$ at α in a way equivalent to $P(\alpha) = 0$ if and only if $(z - \alpha) \mid P(z)$, and this has enabled him to develop a rich theory of factorizations of such polynomials. This works because these twisted polynomial rings are Euclidean rings on the left with degree giving the Euclidean norm just as for regular polynomials. Thus for $\alpha \in K$, every polynomial $P(z) = Q(z)(z - \alpha) + \beta$ for some $\beta \in K$, and this β is $P(\alpha)$. There is no way to get such linear factors in the context in which Gelfand, Retakh et al. work since a product of a linear combination of positive powers of z multiplied on the right by $(z - \alpha)$ is not a linear combination of powers of z.

2. Quasideterminants

2.1. *The definition of quasideterminants*

Recall from the introduction that a (Gelfand-Retakh et al.) *quasideterminant* of a square matrix over a division ring is the last pivot in a successful Gaussian elimination sequence of elementary row and column operations which row or column reduces the matrix to the identity. Let us flesh out that idea.

Look at a square $n \times n$ matrix $\mathbf{A}$ over a not necessarily commutative ring. Partition it into an $(n-1) \times (n-1)$ submatrix $\mathbf{V}$ in the upper left corner, a single element α in the lower right corner, and an $(n-1)$ row vector $\mathbf{r}$ and $(n-1)$ column vector $\mathbf{c}$, namely

$$\mathbf{A} = \left[\begin{array}{c|c} \mathbf{V} & \mathbf{c} \\ \hline \mathbf{r} & \alpha \end{array}\right].$$

Now assume that $\mathbf{V}$ is invertible, with inverse $\mathbf{V}^{-1}$. Then $\mathbf{A}$ is invertible if and only if the matrix

$$\left[\begin{array}{c|c} \mathbf{V} & \mathbf{c} \\ \hline \mathbf{r} & \alpha \end{array}\right] \left[\begin{array}{c|c} \mathbf{V}^{-1} & \mathbf{0} \\ \hline \mathbf{0} & 1 \end{array}\right] = \left[\begin{array}{c|c} \mathbf{I} & \mathbf{c} \\ \hline \mathbf{r}\mathbf{V}^{-1} & \alpha \end{array}\right] \tag{$*$}$$

---

[b]Note that a single inversion is required in these rational expressions for the roots ρ and σ. In general, for a monic polynomial of degree n, there will be precisely $n-1$ inversions required to express the coefficients in terms of the right roots if the right roots generate an invertible van der Monde matrix.

is invertible. And this matrix is invertible if and only if when you do elementary row operations to make the vector portion of the last row zero (premultiply by elementary matrices) or the vector portion of the last column zero (postmultiply by elementary matrices), the entry in the n, n position is a unit. That new last entry in both cases is $\alpha - \mathbf{r}\mathbf{V}^{-1}\mathbf{c}$, and this expression is the n, n quasideterminant of $\mathbf{A}$, denoted

$$|\mathbf{A}|_{n,n} = \left|\begin{array}{c|c} \mathbf{V} & \mathbf{c} \\ \hline \mathbf{r} & \alpha \end{array}\right|_{n,n} = \alpha - \mathbf{r}\mathbf{V}^{-1}\mathbf{c}.$$

Recapping, in the case that $\mathbf{V}$ is invertible, $\mathbf{A}$ is invertible if and only if $|\mathbf{A}|_{n,n}$ is a unit.

What is the significance of the fact that our quasideterminant is in the n, n position. Basically it is only that it is much more convenient to picture things in that manner. If you have an inverse of the matrix $\mathbf{A}^{i,j}$ obtained by omitting row i and column j, you can do elementary row operations to clear the remaining column j entries missing the i, j entry or the remaining row i entries missing the i, j entry. In that case we call the pivot remaining after this Gaussian elimination $|\mathbf{A}|_{i,j}$. And just as in the illustration for $|\mathbf{A}|_{n,n}$,

$$|\mathbf{A}|_{i,j} = a_{i,j} - \mathbf{r}^{j}(\mathbf{A}^{i,j})^{-1}\mathbf{c}^{i} \qquad (**)$$

where $\mathbf{r}^{j}$ is row i with the column j entry removed, and $\mathbf{c}^{i}$ is column j with the row i entry removed. In case the submatrix $\mathbf{A}^{i,j}$ is not invertible, the quasideterminant $|\mathbf{A}|_{i,j}$ is not defined because there is no last pivot.

Note that, for any ring, if a quasideterminant $|\mathbf{A}|_{i,j}$ is defined, then $\mathbf{A}$ is invertible if and only this quasideterminant of $\mathbf{A}$ is a unit.

2.2. *Some elementary properties of quasideterminants*

There are some obvious consequences of the definition of quasideterminant as a pivot resulting from adding right multiples of columns to other columns or adding left multiples of rows to other rows once you have inverted $\mathbf{A}^{i,j}$. Not unsurprisingly, quasideterminants behave very nicely with respect to the elementary column (or row) operations of adding a multiple of one column (row) to another, multiplying a column (row) by a unit, or permuting two columns (rows). Doing column operations involves postmultiplication by elementary matrices, and row operations involve premultiplications.

Remark 1. If you add a (right) multiple of one column of $\mathbf{B}$ to a different column of $\mathbf{B}$ to get a matrix $\mathbf{A}$, no quasideterminant that remains defined is changed. The inverse of the new minor $\mathbf{A}^{i,j}$, if it exists, is an elementary matrix (perhaps the identity) times the inverse of the old minor $\mathbf{B}^{i,j}$. The elementary column operations used in computing the quasideterminant $|\mathbf{A}|_{p,q}$ will simply undo the elementary column operation performed to get $\mathbf{A}$ from $\mathbf{B}$. The same is true for adding a multiple of one row to another. The caveat about remaining defined is there only to cover the case where the column (or row) being added to or subtracted from another one is the location of the final pivot and so so may make the matrix $\mathbf{A}^{i,j}$ singular. For example, if $\mathbf{A} = \begin{bmatrix} 1 & 1 \\ 3 & 2 \end{bmatrix}$ subtracting column 2 from column 1 makes $|\mathbf{A}|_{2,2}$ undefined. Subtracting row 1 from row 2 leads to

$$\begin{vmatrix} 1 & 1 \\ 2 & 1 \end{vmatrix}_{2,2} = -1 = |\mathbf{A}|_{2,2}.$$

Remark 2. If you postmultiply column j of a matrix $\mathbf{A}$ by a scalar α, then

$$|\mathbf{A}|_{p,q} = \begin{cases} |\mathbf{A}|_{p,j}\alpha & \text{if } q = j \\ |\mathbf{A}|_{p,q} & \text{if } q \neq j \text{ and } \alpha \text{ is invertible} \\ \text{undefined} & \text{if } q \neq j \text{ and } \alpha \text{ is not invertible} \end{cases}$$

and the transpose of this holds for rows.

Remark 3. If you permute columns i and j, then

$$|\mathbf{A}|_{p,q} = \begin{cases} |\mathbf{A}|_{p,q} & \text{if } q \notin \{i,j\} \\ |\mathbf{A}|_{p,u} & \text{if } q \in \{i,j\} \text{ and } u \in \{i,j\} \setminus \{q\} \end{cases}$$

Remark 4. The quasideterminant $|\mathbf{A}|_{i,j}$ is independent of whether you do elementary row operations or elementary column operations to get the final pivot or a combination of them, provided you take into account operations that may change the quasideterminant as in Remark 2.

Remark 5. If $\mathbf{A}$ is a square matrix over a division ring, then $\mathbf{A}$ is invertible if and only if there exists a quasideterminant $|\mathbf{A}|_{i,j}$ which is defined and nonzero. This is because you can always invert the matrix by elementary column or elementary row operations.

Remark 6. Over a division ring there is an analog of Cramer's rule. One solves a system of linear equations with unknowns $\{x_i\}$ written on the left (so multiplying by elementary matrices is on the right) by column reducing the augmented matrix. If you are looking for variable j and the k,j minor

$\mathbf{A}^{k,j}$ is invertible, then you can do Gaussian elimination on all rows but the k-th with pivots in all columns except the j-th. The next reduction yields the equation $x_j|\mathbf{A}|_{k,j} = |\mathbf{B}|_{k,j}$ so $x_j = |\mathbf{B}|_{k,j}(|\mathbf{A}|_{k,j})^{-1}$ where $\mathbf{B}$ is the matrix obtained by replacing row j of $\mathbf{A}$ by the row of constants, provided the quasideterminant $|\mathbf{A}|_{k,j}$ is defined and invertible. Our illustration is for the case $k = j = n$ where the matrix $\mathbf{A} = \left[\begin{array}{c|c} \mathbf{V} & \mathbf{c} \\ \hline \mathbf{r} & \alpha \end{array}\right]$ is $n \times n$:

$$\begin{bmatrix} \mathbf{V} & \mathbf{c} \\ \mathbf{r} & \alpha \\ \mathbf{b} & \sigma \end{bmatrix} \rightsquigarrow \begin{bmatrix} \mathbf{I} & \mathbf{c} \\ \mathbf{r}\mathbf{V}^{-1} & \alpha \\ \mathbf{b}\mathbf{V}^{-1} & \sigma \end{bmatrix} \rightsquigarrow \begin{bmatrix} \mathbf{I} & \mathbf{0} \\ \mathbf{r}\mathbf{V}^{-1} & \alpha - \mathbf{r}\mathbf{V}^{-1}\mathbf{c} \\ \mathbf{b}\mathbf{V}^{-1} & \sigma - \mathbf{r}\mathbf{V}^{-1}\mathbf{c} \end{bmatrix}.$$

Remark 7. Every invertible square matrix $\mathbf{A}$ over a division ring has factorizations

$$\mathbf{A} = \mathbf{P}_1\mathbf{L}_1\mathbf{D}_1\mathbf{U}_1$$
$$= \mathbf{P}_2\mathbf{U}_2\mathbf{D}_2\mathbf{L}_2$$

where the $\mathbf{L}_i$ are lower triangular with 1's on the diagonal, the $\mathbf{U}_i$ are upper triangular with 1's on the diagonal, the $\mathbf{P}_i$ are permutation matrices, and the n, n entry of the diagonal matrix $\mathbf{D}$ is a quasideterminant of $\mathbf{A}$. For the first case, simply do a forward pass of Gaussian elimination, which eliminates starting from column n and so corresponds to post multiplication by a lower triangular matrix with 1's on the diagonal, permute rows (which corresponds to premultiplication by a permutation matrix) so that the result becomes upper triangular, and factor the pivots out of the rows of the upper triangular matrix that remains. For the second equation, start your Gaussian elimination in column 1 and permute to make what is left lower triangular. If you can pivot on the diagonal, that is, if you can factor $\mathbf{A} = \mathbf{LDU}$ with identity permutation, then the three matrices $\mathbf{L}$, $\mathbf{D}$, and $\mathbf{U}$ are unique and independent of the method employed to get the factorization (such as the order of doing elementary row operations below the diagonal to clear columns).

Remark 8. In Remark 7, the diagonal elements $d_{i,i}$ of the matrix are quasideterminants of an $i \times i$ submatrix of $\mathbf{A}$. In particular, if the underlying division ring is commutative, the determinant of $\mathbf{A}$ is a product of quasideterminants of $i \times i$ submatrices of $\mathbf{A}$. If pivoting is on the diagonal, then these submatrices have first entry at position $1, 1$ and last entry at position i, i for $1 \le i \le n$.

2.2.1. *Heredity*

The advantage of this definition of quasideterminant is that it can be used for block matrices in situations where all of the blocks need not be square. Over any ring, if the minor $\mathbf{A}^{i,j}$ of the matrix $\mathbf{A}$ is invertible, one can use $(**)$ to define the i, j quasideterminant of $\mathbf{A}$. If $\mathbf{A}$ is written in block form as

$$\mathbf{A} = \begin{bmatrix} \mathbf{B}_{1,1} & \cdots & \mathbf{B}_{1,m} \\ \vdots & \ddots & \vdots \\ \mathbf{B}_{m,1} & \cdots & \mathbf{B}_{m,m} \end{bmatrix}$$

where the $\mathbf{B}_{i,i}$ are square matrices, not necessarily of the same size, and if the minor

$$\mathbf{A}^{(i,j)_b} = \begin{bmatrix} \mathbf{B}_{1,1} & \cdots & \mathbf{B}_{1,m} \\ \vdots & \ddots & \vdots \\ \mathbf{B}_{m,1} & \cdots & \mathbf{B}_{m,m} \end{bmatrix}^{(i,i)_b}$$

obtained by deleting the i-th blocked row and j-th blocked column is square and invertible, then the i, j quasiminor of $\mathbf{A}$ with respect to this block decomposition is

$$|\mathbf{A}|^{(i,j)_b} = \mathbf{B}_{i,j} - \begin{bmatrix} \mathbf{B}_{i,1} & \cdots & \mathbf{B}_{i,m} \end{bmatrix}^{j_b} \left(\mathbf{A}^{(i,i)_b} \right)^{-1} \begin{bmatrix} \mathbf{B}_{1,m} \\ \vdots \\ \mathbf{B}_{m,m} \end{bmatrix}^{i_b} .$$

This gives a property, called *heredity*, namely that if required submatrices are invertible in this block decomposition, a quasideterminant $|\mathbf{B}|_{u,v}$ of a square $\mathbf{B} = \mathbf{B}_{i,j}$ is the quasideterminant $\left| |\mathbf{A}|^{(i,j)_b} \right|_{u,v}$.

3. The division ring of rational functions

In working with division rings, it is very helpful to be able to adjoin elements in a given set with no extra relations, and get some division ring in which they are invertible. The relevant theory was developed by Amitsur, Bergman, and Cohn. See for example, P. M. Cohn, [2]. It provides a general setting for doing computations such as those in the rest of this paper.

Given a set $X = \{x_i : 1 \le i \le n\}$, let $\mathcal{F}(X)$ be the free algebra over $\mathbb{Q}$ generated from elements of X by the operations of addition, subtraction, multiplication, division, and writing down formal inverses. No relations are imposed. Any map σ from X to a $\mathbb{Q}$-algebra R extends to a homomorphism on some subset $E(\sigma) \subseteq \mathcal{F}(X)$. Every $f \in \mathcal{F}(X)$, that is, every one of these words involving the x_i, $+$, $-$, $\cdot$, $^{-1}$, and parentheses or other grouping indicators, determines a subset $\mathrm{dom}(f)$, perhaps empty, of maps σ such that $f \in E(\sigma)$. If f and g have nonempty domains, then $\mathrm{dom}(f) \cap \mathrm{dom}(g) \neq \emptyset$, so we can define an equivalence relation on rational formulas with nonempty domains (nondegenerate) and if R is a division ring, the resulting equivalence classes form a skewfield $\mathfrak{F}_R(X)$. If R is infinite dimensional over its center, this skewfield does not depend on R, and we can just call it $\mathfrak{F}(X)$, the free skewfield in X over the center of R.

If R is a commutative field (of characteristic 0 unless you change the $\mathbb{Q}$ above to $\mathbb{Z}$), and $R\langle X \rangle$ the R-algebra of polynomials in the noncommuting indeterminants X, then the identity map on X extends to an embedding of $R\langle X \rangle$ into $\mathcal{F}(X)$ which is universal with respect to maps from X to R-algebras K extending to maps from some maximal subring $S \subseteq \mathcal{F}(X)$ to K where $S \supseteq R\langle X \rangle$. For such an S, if $\alpha \in S$ maps to an invertible element in K, then α must be invertible in S. These rings S are in the background in much of the discussion of quasideterminants. They lead to the concept of inversion height of an element of $\mathcal{F}(X)$, namely the minimal number of inverses necessary to construct that element. For example, the inversion height of a 'generic' quasideterminant of an $n \times n$ invertible matrix is $n-1$, since Gaussian elimination computes the last pivot by taking inverses of the $n - 1$ previous pivots, and 'generic' means you cannot avoid any of these inverses.

4. Van der Monde quasideterminants

4.1. *Independent sets of elements*

From here on we will be working over a division ring F. The van der Monde matrix on the set $\{x_i : 1 \leq i \leq n\} \subseteq F$ is the matrix

$$\mathbf{V}[x_1, \ldots, x_n] = \begin{bmatrix} 1 & 1 & \ldots & 1 & 1 \\ x_1 & x_2 & \ldots & x_{n-1} & x_n \\ \vdots & \vdots & \ddots & \vdots & \vdots \\ x_1^{n-2} & x_2^{n-2} & \ldots & x_{n-1}^{n-2} & x_n^{n-2} \\ x_1^{n-1} & x_2^{n-1} & \ldots & x_{n-1}^{n-1} & x_n^{n-1} \end{bmatrix}.$$

The $\{x_i : 1 \leq i \leq n\}$ are called *independent* if $\mathbf{V}[x_1, \ldots, x_n]$ is invertible.

Proposition 1. *The $\{x_i : 1 \leq i \leq n\}$ are independent if and only if for every i with $2 \leq i \leq n$, the quasideterminant $\mathbf{V}(x_1, \ldots, x_i) = |\mathbf{V}[x_1, \ldots, x_i]|_{i,i}$*

is defined and invertible. That is, $\mathbf{V}[x_1, \ldots, x_n]$ *is invertible if and only if you can invert it by pivoting on the diagonal.*

Proof. Clearly if the quasideterminant $\mathbf{V}(x_1, \ldots, x_i)$ is defined and invertible then Gaussian elimination pivoting on the diagonal will invert the matrix $\mathbf{V}[x_1, \ldots, x_n]$.

Now assume that some $\mathbf{V}(x_1, \ldots, x_i)$ is not both defined and invertible, and take i minimal with this property. Then the first $i - 1$ rows of $\mathbf{V}[x_1, \ldots, x_n]$ are independent, and you can row reduce them by pivoting on the diagonal. However, the i-th row is a linear combination of those first $i - 1$ rows, that is, for $1 \le j \le i$, there is a monic polynomial $p(x) = x^{i-1} + \sum_{j=0}^{i-2} a_j x^j$ such that $p(x_k) = 0$ for $1 \le k \le i$. Then $x_k^{i-1} = -\sum_{j=0}^{i-2} a_j x_k^j$ so every power of x_k^m with $m \ge i - 1$ is equal to $q(x_k)$ for some polynomial q of degree $\le i - 2$ independent of k. That is, in your (row) Gaussian elimination there cannot be any pivot in column i, so $\mathbf{V}[x_1, \ldots, x_n]$ is not invertible. $\qquad\square$

Van der Monde matrices with $n > 2$ need not be invertible. We will illustrate with our example of roots of a polynomial. Let $R = \mathbb{H}$. Let $p(x)$ be the polynomial $p(x) = x^3 + x^2 + x + 1$. Then p has an infinite number of right roots in $\mathbb{H}$, since

$$(a + bi + cj + dk)^2 = a^2 - b^2 - c^2 - d^2 + 2abi + 2acj + 2adk$$

so if $a = 0$ the square is a negative real. Dividing by $\sqrt{b^2 + c^2 + d^2}$ gives a square root of -1 and so a root of $p(x) = 0$ for any 4-tuple $(0, b, c, d)$ not all zero. We look at a particular set of four roots, namely $\{-1, i, j, k\}$, and the van der Monde matrices generated by the first three and the last three and do Gaussian elimination on them.

$$\begin{bmatrix} 1 & 1 & 1 \\ -1 & i & j \\ 1 & -1 & -1 \end{bmatrix} \rightsquigarrow \begin{bmatrix} 1 & 1 & 1 \\ 0 & 1+i & 1+j \\ 0 & -2 & -2 \end{bmatrix} \rightsquigarrow \begin{bmatrix} 1 & 1 & 1 \\ 0 & 1 & \dfrac{(1-i)(1+j)}{2} \\ 0 & -2 & -2 \end{bmatrix}$$

$$\rightsquigarrow \begin{bmatrix} 1 & 1 & 1 \\ 0 & 1 & \dfrac{1-i+j-k}{2} \\ 0 & 0 & -1-i+j-k \end{bmatrix}$$

so this matrix is invertible, and $\{1, i, j\}$ is independent The matrix

$$\mathbf{V}[i, j, k] = \begin{bmatrix} 1 & 1 & 1 \\ i & j & k \\ -1 & -1 & -1 \end{bmatrix}$$

is clearly not invertible; although the matrix $\mathbf{V}^{3,3}$ is invertible, the quasideterminant $|\mathbf{V}|_{3,3}$ is zero.

Remark 9. The proof of Proposition 1 shows that this example encapsulates the only way a van der Monde matrix can fail to be invertible. Indeed, $\mathbf{V}[x_1, x_2, \ldots, x_n]$ is invertible if and only if for all subsets $S \subseteq \{1, 2, \ldots, n\}$ with the cardinality of S equal to i, any nonzero polynomial $p(x)$ of degree $< i$ has $p(x_s) \neq 0$ for some $s \in S$.

4.2. *Factorization of the van der Monde quasideterminant*

Now let x_1, x_2, $\ldots$, x_{n-1}, x_n and x_1, x_2, $\ldots$, x_{n-1}, z be independent sets of elements in a division ring F where $n \geq 2$. We wish to calculate the $(n+1) \times (n+1)$ quasideterminant

$$
\mathbf{V}(x_1, x_2, \ldots, x_{n-1}, x_n, z) =
\begin{vmatrix}
1 & 1 & \cdots & 1 & 1 & 1 \\
x_1 & x_2 & \cdots & x_{n-1} & x_n & z \\
x_1^2 & x_2^2 & \cdots & x_{n-1}^2 & x_n^2 & z^2 \\
\vdots & & \ddots & & & \\
x_1^{n-1} & x_2^{n-1} & \cdots & x_{n-1}^{n-1} & x_n^{n-1} & z^{n-1} \\
x_1^n & x_2^n & \cdots & x_{n-1}^n & x_n^n & z^n
\end{vmatrix}_{n,n} .
$$

To get an idea what to expect, and to start our induction, let us look at the case $n = 2$ with a slight change of notation to avoid unnecessary clutter in this case. Elementary row operations correspond to multiplications on the left and elementary column operations correspond to multiplications on the right. Since we wish to have the first row be row 1 just as the first column is column 1, there is a shift in row index, and the (i, j) entry is actually $x_j{}^{i-1}$. We start with the matrix

$$
\begin{bmatrix}
1 & 1 & 1 \\
\rho & \sigma & z \\
\rho^2 & \sigma^2 & z^2
\end{bmatrix}
$$

and reduce to an upper triangular matrix in a somewhat atypical way using both row and column operations.

$$\begin{bmatrix} 1 & 1 & 1 \\ \rho & \sigma & z \\ \rho^2 & \sigma^2 & z^2 \end{bmatrix} \xrightarrow[R_3 \leftarrow R_3 - \rho R_2]{} \begin{bmatrix} 1 & 1 & 1 \\ \rho & \sigma & z \\ 0 & \sigma^2 - \rho\sigma & z^2 - \rho z \end{bmatrix}$$

$$\xrightarrow[R_2 \leftarrow R_2 - \rho R_1]{} \begin{bmatrix} 1 & 1 & 1 \\ 0 & \sigma - \rho & z - \rho \\ 0 & \sigma^2 - \rho\sigma & z^2 - \rho z \end{bmatrix}$$

$$\xrightarrow[C_2 \leftarrow C_2 - C_1]{} \begin{bmatrix} 1 & 0 & 1 \\ 0 & \sigma - \rho & z - \rho \\ 0 & \sigma^2 - \rho\sigma & z^2 - \rho z \end{bmatrix}$$

$$\xrightarrow[C_3 \leftarrow C_3 - C_1]{} \begin{bmatrix} 1 & 0 & 0 \\ 0 & \sigma - \rho & z - \rho \\ 0 & \sigma^2 - \rho\sigma & z^2 - \rho z \end{bmatrix}$$

$$\xrightarrow[C_2 \leftarrow C_2 (\sigma - \rho)^{-1}]{} \begin{bmatrix} 1 & 0 & 0 \\ 0 & 1 & z - \rho \\ 0 & (\sigma - \rho)\sigma(\sigma - \rho)^{-1} & z^2 - \rho z \end{bmatrix}$$

$$\xrightarrow[C_3 \leftarrow C_3 (z - \rho)^{-1}]{} \begin{bmatrix} 1 & 0 & 0 \\ 0 & 1 & 1 \\ 0 & (\sigma - \rho)\sigma(\sigma - \rho)^{-1} & (z - \rho)z(z - \rho)^{-1} \end{bmatrix}$$

$$\xrightarrow[R_3 \leftarrow R_3 - (\sigma - \rho)\sigma(\sigma - \rho)^{-1} R_2]{} \begin{bmatrix} 1 & 0 & 0 \\ 0 & 1 & 1 \\ 0 & 0 & (z - \rho)z(z - \rho)^{-1} - (\sigma - \rho)\sigma(\sigma - \rho)^{-1} \end{bmatrix}.$$

What is $\mathbf{V}(\rho, \sigma, z)$? There is only one elementary operation performed in this reduction that affects this quasideterminant, and that is multiplication of column 3 by $(z - \rho)^{-1}$, since the quasideterminant is the pivot in the third column. This multiplication also multiplies the quasideterminant by $(z - \rho)^{-1}$ by Remark 2. So to get $\mathbf{V}(\rho, \sigma, z)$ we must multiply that last pivot by $((z - \rho)^{-1})^{-1}$. As we saw in the preliminary remarks on right roots, $\mathbf{V}(u, v) = v - u$, so $\mathbf{V}(\rho, \sigma, z)$ can be written as

$$\mathbf{V}(\rho, \sigma, z) = (\mathbf{V}(\rho, z)z\mathbf{V}(\rho, z)^{-1} - \mathbf{V}(\rho, \sigma)\sigma\mathbf{V}(\rho, \sigma)^{-1})(z - \rho)$$

where the 2×2 quasideterminants exist and are not zero by the independence hypothesis.

For the general case, set

$$y_1 = x_1, \qquad y_k = \mathbf{V}(x_1, x_2, \ldots, x_{k-1}, x_k) x_k \mathbf{V}(x_1, x_2, \ldots, x_{k-1}, x_k)^{-1}$$

$$z_1 = z, \qquad z_k = \mathbf{V}(x_1, x_2, \ldots, x_{k-1}, z) z \mathbf{V}(x_1, x_2, \ldots, x_{k-1}, z)^{-1}$$

Theorem 1. *Let $\{x_i : 1 \le i \le n\}$ and $\{x_i : 1 \le i \le n-1\} \cup \{z\}$ be independent sets where $n \ge 2$. Then*

$$\mathbf{V}(x_1, x_2, \ldots, x_n, z) = (z_n - y_n)(z_{n-1} - y_{n-1}) \ldots (z_2 - y_2)(z_1 - y_1).$$

Proof. We use induction. The previous discussion is both the basis step and the template for the induction step. We start with the matrix

$$\begin{bmatrix} 1 & 1 & \ldots & 1 & 1 & 1 \\ x_1 & x_2 & \ldots & x_{n-1} & x_n & z \\ x_1^2 & x_2^2 & \ldots & x_{n-1}^2 & x_n^2 & z^2 \\ \vdots & \vdots & \ddots & \vdots & \vdots & \vdots \\ x_1^{n-1} & x_2^{n-1} & \ldots & x_{n-1}^{n-1} & x_n^{n-1} & z^{n-1} \\ x_1^n & x_2^n & \ldots & x_{n-1}^n & x_n^n & z^n \end{bmatrix}$$

and subtract x_1 times row i from row $i+1$ for i going from n to 1 (bottom to top) to get

$$\begin{bmatrix} 1 & 1 & \ldots & 1 & 1 & 1 \\ 0 & x_2 - x_1 & \ldots & x_{n-1} - x_1 & x_n - x_1 & z - x_1 \\ 0 & x_2^2 - x_1 x_2 & \ldots & x_{n-1}^2 - x_1 x_{n-1} & x_n^2 - x_1 x_n & z^2 - x_1 z \\ \vdots & \vdots & \ddots & \vdots & \vdots & \vdots \\ 0 & x_2^{n-1} - x_1 x_2^{n-2} & \ldots & x_{n-1}^{n-1} - x_1 x_{n-1}^{n-2} & x_n^{n-1} - x_1 x_n^{n-2} & z^{n-1} - x_1 z^{n-2} \\ 0 & x_2^n - x_1 x_2^{n-1} & \ldots & x_{n-1}^n - x_1 x_{n-1}^{n-1} & x_n^n - x_1 x_n^{n-1} & z^n - x_1 z^{n-1} \end{bmatrix}.$$

By Remark 1 this does not change the quasideterminant. Now observe that the i, j entry has a factor of x_j^{i-2} on the right and look at the submatrix

$$\mathbf{B} = \begin{bmatrix} x_2 - x_1 & \ldots & x_{n-1} - x_1 & x_n - x_1 & z - x_1 \\ (x_2 - x_1)x_2 & \ldots & (x_{n-1} - x_1)x_{n-1} & (x_n - x_1)x_n & (z - x_1)z \\ \vdots & \ddots & \vdots & \vdots & \vdots \\ (x_2 - x_1)x_2^{n-2} & \ldots & (x_{n-1} - x_1)x_{n-1}^{n-2} & (x_n - x_1)x_n^{n-2} & (z - x_1)z^{n-2} \\ (x_2 - x_1)x_2^{n-1} & \ldots & (x_{n-1} - x_1)x_{n-1}^{n-1} & (x_n - x_1)x_n^{n-1} & (z - x_1)z^{n-1} \end{bmatrix}.$$

By independence of $\{x_i\}$, we can pivot on the diagonal, and we have already used Gaussian elimination with pivots in the first column to clear that column of the van der Monde matrix below the $1, 1$ position. Hence

$$\mathbf{V}(x_1, x_2, \ldots, x_n, z) = |\mathbf{B}|_{n,n}.$$

Multiply the column of $\mathbf{B}$ involving x_j by $(x_j - x_1)^{-1}$ for $2 \leq j \leq n$, and the last column by $(z - x_1)^{-1}$. By Remark 2, the first $n-1$ of these operations do not change $|\mathbf{B}|_{n,n}$ but the last multiplies it by $(z - x_1)^{-1}$. This gives us a matrix $\mathbf{C} =$

$$\begin{bmatrix} 1 & \cdots & 1 & 1 \\ (x_2 - x_1)x_2(x_2 - x_1)^{-1} & \cdots & (x_2 - x_1)x_n(x_2 - x_1)^{-1} & (z - x_1)z(z - x_1)^{-1} \\ \vdots & \ddots & \vdots & \vdots \\ (x_2 - x_1)x_2^{n-2}(x_2 - x_1)^{-1} & \cdots & (x_n - x_1)x_n^{n-2}(x_n - x_1)^{-1} & (z - x_1)z^{n-2}(z - x_1)^{-1} \\ (x_2 - x_1)x_2^{n-1}(x_2 - x_1)^{-1} & \cdots & (x_n - x_1)x_n^{n-1}(x_n - x_1)^{-1} & (z - x_1)z^{n-1}(z - x_1)^{-1} \end{bmatrix}$$

which we recognize as van der Monde matrix which has $\mathbf{C}^{n,n}$ and $\mathbf{C}^{n,n-1}$ invertible because they are obtained from invertible submatrices of $\mathbf{V}(x_1, x_2, \ldots, x_n, z)$ by elementary row and column operations. We can thus apply induction to get

$$|\mathbf{C}|_{n,n} = \mathbf{V}((x_2 - x_1)x_2(x_2 - x_1)^{-1}, \ldots, (x_n - x_1)x_n(x_n - x_1)^{-1}, (z - x_1)z(z - x_1)^{-1})$$

so

$$\mathbf{V}(x_1, x_2, \ldots, x_n, z) = |\mathbf{B}|_{n,n} =$$
$$\mathbf{V}((x_2 - x_1)x_2(x_2 - x_1)^{-1}, \ldots, (x_n - x_1)x_n(x_n - x_1)^{-1}, (z - x_1)z(z - x_1)^{-1})(z - x_1).$$

Set $\hat{x}_i = (x_i - x_1)x_i(x_i - x_1)^{-1}$ and $\hat{z} = (z - x_1)z(z - x_1)^{-1}$. By ignoring the appropriate rows and columns in this argument we see that $\mathbf{V}(x_1, x_2, \ldots, x_k) = \mathbf{V}(\hat{x}_2, \hat{x}_3, \ldots, \hat{x}_k)(x_k - x_1)$ so

$$\mathbf{V}(x_1, x_2, \ldots, x_k)x_k\mathbf{V}(x_1, x_2, \ldots, x_k)^{-1}$$
$$= \mathbf{V}(\hat{x}_2, \hat{x}_3, \ldots, \hat{x}_k)(x_k - x_1)x_k(\mathbf{V}(\hat{x}_2, \hat{x}_3, \ldots, \hat{x}_k)(x_k - x_1))^{-1}$$
$$= \mathbf{V}(\hat{x}_2, \hat{x}_3, \ldots, \hat{x}_k)\hat{x}_k\mathbf{V}(\hat{x}_2, \hat{x}_3, \ldots, \hat{x}_k)^{-1}$$

and

$$\mathbf{V}(x_1, \ldots, x_{k-1}, z)z\mathbf{V}(x_1, \ldots, x_{k-1}, z)^{-1} =$$
$$\mathbf{V}(\hat{x}_2, \ldots, \hat{x}_{k-1}, \hat{z})\hat{z}\mathbf{V}(\hat{x}_2, \ldots, \hat{x}_{k-1}, \hat{z})^{-1}$$

so our induction hypothesis completes the proof. $\qquad\square$

Theorem 1 leads to a formulation that connects this determinant with a polynomial.

Theorem 2. *If $\{x_1, x_2, \ldots, x_n, z\}$ are independent, then*

$$\mathbf{V}(x_1, x_2, \ldots, x_n, z) = z^n + a_{n-1}z^{n-1} + \cdots + a_1 z + a_0$$

where

$$a_{n-k} = (-1)^k \sum_{1 \leq i_1 < i_2 < \cdots < i_k \leq n} y_{i_k} \cdot y_{i_{k-1}} \cdots \cdots y_{i_1}.$$

In particular the coefficient of z^{n-1} is $-(y_1 + y_2 + \cdots + y_n)$ and the constant term is $(-1)^n y_n \cdots \cdots y_1$.

Proof. By Theorem 1 and the (obvious) induction hypothesis,

$$
\begin{aligned}
\mathbf{V}(x_1, x_2, \ldots, x_n, z) &= \\
&= (z_n - y_n)\mathbf{V}(x_1, x_2, \ldots, x_{n-1}, z) \\
&= \mathbf{V}(x_1, x_2, \ldots, x_{n-1}, z)z - y_n \mathbf{V}(x_1, x_2, \ldots, x_{n-1}, z) \\
&= (z^n + b_{n-2}z^{n-1} + \cdots + b_0 z) - y_n(z^{n-1} + b_{n-2}z^{n-2} + \cdots + b_0) \\
&= z^n + (b_{n-2} - y_n)z^{n-1} + (b_{n-3} - y_n b_{n-2})z^{n-2} + \cdots + (b_0 - y_n b_1)z - y_n b_0 \\
&= z^n + a_{n-1}z^{n-1} + \cdots + a_1 z + a_0
\end{aligned}
$$

where the b_i have the appropriate form, and, by inspection of the form, since $a_i = b_{i-1} - y_n b_i$ for intermediate a_i, so do the a_i. $\qquad\square$

4.3. *The case of n independent right roots of a polynomial of degree n*

Let us assume that the polynomial

$$P(x) = x^n + a_{n-1}x^{n-1} + \cdots + a_1 x + a_0$$

with coefficients in a division ring has a set of n independent (right) roots $\{x_i : 1 \leq i \leq n\}$. The variable x is not assumed to commute with the coefficients. From the hypothesis $P(x_i) = 0$ for all i, we get a system of equations

$$
\begin{aligned}
-x_1^n &= a_{n-1}x_1^{n-1} + a_{n-2}x_1^{n-2} + \cdots + a_1 x_1 + a_0 \\
-x_2^n &= a_{n-1}x_2^{n-1} + a_{n-2}x_2^{n-2} + \cdots + a_1 x_2 + a_0 \\
&\quad\vdots \\
-x_{n-1}^n &= a_{n-1}x_{n-1}^{n-1} + a_{n-2}x_{n-1}^{n-2} + \cdots + a_1 x_{n-1} + a_0 \\
-x_n^n &= a_{n-1}x_n^{n-1} + a_{n-2}x_n^{n-2} + \cdots + a_1 x_n + a_0
\end{aligned}
$$

or

$$[-x_1^n \; -x_2^n \; \ldots \; -x_{n-1}^n \; -x_n^n] = [a_{n-1} \; a_{n-2} \; \ldots \; a_1 \; a_0]\mathbf{V}[x_1, x_2, \ldots, x_n].$$

By Cramer's rule (Remark 6), each a_i is a quotient of two quasideterminants of $n \times n$ submatrices of the $(n+1) \times n$ block matrix

$$\left[\frac{\mathbf{V}[x_1, x_2, \ldots, x_n]}{-x_1^n \; -x_2^n \; \ldots \; -x_{n-1}^n \; -x_n^n} \right],$$

where independence of the $\{x_i\}$ insures that $\mathbf{V}[x_1, x_2, \ldots, x_n]$ is invertible. Hence the a_i are rational functions of the roots, and because the a_i are independent of the order of the $\{x_i\}$, those rational functions must be symmetric in the $\{x_i\}$. Our work on quasideterminants gives us as much better way of determining this polynomial, as $\mathbf{V}(x_1, x_2, \ldots, x_n, x)$ is another monic polynomial of degree n with these same roots by Theorems 1 and 2, and $P(x) - \mathbf{V}(x_1, x_2, \ldots, x_n, x)$ is a polynomial of degree $< n$ with n independent roots. It must be 0, so Theorem 2 gives us another way to express the coefficients of $P(x)$.

5. Symmetric functions

In the commutative case, it is well known that every symmetric polynomial in variables $\{x_1, x_2, \ldots, x_n\}$ is a polynomial in the elementary symmetric functions (the coefficients of $\prod_{i=1}^n (x - x_i)$). If the commutative assumption is dropped, one can ask if there is a comparable theorem. Bergman and Cohn [1] show that the symmetric polynomials in $R\langle X \rangle$ for R a field and X a set of at least two variables is free but infinitely generated. For example, if $X = \{x, y\}$, then a free generating set for the symmetric polynomials in $R\langle X \rangle$ is $\{x^n + y^n : n \in \omega\}$. Now we look at symmetric elements in $\mathcal{F}(X)$.

The commutative elementary symmetric functions have a precise analog in the expansion of the van der Monde quasideterminant $\mathbf{V}(x_1, x_2, \ldots, x_n)$, even though the functions arising in it do not appear symmetric just by looking at them, and they are not polynomials. In Theorem 2 we set

$$y_1 = x_1, \quad y_k = \mathbf{V}(x_1, x_2, \ldots, x_k) x_k \mathbf{V}(x_1, x_2, \ldots, x_k)^{-1}$$

and showed that, if $\{x_1, x_2, \ldots, x_n, z\}$ are independent, then

$$\mathbf{V}(x_1, x_2, \ldots, x_n, z) = z^n + a_{n-1} z^{n-1} + \cdots + a_1 z + a_0.$$

Gelfand, Retakh et al. study noncommutative symmetric functions using these (unsigned) coefficients as the elementary symmetric functions in

$\mathcal{F}(X)$, that is, for $\Lambda_k(X)$ the elementary symmetric function of degree k they define

$$\Lambda_1(X) = y_1 + y_2 + \ldots y_n$$
$$\Lambda_2(X) = \sum_{i<j} y_j y_i$$
$$\ldots$$
$$\Lambda_n(X) = y_n \cdot y_{n-1} \cdots \cdots y_1$$

I repeat that the y_i are *not* polynomials in the $\{x_i\}$, but instead rational functions in $\mathcal{F}(X)$. The $\Lambda_i(X)$ are rational functions of the $\{x_i\}$ which are symmetric in the $\{x_i\}$ but not in the $\{y_i\}$ since order matters here.

For example, with $n = 2$,

$$y_1 = x_1,$$
$$y_2 = (x_2 - x_1)x_2(x_2 - x_1)^{-1}.$$

Our theory says $y_2 y_1$ is symmetric, and to give an idea of how nonobvious this may be, we can confirm that by noting

$$
\begin{aligned}
(x_2 - x_1)&x_2(x_2 - x_1)^{-1}x_1 = \\
&= (x_2 - x_1)x_2(x_2 - x_1)^{-1}x_1 \\
&\quad - (x_2 - x_1)x_2(x_2 - x_1)^{-1}x_2 + (x_2 - x_1)x_2(x_2 - x_1)^{-1}x_2 \\
&= (x_2 - x_1)x_2(x_2 - x_1)^{-1}(x_1 - x_2) \\
&\quad + (x_2 - x_1)x_2(x_2 - x_1)^{-1}x_2 \\
&= -(x_2 - x_1)x_2 + (x_2 - x_1)x_2(x_2 - x_1)^{-1}x_2 \\
&\quad - (x_2 - x_1)x_1(x_2 - x_1)^{-1}x_2 + (x_2 - x_1)x_1(x_2 - x_1)^{-1}x_2 \\
&= -(x_2 - x_1)x_2 + (x_2 - x_1)^2(x_2 - x_1)^{-1}x_2 \\
&\quad + (x_2 - x_1)x_1(x_2 - x_1)^{-1}x_2 = (x_2 - x_1)x_1(x_2 - x_1)^{-1}x_2 \\
&= (-1)^2(x_1 - x_2)x_1(x_1 - x_2)^{-1}x_2.
\end{aligned}
$$

On the other hand, in $\mathbb{H}$ if $x_1 = i$ and $x_2 = j$, then $y_1 = i$ and $y_2 = (j - i)i(i - j)/\sqrt{2}$ and $i(j - i)i(i - j)/\sqrt{2} = (k + 1)(-1 - k)/\sqrt{2}$ whereas $j(i - j)j(j - i)/\sqrt{2} = (-k + 1)(-1 + k)/\sqrt{2}$ and these are not equal, so preimages in $\mathcal{F}(x_1, x_2)$ cannot be equal.

The major theorem in the theory of noncommutative symmetric functions in $\mathcal{F}(X)$ was proved in Robert Lee Wilson, [17].

Theorem 3 (Wilson, 2001). *Let a polynomial $P(y_1, y_2, \ldots, y_n)$ be symmetric as a function in x_1, x_2, $\ldots$, x_n. Then $P(y_1, y_2, \ldots, y_n) =$*

$Q(\Lambda_1, \Lambda_2, \ldots, \Lambda_n)$ *where Q is a polynomial over $\mathbb{Q}$ in n noncommuting indeterminants.*

Since the Λ_k are rational functions in the $\{x_i\}$, this says that symmetric polynomials are in general not polynomials in the natural variables, only in these special rational functions. This theorem is a generalization of the well known theorem in the commutative case in the sense that if all of the appropriate variables commute, it reduces precisely to that commutative theorem.

5.1. *A side remark*

Lam and Leroy, in [4], using entirely different techniques, found a polynomial of degree n whose coefficients were candidates for a theory of symmetric functions. In the Euclidean domain $K[t]$, where t commutes with all the elements of K, the polynomials $\{t - x_i\}$ have a least common multiple, and independence forces it to be of degree n. By Theorem 1, since the order of the $\{x_i\}$ is irrelevant in computing the van der Monde quasideterminant, each of the polynomials in $\{t - x_i\}$ is a right factor of the van der Monde quasideterminant. Since this quasideterminant is monic, the van der Monde quasideterminant is the least common multiple of these linear polynomials. It is rather interesting that the same polynomial arises with two dramatically different interpolations. Wilson's Theorem, Theorem 3 showed that their remark that the coefficients of the least common multiple of the polynomials could lead to a theory of symmetric functions is correct.

6. A very brief introduction to the algebra Q_n

6.1. *The definition of Q_n*

For the remainder of this paper, we fix the following notation. Let F be a division algebra over a commutative field K, and let t be an indeterminate which commutes with all of the elements of K. Let $P(t) = t^n + a_{n-1}t^{n-1} + \ldots a_1 t + a_0 \in K[t]$ have some independent set $\{x_i : 1 \leq i \leq n\}$ of right roots. In this case we have seen in section 4.3 that $P(t) = \mathbf{V}(x_1, x_2, \ldots, x_n, t)$.

By Theorem 1, $P(t)$ has a factorization

$$P(t) = (t - y_n)(t - y_{n-1}) \ldots (t - y_1)$$

since all of the z_i are equal to t in this case. Every permutation σ of the $\{x_i : 1 \leq i \leq n\}$ of n independent right roots of $P(t)$ gives rise to a set

$\{y_{\sigma,k} : 1 \leq k \leq n\}$ which appear in such a factorization. The problem being investigated is how to capture information about the set $\{y_{\sigma,k} : \sigma \in S_n, 1 \leq k \leq n\}$. References for this include [15] and the survey article [20].

Initially we work in a generic situation independent of the polynomial P. We introduce special indeterminants which do not commute with the coefficients or each other and which capture the defining properties of the $y_{\sigma,k}$.

For any unordered subset $A = \{\rho_1, \ldots, \rho_{k-1}\} \subseteq \{1, 2, \ldots, n\}$ and every $i \in \{1, 2, \ldots, n\} \setminus A$, we define elements $\mathbf{x}_{A,i} \in F$ by

$$\mathbf{x}_{\emptyset,i} = x_i = y_i;$$

$$\mathbf{x}_{A,i} = \mathbf{V}(\{i\} \cup A)x_i\mathbf{V}(\{i\} \cup A)^{-1} = y_{\sigma,k} \text{ for the set } x_{\rho_1}, \ldots, x_{\rho_{k-1}}, x_i.$$

The order of the elements of A does not matter; that is, since the quasideterminant is the pivot in the column corresponding to x_i, by Remark 3, this is independent of the order on A. Moreover, by Theorem 2, we have the coefficients of the van der Monde quasideterminant $\mathbf{V}(x_1, \ldots, x_n, t) = P(t) = t^n + a_{n-1}t^{n-1} + \cdots + a_0$ of the form

$$a_{n-m} = (-1)^m \sum_{1 \leq i_1 < i_2 < \cdots < i_m \leq n} y_{i_m} \cdot y_{i_{m-1}} \cdots \cdots y_{i_1}$$

and these a_i are symmetric in the $\{x_i\}$. In particular

$$a_{n-1} = -(y_1 + y_2 + \cdots + y_n)$$

$$a_0 = (-1)^n y_n \cdots \cdots y_1$$

are symmetric in the $\{x_i\}$. That means, since you can permute x_i and x_j and cancel all the y_{a_m} corresponding only to elements in A, that

$$\mathbf{x}_{A\cup\{i\},j} + \mathbf{x}_{A,i} = \mathbf{x}_{A\cup\{j\},i} + \mathbf{x}_{A,j} \tag{\dag}$$
$$\mathbf{x}_{A\cup\{i\},j} \cdot \mathbf{x}_{A,i} = \mathbf{x}_{A\cup\{j\},i} \cdot \mathbf{x}_{A,j}$$

for all $A \subseteq \{1, 2, \ldots, n\}$ and $i \neq j \in \{1, 2, \ldots, n\} \setminus A$. It turns out that these strange looking symmetry relations are extremely powerful.

For the field K, we define the K-algebra Q_n as the K-algebra generated by formal noncommuting variables

$$\mathfrak{X} = \{z_{A,i} : A \subseteq \{1, 2, \ldots, n\}, i in \{1, 2, \ldots, n\} \setminus A\}$$

subject only to the relations $(\dag)$ with $\mathbf{x}$ replaced by z. That is, if $\mathcal{F}_n(\mathfrak{X})$ denotes the free associative K-algebra on the set $\mathfrak{X}$ and $\mathcal{J}_n(\mathfrak{X})$ denotes the ideal of $\mathcal{F}_n(\mathfrak{X})$ generated by

$$\{z_{A\cup\{i\},j} + z_{A,i} - z_{A\cup\{j\},i} - z_{A,j}, z_{A\cup\{i\},j} \cdot z_{A,i} - z_{A\cup\{j\},i} \cdot z_{A,j}\},$$

then $Q_n = \mathcal{F}_n(\mathfrak{X})/\mathcal{J}_n(\mathfrak{X})$.

6.2. *Some properties of Q_n*

There is no polynomial P in the formal definition of Q_n. Hence Q_n is universal for any factorization of a polynomial $\hat{P}$ of degree n with n independent right roots into a product of linear factors over any K-algebra division ring of coefficients. Indeed given any independent set of right roots $\{x_i : 1 \le i \le n\}$ each permutation σ of $\{1, \ldots, n\}$ gives rise to a factorization

$$\hat{P}(t) = (t - y_n)(t - y_{n-1}) \ldots (t - y_1)$$

where t is an indeterminate commuting with all elements of the underlying division ring. There is a homomorphism from Q_n to the coefficient division ring given by

$$z_{\{\sigma(1),\sigma(2),\ldots,\sigma(i)\},\sigma(i+1)} \mapsto y_i.$$

It turns out that these algebras have a deep structure. For example, they contain a copy of the free algebra on n elements generated by $\{z_{\emptyset,i}\}$, and they have a quotient isomorphic to polynomials in n commuting variables. It is possible to construct a basis of Q_n that can be used to get information about it. Its Hilbert series has been computed in [18] as

$$H(Q_n, t) = \frac{1 - t}{1 - t(2 - t)^n}.$$

It has been shown to be a Koszul algebra in [21], with all the structure that that implies.

References

1. G. M. Bergman and P. M. Cohn, *Symmetric elements in free powers of rings*, J. London Math. Soc. (2) **1** (1969), 525–534. MR 40 #4301
2. P. M. Cohn, *Skew field constructions*, Cambridge University Press, Cambridge, 1977, London Mathematical Society Lecture Note Series, No. 27. MR 57 #3190
3. T. Y. Lam, *A general theory of Vandermonde matrices*, Exposition. Math. **4** (1986), no. 3, 193–215. MR **88j**:16024
4. T. Y. Lam and A. Leroy, *Vandermonde and Wronskian matrices over division rings*, J. Algebra **119** (1988), no. 2, 308–336. MR **90f**:16005
5. I. M. Gelfand and V. S. Retakh, *Determinants of matrices over noncommutative rings*, Funktsional. Anal. i Prilozhen. **25** (1991), no. 2, 13–25, 96. MR **92k**:15018
6. I. M. Gelfand and V. S. Retakh, *Theory of noncommutative determinants, and characteristic functions of graphs*, Funktsional. Anal. i Prilozhen. **26** (1992), no. 4, 1–20, 96. MR **94b**:15003

7. V. S. Retakh and V. N. Shander, *The Schwarz derivative for noncommutative differential algebras*, Unconventional Lie algebras, Adv. Soviet Math., vol. 17, Amer. Math. Soc., Providence, RI, 1993, pp. 139–154. MR **94k**:12008

8. I. M. Gelfand, D. Krob, A. Lascoux, B. Leclerc, V. S. Retakh, and J.-Y. Thibon, *Noncommutative symmetric functions*, Adv. Math. **112** (1995), no. 2, 218–348. MR **96e**:05175

9. I. Gelfand and V. Retakh, *Noncommutative Vieta theorem and symmetric functions*, The Gelfand Mathematical Seminars, 1993–1995, Gelfand Math. Sem., Birkhäuser Boston, Boston, MA, 1996, pp. 93–100. MR 1 398 918

10. P. Etingof, I. Gelfand, and V. Retakh, *Factorization of differential operators, quasideterminants, and nonabelian Toda field equations*, Math. Res. Lett. **4** (1997), no. 2-3, 413–425. MR **98d**:58081

11. I. Gelfand and V. Retakh, *Quasideterminants. I*, Selecta Math. (N.S.) **3** (1997), no. 4, 517–546. MR **2000c**:05150

12. P. Etingof, I. Gelfand, and V. Retakh, *Nonabelian integrable systems, quasideterminants, and Marchenko lemma*, Math. Res. Lett. **5** (1998), no. 1-2, 1–12. MR **99h**:58083

13. P. Etingof and V. Retakh, *Quantum determinants and quasideterminants*, Asian J. Math. **3** (1999), no. 2, 345–351. MR **2001j**:17025

14. V. Retakh, Ch. Reutenauer, and A. Vaintrob, *Noncommutative rational functions and Farber's invariants of boundary links*, Differential topology, infinite-dimensional Lie algebras, and applications, Amer. Math. Soc. Transl. Ser. 2, vol. 194, Amer. Math. Soc., Providence, RI, 1999, pp. 237–246. MR **2000h**:57024

15. I. Gelfand, V. Retakh, and R. L. Wilson, *Quadratic linear algebras associated with factorizations of noncommutative polynomials and noncommutative differential polynomials*, Selecta Math. (N.S.) **7** (2001), no. 4, 493–523. MR **2002i**:16026

16. I. Gelfand, S. Gelfand, and V. Retakh, *Noncommutative algebras associated to complexes and graphs*, Selecta Math. (N.S.) **7** (2001), no. 4, 525–531. MR **2002i**:16027

17. R. L. Wilson, *Invariant polynomials in the free skew field*, Selecta Math. (N.S.) **7** (2001), no. 4, 565–586.

18. I. Gelfand, S. Gelfand, V. Retakh, S. Serconek, and R. L. Wilson, *Hilbert series of quadratic algebras associated with pseudo-roots of noncommutative polynomials*, J. Algebra **254** (2002), no. 2, 279–299. MR 1 933 871

19. I. Gelfand, V. Retakh, and R. L. Wilson, *Quaternionic quasideterminants and determinants*, Lie groups and symmetric spaces, Amer. Math. Soc. Transl. Ser. 2, vol. 210, Amer. Math. Soc., Providence, RI, 2003, pp. 111–123. MR 2 018 356

20. I. Gelfand, S. Gelfand, V. Retakh, and R. L. Wilson, *Quasideterminants*, preprint from arXiv:math.QA/0208146 v3, 26 Jan 2004, http://arXiv.org/abs/math/0208146

21. S. Serconek and R. L. Wilson, *Quadratic algebras associated with decompositions of noncommutative polynomials are Koszul algebras*, J. Algebra (2004), to appear.

22. Z. Škoda, *Noncommutative localization in noncommutative geometry*, preprint from arXiv:math.QA/0403276 v1, 16 March 2004, http://arXiv.org/abs/math/0403276

CERTAINS RÉSULTATS SUR UNE EXTENSION MINIMALE

MOHAMED OUKESSOU

Département de Mathématiques
Faculté des Sciences et Techniques
BP523 Beni-Mellal, Maroc

Ce papier comporte une suite de l'étude qui a été faite sur les suranneaux minimaux dans [5], [6]. On rappelle qu'un domaine T est dit un suranneau minimal de R, si $R \subset T$ tels que R n'est pas un corps et l'extension $R \subset T$ n'admet pas d'anneaux intermédiaires propres.

Dans ce travail nous allons comparer la dimension de Krull de R et celle de T dans le cas où $R \subset T$ est une extension minimale et T est R-plat.

Soit T un suranneau minimal de R tel que T est local et T est R-plat, si R n'est pas local, alors R admet exactement deux idéaux maximaux η, η' et tout idéal premier non maximal contenu dans η est aussi contenu dans η'. Nous allons achever ce travail par l'étude du transfert des domaines de pseudo-valuation et des domaines divisés entre R et T où $R \subset T$ est une extension minimale.

Définition 1. Soient R, T deux domaines tel que $R \subset T$. On dit que T est un *suranneau minimal* de R, si R n'est pas un corps et si l'extension $R \subset T$ n'admet pas d'anneaux intermédiaires propres. On dit aussi que l'*extension $R \subset T$ est minimale*.

Dans ce papier on désigne par K le corps des fractions de R et par $\mathrm{Spec}(R)$, $\dim(R)$ respectivement l'ensemble des idéaux premiers de R et la dimension de Krull de R, enfin par $\subset$ et $\subseteq$ les inclusions stricte et large respectivement.

On commencera par rappeler quelques propriétés fondamentales des suranneaux minimaux.

Propriétés 1. *Soit $R \subset T$ une extension minimale. Alors :*

(i) *T est contenu dans le corps des fractions de R [8, lemme 2].*

(ii) *Ou bien T est entier sur R ou bien T est R-plat [5, lemme 3.1].*

(iii) *Il existe un idéal maximal η de R tel que $R_p = T_p$ pour tout idéal premier p de R distinct de η, de plus $\eta T = T$ si T est R-plat et*

$\eta T = \eta$ *si* T *est entier sur* R *[3, théorème 2.2].*

Dans le théorème suivant nous allons examiner la relation liant la dimension de Krull de R et celle de T, pour cela nous avons besoin des deux lemmes suivants.

Lemme 1. *Soit* $R \subset T$ *une extension minimale tels que* $T \neq K$ *et* T *est* R*-plat.*

(1) *Si* $(R : T) = J = 0$, *alors* $\dim(R) = \dim(T)$.

(2) *Si* $(R : T) = J \neq 0$ *tel que* $\dim(\frac{R}{J}) \leq \dim(\frac{T}{J})$, *alors* $\dim(R) = \dim(T)$.

La démonstration de ce lemme est exposée dans [5, théorème 3.3 et proposition 3.5].

Définition 2. Soient A, B deux anneaux tels que $A \subset B$ et partageant un idéal commun I non nul. On note par X l'ensemble des idéaux premiers de B contenant I. On dit que le couple (A, B) est *presque simple* si tout idéal premier de B contenant I est maximal.

Le lemme suivant est exactement le corollaire 2 de [1].

Lemme 2. *Si le couple* (A, B) *est presque simple, alors* $\dim(A) = \mathrm{Sup}_{Q \in X}\{\dim(B), \mathrm{ht}_B(Q) + \dim(A/Q \cap A)\}$.

Théorème 1. *Soit* T *un suranneau minimal de* R *tel que* T *est* R*-plat.*

(1) *Si* $J = (R : T)$ *n'est pas un idéal maximal de* T *alors* $\dim(R) = \dim(T)$.

(2) *On suppose que* $J = (R : T)$ *est un idéal maximal de* T.

    (a) *Si* $\mathrm{ht}_T(J) < \dim(T)$, *alors* $\dim(R) = \dim(T)$.

    (b) *Si* $\mathrm{ht}_T(J) = \dim(T)$, *alors* $\dim(R) = \dim(T) + 1$.

Preuve. (1) On suppose que $(R : T) = J$ n'est pas un idéal maximal de T. Si $(R : T) = J = 0$ on applique (1) du lemme 1.1. Supposons que $(R : T) = J \neq 0$, si J est un idéal maximal de R alors $\dim(\frac{R}{J}) = 0 \leq \dim(\frac{T}{J})$ puis on applique (2) du lemme 1.1. Si J n'est pas un idéal maximal de R, alors d'après [5, proposition 2.2] $\frac{R}{J} \subset \frac{T}{J}$ est une extension minimale et $\frac{T}{J}$ est $\frac{R}{J}$-plat, comme $(\frac{R}{J} : \frac{T}{J}) = 0$ et $\frac{T}{J}$ est distinct du corps des fractions de $\frac{R}{J}$, d'après ce qui précède on a $\dim(\frac{R}{J}) = \dim(\frac{T}{J})$ et en utilisant (2) du lemme 1.1 on déduit que $\dim(R) = \dim(T)$.

(2) On suppose que $J = (R : T)$ est un idéal maximal de T, dans ce cas l'ensemble X défini précédemment se réduit à $\{J\}$, et d'après le lemme précédent on a $\dim(R) = \mathrm{Sup}(\dim(T), \mathrm{ht}_T(J) + \dim(\frac{R}{J}))$.

Si J est un idéal maximal de R, alors $\dim(\frac{R}{J}) = \dim(\frac{T}{J})$ puis on applique le lemme 1.1. Supposons que J n'est pas un idéal maximal de R, donc

$\frac{R}{J} \subset \frac{T}{J}$ est une extension minimale, comme $\frac{T}{J}$ coïncide avec le corps des fractions de $\frac{R}{J}$ alors d'après [6, remarque 1.1] $\frac{R}{J}$ est un anneau de valuation de dimension 1.

(a) Si $\mathrm{ht}_T(J) < \dim(T)$, alors $\mathrm{ht}_T(J) + 1 \leq \dim(T)$, par suite $\dim(R) = \mathrm{Sup}(\dim(T), \mathrm{ht}_T(J) + \dim(\frac{R}{J})) = \dim(T)$.

(b) Si $\mathrm{ht}_T(J) = \dim(T)$, alors $\dim(R) = \mathrm{ht}_T(J) + 1 = \dim(T) + 1$. $\square$

Corollaire 1. *Soit R un anneau de valuation de dimension 2, alors R admet un unique suranneau minimal T de dimension 1.*

Preuve. On a $\mathrm{Spec}(R) = \{0, p, \eta\}$ où η est l'idéal maximal de R, avec $0 \subset p \subset \eta$ c'est-à-dire $\mathrm{ht}(p) = 1$. On vérifie aisément que $T = R_p$ est un suranneau minimal de R. Soit T' un suranneau minimal de R, alors T' est local d'idéal maximal $M = (R : T')$ (voir [6, proposition 1.1]) et on a $T'_M = T' = R_{M \cap R}$. Or d'après (iii) de propriétés 1.1 on a $M \cap R \neq \eta$ par suite $M \cap R = p$ ainsi $T' = T = R_p$ et $\dim(T) = \mathrm{ht}(p) = 1$. $\square$

Proposition 1. *Soit T un suranneau minimal de R tels que T est R-plat et R est local d'idéal maximal η.*

(i) *Si $\dim(R) = 1$, alors R est de valuation et $T = K$.*

(ii) *Pour tout idéal premier p de R tel que $p \neq \eta$ on a $p \subseteq (R : T) = J$ et si de plus $p \neq J$, alors $JR_p = R_p$.*

Preuve. (i) On a $J = (R : T) \in \mathrm{Spec}(R)$ ceci entraîne que $J = \eta$ ou $J = 0$. Comme $\eta T = T$, alors $J = 0$. D'après [3, proposition 3.3] $R_\eta = R$ est un anneau de valuation et en utilisant [6, remarque 1.1] on déduit que $T = K$.

(ii) D'après [6, proposition 1.1] T est local d'idéal maximal $M = (R : T)$. Soit $p \in \mathrm{Spec}(R)$ tel que $p \neq \eta$, d'après [5, proposition 3 ;2] il existe $Q \in \mathrm{Spec}(T)$ tel que $Q \cap R = p$ donc $Q \subseteq M$ ainsi que $Q \cap R = p \subseteq M \cap R = J$. Supposons que $p \subset J$ et soit $a \in J \setminus p$; alors $a \in R \setminus p$ par suite $1/a \in R_p$ et $1 = a/a \in JR_p$ donc $JR_p = R_p$ $\square$

Théorème 2. *Soit T un suranneau minimal de R tel que T est local d'idéal maximal M et T est R-plat.*

(1) *Si R n'est pas local, alors R admet exactement deux idéaux maximaux η, η' et $\eta' = M \cap R$.*

(2) *Tout idéal premier non maximal de R contenu dans η est aussi contenu dans η'.*

Preuve. 1er *cas* : $J = (R : T) = 0$. Supposons que R n'est pas local et soient η, η' deux idéaux maximaux de R. D'après [5, proposition 3.2] η ou η' se relève à T. Considérons Q un idéal premier de T tel que $Q \cap R = \eta'$, en utilisant [5, théorème 3.2], on déduit que Q est un idéal maximal de T, par suite $M = Q$, donc nécessairement η et η' sont les seuls idéaux maximaux de R.

$2^{\grave{e}me}$ *cas* : $J = (R : T) \neq 0$. Si J n'est pas un idéal maximal de R, alors $\frac{R}{J} \subset \frac{T}{J}$ est une extension minimale et $\frac{T}{J}$ est local d'idéal maximal $\frac{M}{J}$ tel que $\frac{T}{J}$ est $\frac{R}{J}$-plat. Comme $(\frac{R}{J} : \frac{T}{J}) = 0$, d'après le premier cas $\frac{R}{J}$ admet au plus deux idéaux maximaux $\frac{\eta}{J}$ et $\frac{\eta'}{J}$ où η et η' sont deux.idéaux maximaux de R contenant J. Supposons qu'il existe un idéal maximal η'' de R distinct de η et η', donc η'' ne contient pas J par suite $\eta'' + J = R$. Or η'' se relève à T, soit $Q \in \mathrm{Spec}(T)$ tel que $Q \cap R = \eta''$. Posons $Q + J = I$; alors $I \cap R = R$ d'où $I = T$ par suite $Q + J = T$ et $M + J = M = T$ ce qui est impossible.

Si $J = (R : T)$ est un idéal maximal de R, supposons qu'il existe deux idéaux maximaux η, η' distincts de J. On a $\eta T = T$ et $\eta' = Q \cap R$ où $Q \in \mathrm{Spec}(T)$. D'autre part $Q \subset M \Rightarrow M \cap R = \eta'$, or $J \subseteq M \Rightarrow J = \eta'$ ceci est absurde, on conclut donc que R admet exactement deux idéaux maximaux.

(2) Si R n'est pas local alors d'après (1) R admet exactement deux idéaux maximaux. Supposons que $M \cap R$ n'est pas maximal, donc $M \cap R \subset \eta$ où η est un idéal maximal de R, par suite on a $R \subset R_\eta \subset R_{M \cap R} = T$ ceci est absurde, d'où $M \cap R = \eta'$ est un idéal maximal de R. Soit p un idéal premier non maximal de R contenu dans η. Si p n'est pas contenu dans η', nous allons montrer que $R \subset R_{R \setminus p \cup \eta'} \subset R_{\eta'} = T$. En effet, si $R_{R \setminus p \cup \eta'} = R_{\eta'}$ alors $p \cup \eta' = \eta'$ par suite $p \subset \eta'$ ceci est contraire à l'hypothèse. Supposons que $R_{R \setminus p \cup \eta'} = R$. Il est clair que η n'est pas contenu dans $p \cup \eta'$ donc il existe $s \in \eta \setminus p \cup \eta'$ c'est-à-dire $\frac{1}{s} \in R_{R \setminus p \cup \eta'} = R$ donc $\frac{1}{s} = a \Rightarrow as = 1$ avec $a \in R$ ce qui montre que s est inversible. D'où $R \subset R_{R \setminus p \cup \eta'} \subset R_{\eta'} = T$ ceci est contraire à la définition de la minimalité par suite p est nécessairement contenu dans η'. $\qquad\square$

Remarque 1. (a) Soit R un domaine noethérien de dimension 1, alors d'après [8, théorème 3] R admet un suranneau minimal T. Si T est R-plat, d'après [5, proposition 3.2] T est de la forme $\bigcap_p R_p$ où $p \in \mathrm{Spec}(R)$ et $p \neq \eta$ avec η un idéal maximal de R. Si T est entier sur R, alors T n'est pas nécessairement de la forme précédente. En effet il suffit de considérer $R = F[[X^2, X^7]]$ et $T = F[[X^2, X^5]]$ où F est un corps. On a R est un domaine noethérien de dimension 1 et d'après [7, exemple 3.2] $R \subset T$ est une extension minimale entière, de plus R et T sont locaux d'idéaux maximaux respectivement $X^2 R + X^7 R$ et $X^2 T + X^5 T$.

(b) Soit T un suranneau minimal de R tel que T est entier sur R c'est-à-dire $R \subset T \subseteq \overline{R}$. L'exemple suivant montre que T peut coïncider avec $\overline{R}$. Soient $R = F[[X^2, X^3]]$ et $T = F[[X]]$ avec F un corps quelconque, alors on vérifie facilement que $T = \overline{R}$, d'autre part $\overline{R} = F[[X]]$ est de Prüfer, et en utilisant [4, théorème 2.4] on déduit que $F[[X]]$ est l'unique suranneau minimal de $F[[X^2, X^3]]$.

Définition 3. On dit qu'un domaine R est *de pseudo-valuation* si pour tout idéal premier p de R satisfait la propriété suivante : $\forall x, y \in K$ tels que $xy \in p$ entraîne $x \in p$ ou $y \in p$. Tout anneau de valuation est de pseudo-valuation.

Proposition 2. (1) *Soit* (R, η) *un domaine de pseudo-valuation qui n'est pas de valuation. Si R admet un suranneau minimal T, alors T est entier sur R.*

(2) *Soit* (R, η) *un domaine local noethérien admettant un suranneau minimal T tel que T est R-plat, alors R est un anneau de valuation discrète.*

Preuve. (1) Soit $V = (\eta : \eta)_K$ l'anneau de valuation associé à R. D'après [7, p. 564] T et $(\eta : \eta)_K$ sont comparables, donc nécessairement $T \subseteq (\eta : \eta)_K$. Si T est R-plat, on aura $\eta T = \eta$ ce qui est absurde d'où T est entier sur R.

(2) D'après [6, proposition 1.1] T est local d'idéal maximal $(R : T)$, comme R est noethérien et T est R-plat, alors $(R : T) = 0$ par suite T est un corps et d'après [6, remarque 1.1] R est un anneau de valuation de dimension 1. $\qquad \square$

Définition 4. Un domaine R est dit *divisé*, si pour tout idéal premier p de R satisfait $p = pR_p$. D'après [7, p. 563] R est un domaine divisé si et seulement si pour tout idéal premier p de R et pour tout élément $r \in R$ on a $rR \subseteq p$ ou $p \subseteq rR$.

Soient T un suranneau de R et $p \in \mathrm{Spec}(R)$. On dit que p est *T-fort* si pour tout $x, y \in T$ satisfaisant $xy \in p$ alors $x \in p$ ou $y \in p$. Si tout idéal premier de R est T-fort, on dit que $R \subset T$ *est une extension forte*.

Examinons cette structure dans une extension minimale.

Proposition 3. *Soit T un suranneau minimal de R.*

(a) *On suppose que T est entier sur R.*

    Si T est divisé alors R l'est aussi.

    Si R est divisé et T est local alors T est aussi divisé.

(b) *On suppose que T est R-plat.*

    Si R est divisé alors T l'est également.

    Si T est divisé et R est local alors R est divisé.

Preuve. (a) Comme $R \subset T$ est une extension minimale entière, alors $R \subset T$ est une extension très finie et en appliquant [2, corollaire 2.2] on déduit que R est divisé si T est divisé.

Montrons que $QT_Q = Q$ pour tout $Q \in \operatorname{Spec}(T)$. Si $Q \cap R \neq (R : T)$, d'après [5, lemme 3.3] on a $T_Q = R_p$ où $p = Q \cap R$ donc $QT_Q = pR_p$ et comme R est divisé alors $pR_p = p$ par suite $QT_Q = Q$. Supposons que $Q \cap R = (R : T)$, d'après [5, proposition 3.2] $(R : T)$ est un idéal maximal donc Q est l'idéal maximal de T par suite $QT_Q = QT = Q$ par conséquent T est divisé.

(b) Si T est R-plat alors on a $T_Q = R_p$, donc si R est divisé alors $QT_Q = pR_p = p = Q$ où $p = Q \cap R$, ce qui montre que T est aussi divisé.

On suppose que T est divisé et R est local d'idéal maximal η. D'après [6, proposition 1.1] T est d'idéal maximal $M = (R : T)$.

Soit $p \in \operatorname{Spec}(R)$ et montrons que $pR_p = p$. Si $p = \eta$ c'est évident. Supposons $p \neq \eta$, d'après [5, proposition 3.2] il existe $Q \in \operatorname{Spec}(T)$ tel que $Q \cap R = p$ et comme $T_Q = R_p$ et $QT_Q = Q$ alors $QT_Q = pR_p = Q \subseteq (R : T) \subseteq R$ par suite $pR_p = p$. $\square$

Corollaire 2. *Si R est un domaine divisé et T est minimal sur R tel que T est R-plat, alors $R \subset T$ est une extension forte.*

Preuve. Montrons que pour tout $p \in \operatorname{Spec}(R)$ on a p est T-fort. Si $p \neq \eta$ alors $p = Q \cap R$ où $Q \in \operatorname{Spec}(T)$ et $T_Q = R_p$. D'après [7, théorème 2.2] p est R_p-fort donc p est T-fort. Supposons que $p = \eta$ où η est l'idéal maximal de R et soient $x, y \in T$ tels que $xy \in \eta$. Si $x, y \in R$ c'est évident. Si $x \in T \setminus R$ d'après [6, proposition 1.1] $x^{-1} \in R$ par suite $y = x^{-1}xy \in \eta$. $\square$

Remarque 2. Si R est divisé et T est R-plat alors pour tout $Q \in \operatorname{Spec}(T)$ on a montré que $Q = p$ où $p = Q \cap R$ et comme $\operatorname{Spec}(T) \to \operatorname{Spec}(R)$ est injective et tous les idéaux premiers de R se relèvent à T sauf un idéal maximal η de R (voir [5, proposition 3.2]), alors $\operatorname{Spec}(R) = \operatorname{Spec}(T) \cup \{\eta\}$.

Bibliographie

1. J. P. Cahen, *Couple d'anneaux partageant un idéal*, Arch. Math. (Basel) **51** (1988), 505–514.

2. D. E. Dobbs, *Coherence, ascent of going down, and pseudo-valuation domains*, Houston J. Math. **4** (1978), 551–567.

3. D. Ferrand, J. P. Olivier, *Homomorphismes minimaux d'anneaux*, J. Algebra, **16** (1970), 461–471.

4. R. Gilmer, W. J. Heinzer, *Intersection of quotient rings of an integral domain*, J. Math. Kyoto. Univ, **7**(2), 133–150, (1967).

5. M. Oukessou, A. Miri, *Sur les suranneaux minimaux*, Extracta Math. **14** (1999), 333–347.

6. M. Oukessou, A. Miri, *Sur les suranneaux minimaux II*, Comm. Algebra, **31** (2003), 5683–5692.

7. A. Okabe, *Some ideal-theoretical characterizations of divided domains*, Houston J. Math., **12** (1986), 563–577.

8. J. Sato, T. Sugatani, K. I. Yoshida, *On minimal overrings of a Noetherian domain*, Comm. Algebra, **20** (1982), 1735–1746.

*-IDENTITIES IN MATRIX SUPERALGEBRAS WITH SUPERINVOLUTION *

TSETSKA GRIGOROVA RASHKOVA*

Centre of Mathematics and Informatics
University of Rousse "A. Kanchev"
7017 Rousse, Bulgaria
E-mail: tcetcka@ami.ru.acad.bg

In the paper the notion of superinvolution for superalgebras is considered and identities of a special kind in skew-symmetric variables with respect to the superinvolution are investigated. The new results are interpreted in connection with analogous investigations for matrix algebras with symplectic involution.

1. Basic notions

Definition 1.1. Let F be a field of characteristic different from 2. A *superalgebra* is a Z_2-graded F-algebra A such that

$$A = A_0 \oplus A_1, \quad A_\alpha A_\beta \subseteq A_{\alpha+\beta} \qquad (\alpha, \beta \in Z_2).$$

If $a \in A_\alpha$, then a is *homogeneous of degree* α, and we write $\bar{a} = \alpha$. Elements from A_0 are called *even* and elements from A_1 are called *odd*.

If $A = A_0 \oplus A_1$ is a superalgebra, then A_0 is a supersubalgebra of A. The superalgebra A is said to be *trivial* if $A_1 = 0$.

A superalgebra is said to be *simple* if it does not have nontrivial proper (graded) ideals and the multiplication is not trivial.

An *associative superalgebra* is just an associative Z_2-graded algebra. Every associative superalgebra will be supposed to be *nontrivial*, and *unital*, in which case the unit is an even element (in particular, A_0 is also unital).

The following example is a basic one of a nontrivial unital central simple associative superalgebra.

---

*Partially supported by Grant MM1106/2001 of the Bulgarian Foundation for Scientific Research.

Example 1.2. Let $V = V_0 \oplus V_1$ be a *vector space*. Then the associative algebra $\text{End}(V)$ is provided with the induced Z_2-grading $\text{End}(V) = \text{End}(V)_0 \oplus \text{End}(V)_1$, in which

$$\text{End}(V)_\alpha = \{\alpha \in \text{End}(V) \mid \alpha(V_\beta) \subseteq V_{\beta+\alpha}\}.$$

Suppose that $\dim V_0 = r \geq 1$ and $\dim V_1 = s \geq 1$. Using the language of matrices we get a superalgebra $M(r \mid s)$, whose underlying algebra is that of square matrices of order $r + s$ and whose Z_2-grading is determined in the following way:

$$M(r \mid s)_0 = \left\{ \begin{bmatrix} A & 0 \\ 0 & D \end{bmatrix} \,\middle|\, A \in M_r(F), D \in M_s(F) \right\},$$

$$M(r \mid s)_1 = \left\{ \begin{bmatrix} 0 & B \\ C & 0 \end{bmatrix} \,\middle|\, B \in M_{r,s}(F), C \in M_{s,r}(F) \right\}.$$

In the next section we focus our considerations on the algebra $M(r) = M(r \mid r)$ for $r = 2$.

We recall that $G = G_0 \oplus G_1$ is the *Grassmann superalgebra* with the standard grading. If $A = A_0 \oplus A_1$ is a superalgebra, the algebra $G(A) = A_0 \otimes G_0 + A_1 \otimes G_1$, which is a subalgebra of the tensor product $G \otimes A$, is called the *Grassmann envelope* of A.

The next two propositions show the importance of the superalgebras in the theory of algebras with (ordinary) polynomial identities over a field of characteristic 0.

Proposition 1.3 ([5, Theorem 1.1]). *An arbitrary nontrivial variety of associative algebras over a field of characteristic zero is generated by the Grassmann envelope of some finitely-generated PI-superalgebra.*

Proposition 1.4 ([5, Theorem 2.2]). *A variety of superalgebras generated by a finitely-generated PI-superalgebra is generated by a finite-dimensional superalgebra.*

Definition 1.5. Let A be an associative superalgebra. A *superinvolution* on A is a Z_2-graded linear map $* : A \to A$ such that, for all $a, b \in A$, $(a^*)^* = a$ and $(ab)^* = (-1)^{\bar{a}\bar{b}} b^* a^*$, where $\bar{x}$ means the parity of x; $\bar{x} = i$ if $x \in A_i$, $i = 0, 1$.

If $*$ is a superinvolution on A, then the restriction of $*$ to A_0 is an involution on A_0.

Example 1.6. Let $V = V_0 \oplus V_1$ be a vector space with $\dim V_0 = r \geq 1$ and $\dim V_1 = s \geq 1$. Let $(\,,\,)$ be a *nondegenerate super-symmetric bilinear form on V*; i.e., the restriction of $(\,,\,)$ to V_0 is symmetric, the restriction of $(\,,\,)$ to V_1 is skew-symmetric (in particular, $s = 2t$ is even), and V_0 and V_1 are orthogonal. The adjoint endomorphism a^* of $a \in \mathrm{End}(V)$ is given by

$$(a(v), w) = (-1)^{\bar{a}\bar{v}}(v, a^*(w)), \qquad v \in V_0, w \in V_1.$$

The map $a \to a^*$ defines a superinvolution on $\mathrm{End}(V)$, which we call *orthosymplectic superinvolution* and will denote it by *osp*.

In the language of matrices let $H \in M_r(F)$ (respectively, $K \in M_s(F)$) be the matrix associated with the restriction of $(\,,\,)$ to V_0 (respectively to V_1). Notice that H is a symmetric matrix, K is a skew-symmetric matrix, and both H and K are invertible. It is easily checked that the orthosymplectic superinvolution on $M(r \mid s)$ is given by

$$\begin{bmatrix} A & B \\ C & D \end{bmatrix}^{osp} = \begin{bmatrix} H & 0 \\ 0 & K \end{bmatrix}^{-1} \begin{bmatrix} A & -B \\ C & D \end{bmatrix}^{t} \begin{bmatrix} H & 0 \\ 0 & K \end{bmatrix},$$

where t denotes the usual matrix transposition.

Example 1.7. Let us consider the superalgebra $M(r)$. We will call the following superinvolution defined on $M(r)$ a *transposition superinvolution*, and we will denote it by *trp*:

$$\begin{bmatrix} A & B \\ C & D \end{bmatrix}^{trp} = \begin{bmatrix} D^t & -B^t \\ C^t & A^t \end{bmatrix}.$$

Theorem 1.8 ([1, Theorem 3.2]). *The only superinvolutions on $M(1)$ are trp and $(trp)p$, where p is the automorphism of $M(1)$ given by $p(a_0 + a_1) = a_0 - a_1$ (the parity automorphism).*

The two possibilities for $*$ in this case are:

$$\begin{bmatrix} a & b \\ c & d \end{bmatrix}^{*} = \begin{bmatrix} d & -b \\ c & a \end{bmatrix};$$

$$\begin{bmatrix} a & b \\ c & d \end{bmatrix}^{*} = \begin{bmatrix} d & b \\ -c & a \end{bmatrix}.$$

2. The algebra $M(2)$

Let A be an associative superalgebra with superinvolution $*$. An element $a \in A$ is called *symmetric* if $a^* = a$ and *skew-symmetric* if $a^* = -a$. Both the set $H = H(A, *) = \{a \in A \mid a^* = a\}$ of symmetric elements and the set

$K = K(A, *) = \{a \in A \mid a^* = -a\}$ of skew elements could be interpreted as graded subspaces of A. Since the characteristic of F is different from 2, it is obvious that $H \cap K = 0$. Moreover, given $a \in A$, $a = (a+a^*)/2 + (a-a^*)/2$, with $(a + a^*)/2 \in H$ and $(a - a^*)/2 \in K$. Hence $A = H \oplus K$.

We define two new multiplications on A by the following expressions:

$$[a, b] = ab - (-1)^{\bar{a}\bar{b}} ba;$$
$$a \circ b = ab + (-1)^{\bar{a}\bar{b}} ba.$$

We obtain in this way two new superalgebras A^- and A^+ with the same gradings as A. It is easily checked that $H \circ H \subseteq H$, $H \circ K \subseteq K$ and $K \circ K \subseteq H$. Thus $A = H \oplus K$ is a superalgebra (under the product $\circ$). In particular, K is a subalgebra of A^-; hence it is a Lie superalgebra. And H is a subalgebra of A^+; hence it is a Jordan algebra.

We consider superinvolutions in $M(2)$ as defined in Examples 1.7 and 1.6 and the algebra K in any of the cases. The dimension of this algebra is less than the dimension of the analogous algebra for the ordinary symplectic involution $(\dim(K, trp) = \dim(K, osp) = 8$ while in the ordinary one it is 10).

Now we give a basis of K in the case of trp involution. According to Example 1.7 a skew-symmetric matrix $k \in K$ is of the form

$$\begin{bmatrix} a & c & x & d \\ e & b & d & y \\ 0 & f & -a & -e \\ -f & 0 & -c & -b \end{bmatrix}.$$

Thus the basis of K consists of the matrices

$$l_1 = e_{11} - e_{33}, \qquad l_2 = e_{22} - e_{44},$$
$$l_3 = e_{14} + e_{23}, \qquad l_4 = e_{32} - e_{41},$$
$$l_5 = e_{12} - e_{43}, \qquad l_6 = e_{21} - e_{34},$$
$$l_7 = e_{13}, \qquad l_8 = e_{24}.$$

In the case of osp superinvolution a skew-symmetric matrix is of the form

$$\begin{bmatrix} a & 0 & c & d \\ 0 & -a & e & f \\ -f & -d & b & x \\ e & c & y & -b \end{bmatrix}$$

and the basis of K is the following:

$$m_1 = e_{11} - e_{22}, \qquad m_2 = e_{33} - e_{44},$$
$$m_3 = e_{13} + e_{42}, \qquad m_4 = e_{14} - e_{32},$$
$$m_5 = e_{23} + e_{41}, \qquad m_6 = e_{24} - e_{31},$$
$$m_7 = e_{34}, \qquad m_8 = e_{43}.$$

We are interested in defining Bergman type identities for the superalgebra $M(n)$. These polynomials are considered in the ordinary case by Formanek [4] and Bergman [2].

To a polynomial in commuting variables

$$g(t_1, \ldots, t_{n+1}) = \sum \alpha_p t_1^{p_1} \ldots t_{n+1}^{p_{n+1}} \in F[t_1, \ldots, t_{n+1}] \qquad (1)$$

we relate a polynomial $v(g)$ from the free associative algebra $F\langle x, y_1, \ldots, y_n \rangle$

$$v(g) = v(g)(x, y_1, \ldots, y_n) = \sum \alpha_p x^{p_1} y_1 x^{p_2} y_2 \ldots x^{p_n} y_n x^{p_{n+1}}. \qquad (2)$$

Any polynomial $f(x, y_1, \ldots, y_n)$ multilinear in $y_1, \ldots, y_n$ can be written as

$$f(x, y_1, \ldots, y_n) = \sum v(g^{(i)})(x, y_{i_1}, \ldots, y_{i_n}), \quad g^{(i)} \in F[t_1, \ldots, t_{n+1}]. \qquad (3)$$

What we have in the ordinary case and for matrix algebras with the symplectic involution (we denote it by $**$) could be expressed in the following three propositions.

We recall first that in the matrix algebra $M_{2n}(F, **)$ over a field F of characteristic zero the symplectic involution $**$ is defined by

$$\begin{bmatrix} A & B \\ C & D \end{bmatrix}^{**} = \begin{bmatrix} D^t & -B^t \\ -C^t & A^t \end{bmatrix},$$

where A, B, C, D are $n \times n$ matrices and t is the usual transpose.

Proposition 2.1 ([2, Section 6, (27)]). (i) *The polynomial $v(g^{(i)})$ from (2) is an identity for $M_n(F)$ if and only if*

$$g_n = \prod_{1 \leq p < q \leq n+1} (t_p - t_q)$$

divides $g^{(i)}(t_1, \ldots, t_{n+1})$ for all $i = (i_1, \ldots, i_n)$.

(ii) *The polynomial $f(x, y_1, \ldots, y_n)$ from (1) is an identity for $M_n(F)$ if and only if every summand $v(g^{(i)})$ is also an identity for $M_n(F)$.*

Proposition 2.2 ([7, Theorem 1]). *Let the polynomial $f(x, y_1, \ldots, y_n)$ of type (3) be a $**$-identity in skew-symmetric variables for $M_{2n}(F, **)$. Then*

$$g_{2n,0} = \prod_{\substack{1 \leq p < q \leq n+1 \\ (p,q) \neq (1,n+1)}} (t_p^2 - t_q^2)(t_1 - t_{n+1})$$

divides the polynomials $g^{(i)}$ from (1) for all $i = (i_1, \ldots, i_n)$.

The notations g_n and $g_{2n,0}$ will be kept in the sequel.

According to [3] the polynomial $f(x, y_1, y_2) = \sum_{(i_1,i_2) \in \mathrm{Sym}(2)} \cdot v(g_2)(x, y_{i_1}, y_{i_2})$ is a $**$-identity in skew variables for $M_4(F, **)$ (the summands $v(g_2)(x, y_{i_1}, y_{i_2})$ themselves are not such identities).

Proposition 2.3 ([8, Theorem 3]). *Considered in $M_{2n}(F, **)$, the polynomial f from (3) satisfies $f(a, r_1, \ldots, r_n) = 0$ for any skew-symmetric matrix a and all matrices $r_1, \ldots, r_n$ if and only if*

$$(t_1 + t_{n+1})g_{2n,0} = \prod_{1 \leq p < q \leq n+1} (t_p^2 - t_q^2)$$

divides the polynomials $g^{(i)}(t_1, \ldots, t_{n+1})$ for all $i = (i_1, \ldots, i_n)$.

These propositions show that the ordinary case is the basis for all further generalizations. A confirmation of this are the obvious result that an identity either in skew-symmetric or symmetric variables in the matrix algebra with symplectic involution $M_{2n}(F, **)$ is an ordinary identity for $M_n(F)$ and the following well known result.

Proposition 2.4. *Every identity either in A_0 or A_1 for the superalgebra $A = M(n) = A_0 \oplus A_1$ is an ordinary identity for the matrix algebra $M_n(F)$.*

Proof. For the algebra A_0 we consider the matrices

$$a_i = \begin{bmatrix} A_i & 0 \\ 0 & D_i \end{bmatrix},$$

while, for A_1, the matrices

$$b_i = \begin{bmatrix} 0 & B_i \\ C_i & 0 \end{bmatrix}.$$

We have

$$f(a_1, \ldots, a_m) = \begin{bmatrix} f(A_1, \ldots, A_m) & 0 \\ 0 & f(D_1, \ldots, D_m) \end{bmatrix} = 0$$

and

$$f(b_1,\ldots,b_m) = \begin{bmatrix} f(B_1,\ldots,B_m) & 0 \\ 0 & f(C_1,\ldots,C_m) \end{bmatrix} = 0 \quad \text{or}$$

$$f(b_1,\ldots,b_m) = \begin{bmatrix} 0 & f(B_1,\ldots,B_m) \\ f(C_1,\ldots,C_m) & 0 \end{bmatrix} = 0,$$

depending on the degree of the polynomial f.

Thus $f(x_1,\ldots,x_m)$ is an ordinary polynomial identity for $M_n(F)$. $\square$

Remark 2.5. As some of the skew-symmetric matrices to both of the considered superinvolutions are matrices of type a_i the above result is valid for the presentation $M(n) = H \oplus K$ as well.

Considering the superalgebra $M(2)$ we form the polynomial $g_{2n,0}$ for $n = 2$ $(g_{4,0} = (t_1^2 - t_2^2)(t_2^2 - t_3^2)(t_1 - t_3))$ and the corresponding associative polynomial f of type (3), i.e.,

$$P_7(x, y_1, y_2) = f(x, y_1, y_2) = \sum_{(i_1, i_2) \in \mathrm{Sym}(2)} v(g_{4,0})(x, y_{i_1}, y_{i_2}).$$

A straightforward corollary of Proposition 2 [6] gives that the polynomial P_7 is an identity for $M_4(F, **)$ in skew-symmetric variables, where $**$ is the symplectic involution. But the supercase (both $(M(2), osp)$ and $(M(2), trp)$) is different. By the system for computer algebra *Mathematica* we calculate

$$P_7(l_1 - 3l_2, l_3 - 2l_5, l_4 + 4l_6) = -1536e_{43};$$
$$P_7(m_2, m_3, m_5) = 4e_{43}.$$

Now we begin investigating Bergman type identities for superalgebras with superinvolutions.

I. Due to Proposition 2.4 and Proposition 2.1 we begin with

$$\begin{aligned}
P_5(x, y_1, y_2) &= v(g_2)(x, y_1, y_2) + v(g_2)(x, y_2, y_1) \\
&= a(x^2 y_1 x y_2 - x^2 y_1 y_2 x - x y_1 x^2 y_2 + x y_1 y_2 x^2 - y_1 x y_2 x^2 + y_1 x^2 y_2 x) \\
&\quad + b(x^2 y_2 x y_1 - x^2 y_2 y_1 x - x y_2 x^2 y_1 + x y_2 y_1 x^2 - y_2 x y_1 x^2 + y_2 x^2 y_1 x).
\end{aligned}$$

We calculate

$$f(l_1, l_3, l_5) = (-2a + 2b)e_{13};$$
$$f(l_1, l_3, l_8) = 2ae_{13};$$
$$f(m_2, m_4, m_5 - 2m_6) = 4(a + b)e_{34};$$
$$f(m_2 - 2m_1, m_1 - 3m_4, m_3 - 2m_4 + m_5 - m_6 + 3m_7 - 7m_8) =$$
$$36(a - b)e_{12} - 126ae_{13} + 18(a + b)e_{34} + 126be_{42}.$$

These relations show that there are no Bergman type identities of degree 5 for K in both cases $(M(2), osp)$ and $(M(2), trp)$.

II. The next degree is treated analogously. Due to Proposition 2.4 and Proposition 2.1 the commutative polynomials are

$$g^{(i)}(t_1, t_2, t_3) = (t_1 - t_2)(t_2 - t_3)(t_1 - t_3)(a_i t_1 + b_i t_2 + c_i t_3), \quad i = 1, 2$$

and the Bergman type polynomial has the form

$$f(x, y_1, y_2) = v(g^{(1)})(x, y_1, y_2) + v(g^{(2)})(x, y_2, y_1).$$

We get

$$f(l_1, l_3, l_5) = (-2a_1 + 2a_2 + 2c_1 - 2c_2)e_{13};$$
$$f(l_2, l_4, l_5) = (-2a_1 - 2a_2 + 2c_1 + 2c_2)e_{42};$$
$$f(l_1, l_7, l_4) = (-2a_1 + 2b_1)e_{12} + (2b_2 - 2c_2)e_{43}.$$

This leads to $a_1 = b_1 = c_1 = a, a_2 = b_2 = c_2 = b$. The next evaluation

$$f(l_2 - 2l_1, l_6 - 3l_5, l_7 + 4l_3) = 144(a + b)e_{13} + 12be_{14} + 12ae_{23} + 48(a + b)e_{24}$$

shows that there is no Bergman type identity of degree 6 in the case of *trp* superinvolution.

In the case of *osp* superinvolution we calculate

$$f(m_2, m_4, m_5 - 2m_6) = (2a_1 + 2a_2 + 2(a_1 - c_1) - 2c_1 + 2(a_2 - c_2) - 2c_2)e_{34}.$$

This gives that $a_1 + a_2 = c_1 + c_2$. Substituting c_2 from it, we get

$$f(m_2 - 2m_1, m_6 - 3m_5, m_7 - 2m_6) = (-288a_1 - 72b_1 - 72b_2 + 288c_1)e_{21}$$
$$+ (-36a_1 - 18b_1 + 18c_1)e_{24} + (36a_1 + 18a_2 + 18b_2 - 36c_1)e_{31};$$
$$f(m_2 - 2m_1, m_1 - 3m_4, m_3 - 2m_4 + m_5 - m_6 + 3m_7 - 7m_8)$$
$$= (-144a_1 - 36b_1 - 36b_2 + 144c_1)e_{11} + (252a_1 + 126b_1 - 126c_1)e_{13}$$
$$+ (36b_1 - 36b_2)e_{34} + (252a_1 + 126a_2 + 126b_2 - 252c_1)e_{42}.$$

The system

$$\begin{cases} -4a_1 - b_1 - b_2 + 4c_1 = 0 \\ -2a_1 - b_1 + c_1 = 0 \\ 2a_1 + a_2 + b_2 - 2c_1 = 0 \\ b_1 = b_2 \end{cases}$$

gives $c_1 = 0$, $b_1 = -2a_1$, $b_2 = -2a_1$, $a_2 = 0$ and thus we reduce the case to one unknown coefficient a_1. Calculating now the polynomial f for $x = m_1 - 2m_2$, $y_1 = m_1 + 3m_2 + m_4$ and $y_2 = m_5 - m_6 + 3m_8$, we get that its value is $108a_1(e_{13} + e_{42})$, leading to $a_1 = 0$.

So there are no Bergman type identities of degree 6 for K in this case as well.

III a. Now we consider the next degree for $(M(2), trp)$. The commutative polynomials are of type

$$g^{(i)}(t_1, t_2, t_3) = (t_1 - t_2)(t_2 - t_3)(t_1 - t_3)(a_i t_1^2 + b_i t_2^2 + c_i t_3^2 + d_i t_1 t_2 + e_i t_1 t_3 + f_i t_2 t_3),$$

$$i = 1, 2.$$

We denote the corresponding Bergman type polynomial by $f(x, y_1, y_2)$ and calculate

$$\begin{aligned}
f(l_1, l_3, l_5) &= (-2a_1 + 2a_2 - 2c_1 + 2c_2 + 2e_1 - 2e_2)e_{13}; \\
f(l_1, l_3, l_8) &= (2a_1 + 2c_1 - 2e_1)e_{13}; \\
f(l_2, l_4, l_5) &= (2a_1 + 2a_2 + 2c_1 + 2c_2 - 2e_1 - 2e_2)e_{42}; \\
f(l_1, l_7, l_4) &= (-2a_1 - 2b_1 + 2d_1)e_{12} + (2b_2 + 2c_2 - 2f_2)e_{43}; \\
f(l_2 - 2l_1, l_6, l_3) &= (-4a_2 - 32b_1 + 32b_2 + 4c_1 + c_2 - 8(b_1 - d_1) + 4(-a_1 + d_1) \\
&\quad - 8(b_2 - d_2) - 4(-a_2 + d_2) - 2e_1 + 2(-d_1 + e_1) \\
&\quad + 4(b_1 - c_1 - d_1 + e_1) + 2e_2 + 2(-d_2 + e_2) \\
&\quad - 4(b_2 - c_2 - d_2 + e_2) - 4(c_1 - f_1) \\
&\quad + 8(d_1 - f_1) - 2(-a_1 + c_1 + d_1 - f_1) - 8(-b_1 + f_1) \\
&\quad + 2(-e_1 + f_1) - 4(a_1 - b_1 - e_1 + f_1) \\
&\quad + 4(c_2 - f_2) + 8(d_2 - f_2) - 2(-a_2 + c_2 + d_2 - f_2) \\
&\quad - 8(-b_2 + f_2) + 2(-e_2 + f_2) + 4(a_2 - b_2 - e_2 + f_2))e_{24}.
\end{aligned}$$

Equating the coefficients to zero, we get the system

$$\begin{cases} a_1 + c_1 = e_1 \\ a_2 + c_2 = e_2 \\ a_1 + b_1 = d_1 \\ b_2 + c_2 = f_2 \\ a_1 = b_1 + f_1 - b_2 - d_2 + c_2 \\ d_1 = 2b_1 + f_1 - b_2 - d_2 + c_2 \\ e_1 = b_1 + f_1 + c_1 - b_2 - d_2 + c_2. \end{cases}$$

Taking into account the relations for a_1, d_1, e_1, e_2 and f_2, we calculate

$$\begin{aligned} f(l_2 - 2l_1, l_6, l_7 + 4l_3) &= (-72a_2 - 24b_2 - 12c_2 + 48d_2)e_{14} \\ &+ (8b_1 - 12b_2 + 44c_1 - 4(-b_1 + c_1) + 12c_2 - 12d_2 \\ &+ 32(c_1 - f_1) + 28f_1 - 32(-b_1 + f_1))e_{23} \\ &+ (-40b_1 + 8c_1 + 8(-b_1 + c_1) - 16(c_1 - f_1) + 16f_1 - 32(-b_1 + f_1))e_{24}; \\ f(l_2 - 2l_1, l_5, l_3) &= \\ & (32b_1 - 36b_2 + 8c_1 - 16(-b_1 + c_1) + 8(c_1 - f_1) + 4f_1 + 4(-b_1 + f_1))e_{13}. \end{aligned}$$

Thus we have

$$\begin{aligned} c_2 &= 4d_2 - 6a_2 \\ f_1 &= 2c_1 - 2a_2 + d_2 \\ b_1 &= b_2 = 0. \end{aligned}$$

Considering now the polynomial f in 3 variables c_1, a_2 and d_2 and calculating $f(l_2 - 2l_1, l_5, l_7 - l_3) = (160a_2 - 64d_2)e_{13}$ and $f(l_2 - 2l_1, l_6 - l_8, l_4 + l_5) = 32a_2e_{11} + (-4a_2 + 2d_2)e_{22} + (384a_2 - 208d_2)e_{31} + (-128a_2 + 64d_2)e_{33} + (-68a_2 + 34d_2)e_{42} + 4a_2e_{44}$, we get $a_2 = d_2 = 0$. Thus we come to the solution $a_2 = b_2 = c_2 = d_2 = f_2 = e_2 = 0$, $a_1 = d_1 = f_1 = 2c_1$, $e_1 = 3c_1$ and the polynomial has to be of type (2) and associated to the commutative polynomial

$$(t_1 - t_2)(t_1 - t_3)(t_2 - t_3)(2t_1^2 + 2t_1t_2 + 3t_1t_3 + 2t_2t_3 + t_3^2).$$

Now $f(l_1, l_3 - l_4, l_5 + 2l_7 - l_8) = 4c_1e_{43}$ meaning that there is no Bergman type identity of degree 7 for K.

III b. Now we consider the seventh degree in the orthosymplectic case.

We calculate

$$f(m_1, m_3, m_4) = (-2a_1 + 2a_2 - 2c_1 + 2c_2 + 2e_1 - 2e_2)e_{22};$$
$$f(m_2, m_3, m_5) = (-2a_1 - 2a_2 - 2c_1 - 2c_2 + 2e_1 + 2e_2)e_{43};$$
$$f(m_2, m_4, m_8) = (2b_1 + 2c_1 - 2f_1)e_{13} + (-2a_2 - 2b_2 + 2d_2)e_{42}.$$

Thus we get

$$e_1 = a_1 + c_1, \quad e_2 = a_2 + c_2, \quad f_1 = b_1 + c_1, \quad d_2 = a_2 + b_2. \tag{4}$$

Substituting all this in the polynomial, we make some more calculations, namely

$$\begin{aligned}
f(m_2 &- 2m_1, m_6, m_8 + m_3) = \\
&(-76a_2 + 8b_2 + 16c_2 - 16(-b_2 + c_2) - 4(a_2 + b_2 - f_2) + 8(c_2 - f_2) \\
&\quad - 16(b_2 + c_2 - f_2) - 4(-b_2 + f_2) - 16(-a_2 - c_2 + f_2) + 8(-b_2 - c_2 + f_2))e_{11} \\
&\quad + (-32a_1 - 4b_1 - 16(-a_1 + b_1) - 4(b_1 - d_1) + 8(a_1 + b_1 - d_1) \\
&\quad + 12d_1 + 8(-a_1 + d_1) - 16(-a_1 - b_1 + d_1))e_{22} \\
&\quad + (-36a_1 - 2(-a_1 + b_1) - 12c_1 - 4(b_1 - d_1) + 2(a_1 + b_1 - d_1) + 6d_1 \\
&\quad + 8(-a_1 + d_1) - 4(-a_1 - b_1 + d_1))e_{23} \\
&\quad + (4a_1 - 4b_1 + 2(-a_1 + b_1) + 8(b_1 - d_1) - 4(a_1 + b_1 - d_1) + 6d_1 \\
&\quad - 4(-a_1 + d_1) + 2(-a_1 - b_1 + d_1))e_{33} \\
&\quad + (-10a_2 + 2b_2 + 28c_2 - 2(-b_2 + c_2) - 2(a_2 + b_2 - f_2) - 4(c_2 - f_2) \\
&\quad + 2(b_2 + c_2 - f_2) + 8(-b_2 + f_2) + 2(-a_2 - c_2 + f_2) - 4(-b_2 - c_2 + f_2))e_{41} \\
&\quad + (-4a_2 - 4b_2 - 2c_2 + 2(-b_2 + c_2) + 8(a_2 + b_2 - f_2) - 4(c_2 - f_2) \\
&\quad + 2(b_2 + c_2 - f_2) + 8(-b_2 + f_2) + 2(-a_2 - c_2 + f_2) - 4(-b_2 - c_2 + f_2))e_{44}; \\
f(m_1 &- m_2, m_3, m_5) = 2a_2 e_{43}.
\end{aligned}$$

We come to

$$a_2 = 0, \quad f_2 = 3c_2, \quad d_1 = 3a_1 + c_1. \tag{5}$$

Substituting the last relations in the polynomial, we calculate

$$f(m_2 - 2m_1, m_3, m_4) = (-72a_1 + 36b_1 - 20b_2 + 56c_2 + 16(-b_2 + c_2))e_{12};$$
$$f(m_2 - 2m_1, m_3, m_5) = (-36a_1 + 36b_1 + 38b_2 - 38c_2 + 2(-b_2 + c_2))e_{43}$$

getting $a_1 = -2b_2 + 3c_2$ and $b_1 = -3b_2 + 4c_2$.

Taking into account these relations, we make the last calculation $f(m_2 - 2m_1, m_4, m_8) = (96b_2 - 144c_2)e_{13} + 48c_2 e_{42}$. This shows $b_2 = c_2 = 0$ and relations (4) and (5) give $c_1 = d_1 = e_1 = f_1$. This means that the Bergman type identity has to be of type (2) and its corresponding commutative polynomial is

$$g^{(i)}(t_1, t, t_3) = (t_1 - t_2)(t_2^2 - t_3^2)(t_1^2 - t_3^2).$$

Hence the following two theorems are valid.

Theorem 2.6. *The least possible degree of a Bergman type identity of type* (3) *for K in the superalgebra* $(M(2), trp)$ *is 8.*

Theorem 2.7. *If there is a Bergman type identity for K in the superalgebra* $(M(2), osp)$ *of degree 7 (the least possible) it could be a polynomial $v(g)$ of type* (2), *in which $g = g1 = (t_1 - t_2)(t_2^2 - t_3^2)(t_1^2 - t_3^2)$.*

For Bergman type polynomials of type (2) we could say a bit more.

Lemma 2.8. *If a polynomial $v(g)$ of type* (2) *is an identity for K in* $(M(2), trp)$ *then the commutative polynomial $(t_1 + t_3)g_{4,0}$ divides the polynomial g.*

Proof. Let $x^{a_1} y_1 x^{a_2} y_2 x^{a_3}$ be a monomial of the polynomial $v(g)$. Substituting $x = \rho_1 l_1 + \rho_2 l_2$ for the algebraically independent variables ρ_1 and ρ_2, $y_1 = l_5$ and $y_2 = l_3$, we get $\rho_1^{a_1} \rho_2^{a_2} (-\rho_1)^{a_3}$. This leads to $g(\rho_1, \rho_2, -\rho_1) = 0$, meaning that $(t_1 + t_3)$ divides the polynomial g.

Substituting $x = \rho_1 l_1 + \rho_2 l_2$, $y_1 = l_5$ and $y_2 = l_8$, we get $\rho_1^{a_1} \rho_2^{a_2} (-\rho_2)^{a_3}$. This leads to $g(\rho_1, \rho_2, -\rho_2) = 0$, meaning that $(t_2 + t_3)$ divides g.

At the end, substituting $x = \rho_1 l_1 + \rho_2 l_2$, $y_1 = l_7$ and $y_2 = l_4$, we get $\rho_1^{a_1} (-\rho_1)^{a_2} \rho_2^{a_3}$. This leads to $g(\rho_1, -\rho_1, \rho_2) = 0$, meaning that $(t_1 + t_2)$ divides the polynomial g.

Applying Proposition 2.1 and Remark 2.5 we have $(t_1 - t_2)(t_1 - t_3)(t_2 - t_3)$ dividing g. Thus we get the desired result. $\qquad\square$

Lemma 2.9. *If a polynomial $v(g)$ of type* (2) *is an identity in skew-symmetric variables with respect to the orthosymplectic superinvolution for $M(2)$ then the commutative polynomial $g1 = (t_1 - t_2)(t_1^2 - t_3^2)(t_2^2 - t_3^2)$ divides the polynomial g.*

Proof. Let $x^{a_1} y_1 x^{a_2} y_2 x^{a_3}$ be a monomial of the polynomial $v(g)$. Substituting $x = \rho_1 m_1 + \rho_2 m_2$, $y_1 = m_3$ and $y_2 = m_7$, we get $\rho_1^{a_1} \rho_2^{a_2} (-\rho_2)^{a_3}$. This leads to $g(\rho_1, \rho_2, -\rho_2) = 0$, meaning that $(t_2 + t_3)$ divides g.

Substituting $x = \rho_1 m_1 + \rho_2 m_2$, $y_1 = m_3$ and $y_2 = m_4$, we get $-\rho_1^{a_1} \rho_2^{a_2} (-\rho_1)^{a_3}$. This leads to $-g(\rho_1, \rho_2, -\rho_1) = 0$, meaning that $(t_1 + t_3)$ divides the polynomial g.

Using Proposition 2.1 and Remark 2.5 as in Lemma 2.8, we get the desired result. $\qquad\square$

Corollary 2.10. *The least degree of a polynomial of type (2), which is an identity for K in the superalgebra $(M(2), trp)$ is 8.*

Proof. Due to Lemma 2.8 we consider $P_8(x, y_1, y_2) = v((t_1 + t_3)g_{4,0})$, namely

$$P_8(x, y_1, y_2) = x^4 y_1 x^2 y_2 - x^4 y_1 y_2 x^2 + x^2 y_1 y_2 x^4 - x^2 y_1 x^4 y_2 + y_1 x^4 y_2 x^2 - y_1 x^2 y_2 x^4.$$

In order to prove that it is an identity for K we use [9, App. A]. $\qquad\square$

Let

$$f_\rho(e_{i_1 j_1}, e_{i_2 j_2}, \ldots, e_{i_{m-1} j_{m-1}}, \sum_{i=1}^m \xi_i e_{ii})$$
$$= \delta_{j_1 i_2} \ldots \delta_{j_{m-2} i_{m-1}} \rho(\xi_{i_1}, \ldots, \xi_{i_{m-1}}, \xi_{j_{m-1}}) e_{i_1 j_{m-1}}$$

and $\rho'(\xi_1, \ldots, \xi_n) = \rho(\xi_1, \ldots, \xi_n, \xi_1)$. The following is true.

Lemma 2.11 ([9, Lemma A.7]). *Suppose $\rho(\xi_1, \ldots, \xi_{n+1})$ has the following two properties:*

(i) *$(\xi_i - \xi_j)$ divides ρ for every $i < j$ except possibly for $(i, j) = (1, n+1)$,*
(ii) *$\rho'(\xi_1, \ldots, \xi_n)$ is symmetric in $\xi_1, \ldots, \xi_n$.*

Then $f_\rho(e_{i_1 j_1}, e_{i_2 j_2}, \ldots, e_{i_n j_n}, \sum_{i=1}^n \xi_i e_{ii}) = 0$ unless $\{e_{i_1 j_1}, \ldots, e_{i_n j_n}\}$ is a cycle, in which case $f_\rho(e_{i_1 j_1}, e_{i_2 j_2}, \ldots, e_{i_n j_n}, \sum_{i=1}^n \xi_i e_{ii}) = \rho'(\xi_1, \ldots, \xi_n) e_{i_1 i_1}$.

We apply the lemma for the polynomial $\rho(t_1, t_2, t_3) = (t_1 + t_3)g_{4,0}$ and get that $P_8(x, y_1, y_2)$ is a Bergman type polynomial for K in the superalgebra $(M(2), trp)$.

Corollary 2.12. (i) *A Bergman type identity of minimal degree for K in the superalgebra $(M(2), osp)$ is a polynomial of type $v(g)$, in which*

$$g = g1 = (t_1 - t_2)(t_2^2 - t_3^2)(t_1^2 - t_3^2).$$

(ii) *All Bergman type polynomials of type (2), in which*

$$g = t_1^\alpha (t_1 + t_2)^\beta (t_2 + t_3)^\gamma t_3^\delta (t_1 - t_2)(t_2^2 - t_3^2)(t_1^2 - t_3^2)$$

are identities for K in $(M(2), osp)$.

Proof. (i) We consider $P_7(x, y_1, y_2) = v(g1)(x, y_1, y_2)$, namely

$$P_7 = x^3 y_1 x^2 y_2 - x^3 y_1 y_2 x^2 - x y_1 x^2 y_2 x^2 + x y_1 y_2 x^4$$
$$+ x^2 y_1 x^3 y_2 - x^2 y_1 x y_2 x^2 - y_1 x^3 y_2 x^2 + y_1 x y_2 x^4$$

and apply Lemma 2.11 for $\rho = g1$.

(ii) Bergman type polynomials of type (2), in which $g = t_1^\alpha (t_1 + t_2)(t_2 + t_3) t_3^\delta g1$ are consequences of P_7 of degree $\alpha + \delta + 2$, namely

$$x^\alpha v(g1)(x, y_1 = y_1 x + x y_1, y_2 = y_2 x + x y_2) x^\delta.$$

If we substitute $y_1 = y_1 x + x y_1$ α times and $y_2 = y_2 x + x y_2$ β times we get polynomials of the type considered.

We note that $y_1 x + x y_1$ and $y_2 = y_2 x + x y_2$ are skew-symmetric matrices with respect to the superinvolution. By the definition of the operation $[a, b]$ at the beginning of part II of the paper, $y_1 x + x y_1 = [y_1, x]$ and $[K, K] \subset K$.

Using (i), this ends the proof. $\qquad\square$

The analogue of Theorem 2.6 in the symplectic case [3] gives least degree 7, while the analogue of Corollary 2.10 and Corollary 2.12 (namely Proposition 2.3) gives least degree 8.

This shows that the supercase converts in some sense the results of the ordinary one at least for small n.

Acknowledgments

The author is very thankful to the organizers of the International Conference on Algebras, Modules and Rings, Lisbon 2003 for the possibility to take part in it and give a talk being the initial version of this paper.

References

1. C. G. Ambrozi, I. P. Shestakov, *On the Lie structure of the skew elements of a simple superalgebra with superinvolution* J. Algebra **208** (1998), 43–71.
2. G. M. Bergman, *Wild automorphisms of free P.I. algebras and some new identities* (1981), preprint.
3. A. Giambruno, A. Valenti, *On minimal ∗-identities of matrices*, Linear Multilin. Algebra **39** (1995), 309–323.
4. E. Formanek, *Central polynomials for matrix rings*, J. Algebra **23** (1972), 129–132.
5. Al. R. Kemer, *Ideals of Identities of Associative Algebras*, Translations of Mathematical Monographs, vol. 87, 1991, AMS, Providence, Rhode Island.
6. Ts. G. Rashkova, *Central polynomials for low order matrix algebras with involution*, Comm. Algebra **28** (10) (2000), 4879–4887.

7. Ts. G. Rashkova, *Identities in algebras with involution*, Bull. Austral. Math. Soc. **60** (1999), 469–477.

8. Ts. G. Rashkova, V. Drensky, *Identities of representation of Lie algebras and *-polynomial identities*, Rend. Circ. Mat. Palermo (2) **XLVIII** (1999), 153–162.

9. L. H. Rowen, *"Polynomial Identities in Ring Theory"*, Acad. Press, New York, 1980.

STRUCTURE AND REPRESENTATIONS OF CONFORMAL ALGEBRAS

ALEXANDER RETAKH*

*Department of Mathematics
Massachusetts Institute of Technology
Cambridge, MA 02139
E-mail: retakh@math.mit.edu*

Introduction

0.1. In the last two decades vertex algebras have been an important tool in such diverse subjects as representations of infinite-dimensional algebras and the theory of finite groups [15, 19].

Roughly speaking, a vertex algebra is a space V such that to each element of V there corresponds a *formal distribution*, i.e., an element of $\operatorname{End} V[[z^{\pm 1}]]$. (Note that any algebraic operation performed on the space of formal distribution will have to involve more than one variable, as the distributions are power series in both z and z^{-1}.) Two distributions $a(z)$ and $b(w)$ must be *local*, that is, commute outside the diagonal of the zw-plane.

0.2. The first definition of vertex algebras was given in [6] and is rather involved. With time a need for an algebraic formalism for vertex algebras became clear. Since local formal distributions are in some sense meromorphic, it is reasonable to look at the "singular" part first. Such an approach was emphasized in [26, 27] and especially [19], where this theory was fully developed (see also [4, 23] for geometric counterparts).

This setting is quite general; for an algebra A, consider the following operation on formal distributions over A:

$$a(z) \, \widehat{(n)} \, b(z) = \operatorname{Res}_{w=0} a(w)b(z)(w - z)^n, \quad a(z), b(z) \in A[[z^{\pm 1}]], n \in \mathbb{Z}_{\geq 0}. \tag{0.1}$$

---

*Partially supported by the NSF

290

Two formal distributions are local if a finite number of their products (0.1) is nonzero. The formalization of this definition leads to the concept of a *conformal algebra* (Definition 1.1). Then a vertex algebra is defined as having two related structures: that of a Lie conformal algebra and a left symmetric differentiable algebra [2].

0.3. Conformal algebras also have an intriguing connection to Hamiltonian formalism in the theory of nonlinear evolutionary equations [1, 34]. In fact the first appearance in the literature of conformal-like structures predates the discovery of vertex algebras and comes from the calculus of variations [17]. However, this subject is outside the scope of this survey.

0.4. This survey is dedicated to the study of conformal algebras and their representations. Other expository papers on the subject appeared in the past, in particularly, [20] and [35]; however, there have been new developments in the field since their publication.

We begin by defining conformal algebras and discussing their general properties in Section 1. It also contains several basic examples. Section 2 develops the theory of representations of conformal algebras. The bulk of this survey, Section 3, is dedicated to the very important conformal algebras gc_n and Cend_n and their subalgebras (roughly, they are the analogs of matrix algebras). We conclude with conjectures and open questions in Section 4.

Most of the results here appeared elsewhere; however, several remarks are new.

Throughout this survey the base field is $\mathbb{C}$.

1. Basic Definitions and Examples

1.1. We begin with the formal definition of a conformal algebra.

Definition 1.1 ([19]). A *conformal algebra* C is a $\mathbb{C}[\partial]$-module endowed with bilinear operations $\underline{n}$: $C \otimes C \to C$, $n \in \mathbb{Z}_{\geq 0}$, such that for any $a, b \in C$

    (1) (locality axiom) $a \underline{n} b = 0$ for $n > N(a,b)$ ($N(a,b)$ is called the *order of locality* of a and b);

    (2) (Leibniz rule) $\partial(a \underline{n} b) = (\partial a) \underline{n} b + a \underline{n} (\partial b)$;

    (3) $(\partial a) \underline{n} b = -na \underline{n-1} b$.

A more succinct way to present operations in a conformal algebra C is via the so-called λ-*product*. Let λ be a formal variable. Define the map

$C \otimes C \to \mathbb{C}[\lambda] \otimes C$ as

$$a_\lambda b = \sum_n \frac{\lambda^n}{n!} a \, \textcircled{n} \, b, \quad a, b \in C, n \in \mathbb{Z}_{\geq 0}. \tag{1.1}$$

We then arrive at an alternative definition of a conformal algebra: this is a $\mathbb{C}[\partial]$-module with a bilinear λ-product satisfying analogs of axioms (2) and (3) (a quick exercise is to deduce them explicitly!). Clearly locality is automatic here; after all, the λ-product produces a polynomial in λ. The λ-product works extremely well in calculations; we will see some evidence of this below.

1.2. A set of mutually local formal distributions closed with respect to product $\textcircled{n}$ and ∂_z is a conformal algebra. Thus conformal algebras provide an algebraic formalism for algebras of local formal distributions. Conversely, every conformal algebra can be made into an algebra of formal distributions. This is the procedure:

Let C be a conformal algebra. For each integer n, consider the linear space $\hat{A}(n)$ isomorphic to C. The element corresponding to $a \in C$ is denoted $\hat{a}(n)$. Let $\hat{A} = \bigoplus_{n \in \mathbb{Z}} \hat{A}(n)$ and let E be the subspace of $\hat{A}$ spanned by the elements of the form $(\partial a)(n) + na(n-1)$ for all $a \in C, n \in \mathbb{Z}$. The quotient space $\mathrm{Coeff}\, C = \hat{A}/E$ is the *coefficient algebra* of C. The image $\hat{a}(n)$ in $\mathrm{Coeff}\, C$ is denoted $a(n)$.

It remains to introduce the operation on $\mathrm{Coeff}\, C$. The following formula holds for any two formal distributions a and b:

$$a(m)b(n) = \sum_{j \geq 0} \binom{m}{j} (a \, \textcircled{j} \, b)(m + n - j).$$

Now we take it as the definition of the product in $\mathrm{Coeff}\, C$. It follows that C is isomorphic to an algebra of formal distributions over $\mathrm{Coeff}\, C$ (and ∂ acts properly because we factored out E).

Remark 1.2. The above construction is taken from [31]. [19] follows a somewhat different approach by introducing a conformal structure on $\hat{A}$. (This is more similar to the original construction for vertex algebras in [6].)

$\mathrm{Coeff}\, C$ is universal among all algebras of possible coefficients. Namely, let B be an algebra such that there exists a homomorphism $C \to B[[z, z^{-1}]]$. Then there exists a unique homomorphism $\phi : \mathrm{Coeff}\, C \to B$ such that the

292

diagram

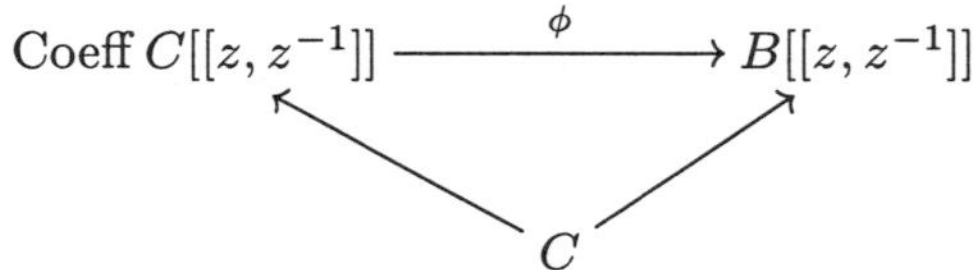

commutes.

1.3. Now we can define Lie and associative conformal algebras.

Let $\mathcal{X}$ be a variety of algebras (Lie, associative, commutative, etc.). Then we say that C is $\mathcal{X}$ *conformal* if $\operatorname{Coeff} C$ is $\mathcal{X}$ (or more rigorously lies in $\mathcal{X}$).

This definition is in a sense unsatisfactory: it refers to another object and a non-conformal one at that. An improvement would be to define a variety of conformal algebras directly. So, assume that $\mathcal{X}$ is defined by identities $\{f_\alpha\}$ (e.g., the Jacobi identity and anti-commutativity for Lie algebras). Then one can produce identities $\{g_\alpha\}$ such that a conformal algebra C satisfies them if and only if C is $\mathcal{X}$ conformal. The algorithm for the construction of such identities g_α can be found in [19]; a more detailed discussion appears in [24].

Conformal identities look much better in their λ-form. To provide a few examples, the conformal law of associativity is (for any $a, b, c \in C$)

$$a_\lambda(b_\mu c) = (a_\lambda b)_{\lambda+\mu} c, \tag{1.2}$$

and anti-commutativity and the Jacobi identity are, respectively,

$$[a_\lambda b] = -[b_{-\partial-\lambda} a], \tag{1.3}$$

$$[a_\lambda [b_\mu c]] = [[a_\lambda b]_{\lambda+\mu} c] + [b_\mu [a_\lambda c]] \tag{1.4}$$

(we denote the λ-product in a Lie conformal algebra as $[\ _\lambda\ ]$ to emphasize its relation to the ordinary Lie bracket. That is, for a Lie conformal algebra C we have $[\ _\lambda\ ] : C \otimes C \to C[\lambda]$).

Remark 1.3. An associative conformal algebra can always be turned into a Lie conformal one. The λ-bracket is defined as

$$[a_\lambda b] = a_\lambda b - b_{-\lambda-\partial} a.$$

This allows us to abuse the language sometimes and speak of Lie subalgebras of an associative conformal algebra C without mentioning that we first turn C into a Lie conformal algebra as above. The same goes for embeddings of Lie conformal algebras into associative ones etc.

At this point a reasonable question would be, If we have something like "conformal varieties", can we also get free conformal algebras? In such generality, the answer is negative. Indeed, a free object must map onto any other, and for conformal algebras this means that the orders of locality for generators in a free conformal algebra would be unbounded. However, if we restrict the orders of locality for generators beforehand, we can construct a free conformal algebra. Namely, let S be a set of letters with a given function $N : S \times S \to \mathbb{Z}_{\geq 0}$. Consider the category of $\mathcal{X}$ conformal algebras generated by S such that for any $a, b \in S$ and $n > N(a, b)$, $a\,{}_{(n)}\,b = 0$. This category does possess the universal object called the *free conformal algebra* corresponding to the function N. For details, see [31] or [5].

The above discussion shows that a conformal algebra can be described as universal in some category only if all objects in this category satisfy given locality conditions. With this in mind, one can ask if there exist universal enveloping algebras for Lie conformal algebras. This turns out to be true for some algebras (e.g., finite) but in general there exist Lie conformal algebras that can not be embedded into associative conformal algebras (in the sense of Remark 1.3), see [32].

1.4. The paragraphs above dealt with "universal algebra"; we continue with the definitions from the structure theory.

A *subalgebra* of a conformal algebra C is a $\mathbb{C}[\partial]$-submodule of C closed with respect to all the operations ${}_{(n)}$. An *ideal* of C is a $\mathbb{C}[\partial]$-submodule I such that for all n, $C\,{}_{(n)}\,I \subset C$ and $I\,{}_{(n)}\,C \subset C$.

A conformal algebra C is *simple* if its only ideals are C and 0.

A conformal algebra C is *nilpotent* of order d if for any $n_1, n_2, \ldots, n_{d-1} \in \mathbb{Z}_{\geq 0}$, $C\,{}_{(n_1)}\,C \ldots {}_{(n_{d-1})}\,C = 0$. (There is an abuse of notation going on: the product of d elements of C is not defined unless we say how the brackets are inserted. Here we mean that the product is 0 for any insertion of brackets.)

An associative conformal algebra C is *semisimple* if it has no nonzero nilpotent ideals.

The *derived conformal algebra* of a Lie conformal algebra C is $C' = \sum_n C\,{}_{(n)}\,C$. As usual we set $C^{(1)} = C'$, $C^{(m+1)} = (C^{(m)})'$, $m \geq 1$, and say that C is *solvable* if $C^{(m)} = 0$ for some m. A Lie conformal algebra is *semisimple* if it has no nonzero solvable ideals.

One can go on and define, e.g., prime conformal algebras, various conformal radicals, and so on, but we do not need them in this survey.

1.5. Now we turn to examples of conformal algebras.

The first example shows that every "ordinary" algebra can be made conformal.

Example 1.4. Let B be an algebra. Consider the affinization $B[t^{\pm 1}]$ of B and the collection $\mathcal{F} \subset B[t^{\pm 1}][[z, z^{-1}]]$ of formal distributions of the form

$$\widetilde{b} = \sum_{m \in \mathbb{Z}} bt^m z^{-m-1}, \quad b \in B. \tag{1.5}$$

We claim that the module $\operatorname{Cur} B = \mathbb{C}[\partial] \otimes \mathcal{F}$ is a conformal algebra. For this we have only to show that all elements of $\operatorname{Cur} B$ are pairwise mutually local.

It is easy to see that for any $a, b \in B$,

$$\widetilde{a} \textcircled{0} \widetilde{b} = \widetilde{ab}, \qquad \widetilde{a} \textcircled{n} \widetilde{b} = 0, \ n > 0$$

(where the products $\textcircled{n}$ are understood in the sense of (0.1)), and as for the rest of the elements of $\operatorname{Cur} B$, their mutual locality follows from axioms (2) and (3).

$\operatorname{Cur} B$ is called the *current algebra* over B.

Remark 1.5. It was easy to check mutual locality for all elements of $\operatorname{Cur} B$; however, this might not be so for an arbitrary collection of formal distributions. Fortunately, there is a way out for Lie and associative algebras.

Lemma 1.6 *(Dong's Lemma). Let $a, b,$ and c be pairwise mutually local formal distributions over a Lie or associative algebra. Then for all $n \in \mathbb{Z}_{\geq 0}$ the formal distributions $a \textcircled{n} b$ and c are mutually local.*

In particular, Dong's lemma implies that for Lie and associative conformal algebras we have only to check that the generators are local.

For associative algebras, there exists a generalization of current algebras.

Example 1.7. Let B be an associative algebra with a locally nilpotent derivation δ. Consider its (localized) Ore extension $B[t^{\pm 1}; \delta]$ and the collection $\mathcal{F} \subset B[t^{\pm 1}; \delta][[z, z^{-1}]]$ of formal distributions of the type (1.5). For $a, b \in B$, we have

$$\widetilde{a} \textcircled{n} \widetilde{b} = (-1)^n \widetilde{a\delta^n(b)}$$

(or $\widetilde{a}_\lambda \widetilde{b} = \widetilde{ae^{-\lambda\delta}b}$ in the λ-notation).

The conformal algebra $\operatorname{Diff} B = \mathbb{C}[\partial] \otimes \mathcal{F}$ is called the *differential algebra* over B.

In the Lie case, the smallest non-current conformal algebra is the Virasoro conformal algebra.

Example 1.8. Consider the Lie algebra $\mathrm{Vect}\,\mathbb{C}^{\times}$ of regular vector fields on $\mathbb{C}^{\times}$. It is well known that the fields $L_n = -t^{n+1}\partial_t$, $n \in \mathbb{Z}$, form a basis of $\mathrm{Vect}\,\mathbb{C}^{\times}$. The formal distribution $L(z) = \sum_n L_n z^{-n-2}$ is local with itself:

$$[L_\lambda L] = (\partial + 2\lambda)L.$$

The conformal algebra $\mathrm{Vir} = \mathbb{C}[\partial] \otimes L$ is called the *Virasoro conformal algebra*.

Remark 1.9. $\mathrm{Vect}\,\mathbb{C}^{\times}$ is the algebra of infinitesimal conformal transformations of $\mathbb{C}^{\times}$. This explains the choice of the term "conformal" since Vir is the smallest non-trivial (i.e., non-current) conformal algebra.

Of course, the "ordinary" Virasoro algebra is not $\mathrm{Vect}\,\mathbb{C}^{\times}$ but its central extension. However, one can also construct the central extension $\widehat{\mathrm{Vir}}$ of Vir such that $\mathrm{Coeff}\,\widehat{\mathrm{Vir}}$ is the Virasoro algebra.

Example 1.10. As a vector space $\widehat{\mathrm{Vir}} = \mathrm{Vir} \oplus \mathbb{C}\mathsf{c}$ (here $\partial\mathsf{c} = 0$) and the λ-brackets are

$$[L_\lambda L] = (\partial + 2\lambda)L + \frac{\lambda^3}{12}\mathsf{c}, \quad [L_\lambda \mathsf{c}] = 0, \quad [\mathsf{c}_\lambda \mathsf{c}] = 0.$$

In the same vein, for a simple finite-dimensional Lie algebra $\mathfrak{g}$, one can construct the central extension $\widehat{\mathrm{Cur}\,\mathfrak{g}}$ of $\mathrm{Cur}\,\mathfrak{g}$ such that $\mathrm{Coeff}\,\widehat{\mathrm{Cur}\,\mathfrak{g}} = \widehat{\mathfrak{g}}$. Namely, $\widehat{\mathrm{Cur}\,\mathfrak{g}}$ is generated by elements $\widetilde{g}$, $g \in \mathfrak{g}$, and c (again, $\partial\mathsf{c} = 0$) such that

$$[\widetilde{g}_\lambda \widetilde{h}] = \widetilde{[g,h]} + \lambda(g|h)\mathsf{c}, \quad [\widetilde{g}_\lambda \mathsf{c}] = 0, \quad [\mathsf{c}_\lambda \mathsf{c}] = 0.$$

1.6. For brevity we say that a conformal algebra is *finite* if it is finite as a $\mathbb{C}[\partial]$-module.

A current algebra $\mathrm{Cur}\,B$ is simple (respectively, finite) if and only if B is simple (respectively, finite). Thus, we already know a number of (admittedly not very interesting) examples of finite simple conformal algebras. Are there others?

In the Lie case, the Virasoro conformal algebra is also simple and finite but this is it.

Theorem 1.11. (1) *[11] Let C be a finite simple Lie conformal algebra. Then C is isomorphic to either Vir or a current algebra $\mathrm{Cur}\,\mathfrak{g}$ over a simple finite-dimensional Lie algebra $\mathfrak{g}$.*

(2) *[20] Let C be a finite simple associative conformal algebra. Then C is isomorphic to a current algebra $\mathrm{Cur}\,\mathrm{Mat}_n(\mathbb{C})$.*

The first part of Theorem 1.11 is proved by the careful study of the Lie algebra of non-negative coefficients of C. When completed with respect to its natural topology, this algebra becomes linearly compact and then, after some additional work, one applies the Cartan-Guillemin theorem to obtain the complete classification.

The second part follows from the first via standard algebraic techniques.

Remark 1.12. As we have just seen, in the conformal setting the (centerless) Virasoro algebra and the Laurent extensions of finite-dimensional simple Lie algebras appear as coefficient algebras of finite simple Lie conformal algebras. In the non-conformal universe, however, these belong to two very distinct worlds of Cartan and affine Kac-Moody algebras.

We can extend Theorem 1.11 to the semisimple case but it is not as straightforward as one may think.

Example 1.13. For a finite-dimensional Lie algebra $\mathfrak{g}$, the $\mathbb{C}[\partial]$-module $\mathrm{Vir} \oplus \mathrm{Cur}\,\mathfrak{g}$ carries the following conformal structure:

$$[L_\lambda L] = (\partial + 2\lambda)L, \quad [\widetilde{g}_\lambda \widetilde{h}] = \widetilde{[g, h]}, \quad [L_\lambda \widetilde{g}] = (\partial + \lambda)\widetilde{g}.$$

So, we obtain the semidirect product of Vir and $\mathrm{Cur}\,\mathfrak{g}$. When $\mathfrak{g}$ is semisimple, this semidirect product is semisimple too.

This, however, is the only surprise in the classification of finite semisimple Lie conformal algebras.

Theorem 1.14. (1) *[11] Let C be a finite semisimple conformal Lie algebra. Then C is isomorphic to a direct sum of copies of finite simple Lie conformal algebras and semidirect products of* Vir *and* $\mathrm{Cur}\,\mathfrak{g}$ *for semisimple finite-dimensional $\mathfrak{g}$'s.*

(2) [20] A finite semisimple associative conformal algebra is isomorphic to a direct sum of copies of simple associative conformal algebras.

The second part of this theorem seems to be a weak version of the Artin-Wedderburn theorem: in the end, all we get is matrices. However, from the representation-theoretic point of view, current algebras over matrices and, in general, finite conformal algebras are not the right analogs of ordinary matrices. We discuss this in the next section.

2. Representation Theory of Conformal Algebras

Defining a module M over a conformal algebra C is easy and we now have a choice of two approaches: either modify Definition 1.1 by considering the products $C \otimes M \to M$ or define a module of formal distributions over an algebra of formal distributions imitating (0.1). However, it is even better to take a more general approach and start with the definition of a conformal linear map. In particular, this will help us to construct certain representations later on.

2.1. Let M and N be two $\mathbb{C}[\partial]$-modules. A *conformal linear map* from M to N is a $\mathbb{C}$-linear map $\phi : M \to \mathbb{C}[\lambda] \otimes N$, denoted $\phi_\lambda : M \to N$, such that $\partial\phi_\lambda - \phi_\lambda\partial = -\lambda\phi_\lambda$. As above, we can define the operators $\phi \textcircled{n}$ via $\phi_\lambda m = \sum_n (\lambda^n/n!)\phi \textcircled{n} m$.

We can "differentiate" conformal linear maps by putting $(\partial\phi)_\lambda = -\lambda\phi_\lambda$.

The space of all conformal linear maps from M to N is denoted $\mathrm{Chom}(M, N)$. It carries a natural structure of a $\mathbb{C}[\partial]$-module. When $M = N$, we can also define the products $\psi \textcircled{n} \phi$ of conformal linear maps by applying the conformal law of associativity (1.2). However, in general this does not make $\mathrm{Chom}(M, M)$ into a conformal algebra as the locality condition fails. For instance, let M be of infinite rank and $\phi \in \mathrm{Chom}(M, M)$ be such that for any n there exists $u_n \in M$ whose order of locality with ϕ is n. Put $v_n = \phi \textcircled{n} u_n$. Let $\psi \in \mathrm{Chom}(M, M)$ be such that $\psi \textcircled{0} v_n \neq 0$ for all n. Then for any n, $(\psi \textcircled{n} \phi) \textcircled{n} u_n \neq 0$ and thus ψ and ϕ are not local.

On the other hand, if M is of finite rank, any two elements of $\mathrm{Chom}(M, M)$ are local with each other. In this case we denote the associative conformal algebra $\mathrm{Chom}(M, M)$ as $\mathrm{Cend}\, M$.

$\mathrm{Cend}\, M$ with a Lie conformal bracket (see Remark 1.3) is denoted $\mathrm{gc}\, M$.

To simplify notations, we also denote $\mathrm{Cend}\, \mathbb{C}[\partial]^n$ and $\mathrm{gc}\, \mathbb{C}[\partial]^n$ as Cend_n and gc_n, respectively.

2.2. Now we will define a module over a conformal algebra.

Definition 2.1. A *module M* over a Lie or associative conformal algebra C is a $\mathbb{C}[\partial]$-module endowed with an operation $C \otimes M \to \mathbb{C}[\lambda] \otimes M$ such that for any $a, b \in C$ and $v \in M$,

(1) $(\partial a)_\lambda v = -\lambda a_\lambda v, \quad a_\lambda \partial v = (\partial + \lambda)(a_\lambda v)$;
(2) $a_\lambda(b_\mu v) = [a_\lambda b]_{\lambda+\mu} v + b_\mu(a_\lambda v)$ if C is Lie; $a_\lambda(b_\mu v) = (a_\lambda b)_{\lambda+\mu} v$ if C is associative.

Simply put, M is a C-module if there exists a map $C \to \mathrm{Chom}(M, M)$ of conformal algebras that satisfies a version of the Jacobi identity or associativity.

As usual we can define a *submodule*, an *irreducible* module (contains no non-trivial submodules), an *indecomposable* module (does not split into a direct sum of non-trivial submodules), etc. We call a module *finite* if it is finite over $\mathbb{C}[\partial]$.

Remark 2.2. If C is a Lie conformal algebra and M and N are modules over C, $\mathrm{Chom}(M, N)$ also carries a natural structure of a C-module. Namely, we set

$$(a_\lambda \phi)_\mu m = a_\lambda(\phi_{\mu-\lambda} m) - \phi_{\mu-\lambda}(a_\lambda m), \quad a \in C, \phi \in \mathrm{Chom}(M, N), m \in M.$$

Then we can define the *contragradient* C-module $U^* = \mathrm{Chom}(M, \mathbb{C})$, where $\mathbb{C}$ stands for the trivial C-module (with the trivial action of ∂).

For a finite M, we also can define $M \otimes N = \mathrm{Chom}(M^*, N)$.

As in Section 1, one can show that a module over a conformal algebra C can be always viewed as a module of formal distributions over a "coefficient module" which, in turn, is a module over $\mathrm{Coeff}\, C$.

2.3. Below we construct modules for finite simple conformal algebras (cf. Theorem 1.11).

Example 2.3. For any $\Delta, \alpha \in \mathbb{C}$ consider the space of densities $F(\Delta, \alpha) = \mathbb{C}[t, t^{-1}] e^{-\alpha t} (dt)^{1-\Delta}$. This is naturally a module over $\mathrm{Vect}\,\mathbb{C}^\times$; it is irreducible whenever $\Delta \neq 0$.

The formal distribution $m(z) = \sum_n (t^n e^{-\alpha t} (dt)^{1-\Delta}) z^{-n-1}$ spans the module $M(\Delta, \alpha)$ over the Virasoro conformal algebra Vir with the action induced from that of $\mathrm{Vect}\,\mathbb{C}^\times$ on $F(\Delta, \alpha)$. Explicitly, $L_\lambda m = (\partial_z + \alpha + \Delta\lambda)m$. Again, whenever $\Delta \neq 0$, $M(\Delta, \alpha)$ is irreducible.

In fact, the following is true.

Theorem 2.4 ([9]). *Any non-trivial irreducible finite module over the Virasoro conformal algebra is isomorphic to $M(\Delta, \alpha)$ with $\Delta \neq 0$.*

Finite indecomposable modules over Vir were studied in [10]. Complete reducibility does not hold here. In fact, by classifying finite central extensions of $M(\Delta, \alpha)$, one can arrive directly at the classification of central extensions of certain modules over regular vector fields on $\mathbb{C}$. This is an example of a connection between the cohomology of conformal algebras [3] and the cohomology of infinite-dimensional Lie algebras [14, 13, 18].

Example 2.5. It is easy to construct a module over a current algebra. Let A be an algebra and U an A-module. Then $\mathrm{Cur}\, A$ acts on the module $\widetilde{U} = \mathbb{C}[\partial] \otimes U$ with the natural action $\widetilde{a}_\lambda(1 \otimes u) = 1 \otimes au$, $a \in A$, $u \in U$.

A companion result to Theorem 2.4, also proved in [9], states that a non-trivial irreducible finite module over $\mathrm{Cur}\,\mathfrak{g}$ for a finite-dimensional semisimple Lie algebra $\mathfrak{g}$ is of the form $\widetilde{U}$ for a non-trivial irreducible finite-dimensional $\mathfrak{g}$-module U. We will see below (Theorem 3.10) that an even more general result holds for associative unital algebras: every module over such an algebra is of the form $\widetilde{U}$ (and $\widetilde{U}$ is irreducible if and only if U is).

3. Cend_n and gc_n

3.1. Here we present the explicit constructions of the conformal algebras Cend_n and gc_n.

Every conformal endomorphism of the module $\mathbb{C}[\partial]$ is determined by the image of ∂, thus roughly speaking, Cend_1 is isomorphic to $\mathbb{C}[\partial] \otimes_{\mathbb{C}} \mathbb{C}[\partial]$ (where the second component is responsible for the image and the first, for the $\mathbb{C}[\partial]$-module structure of this conformal algebra). In the case of $\mathbb{C}[\partial]^n = \mathbb{C}[\partial] \otimes \mathbb{C}^n$ we have also to account for the action of $\mathrm{Mat}_n(\mathbb{C})$ on $\mathbb{C}^n$.

However, to get more explicit expressions, it is perhaps better to go down to the level of coefficients.

The coefficient algebra of Cend_n or gc_n is the algebra $\mathrm{Mat}_n(\mathrm{Vect}\,\mathbb{C}^\times)$ viewed as either an associative or Lie algebra. The formal distributions that span Cend_n as a $\mathbb{C}[\partial]$-module (here and further, we simply write ∂ for ∂_z) are

$$J_A^m = \sum_{n \in \mathbb{Z}} At^n(-\partial_t)^m z^{-n-1}, \quad A \in \mathrm{Mat}_n(\mathbb{C}), m \in \mathbb{Z}_{\geq 0}.$$

The action on $\mathbb{C}[\partial]^n$ arises from the standard action of $\mathrm{Mat}_n(\mathrm{Vect}\,\mathbb{C}^\times)$ on the space $\mathbb{C}^n[t^{\pm 1}]$. Namely, for $v \in \mathbb{C}^n$ let $\tilde{v} = \sum_n vt^n z^{-n-1}$. Then

$$J_A{}^m{}_\lambda \tilde{v} = (\partial + \lambda)^m Av. \tag{3.1}$$

We call this action *canonical*.

Remark 3.1. We can tweak the above construction by putting

$$J_A{}^m{}_\lambda \tilde{v} = (\partial + \lambda + \alpha)^m Av, \quad \alpha \in \mathbb{C}.$$

More explicitly, we consider the action of $\mathrm{Mat}_n(\mathrm{Vect}\,\mathbb{C}^\times)$ on the space $\mathbb{C}^n[t^{\pm 1}]e^{-\alpha t}$ and then pass to the formal distributions. Notice, though, that we still get an action of Cend_n on $\mathbb{C}[\partial]^n$. We denote this representation E_n^α.

In fact, this action comes from the automorphism of Cend_n sending the generator J_A^1 to $J_A^1 + \alpha J_{\mathrm{Id}}^0$, where Id is the identity matrix.

Another way to obtain the explicit presentation of Cend_n is to use the language of differential conformal algebras (see Example 1.7). (This is more in tune with our statement that $\mathrm{Cend}_1 = \mathbb{C}[\partial] \otimes \mathbb{C}[\partial]$.) Here we have $\mathrm{Cend}_n = \mathrm{Diff}(\mathbb{C}[\partial_t] \otimes \mathrm{Mat}_n(\mathbb{C}))$ for the derivation $\mathrm{ad}\, t = [t, \cdot]$.

Actually, an easy computation shows that both explicit presentations above produce the same algebra of formal distributions; we just get two different bases, $\{J_A^m\}$ and $\{\widetilde{f(\partial_t)}\}$. Though the formula for the action of the second basis on $\mathbb{C}[\partial]^n$ does not look as good as (3.1), it still has merits (see below).

Remark 3.2. The coefficient algebra of gc_n can be embedded into the algebra $\widetilde{\mathrm{gl}_\infty}$ (a central extension of the algebra of infinite-dimensional matrices with finitely many nonzero diagonals). This relation is used in the study of finite modules of gc_n but its precise nature—as well as its connection to vertex algebras—is outside the scope of this survey.

3.2. It is clear from the construction that Cend_n contains $\mathrm{Cur}\,\mathrm{Mat}_n(\mathbb{C})$, i.e., that every simple finite conformal algebra is contained in Cend_n for some n. And, for every simple finite-dimensional Lie algebra $\mathfrak{g}$, $\mathrm{Cur}\,\mathfrak{g}$ embeds into gc_n for some n.

The construction of Cend_n (and thus gc_n) also implies that gc_n contains the Virasoro conformal algebra Vir. Moreover, for every $\alpha \in \mathbb{C}$, the conformal subalgebra of gc_n generated by $L_\alpha = J_{\mathrm{Id}}^1 + \alpha J_{\mathrm{Id}}^0$ is isomorphic to Vir (cf. Remark 3.1).

It is more useful here to pass to the basis $\{\widetilde{f(\partial_t)}\}$ of gc_n. In the above notation, $L_\alpha = \widetilde{\partial_t} + \widetilde{\partial 1} + \alpha \widetilde{1}$. We can go further and consider the element $L_{\alpha,\beta} = \widetilde{\partial_t} + \beta\widetilde{\partial 1} + \alpha\widetilde{1}$ for any $\beta \in \mathbb{C}$. The subalgebra generated by $L_{\alpha,\beta}$ is also isomorphic to Vir.

This gives us a family of embeddings $\theta_{\alpha,\beta} : \mathrm{Vir} \hookrightarrow \mathrm{gc}_1$. It is not difficult to show that every embedding of Vir into gc_1 is of this form. Moreover, a map $\theta_{\alpha,\beta}$ and the canonical gc_1-action on $\mathbb{C}[\partial]$ establish the isomorphism $\mathbb{C}[\partial] \simeq M(1 - \beta, \alpha)$ (see Example 2.3). Thus we have just classified all Vir modules of rank one.

Remark 3.3. A direct classification of all embeddings of Vir into gc_n for an arbitrary n would be of great interest. Among other things it would imply the classification of Virasoro elements in gc_n and the complete description of finite Vir modules.

3.3. We essentially view conformal algebras Cend_n and gc_n as analogs of matrix algebras. Since the theory of finite associative algebras is much simpler than its Lie counterpart, we should first focus on the study of Cend_n. There are three directions here: the representation theory, a purely algebraic description of Cend_n (i.e., the first cornerstone for the analog of Artin-Wedderburn), and a more detailed study of its subalgebras. The latter approach, of course, is a dead end when ordinary matrix algebras are concerned but we already saw that Cend_n possesses some interesting subalgebras.

3.4. We begin by focusing on a possible analog of the Artin-Wedderburn theorem.

The biggest problem is that unlike $\mathrm{Mat}_n(\mathbb{C})$, Cend_n is not of finite rank. This suggests to develop a concept of growth for conformal algebras. In analogy with the ordinary theory [25], we define the *Gelfand-Kirillov dimension* of a finitely-generated conformal algebra C as

$$\mathrm{GKdim}\, C = \limsup_{r \to \infty} \frac{\log \mathrm{rk}_{\mathbb{C}[\partial]}(\mathbb{C}[\partial]\text{-span of products of} \leq r \text{ generators of } C)}{\log r}.$$

$$(3.2)$$

By locality all the ranks in (3.2) are finite and, since the function we consider here is monotone, $\mathrm{GKdim}\, C$ is well defined. It possesses the standard properties: GKdim is does not depend on the choice of the generating set, GKdim of a subalgebra or a quotient algebra does not exceed that of the algebra, etc.

For any differential conformal algebra $\mathrm{Diff}\, B$, $\mathrm{GKdim}\, \mathrm{Diff}\, B = \mathrm{GKdim}\, B$. In particular, $\mathrm{GKdim}\, \mathrm{Cend}_n = 1$.

Remark 3.4. In general, for an associative conformal algebra C, $\mathrm{GKdim}\, \mathrm{Coeff}\, C \leq \mathrm{GKdim}\, C + 1$ [28] and the inequality is sometimes strict (e.g., when C is torsion). It is still an open question if the equality is always reached for a torsion-free conformal algebra.

So far we can say that Cend_n is simple, of GK-dimension 1, and differential. The latter property, however, refers to the coefficient algebra and we wish to obtain a description of Cend_n in purely conformal terms.

Mimicking the ordinary algebra, we have to start by defining unital conformal algebras. An ordinary unital algebra contains the field (i.e., a subalgebra of dimension one) whose action is nonzero. The only non-trivial conformal algebra of rank one is $\mathrm{Cur}\, \mathbb{C}$ and, moreover, every $\mathrm{Cur}\, \mathbb{C}$ module M splits as $M = M^0 \oplus M^1$, where $\widetilde{1}\,_{\textstyle{①}} = \mathrm{Id}_{M^1}$ and the action on M^0 is zero (both facts are not hard to show). Hence, we will call an associative conformal algebra C unital if it contains $\mathrm{Cur}\, \mathbb{C}$ and if for the resulting action of $\mathrm{Cur}\, \mathbb{C}$, $C = C^1$.

The element $\widetilde{1}$ is called a *conformal identity*.

Remark 3.5. A more rigorous name would be a *left* conformal identity as we have only the left action of $\mathrm{Cur}\,\mathbb{C}$ on C. And, unlike, an ordinary (two-sided) identity, a conformal identity is not unique.

A differential algebra $\mathrm{Diff}\,B$ can be always made unital by adjoining identity to B. (It is still unknown if one can adjoin a conformal identity to a torsion-free conformal algebra.) The converse is almost true.

Let the *left annihilator* of C be the set $\{a \ni C \mid a_\lambda C = 0\}$.

Theorem 3.6 ([28]). *An associative conformal algebra C with the zero left annihilator is differential, $C = \mathrm{Diff}\,B$. Also, if C is finitely generated, then so is B.*

(The main part of the proof is to show that the conformal structure of C is encoded by the zeroth coefficients and $\widetilde{1}(1)$. It immediately follows that the coefficient algebra is an Ore extension.)

Thus a simple unital conformal algebra is always differential. Utilizing the classification of algebras of GK-dimension 1 [33], we can finally obtain the algebraic description of Cend_n:

Theorem 3.7 ([28]). *A simple unital associative conformal algebra of Gelfand-Kirillov dimension 1 is isomorphic to Cend_n.*

Remark 3.8. The proof of Theorem 3.7 can be extended and yield the classification of unital semisimple conformal algebras of GK-dimension 1. Namely, such an algebra always embeds into a direct sum of Cend_n and a current algebra over a semiprime algebra of zero or linear growth [30]. It would be very difficult to classify all such embeddings: for instance, there exists a prime non-current subalgebra of a current algebra of linear growth [30].

3.5. Now we turn to the representation theory of Cend_n.

Just as in the case of $\mathrm{Cur}\,\mathbb{C}$-modules, a module M over a unital conformal algebra C always splits, $M = M^0 \oplus M^1$. We need only to study the structure of M^1. As in the proof of Theorem 3.6, the action of $\widetilde{1}$ gives a certain rigidity to M^1. Namely, M^1 is filtered by the submodules annihilated by $\widetilde{1}\,\textcircled{n}$. If C is differential, i.e., $C = \mathrm{Diff}\,B$ (with a derivation δ), the lowest non-trivial component of this filtration can be made into a B-module that completely determines the structure of M^1.

Conversely, a (unitary) B-module V gives rise to a C-module $\widetilde{V} = \mathbb{C}[\partial] \otimes V$ with the action

$$\widetilde{a}_\lambda \widetilde{v} = \sum_j \partial^j (\widetilde{\delta^j(a)v}), \quad \text{where } \widetilde{v} = 1 \otimes v, v \in V, \text{ and } a \in A. \tag{3.3}$$

Example 3.9. A finite irreducible module E_n^α over Cend_n constructed in Remark 3.1 has the form $E_n^\alpha = \widetilde{\mathbb{C}^n}$, where ∂_t acts on $\mathbb{C}^n$ as α.

This example can be generalized by considering modules $U \otimes \mathbb{C}^n$ over $\mathbb{C}[\partial_t] \otimes \mathrm{Mat}_n(\mathbb{C})$ with ∂_t acting on U as $\alpha \in \mathrm{End}(U)$. Thus we obtain a Cend_n-module $E_n^\alpha(U) = \widetilde{U \otimes \mathbb{C}^n}$.

Call a C-module M *unitary* if $M = M^1$. The discussion above implies that there is a bijection between unitary C-modules and unitary B-modules. We can make this statement more precise: consider the category $\mathrm{Rep}\,C$ whose objects are unitary C-modules and morphisms are homomorphisms that commute with $a\,\widehat{n}$ for every $a \in C, n \in \mathbb{Z}_{\geq 0}$. Then

Theorem 3.10 ([29]). $\mathrm{Rep}\,B \simeq \mathrm{Rep}\,C$.

As the above equivalence is constructed explicitly, we easily deduce that irreducibles correspond to irreducibles, indecomposables to indecomposables, and that an extension of C-modules arises from an extension of corresponding B-modules.

The only concept from representation theory that does not automatically survive is faithfulness: if $\widetilde{V}$ is faithful, we can only conclude that $\mathrm{Ann}\,V$ does not contain any nonzero δ-stable ideals.

Remark 3.11. It would be more useful to define a sort of "conformal category," i.e., to define $\mathrm{Rep}\,C$ for any conformal algebra with the morphisms also carrying a conformal structure. So far the attempts at such definition have been unsuccessful.

Corollary 3.12. *Finite irreducible Cend_n-modules are of the form E_n^α. Finite indecomposable Cend_n-modules are of the form $E_n^\alpha(U)$ for an indecomposable $\alpha \in \mathrm{End}(U)$.*

The above result was first stated in [20]. Another proof was given in [8]. Though more calculation-heavy, it also works for certain non-unital conformal algebras (see below).

3.6. Finite modules over gc_n look similar to Cend_n-modules; however, the classification methods here are entirely different.

We begin by constructing such modules. Since every representation of Cend_n gives rise to a representation of gc_n, we already have the family $E_n^\alpha(U)$ of gc_n-modules (see Example 3.9).

Recall that we can also construct contragradient modules $E_n^\alpha(U)^*$ (see Remark 2.2).

Theorem 3.13 ([20, 7]). *A finite irreducible gc_n-module has the form E_n^α or $(E_n^\alpha)^*$.*

The first step of the proof is to look at certain representations of the coefficient algebra instead. In fact, for a conformal algebra C it suffices to consider modules over the *annihilation algebra* spanned by ∂ and coefficients $a(n)$, $n \geq 0$. Indeed, a C-module M can be viewed as a module over the annihilation algebra. Conversely, a module V over the annihilation algebra such that for each $v \in V$, $a(n)v = 0$ for $n \gg 0$, gives rise to a C-module.

The annihilation algebra of gc_n is the direct sum of a one-dimensional Lie algebra and the algebra of matrices of regular differential operators on the line. Since we are interested in finite gc_n-modules here, we need to consider only modules with finite-dimensional graded components (i.e., *quasifinite*). Therefore, the proof of Theorem 3.13 comes down to classifying quasifinite modules over the Lie algebra of regular differential operators on the line. This was achieved in [7] via the technique developed in [21].

It can be shown that modules over the annihilation algebra of gc_n are completely reducible [22]. Thus all finite indecomposable modules over gc_n are of the form $E_n^\alpha(U)$ or $E_n^\alpha(U)^*$.

3.7. We have already discussed the finite subalgebras of Cend_n and gc_n. Here we present a continuous family of infinite subalgebras of Cend_n acting irreducibly on $\mathbb{C}[\partial]^n$.

Let $P(\partial_t)$ be a matrix in $\mathrm{Mat}_n(\mathrm{Vect}\,\mathbb{C}^\times)$. The formal distributions from Cend_n whose coefficients lie in $\mathrm{Mat}_n(\mathrm{Vect}\,\mathbb{C}^\times)P(\partial_t)$ form a subalgebra (in fact, a left ideal) of Cend_n denoted $\mathrm{Cend}_{n,P}$. When $P(\partial_t)$ is non-degenerate, the conformal algebra $\mathrm{Cend}_{n,P}$ acts irreducibly on $\mathbb{C}[\partial]^n$.

By applying first elementary transformations to P and then by conjugating $\mathrm{Cend}_{n,P}$, we arrive at an isomorphic subalgebra for $P = \mathrm{diag}(p_1(\partial_t), \ldots, p_n(\partial_t))$, where $p_i(\partial_t) \neq 0$ are monic and $p_i|p_{i+1}$. Such p_i's are called elementary divisors. Therefore, we obtain a family of non-isomorphic subalgebras of Cend_n that act irreducibly on $\mathbb{C}[\partial]$ and is parametrized by sequences of elementary divisors.

Conjecture 3.14 ([20]). *A subalgebra of Cend_n that acts irreducibly on* *the standard module* $\mathbb{C}[\partial]^n$ *is conjugate to either* $\operatorname{Cur}\operatorname{Mat}_n(\mathbb{C})$ *or* $\operatorname{Cend}_{n,P}$ *for a non-degenerate* P.*

That all such finite subalgebras are conjugate to $\operatorname{Cur}\operatorname{Mat}_n(\mathbb{C})$ can be deduced from the conformal Cartan-Jacobson theorem [11].

Another particular case of the conjecture is its restriction to unital subalgebras; here the result follows from Theorem 3.7.

It is possible to relax the definition of unitality and consider associative conformal algebras that contain $\operatorname{Cur}\mathbb{C}$ (in this case $\widetilde{1}$ is called a *conformal idempotent*). For such algebras, the conjecture also holds [36].

However, for arbitrary subalgebras, the only settled case is $n = 1$. Moreover, here one can classify all subalgebras of Cend_1 by carefully investigating subalgebra elements of minimal degree (with respect to both bases $\{J^m\}$ and $\{\widetilde{f(\partial_t)}\}$), similar to the classical proof that the algebra $\mathbb{C}[x]$ is principal [8].

3.8. One of the crucial observations of [8] is that for an associative algebra C and an irreducible C-module M, $M \simeq C_{-\partial-\lambda}m|_{\lambda=\alpha}$ for some $m \in M, \alpha \in \mathbb{C}$. Thus we obtain a surjective map from the set of maximal left ideals of C to the set of non-trivial irreducible C-modules (taken up to isomorphism). Hence,

Theorem 3.15 ([8]). *A finite irreducible* $\operatorname{Cend}_{n,P}$*-module is isomorphic to* E_n^α.

However, the category of representations of $\operatorname{Cend}_{n,P}$ is very different from that of Cend_n; it is actually wild for P of high degree [16]. This can be seen by constructing the ext-quiver (adapted to conformal algebras) for the finite-dimensional extensions of irreducible $\operatorname{Cend}_{n,P}$-modules classified in [8].

Remark 3.16. If one could define a conformal category (i.e., a category of conformal objects, where morphism carry a conformal structure as well, see Remark 3.11), then perhaps representations of $\operatorname{Cend}_{n,P}$ could be described as a deformation of $\operatorname{Rep}\operatorname{Cend}_n$. Moreover, an analog of the density theorem for endomorphisms in such a category might help in solving Conjecture 3.14.

3.9. Above we passed from Cend_n to its subalgebras $\operatorname{Cend}_{n,P}$. In the Lie case, we go from $\mathfrak{gc}_n$ to $\mathfrak{gc}_{n,P}$ (either by introducing the Lie bracket on

---

* *Added in proof.* Recently P. Kolesnikov has announced a proof of this conjecture; details are forthcoming.

$\mathrm{Cend}_{n,P}$ or by also considering the subalgebra of formal distributions of gc_n with coefficients divisible on the right by P). Thus, we obtain subalgebras of gc_n acting irreducibly on $\mathbb{C}[\partial]^n$.

But there is more. Since gc_n is simple, it can be viewed as both the analog of gl_n and sl_n. What about analogs of other simple Lie algebras? In particular, what are orthogonal and symplectic conformal Lie algebras?

Consider an anti-involution $*$ on $\mathrm{Vect}\,\mathbb{C}^\times$: $\partial_t^* = -\partial_t, t^* = t$. It can be extended to $\mathrm{Mat}_n(\mathbb{C}) \otimes \mathrm{Vect}\,\mathbb{C}^\times$ by applying $*$ to the second component and a matrix anti-involution to the first. This gives us an anti-involution of Cend_n.

Theorem 3.17 ([8]). *Up to conjugation, all anti-involutions of* Cend_n *are of this form.*

We continue mimicking the constructions of orthogonal and symplectic Lie algebras. Let σ be an anti-involution on Cend_n that arises from a symmetric (resp. skew-symmetric) involution of $\mathrm{Mat}_n(\mathbb{C})$. The fixed points of $-\sigma$ form the *orthogonal conformal algebra* oc_n (resp. *symplectic conformal algebra* spc_n). As gc_n, both oc_n and spc_n are simple.

We can go further and define orthogonal and symplectic subalgebras $\mathrm{oc}_{n,P}$ and $\mathrm{spc}_{n,P}$ of $\mathrm{gc}_{n,P}$. However, this can be done only for hermitian and anti-hermitian P's respectively.

Remark 3.18. There exists another construction of oc_n and spc_n that is more representation-theoretic in spirit. Here we begin by defining a *conformal form* on a $\mathbb{C}[\partial]$-module V: $\langle\,,\,\rangle_\lambda : V \otimes V \to \mathbb{C}[\lambda]$. The form is *bilinear* if

$$\langle \partial v, w \rangle_\lambda = -\lambda \langle v, w \rangle_\lambda = -\langle v, \partial w \rangle_\lambda, \quad v, w \in V$$

and *symmetric* (resp. *skew-symmetric*) if $\langle v, w \rangle_\lambda = \langle w, v \rangle_{-\lambda}$ (resp. $-\langle w, v \rangle_{-\lambda}$).

If V is free over $\mathbb{C}[\partial]$, the form is completely determined by its action on the basis $\{e_i\}$, i.e., by a matrix $P(\lambda)$ with entries $\langle e_i, e_j \rangle_\lambda$.

A conformal bilinear form gives rise to a map $V \to V^*$, which establishes an isomorphism between V and $P(-\partial)V^*$ (for a non-degenerate P).

Thus, though we can not define adjoints with respect to a given form for all operators in Cend_n, there is a well-defined adjoint for elements of $\mathrm{Cend}_{n,P}$. (The details here are rather involved, see [8].) Elements that are skew-symmetric with respect to taking the adjoint form up the Lie conformal algebra $\mathrm{oc}_{n,P}$ (if the corresponding form is symmetric) or $\mathrm{spc}_{n,P}$ (resp. skew-symmetric).

As a companion to Conjecture 3.14, we have

Conjecture 3.19 ([35, 8]). *An infinite subalgebra of gc_n that acts irreducibly on the standard module $\mathbb{C}[\partial]^n$ is conjugate to either $\mathrm{gc}_{n,P}$, $\mathrm{oc}_{n,P}$, or $\mathrm{spc}_{n,P}$ for a non-degenerate P (if defined).*

Remark 3.20. The classification of finite subalgebras of gc_n acting irreducibly on $\mathbb{C}[\partial]^n$ is contained in [11].

Remark 3.21. It is worth to emphasize that here we get only the analogs of classical series A_n, B_n, C_n, and D_n (though the series are not discrete). There is no place for exceptional Lie conformal algebras in this conjecture.

And indeed, so far all attempts to construct such by the analogs of ordinary methods (Tits-Kantor-Koecher construction, sums of representations of algebras of small rank, etc.) have failed.

The existing evidence for this conjecture covers several important cases. In all cases we assume a presence of an important finite subalgebra:

- [36] If C is a simple Lie conformal algebra that has a faithful finite representation and such that $C \supset \mathrm{Cur}\,\mathrm{sl}_2$, then C is one of the subalgebras from Conjecture 3.19.
- [12] If C is a subalgebra of gc_n that acts irreducibly on $\mathbb{C}[\partial]^n$ and is fixed by the action of $L_{(n)}$, $n = 0, 1, 2$, for the Virasoro element $L = \widetilde{\partial}_t + \alpha \partial \widetilde{1}$, then C is one of the subalgebras from Conjecture 3.19.

Remark 3.22. The second statement above can be reformulated: the (centerless) Virasoro Lie algebra contains an sl_2 spanned by L_{-1}, L_0, and L_1. A subalgebra fixed by this sl_2 and acting irreducibly on $\mathbb{C}[\partial]$ is from Conjecture 3.19.

Finally, as far as the study of representations of infinite subalgebras of gc_n goes, only few inroads have been made: [7] has the classification of finite representations of subalgebras of gc_1 containing a Virasoro subalgebra but this is all.

4. Future Developments

4.1. We restate here Conjectures 3.14 and 3.19:

Conjecture 4.1. *Let C be an infinite conformal algebra that acts faithfully and irreducibly on $\mathbb{C}[\partial]$. Then*

- *if C is associative, it is isomorphic to $\mathrm{Cend}_{n,P}$;*

- *if C is Lie, it is isomorphic to either $\mathrm{gc}_{n,P}$, $\mathrm{oc}_{n,P}$, or $\mathrm{spc}_{n,P}$.*

It is well known that any study of infinite-dimensional Lie algebras of linear growth is in general hopeless; fortunately, there is a natural subclass of affine algebras that gives us a controllable rich theory. Since these algebras are finite from the conformal point of view, the subalgebras of gc_n (or, rather, their coefficients) form a good class of algebras of quadratic growth.

More generally, one should study finite subalgebras of gc_n as well: this may lead, for instance, to the classification of Virasoro elements (see Remark 3.3) and other important results.

In view of Theorem 3.7 and its generalization in [36], we also propose a generalization of Conjecture 3.14:

Conjecture 4.2. *Let C be a simple associative algebra of Gelfand-Kirillov dimension 1. Then C is isomorphic to $\mathrm{Cend}_{n,P}$ for some n and P.*

4.2. A closely related issue is the study of representations of $\mathrm{Cend}_{n,P}$, $\mathrm{gc}_{n,P}$, $\mathrm{oc}_{n,P}$ and $\mathrm{spc}_{n,P}$.

We have already mentioned in Remarks 3.11 and 3.16 that the best strategy here might be to define a conformal category with the space of morphisms carrying a conformal structure and then describe this category in terms of some equivalent data. Though successful for unital conformal algebras (see Theorem 3.10), this task is very hard in general. This is not the only obstacle for such a project. For instance, as mentioned above, there is no bijection between the set of maximal left ideals and irreducible modules—i.e., even if one could define a "conformal kernel," it would yield less information than its ordinary analog.

In the Lie case, these questions are closely related to the study of representations of $\widehat{\mathrm{gl}}_\infty$ and its subalgebras. This is a very important subject for infinite-dimensional representation theory and hopefully, the use of conformal language can move it further.

4.3. Representations of Lie algebras are a source for a lot of combinatorics. What about their conformal counterparts?

The study of subalgebras of gc_n that are normalized by the sl_2 part of a Virasoro element (see Remark 3.22) produced a surprising connection with classical Jacobi polynomials [12]. This seems to be the only deep combinatorial result in the conformal algebra field but we can hope for more: for instance, character formulas for gc_n (and, for that matter a good definition of "conformal" characters) and its subalgebras should turn out very interesting from the combinatorial point of view.

4.4. Finally, we should mention a generalization of conformal algebras.

It is clear that instead of $\mathbb{C}[\partial]$-modules in Definition 1.1 we can consider modules over $\mathbb{C}[\partial_1, \ldots, \partial_k]$ (and take n in $\textcircled{n}$ to be a multiindex). A more involved procedure allow us to endow modules over any cocommutative Hopf algebra with a conformal-like structure. Such objects are called *pseudoalgebras*.

So far we have the classification of finite pseudoalgebras, the beginnings of representation theory (both in [1]), and the theory of unital pseudoalgebras [29]. It should be mentioned that simple finite Lie pseudoalgebras arise from either affine algebras or algebras of Cartan type (of any GK dimension), so in a sense this theory unifies the Kac-Moody and the Cartan type sides of infinite-dimensional Lie algebras, cf. Remark 1.12. As far as infinite pseudoalgebras are concerned, some results in [29] (for the associative case only) suggest that their theory is rich and manageable. These objects should be given more attention.

References

1. B. Bakalov, A. D'Andrea, V. G. Kac, *Theory of finite pseudoalgebras*, Adv. Math. **162** (2001), 1–140.
2. B. Bakalov, V. G. Kac, *Field algebras*, Int. Math. Res. Not. (2003), 123–159.
3. B. Bakalov, V. G. Kac, A. Voronov, *Cohomology of conformal algebras*, Comm. Math. Phys. **200** (1999), 561–598.
4. A. Beilinson, V. Drinfeld, *Chiral algebras*, preprint, 1995.
5. L. A. Bokut, Y. Fong, W.-F. Ke, *Gröbner-Shirshov bases and composition lemma for associative conformal algebras: an example*, in "Combinatorial and computational algebra (Hong Kong, 1999)", Contemp. Math. **264**, AMS, Providence, RI, 2000, 63–90.
6. R. Borcherds, *Vertex algebras, Kac-Moody algebras, and the Monster*, Proc. Nat. Acad. Sci. USA **83** (1986), 3068–3071.
7. C. Boyallian, V. G. Kac, J. Liberati, *Finite growth representations of infinite Lie conformal algebras*, J. Math. Phys. **44** (2003), 754–770.
8. C. Boyallian, V. G. Kac, J. Liberati, *On the classification of subalgebras of Cend_N and gc_N*, J. Algebra **260** (2003), 32–63.
9. S.-J. Cheng, V. G. Kac, *Conformal modules*, Asian J. Math. **1** (1997), 181–193; *Erratum*, Asian J. Math. **2** (1998), 153–156.
10. S.-J. Cheng, V. G. Kac, M. Wakimoto, *Extensions of conformal modules*, in "Topological field theory, primitive forms, and related topics (Kyoto, 1996)", Progr. Math. **160**, Birkhäuser, Boston, MA, 1998, 79–129.
11. A. D'Andrea, V. G. Kac, *Structure theory of finite conformal algebras*, Selecta Math. (N. S.) **4** (1998), 377–418.
12. A. De Sole, V. G. Kac, *Subalgebras of gc_N and Jacobi polynomials*, Canad. Math. Bull. **45** (2002), 567–605.

13. D. B. Fuchs, "Cohomology of infinite-dimensional Lie algebras", Contemporary Soviet Mathematics. Consultants Bureau, New York, 1986.

14. B. L. Feigin, D. B. Fuchs, *Homology of the Lie algebra of vector fields on the line*, Funkc. Anal. i Pril. **14**(3) (1980), 45–60 (Russian).

15. I. B. Frenkel, J. Lepowsky, A. Meurman, "Vertex operator algebras and the Monster", Academic Press, Boston, MA, 1988.

16. , V. Futorny, private communication.

17. I. M. Gelfand, I. Ya. Dorfman, *Hamiltonian operators and algebraic structures related to them*, Funct. Anal. and Appl. **13** (1979), 248–262.

18. I. M. Gelfand, D. B. Fuchs, *Cohomologies of the Lie algebra of vector fields on the circle*, Funkc. Anal. i Pril. **2**(4) (1968), 92–93 (Russian).

19. V. G. Kac, "Vertex algebras for beginners", University Lecture Series, **10**. AMS, Providence, RI, 1996. Second edition 1998.

20. V. G. Kac, *Formal distribution algebras and conformal algebras*, in "Proc. XIIth International Congress of Mathematical Physics (ICMP '97) (Brisbane)", Internat. Press, Cambridge, MA, 1999, 80–97.

21. V. G. Kac, A. Radul, *Quasifinite highest weight modules over the Lie algebra of differential operators on the circle*, Comm. Math. Phys. **157** (1993), 429–457.

22. V. G. Kac, A. Radul, M. Wakimoto, *Finite modules over* gc_n, unpublished.

23. M. Kapranov, E. Vasserot, *Vertex algebras and the formal loop space*, preprint, 2001, `math.AG/0107143`.

24. P. Kolesnikov, *Simple Jordan pseudoalgebras*, preprint, 2002, `math.QA/0210264`.

25. G. R. Krause, T. H. Lenagan, "Growth of Algebras and Gelfand-Kirillov Dimension", second edition. AMS, Providence, RI, 2000.

26. B. H. Lian, G. J. Zuckerman, *Commutative quantum operator algebras*, J. Pure Appl. Algebra **100** (1995), 117–139.

27. M. Primc, *Vertex algebras generated by Lie algebras*, J. Pure Appl. Algebra **135** (1999), 253–293.

28. A. Retakh, *Associative conformal algebras of linear growth*, J. Algebra **237** (2001), 769–788.

29. A. Retakh, *Unital associative pseudoalgebras and their representations*, preprint, 2001, `math.QA/0109110`, to appear in J. Algebra.

30. A. Retakh, *Semisimple associative conformal algebras of linear growth*, preprint, 2002, `math.RA/0212168`.

31. M. Roitman, *On free conformal and vertex algebras*, J. Algebra **255** (2002), 297–323.

32. M. Roitman, *Universal enveloping conformal algebras*, Selecta Math. (N. S.) **6**(3) (2000), 319–345.

33. L. Small, J. T. Stafford, R. Warfield, *Affine algebras of Gelfand-Kirillov dimension one are PI*, Math. Proc. Cambridge Phil. Soc. **97** (1985), 407–414.

34. X. Xu, *Equivalence of conformal superalgebras to Hamiltonian superoperators*, Algebra Colloq. **8** (2001), 63–92.

35. E. Zelmanov, *On the structure of conformal algebras*, in "Combinatorial and

computational algebra (Hong Kong, 1999)", Contemp. Math. **264**, AMS, Providence, RI, 2000, 139–153.

36. E. Zelmanov, *Idempotents in conformal algebras*, in "Proceedings of the Third International Algebra Conference (Tainan, 2002)", Kluwer, Dordrecht, Netherlands, 2003, 257–266.

ACTIONS OF TORI AND FINITE FANS

SONIA L. RUEDA

Departamento de Matemáticas. E.T.S. Arquitectura
Universidad Politécnica de Madrid
Madrid, Spain
E-mail: srueda@aq.upm.es

Let k be an algebraically closed field of characteristic 0, $X = k^r \times (k^\times)^s$ and let G be an algebraic torus acting diagonally on X. We construct a fan Δ such that the quotient $Y/\!/G$ is isomorphic to the toric variety determined by Δ and $\mathcal{D}(X) = \mathcal{D}(Y)$, for a distinguished G-invariant open subset Y of k^n. The main goal of this construction is to give necessary and sufficient conditions on Δ for $\mathcal{D}(X)^G$ to have enough simple finite dimensional modules.

Let G be a reductive group acting on the smooth affine variety $X = k^r \times (k^\times)^s$, with k an algebraically closed field k of characteristic 0. We denote the ring of regular functions on X by $\mathcal{O}(X)$ and the ring of differential operators by $\mathcal{D}(X)$. Let $\mathcal{D}(X)^G$ be the subring of $\mathcal{D}(X)$ of invariants under the action of G. There are several papers where actions of tori and finite fans are related, [4], Chapter VI, [5], [11] and more recently [2], [3]. We would like to associate a finite fan of cones to the action of G on X, in such a way that the study of the fan will allow us to get conclusions about the finite dimensional $\mathcal{D}(X)^G$-modules.

Suppose that $n = r + s$. Given a fan (N, Δ), we consider the following open subset of k^n, $Y = \{x \in k^n \mid x_i \neq 0 \text{ for } i \notin \{1, \dots, r\}\}$, where r is the number of one dimensional cones of Δ. We say that a finite fan Δ is *associated to the action* of G on X, if the following conditions hold,

(1) (A1) the quotient variety $Y/\!/G$ is isomorphic to the toric variety determined by the fan Δ, $X(\Delta)$,
(2) (A2) $\operatorname{codim} X \setminus Y \geq 2$.

In this paper, G will be a finite dimensional algebraic torus acting diagonally on X.

Proposition A. There exists a fan Δ associated to the action of the algebraic torus G on X.

The proof of Proposition A is constructive, we give a method to obtain a fan associated to the action of G on X. This result was motivated by the work of I. M. Musson in [11]. Given a finite fan Δ he gives an action of G on Y such that the variety of closed orbits $Y//G$ is isomorphic to the toric variety determined by Δ. A similar result was proved by D. A. Cox in [5]. We consider Proposition A a converse of this results, since our point of departure is the action of G on X. The variety Y is relevant to us because it serves as a bridge between the action of G on X and the fan Δ, condition (A1) explains this connection. In fact, we define Y as in [11]. Furthermore, Y is a toric subvariety of X, which is a toric variety for a dense torus T and G is a subtorus of T. The variety Y admits a good quotient by the action of G. The existence of good quotients of a toric variety by a subtorus action was studied in a recent paper by A. A'Campo-Neuen and J. Hausen, [2]. Also, the open subsets of a normal variety which admit a good quotient by a torus action have been described in [9].

We call $V = k^r, W = (k^\times)^s$ then $X = V \times W \subseteq k^n$, where $n = r + s$. Let us suppose that G acts transitively on X. Let H be the stabilizer of w in W. The connected component H^o of the identity in H is a torus but we may have $H \neq H^o$ and then H/H^o is a finite group. The following result reflects the connection existing between the action of G on X and the action of H on V.

Proposition B. Δ is a fan associated to the action of G on X if and only if Δ is a fan associated to the action of H on V.

The main goal of this paper is to give necessary and sufficient conditions on Δ for $\mathcal{D}(X)^G$ to have a nonzero finite dimensional module. In a recent work with Musson [13], we show that if $\mathcal{D}(X)^G$ has a nonzero finite dimensional module then $\mathcal{D}(X)^G$ has enough simple finite dimensional modules. We say that a k-algebra R has enough simple finite dimensional modules if $\bigcap \operatorname{ann}_R M = 0$, where the intersection is taken over all simple finite dimensional R-modules, [13].

Condition (A2) implies that $\mathcal{D}(X) = \mathcal{D}(Y)$, so we can transfer our attention to the study of $\mathcal{D}(Y)^G$-modules. We say that a finite fan is contained in a half-space if the intersection of its dual cones is not zero.

Proposition C. The $\mathcal{D}(Y)^G$-module $\mathcal{O}(Y)^G$ is finite dimensional if and only if the fan Δ is not contained in a half-space.

This will allow us to prove the following theorem.

Theorem D. The following conditions are equivalent.

(1) $\mathcal{D}(X)^G$ has a nonzero finite dimensional module.
(2) There exists a fan Δ not contained in a half-space and associated to the action of G on X.

When $V^{H^\circ} = 0$, we can modify a fan associated to the action of G on X to get a fan which is not contained in a half-space and it is associated to an action that is different from the original one but gives the same invariant differential operators. This fact allowed us to realize that $V^{H^\circ} = 0$ is a necessary and sufficient condition for $\mathcal{D}(X)^G$ to have a nonzero finite dimensional module, as proved in [13] without the use of fans.

The paper is organized as follows. In § 1, we introduce some notation about actions of tori, finite fans and rings of differential operators. Section 2 contains a method to construct fans that will be proved to be associated to the action of G on X. We prove Proposition B in § 3. In § 4, we prove Proposition C and Theorem D. The last section, contains a description of the members of the family of finite dimensional simple $\mathcal{D}(X)^G$-modules $\{\mathcal{O}(Y)_\chi\}_{\chi \in \mathbb{Z}^m}$, in terms of the fan. This family was proved to have enough members in [13]. We show that the dimension of $\mathcal{O}(Y)_\chi$ is the number of lattice points inside a certain polytope (i.e. a bounded polyhedron). This computation can be done with LattE.

1. Notation

1.1. *Actions of tori*

Set $\mathbb{X}(G) = \mathrm{Hom}(G, k^\times)$, $\mathbb{Y}(G) = \mathrm{Hom}(k^\times, G)$, the groups of characters and one-parameter subgroups of G, respectively.

A *diagonal action* of a torus G on X is an action that extends to a diagonal action on k^n. Such an action is given by an embedding of G into the group T of diagonal matrices in $GL(n)$. Details about this action are given in [13], § 1.1 and the following concepts are described. There exist $\eta_1, \ldots, \eta_n \in \mathbb{X}(G)$ such that *G acts on X with weights* $\eta_1, \ldots, \eta_n$. Identify G with $(k^\times)^m$ and $\mathbb{X}(G)$ with $\mathbb{Z}^m$. We think of $\mathbb{X}(G)$ as a space of column vectors with integer entries. Let L be the $m \times n$ matrix whose i-th column vector is η_i, $i = 1, \ldots, n$. We say that *G acts on X by the matrix L*.

Let $\psi \colon \mathbb{X}(T) \to \mathbb{X}(G)$ be the restriction map. This map is given by multiplication by L. There is a natural bilinear pairing

$$(\_, \_) \colon \mathbb{X}(T) \times \mathbb{Y}(T) \to \mathbb{Z}. \tag{1}$$

defined by the requirement that

$$(a \circ b)(\lambda) = \lambda^{(a,b)} \tag{2}$$

for all $a \in \mathbb{X}(T)$, $b \in \mathbb{Y}(T)$ and $\lambda \in k^{\times}$.

We will assume that G acts faithfully on X. Therefore L has rank m. Let $l = n - m$.

Lemma 1.1. *Assume that $\{\eta_{r+1}, \ldots, \eta_n\}$ are linearly independent. There exist matrices $\Gamma \in GL_m(\mathbb{Z})$, $\Delta \in GL_n(\mathbb{Z})$ such that*

$$\Gamma L \Delta = \begin{bmatrix} b_{11} & \ldots & b_{1l} & d & 0 & \ldots & 0 \\ b_{21} & \ldots & b_{2l} & 0 & d & \ldots & 0 \\ \vdots & \ddots & \vdots & & & \ddots & \\ b_{m1} & \ldots & b_{ml} & 0 & 0 & \ldots & d \end{bmatrix}. \tag{3}$$

where d is a nonzero integer.

Proof. Let $m' = m - s$. Since $\{\eta_1, \ldots, \eta_n\}$ contains m linearly independent vectors, there exist $\eta_{i_1}, \ldots, \eta_{i_{m'}} \in \{\eta_1, \ldots, \eta_n\}$ such that $\eta_{i_1}, \ldots, \eta_{i_{m'}}, \eta_{r+1}, \ldots, \eta_n$ are linearly independent. There exists $\Delta \in GL_n(\mathbb{Z})$ such that the last m' columns of $L\Delta$ are $\eta_{i_1}, \ldots, \eta_{i_{m'}}$. Let Γ' be the $m \times m$ matrix whose i-th column vector is the $(l + i)$-th column of $L\Delta$, $i = 1, \ldots, m$. Then $d := |\det \Gamma'| \neq 0$. Let $\Gamma = d\Gamma'^{-1}$, then the $m \times n$ matrix with integer coefficients $\Gamma L \Delta$ will look like (3). $\square$

If $\{\eta_{r+1}, \ldots, \eta_n\}$ are linearly independent, by Lemma 1.1 and [13], equations (15) and (16), we assume that the matrix L has the special form (3).

1.2. *Finite fans*

As far as possible we follow the notation of [8], Chapter 1. Let $N \simeq \mathbb{Z}^l$ be the l-dimensional lattice. Let (N, Δ) be a fan in N. Recall that each $\sigma \in \Delta$ is a strongly convex rational polyhedral cone in $N_{\mathbb{R}} = N \otimes_{\mathbb{Z}} \mathbb{R}$. Let $M = \mathrm{Hom}_{\mathbb{Z}}(N, \mathbb{Z})$ and $\langle \_, \_ \rangle \colon M \times N \to \mathbb{Z}$ the natural bilinear pairing. For each $\sigma \in \Delta$, let

$$\Lambda_\sigma = M \cap \sigma^{\vee} = \{u \in M \mid \langle u, v \rangle \geq 0 \text{ for all } v \in \sigma\} \tag{4}$$

and $U_\sigma = \mathrm{Spec}\, k[\Lambda_\sigma]$ is a semigroup algebra. By [8], Theorem 1.4 we can glue U_σ to obtain a toric variety $X(\Delta)$.

Denote by $\Delta(1)$ the set of cones of (N, Δ) with dimension one. Given $v \in N$ let $\tau_v = \mathbb{R}_+ v$ be the ray generated by $v \in N$. Let $v, v' \in N$, if $v = cv'$ with $c > 0$ then $\tau_v = \tau_{v'}$. Suppose that $\Delta(1) = \{v_1, \ldots, v_r\}$. Given $\sigma \in \Delta$ we define $[\sigma] = \{i \in \{1, \ldots, r\} \mid \tau_{v_i} \text{ is a face of } \sigma\}$. Then $\sigma = \sum_{i \in [\sigma]} \tau_{v_i}$.

If $u \in M_{\mathbb{R}} = M \otimes_{\mathbb{Z}} \mathbb{R}$, a subset of the form

$$H_u = \{v \in N_{\mathbb{R}} \mid \langle u, v \rangle \geq 0\} \tag{5}$$

with $u \neq 0$ is called a *half-space* in $N_{\mathbb{R}}$, see [12], § 1. We will say that the fan (N, Δ) is *contained in a half-space* if we can find $0 \neq u \in M_{\mathbb{R}}$ such that $\sigma \subseteq H_u$ for all $\sigma \in \Delta$. Equivalently, if the intersection of its dual cones is not zero.

1.3. *Coordinate rings and rings of differential operators*

In this section, we gather some definitions and results from [13], § 2. Note that X is a toric variety with a dense torus $T = (k^\times)^n \subseteq X$. Write Q_i for the character e_i considered as a regular function on T. Then

$$\mathcal{O}(X) = k[Q_1, \ldots, Q_r, Q_{r+1}^{\pm 1}, \ldots, Q_n^{\pm 1}]. \tag{6}$$

We consider the action of G on $\mathcal{O}(T)$ (or $\mathcal{O}(X)$) given by right translation. This convention implies that Q_i has weight η_i. Let $P_i = \partial/\partial Q_i$,

$$\mathcal{D}(X) = k[Q_1, \ldots, Q_r, Q_{r+1}^{\pm 1}, \ldots, Q_n^{\pm 1}, P_1, \ldots, P_n]. \tag{7}$$

If $\lambda = (\lambda_1, \ldots, \lambda_n) \in \mathbb{N}^r \times \mathbb{Z}^s, \mu = (\mu_1, \ldots, \mu_n) \in \mathbb{N}^n$, set $Q^\lambda = Q_1^{\lambda_1} \ldots Q_n^{\lambda_n}$, and $P^\mu = P_1^{\mu_1} \ldots P_n^{\mu_n}$. The elements $Q^\lambda P^\mu \in \mathcal{D}(X)$, with $L\lambda = L\mu$, form a basis of $\mathcal{D}(X)^G$.

Let Y be an open subset of X. Define the codimension of $X \setminus Y$ in k^n, to be

$$\operatorname{codim} X \setminus Y = \dim k^n - \dim X \setminus Y.$$

Proposition 1.2. *If* $\operatorname{codim} X \setminus Y \geq 2$, *then* $\mathcal{O}(X) = \mathcal{O}(Y)$ *and* $\mathcal{D}(X) = \mathcal{D}(Y)$.

Proof. The result follows from [10], Proposition II.2.2. $\square$

2. Fans associated to the action of G

Let us describe Y in detail.

2.1. *The set Y*

Let (N, Δ) be a finite fan. For every $\sigma \in \Delta$ we define $x^{\hat{\sigma}} = \prod_{i \notin [\sigma]} x_i$ and we consider the T-invariant open sets

$$V_\sigma = k^n - Z(x^{\hat{\sigma}}) \tag{8}$$

where $Z(x^{\hat{\sigma}}) = \{x \in k^s \mid x^{\hat{\sigma}} = 0\}$. Let

$$Z = \bigcap_{\sigma \in \Delta} Z(x^{\hat{\sigma}}). \tag{9}$$

Hence Z is closed and T-invariant. We have an open subset

$$Y = k^n - Z = \bigcup_{\sigma \in \Delta} V_\sigma \tag{10}$$

of an affine space k^n. Note that Y might no longer be affine. These sets were introduced in [11], § 1.3. See also [5], Theorem 2.1.

We determine the irreducible components of Z. For $I \subseteq \{1, \ldots, n\}$ set $Z_I = \{x \in k^n \mid x_i = 0 \text{ if } i \in I\}$.

Lemma 2.1. *Any T-invariant irreducible closed set in k^n is some Z_I.*

Proof. See [8], § 3.1. $\qquad\qquad\square$

By Lemma 2.1, Z is a union of irreducible closed subsets Z_I. Observe that when $I \subseteq J$ then $Z_J \subseteq Z_I$ for $I, J \subseteq \{1, \ldots, n\}$. Therefore, the irreducible components that occur in Z are the ones in the family $\mathcal{I}$ of subsets of $\{1, \ldots, n\}$ verifying the following statements.

(1) $Z_I \subseteq Z$ and;
(2) I is minimal verifying the previous condition, i.e. there is no $J \subseteq \{1, \ldots, n\}$, $J \subsetneq I$ such that $Z_J \subseteq Z$.

Thus, $Z = \bigcup_{I \in \mathcal{I}} Z_I$.

2.2. *Construction of the fan associated to the action*

We will use the following lemma to develop our construction.

Lemma 2.2. *There exists an $n \times l$ matrix E that satisfies the following statements.*

(1) The rows of E generate N as a group.
(2) The columns of E are a $\mathbb{Z}$-basis of $\ker \psi$.

Proof. By [1], Theorem 12.4.3, there exist matrices $Q \in GL_m(\mathbb{Z})$ and $P \in GL_n(\mathbb{Z})$ such that

$$L' = QLP = \begin{bmatrix} d_1 & 0 & \cdots & 0 & 0 \cdots 0 \\ 0 & d_2 & & 0 & \\ \vdots & & \ddots & \vdots & \vdots & \vdots \\ 0 & \cdots & & d_m & 0 \cdots 0 \end{bmatrix} \tag{11}$$

with $d_i \neq 0$ for all $i = 1, \ldots m$. Let I_l be the identity $l \times l$ matrix and E' the $n \times l$ matrix with I_l in the last l rows and zeroes in the first m rows. Then $L'E' = 0$. We define $E := PE'$. Let us prove that E satisfies statements 1 and 2.

(1) Let $\bar{P}$ be the matrix obtained by deleting the first m rows of P^{-1}. From the definition of E we get easily that $I_l = \bar{P}E$. This proves that the rows of E generate N as a group.

(2) The columns of E are elements of $\ker \psi$ because $LE = 0$. Given any $\lambda \in \ker \psi$ then $L'P^{-1}\lambda = 0$. The columns of E' are a $\mathbb{Z}$-basis of the kernel of L'. Then there exist $z_1, \ldots, z_l \in \mathbb{Z}$ such that

$$P^{-1}\lambda = E' \begin{bmatrix} z_1 \\ \vdots \\ z_l \end{bmatrix}, \text{ therefore } \lambda = E \begin{bmatrix} z_1 \\ \vdots \\ z_l \end{bmatrix}.$$

This proves that the columns of E generate $\ker \psi$ as a group and since $\ker \psi$ has rank l the result follows. $\qquad\square$

Let E be an $n \times l$ matrix satisfying the statements of Lemma 2.2. We can identify $B = \mathbb{Y}(T)$ with $\mathbb{Z}^n$ and think of it as a space of row vectors with integer entries. Define

$$\varphi \colon B \to N \tag{12}$$

by $\varphi(e) = eE$ for all $e \in B$. By Lemma 2.2(1) φ is onto. Let $K = \ker \varphi$. Then K is a free abelian group of rank m.

Let e_i be the i-th standard basis vector for B and call $v_i = \varphi(e_i)$ the i-th row vector of E, $1 \leq i \leq n$. The matrix E has rank l; hence the subset $\{v_1, \ldots, v_n\}$ of $N = \mathbb{Z}^l$ contains l linearly independent vectors. Observe that $\{v_1, \ldots, v_n\}$ could contain elements that are equal and also the zero element.

Let Δ be any fan in N with $\Delta(1) = \{\tau_{v_i} \mid i = 1, \ldots, r\}$. We will prove that such a fan is associated to the action of G on X.

Example 2.3. Let $r = 4$, $s = 0$, $m = 2$; then $l = 2$. Let

$$L = \begin{bmatrix} 3 & 3 & 2 & 0 \\ 4 & 4 & 0 & 2 \end{bmatrix} \qquad E = \begin{bmatrix} -1 & -2 \\ 1 & 0 \\ 0 & 3 \\ 0 & 4 \end{bmatrix},$$

hence $v_1 = (-1, -2)$, $v_2 = (1, 0)$, $v_3 = (0, 3)$ and $v_4 = (0, 4)$. Then Δ could be the fan with maximal cones $\sigma_1, \sigma_2, \sigma_3$, where $[\sigma_1] = \{1\}$, $[\sigma_2] = \{2\}$ and $[\sigma_3] = \{3, 4\}$.

2.3. *Proof of Proposition A*

(1) Let $K^\perp = \{\lambda \in \mathbb{X}(T) \mid (\lambda, K) = 0\}$. Then $K^\perp = \ker \psi$. There is an isomorphism $w \colon M \to K^\perp$ given by

$$\langle x, \varphi(b) \rangle = (w(x), b) \tag{13}$$

for all $x \in M$, $b \in B$. By equation (13), it can be proved in the same way as [11], Theorem 1 that the variety of closed orbits $Y//G$ is isomorphic to $X(\Delta)$.

(2) Consider the family $\mathcal{I}' = \{I \in \mathcal{I} \mid |I| = 1\}$ and define

$$\hat{Z} := \bigcup_{I \in \mathcal{I}'} Z_I. \tag{14}$$

Since $\bigcup_{\sigma \in \Delta} [\sigma] = \{1, \ldots, r\}$, then $X = k^n - \hat{Z}$ and $X \setminus Y = Z \setminus \hat{Z}$. By (14), $X \setminus Y = \bigcup_{I \in \mathcal{I}''} Z_I$ with $\mathcal{I}'' = \mathcal{I} \setminus \mathcal{I}' = \{I \in \mathcal{I} \mid |I| \geq 2\}$. We also have $\operatorname{codim} \bigcup_{I \in \mathcal{I}''} Z_I = \inf_{I \in \mathcal{I}''} \operatorname{codim} Z_I$ and $\operatorname{codim} Z_I = |I| \geq 2$ for all $I \in \mathcal{I}''$. Therefore $\operatorname{codim} X \setminus Y \geq 2$.

Remark 2.4. There is a canonical morphism $p \colon Y \to X(\Delta)$ such that $X(\Delta)$ is isomorphic to the geometric quotient $Y//G$. We have a covering U_σ of $X(\Delta)$ with U_σ isomorphic to $V_\sigma//G$, for each $\sigma \in \Delta$. Also, $p_{|V_\sigma} \colon V_\sigma \to U_\sigma$ is the categorical quotient of G restricted to V_σ. Therefore, the morphism p is a good quotient as defined in [2], § 3.

3. Fans associated to the action of H on V

Given a finite fan Δ, for each $\sigma \in \Delta$, define $V_\sigma' = \{x \in k^r \mid x_i \neq 0$ if $i \notin [\sigma]\}$. Then $V_\sigma = V_\sigma' \times W$, recall that $n = r + s$.

Suppose G is a torus acting faithfully on V_σ with weights $\eta_1, \ldots, \eta_n$. We assume that G acts transitively on W, then by [13], Lemma 3.1, $\eta_{r+1}, \ldots, \eta_n$ are linearly independent. Let $w = (w_{r+1}, \ldots, w_n)$ be an element of W.

Then $H = G_w = \bigcap_{i=r+1}^{n} \ker \eta_i$. It can be proved in the same way as [13], Lemma 3.2, that the slice representation at w, [7], [15], is isomorphic to (H, V_σ).

Consider the H-invariant open subset of V, $Y' = \bigcup_\sigma V'_\sigma$. This is the variety defined in (10) for the case $n = r$.

Theorem 3.1. *The varieties $Y//G$ and $Y'//H$ are isomorphic.*

Proof. Given $\sigma \in \Delta$. Part of the Luna slice theorem states that there is a closed H-stable subvariety S_σ containing w and a G-equivariant étale map $G \times^H S_\sigma \to V_\sigma$. Taking $S_\sigma = V'_\sigma + w$ we get a G-equivariant isomorphism $\delta_\sigma : G \times^H S_\sigma \to V_\sigma$ and this map induces an isomorphism between $V_\sigma//G$ and $V'_\sigma//H$, this can be proved as [13], Theorem 6.2.

If τ is a face of σ, then $V_\tau \subseteq V_\sigma$, $V'_\tau \subseteq V'_\sigma$ and the isomorphism $V_\sigma//G \cong V'_\sigma//H$ restricts to the isomorphism $V_\tau//G \cong V'_\tau//H$. Thus, we may identify $Y//G = \bigcup_\sigma V_\sigma//G$ with $Y'//H = \bigcup_\sigma V'_\sigma//H$. $\square$

3.1. *Proof of Proposition B*

By Theorem 3.1, $Y//G$ is isomorphic to $X(\Delta)$ if and only if $Y'//H$ is. Let as prove that $\operatorname{codim} X \setminus Y \geq 2$ if and only if $\operatorname{codim} V \setminus Y' \geq 2$.

We have $Y = \{x \in k^n \mid x_i \neq 0 \text{ for } i \notin \bigcup[\sigma]\}$ and $Y' = \{x \in k^r \mid x_i \neq 0 \text{ for } i \notin \bigcup[\sigma]\}$. If $\operatorname{codim} X \setminus Y \geq 2$ then $\mathcal{O}(X) = \mathcal{O}(Y)$, therefore $\bigcup[\sigma] = \{1, \ldots, r\}$. By the proof of Proposition A (2) for the case $n = r$ then $\operatorname{codim} V \setminus Y' \geq 2$. Conversely if $\operatorname{codim} V \setminus Y' \geq 2$, then $\mathcal{O}(V) = \mathcal{O}(Y')$ so $\bigcup[\sigma] = \{1, \ldots, r\}$ and the by proof of Proposition A (2) the result follows.

3.2. $\mathcal{D}(X(\Delta))$-*modules*

Set $\mathfrak{h} = Lie(H) \subseteq \mathfrak{g} = Lie(G)$. For $\lambda \in \mathfrak{g}^*$, $\mu \in \mathfrak{h}^*$ we set

$$\mathcal{B}_\lambda(X) = \mathcal{D}(X)^G/(\mathfrak{g} - \lambda(\mathfrak{g})), \qquad \mathcal{B}_\mu(V) = \mathcal{D}(V)^H/(\mathfrak{h} - \mu(\mathfrak{h})). \tag{15}$$

Here $(\mathfrak{g} - \lambda(\mathfrak{g}))$ is the ideal generated by all elements of the form $x - \lambda(x)$, with $x \in \mathfrak{g}$, and $(\mathfrak{h} - \mu(\mathfrak{h}))$ is defined similarly. Let $i^* : \mathfrak{g}^* \to \mathfrak{h}^*$ be the map obtained from the inclusion $i : \mathfrak{h} \to \mathfrak{g}$.

By [13], Proposition C, there is an injective algebra homomorphism $\mathcal{D}(V)^H \to \mathcal{D}(X)^G$. If $\lambda \in \mathfrak{g}^*$ and $\mu = i^*(\lambda)$, the previous map induces an isomorphism $\mathcal{B}_\mu(V) \cong \mathcal{B}_\lambda(X)$ and by [11], Theorem 5 they are isomorphic to $\mathcal{D}(X(\Delta))$. Note that any simple $\mathcal{D}(X)^G$-module is a $\mathcal{B}_\lambda(X)$-module for some $\lambda \in \mathfrak{g}^*$. So we can reduce the study of finite dimensional simple

$\mathcal{D}(X)^G$-modules to that of finite dimensional simple $\mathcal{D}(V)^H$-modules and also to the study of $\mathcal{D}(X(\Delta))$-modules.

In [14] it is shown that the category of $\mathcal{D}(X(\Delta))$-modules is equivalent to a category of graded $\mathcal{D}(V)$-modules modulo $\mathfrak{b}$-torsion, with $\mathfrak{b} = Z$ defined by equation (9) for $s = 0$.

4. Fans not contained in a half-space

In this section we include some lemmas that will be used to prove Proposition C and Theorem D.

Suppose $I \subseteq \{1, \ldots, r\}$. For $1 \leq i \leq n$, set

$$\varsigma_i = \begin{cases} -\eta_i & \text{if } i \in I, \\ \eta_i & \text{if } i \notin I. \end{cases} \tag{16}$$

Let L_I be the matrix with columns $\varsigma_1, \ldots, \varsigma_n$. Then G_I denotes the m-dimensional torus acting on X by the matrix L_I. By [13], Lemma 5.2, the map $\sigma_I \colon \mathcal{D}(X) \to \mathcal{D}(X)$ defined by

$$\sigma_I(Q_i) = \begin{cases} -P_i & \text{if } i \in I \\ Q_i & \text{if } i \notin I \end{cases} \qquad \sigma_I(P_i) = \begin{cases} Q_i & \text{if } i \in I \\ P_i & \text{if } i \notin I \end{cases} \tag{17}$$

$i = 1, \ldots, n$ is an isomorphism between $\mathcal{D}(X)^G$ and $\mathcal{D}(X)^{G_I}$. Therefore G_I and G have the same invariant differential operators.

Lemma 4.1. *When the matrix L is of the special kind* (3), *then* $v_1, \ldots, v_l$ *are linearly independent.*

Proof. By Lemma 2.2, $LE = 0$ and the rows $v_1, \ldots, v_n$ of E generate N as a group. The equation $LE = 0$ means that for $i = 1, \ldots, m$

$$dv_{l+i} = -\sum_{j=1}^{l} b_{ij} v_j. \tag{18}$$

Thus $v_{l+1}, \ldots, v_n$ belong to the $\mathbb{R}$-span of $v_1, \ldots, v_l$. The result follows from this. $\qquad\square$

Let us suppose that L is of the special kind (3) and let Δ be a fan as in § 2.2. By Lemma 4.1, $\mathcal{B} = \{v_1, \ldots, v_l\}$ is a basis of $N_{\mathbb{R}}$. With respect to $\mathcal{B}$ the vectors $v_{l+1}, \ldots, v_n$ have coordinates

$$v_j = \left(-\frac{1}{d} b_{j-l,1}, \ldots, -\frac{1}{d} b_{j-l,l} \right), \qquad j = l+1, \ldots, n. \tag{19}$$

Let $m' = r - l$. For $i = 1, \ldots, l$, let ρ_i be the vector in $\mathbb{Z}^{m'}$ obtained deleting the last $m - m'$ entries of η_i.

Lemma 4.2. *If $\rho_i = 0$ for some $i \in \{1, \ldots, l\}$, then Δ is contained in a half-space.*

Proof. Consider the basis $\mathcal{B}$ in N. Let $u \in M_{\mathbb{R}}$ such that $\langle u, v_j \rangle = 0$ if $j \neq i$, $j \in \{1, \ldots, l\}$ and $\langle u, v_i \rangle = 1$. Then $\langle u, v_j \rangle = 0$, for all $j = l + 1, \ldots, n$. Therefore Δ is contained in the half-space H_u. $\qquad\square$

Lemma 4.3. *If $\mathcal{O}(Y)^G = k$, then $\eta_{r+1}, \ldots, \eta_n$ are linearly independent.*

Proof. It follows from Proposition 1.2 and [13], Lemma 4.1. $\qquad\square$

4.1. *Proof of Proposition C*

Let

$$\phi_\sigma := \{\lambda \in K^{\perp} \mid (\lambda, e_i) \geq 0 \text{ for all } i \in [\sigma]\}. \tag{20}$$

Then $\mathcal{O}(V_\sigma)^G = k[\phi_\sigma]$. Hence $\mathcal{O}(Y)^G = k$ if and only if $\bigcap_{\sigma \in \Delta} \phi_\sigma = 0$. Furthermore, $w(\Lambda_\sigma) = \phi_\sigma$. Hence $0 \neq u \in \bigcap_{\sigma \in \Delta} \sigma^{\vee}$ if and only if Δ is contained in the half-space H_u. This proves the result.

Remark 4.4. Let us call G' the m-dimensional torus acting on X by a matrix L'. Let Δ' be a fan associated to the action of G'. Suppose that $\mathcal{O}(X)^G = \mathcal{O}(X)^{G'}$. By Proposition C, Δ is contained in a half-space if and only if Δ' is.

4.2. *Proof of Theorem D*

$(1) \Rightarrow (2)$ By [13], Theorem B and Lemma 5.1, there is a subset I of $\{1, \ldots, r\}$ such that $\mathcal{O}(X)^{G_I} = k$. By Proposition A, there exists a fan Δ associated to the action of G_I on X. By Proposition C, Δ is not contained in a half-space.

$\quad(2) \Rightarrow (1)$ By Proposition C, $\mathcal{O}(Y)^G = k$. By Lemma 4.3, Remark 4.4, and Lemma 4.2, $\rho_i \neq 0$ for all $i = 1, \ldots, r$. By [13], Lemma 3.3 and Theorem B the result follows.

4.3. *Construction of an associated fan not included in a half-space*

By [13], Theorem B, if $V^{H^o} = 0$ then $\mathcal{D}(X)^G$ has a nonzero finite dimensional module and by Theorem D there exists a fan Δ associated to the

324

action of G on X and not contained in a half-space. By [13], Lemma 3.3., $V^{H^o} = 0$ if and only if $\rho_i \neq 0$ for all $i = 1, \ldots r$.

Suppose that $\rho_i \neq 0$ for all $i = 1, \ldots, l$, then L is of the special kind (3). We give a construction of a fan associated to the action of G and not contained in a half-space.

Let $v_1^*, \ldots, v_l^*$ be the dual basis of $\mathcal{B}$. Given $j \in \{l+1, \ldots, r\}$, let

$$I_j^0 = \{i \in \{1, \ldots, l\} \mid \langle v_i^*, v_j \rangle = 0\}, \tag{21}$$

$$I_j^+ = \{i \in \{1, \ldots, l\} \mid \langle v_i^*, v_j \rangle > 0\}, \tag{22}$$

$$I_j^- = \{i \in \{1, \ldots, l\} \mid \langle v_i^*, v_j \rangle < 0\}, \tag{23}$$

and

$$I_j = I_j^+ \cup I_j^-. \tag{24}$$

Then there exists $J \subseteq \{l+1, \ldots, r\}$ such that

$$\bigcup_{j \in J} I_j = \{1, \ldots, l\} \tag{25}$$

because $\rho_i \neq 0$,

$$\rho_i = \begin{bmatrix} b_{1i} \\ \vdots \\ b_{m'i} \end{bmatrix} \text{ and } \frac{-1}{d} b_{j-l,i} = \langle v_i^*, v_j \rangle, i = 1, \ldots, l, j = l+1, \ldots, r.$$

Take J to be minimal verifying (25), and let $J = \{j_1, \ldots, j_c\}$ with $c \leq m'$ and

$$|I_{j_h}| \leq |I_{j_{h+1}}| \qquad h = 1, \ldots, c-1. \tag{26}$$

These two assumptions will make the next computation shorter. We take a subset I of $\{1, \ldots, l\}$ in the following way:

$$I := I_{j_1}^+ \bigcup_{h=2}^{c} [(\bigcap_{t=1}^{h-1} I_{j_t}^0) \cap I_{j_h}^+] = \tag{27}$$

$$= I_{j_1}^+ \cup (I_{j_1}^0 \cap I_{j_2}^+) \cup (I_{j_1}^0 \cap I_{j_2}^0 \cup I_{j_3}^+) \cup \ldots \cup (I_{j_1}^0 \cap \ldots \cap I_{j_{c-1}}^0 \cap I_{j_c}^+). \tag{28}$$

Define

$$v_i^I = \begin{cases} -v_i & \text{if } i \in I, \\ v_i & \text{if } i \notin I, \end{cases} \qquad i = 1, \ldots, r. \tag{29}$$

Let Δ_I be a fan in N with $\Delta_I(1) = \{\tau_{v_i^I} \mid i = 1, \ldots, r\}$. This fan is associated to the action of G_I on X.

Proposition 4.5. Δ_I *is not contained in a half-space.*

Proof. Suppose Δ_I is contained in the half-space H_u for some $u \in M_\mathbb{R}$, $u \neq 0$. Then $v_i^I \in H_u$ for all $i = 1, \ldots, r$. Let $u = u_1 v_1^* + \ldots + u_l v_l^*$. Then $u_i \geq 0$ for $i \notin I$ and $u_i \leq 0$ for $i \in I$.

Suppose $1 \leq i \leq r$ and consider three cases:

$$\text{If } i \in I_{j_1}^0, \text{ then } \langle v_i^*, v_{j_1} \rangle = 0.$$
$$\text{If } i \in I_{j_1}^+, \text{ then } \langle v_i^*, v_{j_1} \rangle > 0 \text{ and } u_i \leq 0.$$
$$\text{If } i \in I_{j_1}^-, \text{ then } \langle v_i^*, v_{j_1} \rangle < 0 \text{ and } u_i \geq 0.$$

In all cases we have $u_i \langle v_i^*, v_{j_1} \rangle \leq 0$. Therefore $\langle u, v_{j_1} \rangle \leq 0$. But $v_{j_1} \in H_u$ so $\langle u, v_{j_1} \rangle = 0$. Thus $u_i = 0$ for all $i \in I_{j_1}$.

Analogously we can prove that $\langle u, v_{j_2} \rangle = 0$ and therefore $u_i = 0$ for all $i \in I_{j_2} \setminus I_{j_1}$. Hence $u_i = 0$ for all $i \in I_{j_2} \cup I_{j_1}$. In this way we get that $u_i = 0$ for all $\bigcup_{j \in J} I_j = \{1, \ldots, l\}$. $\qquad\square$

Example 4.6. Let $n = r = 6$ and $m = 2$. The action of G on $kQ_1 + \ldots + kQ_6$ is given by the matrix

$$L = \begin{bmatrix} 0 & -1 & 2 & 0 & 1 & 0 \\ 1 & 0 & -1 & -1 & 0 & 1 \end{bmatrix}. \tag{30}$$

Then $v_1 = (1,0,0,0)$, $v_2 = (0,1,0,0)$, $v_3 = (0,0,1,0)$, $v_4 = (0,0,0,1)$, $v_5 = (0,1,-2,0)$, $v_6 = (-1,0,1,1)$. Also $J = \{5,6\}$ and $I = I_5^+ \cup (I_5^0 \cap I_6^+)$, with $I_5^+ = \{2\}$, $I_5^0 = \{1,4\}$ and $I_6^+ = \{3,4\}$. Therefore $I = \{2,4\}$.

5. Finite polytopes

Let us suppose that $\mathcal{D}(X)^G$ has a nonzero finite dimensional module. We can assume that L is of the special kind (3). Let Δ be a fan associated to the action of G on X and not contained in a half-space. Let Y be as in § 2.1. Define $\Lambda \subseteq \mathbb{Z}^m$ by $\Lambda = \{L\alpha \mid \alpha \in \mathbb{N}^r \times \mathbb{Z}^s\}$. For $\chi \in \Lambda$ define

$$\mathcal{O}(Y)_\chi = \mathrm{span}\{Q^\lambda \in \mathcal{O}(Y) \mid L\lambda = \chi\}. \tag{31}$$

It is easy to see that

$$\mathcal{O}(Y) = \bigoplus_{\chi \in \Lambda} \mathcal{O}(Y)_\chi. \tag{32}$$

For each $\chi = (\chi_1, \ldots, \chi_m) \in \Lambda$, $\mathcal{O}(Y)_\chi$ is a simple $\mathcal{D}(Y)^G$-module by [13], Lemma 4.3 and Lemma 1.2. By [13], Lemma 4.1., $\mathcal{O}(Y)_\chi$ is finite dimensional. Let $\varphi = (\varphi_1, \ldots, \varphi_n) \in \mathbb{N}^r \times \mathbb{Z}^s$ such that $L\varphi = \chi$. Given

$\sigma \in \Delta$ and the $\mathcal{D}(Y)^G$-module $\mathcal{O}(V_\sigma)$, we can easily see that $\mathcal{O}(V_\sigma) = \bigoplus_{\chi \in \Lambda} \mathcal{O}(V_\sigma)_\chi$. Then

$$\mathcal{O}(V_\sigma)_\chi = \text{span}\{Q^\lambda \in \mathcal{O}(V_\sigma) \mid \lambda \in \varphi + K^\perp\}. \tag{33}$$

Let

$$\phi_{\sigma,\chi} := \{\lambda \in \varphi + K^\perp \mid (\lambda, e_i) \geq 0 \text{ for all } i \in [\sigma]\}. \tag{34}$$

We can write

$$\phi_{\sigma,\chi} = \{\varphi + \mu \in \varphi + K^\perp \mid (\mu, e_i) \geq -\varphi_i \text{ for all } i \in [\sigma]\}. \tag{35}$$

Observe that $\mathcal{O}(V_\sigma)_\chi = k[\phi_{\sigma,\chi}]$, by (8) $V_\sigma = \{x \in k^n \mid x_i \neq 0 \text{ for all } i \notin [\sigma]\}$; see also (33) and (34). Therefore

$$\mathcal{O}(Y)_\chi = \bigcap_{\sigma \in \Delta} k[\phi_{\sigma,\chi}] \tag{36}$$

since $Y = \bigcup_{\sigma \in \Delta} V_\sigma$.

Let us consider the following $r \times l$ matrix,

$$P = \begin{bmatrix} -1 & & 0 \\ & \ddots & \\ 0 & & -1 \\ b_{11} & \cdots & b_{1l} \\ \vdots & & \vdots \\ b_{m'1} & \cdots & b_{m'l} \end{bmatrix}.$$

We denote by P_i the i-th row vector of P. Let $b = (b_1, \ldots, b_n) \in \mathbb{N}^r \times \mathbb{Z}^s$ such that

$$b_i = \begin{cases} \varphi_i & \text{if } i \in \{1, \ldots, l\}, \\ d\varphi_i & \text{if } i \in \{l+1, \ldots, n\}. \end{cases} \tag{37}$$

Theorem 5.1. *The dimension of $\mathcal{O}(Y)_\chi$ is the number of lattice points inside the polytope*

$$\{x \in M_\mathbb{R} \mid \langle x, P_i \rangle \leq b_i, i = 1, \ldots, r\}. \tag{38}$$

Proof. Define the sets

$$\psi_{\sigma,\chi} := \{\lambda \in K^\perp \mid (\lambda, e_i) \geq -\varphi_{i,\chi}, \text{ for all } i \in [\sigma]\}. \tag{39}$$

Then $\phi_{\sigma,\chi} = \varphi + \psi_{\sigma,\chi}$ where $\phi_{\sigma,m}$ is the set given in (34). Also $k[\phi_{\sigma,\chi}] = Q^\varphi k[\psi_{\sigma,\chi}]$. Therefore $\mathcal{O}(Y)_\chi = Q^\varphi(\bigcap_\sigma k[\psi_{\sigma,\chi}])$, by (36). Let

$$\Lambda_{\sigma,\chi} := \{x \in M \mid \langle x, v_i \rangle \geq -\varphi_{i,\chi} \text{ for all } i \in [\sigma]\}. \tag{40}$$

Then $\psi_{\sigma,\chi} = w(\Lambda_{\sigma,\chi})$, with w as in (13), and $k[\Lambda_{\sigma,\chi}] \cong k[\psi_{\sigma,\chi}]$. Therefore, the dimension of $\mathcal{O}(Y)_\chi$ is the number of lattice points in the set $\bigcap_{\sigma \in \Delta} \Lambda_{\sigma,\chi}$. Henceforth the dimension of $\mathcal{O}(Y)_\chi$ is the number of lattice points in the polytope

$$\{x \in M_\mathbb{R} \mid \langle x, v_i \rangle \geq -\varphi_i \text{ for all } i = 1, \ldots, r\}.$$

It can be easily seen that this polytope coincides with (38) setting $\mathcal{B}$ as $N_\mathbb{R}$ basis. $\qquad\square$

5.1. *Example*

Assume that $\dim G = 3$ and $X = k^5$. Then $\mathcal{D}(X) = A_5$ is the 5-th Weyl algebra. Let the action of G on X be given by the matrix

$$L = \begin{bmatrix} 2 & 2 & 1 & 0 & 0 \\ 1 & 3 & 0 & 1 & 0 \\ 3 & 1 & 0 & 0 & 1 \end{bmatrix}. \tag{41}$$

We consider the A_5^G-module $\mathcal{O}(Y)_\chi$ with $\chi = (30, 30, 40)$. Then $\dim \mathcal{O}(Y)_\chi = 108$, the number of lattice points inside the polytope $\{(x_1, x_2) \in \mathbb{Z}^2 \mid x_1 \geq 0, x_2 \geq 0, 2x_1 + 2x_2 \leq 30, x_1 + 3x_2 \leq 30, 3x_1 + x_2 \leq 40\}$. The number of points inside the polytope was obtained with LattE, which is a recent computer package for lattice point enumeration [6]. The following picture shows this polytope.

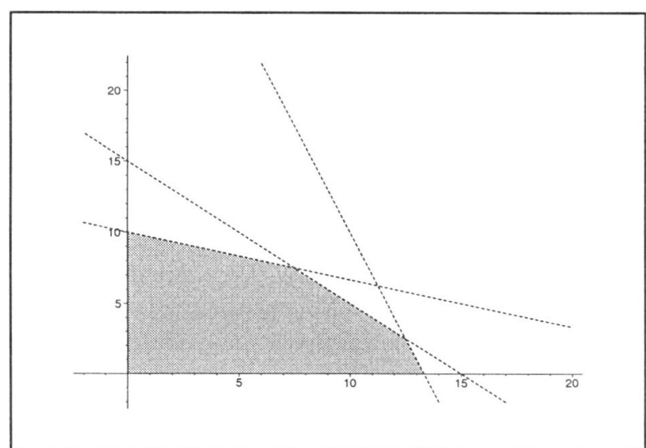

Acknowledgments

This is part of the author's PhD thesis written at the Mathematics Department of the University of Wisconsin-Milwaukee under the supervision of Prof. Ian M. Musson. I would like to thank him for helpful comments on earlier drafts of this paper.

References

1. M. Artin, *Algebra* (Prentice Hall, 1991).
2. A. A'Campo-Neuen and J. Hausen, Quotients of toric varieties by the action of a subtorus, *Tohoku Math. J.* **51** (1999).
3. A. A'Campo-Neuen and J. Hausen, Toric prevarieties and subtorus actions, *Geom. Dedicata* **87** (2001), 35–64.
4. M. Audin, *The Topology of Torus Actions on Symplectic Manifolds*, Progress in Mathematics, Vol. 93 (Birkhäuser, Basel, 1991).
5. D. A. Cox, The homogeneous coordinate ring of a toric variety, *J. Algebraic Geom.* **4** (1995), no. 1, 17–50.
6. J. A. De Loera, R. Hemmecke, J. Tauzer and R. Yoshida, Effective Lattice Point Counting in Rational Convex Polytopes, available via `http://www.math.ucdavis.edu/~latte/theory.html`.
7. D. Luna, Slices étales. Sur les groupes algébriques, *Bull. Soc. Math. France* **33** (Soc. Math. France, Paris, 1973), 81–105.
8. W. Fulton, *Introduction to toric varieties* (Princeton University Press, 1993).
9. J. Hausen, Geometric invariant theory based on Weil divisors. Preprint available at `arXiv:amth.AG/0301204v2`, 2003.
10. T. Levasseur, Anneaux d'opérateurs différentiels, in: P. Dubreil et M.-P. Malliavin, eds., Séminaire d'Algèbre, *Lecture Notes in Mathematics* **867** (Springer, 1981) 157–173.
11. I. M. Musson, Differential operators on toric varieties, *J. Pure Applied Algebra* **95** (1994), 303–315.
12. I. M. Musson, Rings of differential operators on invariant rings of tori, *Trans. Amer. Math. Soc.* **303** (1987), 805–827.
13. I. M. Musson and S. L. Rueda, Finite dimensional representations of invariant differential operators, *Trans. Amer. Math. Soc.*, (accepted for publication). Preprint available at `arXiv:amth.RT/0305279v1`, 2003.
14. M. Mustaţă, G. G. Smith, H. Tsai and U. Walther, $\mathcal{D}$-modules on smooth toric varieties, *J. Algebra* **240** (2001), 744–770.
15. P. Slodowy, Der Scheibensatz für algebraische Transformationsgruppen (pp. 89–113); Algebraische Transformationsgruppen und Invariantentheorie. Edited by H. Kraft, P. Slodowy and T. A. Springer. DMV Seminar, **13**. Birkhäuser Verlag, Basel, 1989.

HOMOLOGICALLY INDUCED $\mathfrak{sl}(1,2)$-MODULES

JOSÉ CARLOS DE SOUSA OLIVEIRA SANTOS

Departamento de Matemática Pura
Rua do Campo Alegre, 687, 4169–007 Porto, Portugal
E-mail: jcsantos@fc.up.pt

The $\mathfrak{sl}(1,2)$-modules that can be obtained from a parabolic subalgebra and a generalized Verma module by (co)homological induction are described. It is proved that, unlike the case of simple Lie algebras, the modules thus obtained starting from a Borel subalgebra depend upon the choice of this subalgebra. It is also proved that every indecomposable module such that the action of a Cartan subalgebra on that module is semisimple can be obtained by (co)homological induction.

Introduction

This article will use the notations and terminology introduced by Victor Kac in his seminal works [2] and [3]. We will start by defining the objects that we will be dealing with. The field we will be working with will be the complex number field. A *super vector space* is a vector space V endowed with a decomposition $V = V_0 \oplus V_1$ (where 0 and 1 should be seen as elements of the group $\mathbb{Z}_2$). A *superalgebra* is an algebra A endowed with super vector space structure $A = A_0 \oplus A_1$ such that, for each $i, j \in \{0, 1\}$, $A_i \cdot A_j \subset A_{i+j}$. For each $i \in \{0, 1\}$ and each $a \in A_i \setminus \{0\}$, let $|a|$ be equal to i. A *Lie superalgebra* is a superalgebra $(\mathfrak{g}, [\,\cdot\,,\,\cdot\,])$ such that

(1) $(\forall X, Y \in (\mathfrak{g}_0 \cup \mathfrak{g}_1) \setminus \{0\})\colon [X, Y] = -(-1)^{|X| \cdot |Y|}[Y, X]$;
(2) $(\forall X, Y, Z \in (\mathfrak{g}_0 \cup \mathfrak{g}_1) \setminus \{0\})\colon [X, [Y, Z]] = [[X, Y], Z] + (-1)^{|X| \cdot |Y|}[Y, [X, Z]]$.

Note that $\mathfrak{g}_0$ is then a Lie algebra and that $\mathfrak{g}_1$ has a natural $\mathfrak{g}_0$-module structure. A *classical Lie superalgebra* is a finite-dimensional simple Lie superalgebra $\mathfrak{g}$ such that the natural action of $\mathfrak{g}_0$ on $\mathfrak{g}_1$ is completely reducible. If $\mathfrak{g}$ is a classical Lie superalgebra, we will say that $\mathfrak{g}$ is *basic classical* if there is some non-degenerate bilinear form $(\,\cdot\,,\,\cdot\,)$ on $\mathfrak{g}$ which is *invariant*, that is, such that

(1) $(\forall X, Y \in (\mathfrak{g}_0 \cup \mathfrak{g}_1) \setminus \{0\})$: $(X, Y) = (-1)^{|X| \cdot |Y|}(Y, X)$;
(2) $(\mathfrak{g}_0, \mathfrak{g}_1) = \{0\}$;
(3) $(\forall X, Y, Z \in \mathfrak{g})$: $([X, Y], Z) = (X, [Y, Z])$.

Of course, every simple finite-dimensional Lie algebra is a basic classical Lie superalgebra.

There are two Lie superalgebras that will be studied in this article. The first one is $\mathfrak{sl}(1, 2)$, whose elements the matrices

$$\begin{pmatrix} a + d & x & y \\ z & a & b \\ t & c & d \end{pmatrix}$$

with $a, b, c, d, x, y, z, t \in \mathbb{C}$. Its superalgebra structure is the one such that

(1) $\mathfrak{sl}(1, 2)_0$ (respectively $\mathfrak{sl}(1, 2)_1$) is the set of those matrices in $\mathfrak{sl}(1, 2)$ such that $x = y = z = t = 0$ (resp. $a = b = c = d = 0$);
(2) if $M, N \in (\mathfrak{sl}(1, 2)_0 \cup \mathfrak{sl}(1, 2)_1) \setminus \{0\}$, then $[M, N] = MN - (-1)^{|M| \cdot |N|} NM$.

The second Lie superalgebra which will be studied in this article is

$$\mathfrak{gl}(1, 1) = \left\{ \begin{pmatrix} a & b \\ c & d \end{pmatrix} \,\middle|\, a, b, c, d \in \mathbb{C} \right\}.$$

In this case, $\mathfrak{gl}(1, 1)_0$ (respectively $\mathfrak{gl}(1, 1)_1$) is the set of those matrices $\begin{pmatrix} a & b \\ c & d \end{pmatrix}$ such that $b = c = 0$ (resp. $a = d = 0$) and the the product is defined as in the case of $\mathfrak{sl}(1, 2)$. As an example of how different the behaviour of these Lie superalgebras can be different from the behaviour of the reductive Lie algebras, it will be enough to observe that, whereas $\mathfrak{g}(2, \mathbb{C})$ has finite-dimensional irreducible representations of any dimension, every finite-dimensional irreducible representation of $\mathfrak{gl}(1, 1)$ has dimension 1 or 2. See the theorem 4.2 for a more precise statement.

Let $\mathfrak{g}$ be a basic classical Lie superalgebra. A *Cartan subalgebra* of $\mathfrak{g}$ is a Cartan subalgebra of $\mathfrak{g}_0$. Let $\mathfrak{h}$ be a Cartan subalgebra of $\mathfrak{g}$. When $\mathfrak{a}$ is a subalgebra of $\mathfrak{g}$ that contains $\mathfrak{h}$, then "$\mathfrak{a}$-module" will mean "$\mathfrak{h}$-semisimple $\mathfrak{a}$-module", unless it is explicitly stated otherwise.

Let $(\cdot, \cdot)$ be a non-degenerate bilinear form on $\mathfrak{h}^*$ induced by an invariant non-degenerate bilinear form on $\mathfrak{g}$. A root α of the pair $(\mathfrak{g}, \mathfrak{h})$ is called an *even* (respectively *odd*) root if $\mathfrak{g}_\alpha \subset \mathfrak{g}_0$ (resp. $\mathfrak{g}_\alpha \subset \mathfrak{g}_1$). Denote by Δ_0 and Δ_1 the set of all even roots and the set of all odd roots respectively and let $\overline{\Delta}_1$ be the set of all *isotropic* roots (that is, the roots α such that $(\alpha, \alpha) = 0$, the subscript 1 being due to the fact that every isotropic root is odd). If $\lambda \in \mathfrak{h}^*$ is a weight, we will say that λ is *typical* when $(\lambda, \alpha) \neq 0$ whenever $\alpha \in \overline{\Delta}_1$. If $\mathfrak{b}$ is a *Borel subalgebra* of $\mathfrak{g}$ (that is, a maximal solvable subalgebra of $\mathfrak{g}$) such that $\mathfrak{h} \subset \mathfrak{g}$, set $\Delta_0(\mathfrak{b})$ (respectively $\Delta_1(\mathfrak{b})$) as the set of all even (resp. odd) roots α such that $\mathfrak{g}_\alpha \subset \mathfrak{b}$ and define

$$\rho_{\mathfrak{b}} = \frac{1}{2} \sum_{\alpha \in \Delta_0(\mathfrak{b})} \alpha - \frac{1}{2} \sum_{\alpha \in \Delta_1(\mathfrak{b})} \alpha.$$

If $\mathfrak{p}$ is a subalgebra of $\mathfrak{g}$ that contains a Borel subalgebra of $\mathfrak{g}$ that contains $\mathfrak{h}$ and if, for every root α of the pair $(\mathfrak{g}, \mathfrak{h})$, $\mathfrak{p}$ contains $\mathfrak{g}_\alpha$ or $\mathfrak{g}_{-\alpha}$, then it will be said that $\mathfrak{p}$ is a *parabolic subalgebra* of $\mathfrak{g}$. Then $\mathfrak{p} = \mathfrak{u} \oplus \mathfrak{s}$, where $\mathfrak{s}$ is a subalgebra of $\mathfrak{p}$ (it is the Levi factor of $\mathfrak{g}$ when $\mathfrak{g}$ is a semisimple Lie algebra) and $\mathfrak{u}$ is the nilradical of $\mathfrak{p}$. If E is an $\mathfrak{s}$-module, then E becomes a $\mathfrak{p}$-module with $\mathfrak{u}$ acting trivially in E. Let

$$M(\mathfrak{p}, E) = \mathcal{U}(\mathfrak{g}) \bigotimes_{\mathcal{U}(\mathfrak{p})} E.$$

Such a module is called a *generalized Verma module* (or simply a *Verma module*, when $\mathfrak{p}$ is a Borel subalgebra). Assume that E is irreducible. It was proved in [7, lemme 2.3] that it has then one and only one irreducible quotient. If $\mathfrak{p}$ is a Borel subalgebra $\mathfrak{b}$ of $\mathfrak{g}$, then $\mathfrak{s} = \mathfrak{h}$ and $E = \mathbb{C}_\lambda$, for some $\lambda \in \mathfrak{h}^*$; in this case, $M(\mathfrak{b}, E)$ will be denoted by $M(\mathfrak{b}, \lambda)$ and its only irreducible quotient will be denoted by $L(\mathfrak{b}, \lambda)$. Every finite-dimensional irreducible $\mathfrak{g}$-module M (distinct from $\{0\}$) is isomorphic to $L(\mathfrak{b}, \lambda)$, for one and only one $\lambda \in \mathfrak{h}^*$. We will say that M is a *typical module* when $\lambda - \rho_{\mathfrak{b}}$ is typical.

If $\mathfrak{s}_0$ is a reductive subalgebra of $\mathfrak{g}_0$, then $\mathcal{HC}(\mathfrak{g}, \mathfrak{s}_0)$ represents the category of $\mathfrak{g}$-modules which, as $\mathfrak{s}_0$-modules, are direct sum of finite-dimensional irreducible modules ($\mathcal{HC}$ stands for Harish-Chandra). A covariant right exact functor

$$\mathcal{L}_0^{\mathfrak{s}_0} : \mathcal{HC}(\mathfrak{g}, \mathfrak{s}_0) \to \mathcal{HC}(\mathfrak{g}, \mathfrak{g}_0)$$

will be defined in section 1 with the following property: when V belongs to $\mathcal{HC}(\mathfrak{g}, \mathfrak{s}_0)$ and is such that, up to isomorphism, there are only a finite number of finite-dimensional irreducible $\mathfrak{g}_0$-modules that are isomorphic to $\mathfrak{g}_0$-submodules of $\mathcal{L}_0^{\mathfrak{s}_0}(V)$, then $\mathcal{L}_0^{\mathfrak{s}_0}(V)$ is the greatest quotient of V in the category $\mathcal{HC}(\mathfrak{g}, \mathfrak{g}_0)$; cf. proposition 1.1 for a more precise statement. In an analogous way (see [7, §4]), a left exact functor

$$\Gamma_{\mathfrak{s}_0}^0 : \mathcal{HC}(\mathfrak{g}, \mathfrak{s}_0) \to \mathcal{HC}(\mathfrak{g}, \mathfrak{g}_0)$$

can be defined such that, when $V \in \mathcal{HC}(\mathfrak{g}, \mathfrak{s}_0)$, $\Gamma^0_{\mathfrak{s}_0}(V)$ is the greatest submodule of V in the category $\mathcal{HC}(\mathfrak{g}, \mathfrak{g}_0)$. The functors $\Gamma^0_{\mathfrak{s}_0}$ and $\mathcal{L}^{\mathfrak{s}_0}_0$ are similar to the Zuckerman functors and the dual Zuckerman functors; cf. [4, chap. II]. For each $i \geqslant 0$, let $\mathcal{L}^{\mathfrak{s}_0}_i$ (respectively $\Gamma^i_{\mathfrak{s}_0}$) denote the i-th derived functor of $\mathcal{L}^{\mathfrak{s}_0}_0$ (resp. $\Gamma^0_{\mathfrak{s}_0}$). It was proved in [7, §6] that, if $\mathfrak{b}$ is a Borel subalgebra of $\mathfrak{g}$, then every finite-dimensional typical $\mathfrak{g}$-module is isomorphic to $\mathcal{L}^{\mathfrak{h}}_i(M(\mathfrak{b}, \lambda))$ for some $i \in \{0, 1, \ldots, \dim(\mathfrak{g}_0/\mathfrak{h})/2\}$ and some weight λ. It was also proved there that, if $\lambda \in \mathfrak{h}^*$ is a typical weight, then $\mathcal{L}^{\mathfrak{h}}_i(M(\mathfrak{b}, \lambda - \rho_\mathfrak{b})) = \{0\}$ except for, at most, one single $i \geqslant 0$ and also that when $\mathcal{L}^{\mathfrak{h}}_i(M(\mathfrak{b}, \lambda - \rho_\mathfrak{b}))$ is different from $\{0\}$, then it is irreducible.

As it will be seen (cf. theorem 2.1), even in such a simple case as when $\mathfrak{g} = \mathfrak{sl}(1, 2)$, the situation changes drastically when one considers all modules of the form $\mathcal{L}^{\mathfrak{h}}_i(M(\mathfrak{b}, \lambda))$. In fact:

(1) Not every finite-dimensional irreducible representation of $\mathfrak{sl}(1, 2)$ is isomorphic to some $\mathcal{L}^{\mathfrak{h}}_i(M(\mathfrak{b}, \lambda))$. Only the typical representations are.

(2) Sometimes $\mathcal{L}^{\mathfrak{h}}_i(M(\mathfrak{b}, \lambda))$ is neither $\{0\}$ nor an irreducible module.

(3) It is not true that there is always, at most; one single $i \geqslant 0$ such that $\mathcal{L}^{\mathfrak{h}}_i(M(\mathfrak{b}, \lambda)) \neq \{0\}$.

It will also be seen (cf. theorem 4.1) that every finite-dimensional irreducible representation of $\mathfrak{sl}(1, 2)$ is isomorphic to $\mathcal{L}^{\mathfrak{s}_0}_i(M(\mathfrak{p}, E))$ for some parabolic subalgebra $\mathfrak{p}$ of $\mathfrak{g}$, some $i \geqslant 0$ and some irreducible $\mathfrak{s}$-module E.

The author expresses his thanks to Jérôme Germoni for making him notice the existence of the short exact sequence (2), from section 2.

1. The functors

The aim of this section is to define the functors $\mathcal{L}^{\mathfrak{s}_0}_i$ and $\Gamma^i_{\mathfrak{s}_0}$ ($i \in \mathbb{Z}_+$). Since $\mathcal{L}^{\mathfrak{s}_0}_i$ (respectively $\Gamma^i_{\mathfrak{s}_0}$) will be defined as the i-th derived functor of the functor $\mathcal{L}^{\mathfrak{s}_0}_0$ (resp. $\Gamma^0_{\mathfrak{s}_0}$), this functor will be defined first. See [7, §4] for further details.

Note that $\mathfrak{sl}(1, 2)_0 \simeq \mathfrak{gl}(2, \mathbb{C})$; we shall identify these Lie algebras. Let $G = SL(2, \mathbb{C}) \times (\mathbb{C}, +)$. Then G is a connected, simply connected complex Lie group such that its Lie algebra is isomorphic to $\mathfrak{gl}(2, \mathbb{C})$. Let $R(G)$ be the vector space generated by the matrix coefficients of the finite-dimensional semisimple representations of G. It turns out that $(\forall f, g \in R(G)): f \cdot g \in R(G)$ and that, therefore, $R(G)$ has a natural structure of an algebra. The group G acts on $R(G)$ by the natural right action, denoted by r and defined as follows: if $g \in G$ and $f \in R(G)$, then

$$(\forall h \in G)\colon (r(g)(f))(h) = f(hg).$$

Note that, under this action, $R(G)$ can be decomposed as a direct sum of finite-dimensional G-modules and that, therefore, the right natural action of G on $R(G)$ induces an action of $\mathfrak{sl}(1,2)_0$. This action will also be denoted by r.

Let $\hat{\mathcal{M}}(G)$ be the dual space of $R(G)$ and let

$$\mathcal{M}(G) = \left\{ \psi \in \hat{\mathcal{M}}(G) \;\middle|\; \#\{\gamma \in \mathfrak{sl}(1,2)_0^{\wedge} \mid \psi(R(G)_\gamma) \neq \{0\}\} < \infty \right\},$$

where $\mathfrak{sl}(1,2)_0^{\wedge}$ stands for the set of equivalent classes of finite-dimensional irreducible representations of $\mathfrak{sl}(1,2)_0$ and $R(G)_\gamma$ is the sum of the submodules of $R(G)$ that belong to the class γ. The multiplication in $R(G)$ induces an action of $R(G)$ in $\mathcal{M}(G)$. Then, if V is an $\mathfrak{sl}(1,2)$-module that belongs to $\mathcal{HC}(\mathfrak{sl}(1,2), \mathfrak{s}_0)$ (where $\mathfrak{s}_0$ is a reductive subalgebra of $\mathfrak{sl}(1,2)_0$), we can define

$$\mathcal{L}_0^{\mathfrak{s}_0}(V) = \mathcal{M}(G) \bigotimes_{\mathcal{U}(\mathfrak{sl}(1,2)_0)} V.$$

An $\mathcal{U}(\mathfrak{sl}(1,2))$-module structure (and, therefore, an $\mathfrak{sl}(1,2)$-module structure) can be defined in $\mathcal{L}_0^{\mathfrak{s}_0}(V)$ in the following way: if $u \in \mathcal{U}(\mathfrak{sl}(1,2))$, choose elements $u_1, u_2, u_3, \ldots \in \mathcal{U}(\mathfrak{sl}(1,2))$ and elements $f_1, f_2, f_2, \ldots \in R(G)$ such that

$$(\forall g \in G)\colon \operatorname{Ad}\left(g^{-1}\right)(u) = \sum_j f_j(g)u_j.$$

Then, if $m \otimes v \in \mathcal{L}_0^{\mathfrak{s}_0}(V)$, the action of u on $m \otimes v$ is given by:

$$u \cdot (m \otimes v) = \sum_j (f_j m) \otimes (u_j v).$$

Let us now define the functor $\Gamma_{\mathfrak{s}_0}^0$. As a vector space,

$$\Gamma_{\mathfrak{s}_0}^0(V) = \left(R(G) \bigotimes V\right)^{\mathfrak{sl}(1,2)_0},$$

where the action of $\mathfrak{sl}(1,2)_0$ on $R(G) \bigotimes V$ is $r \otimes \theta$, θ being the action of $\mathfrak{sl}(1,2)$ on V. In order to define an action of $\mathfrak{sl}(1,2)$ on $\Gamma_{\mathfrak{s}_0}^0(V)$, note that $R(G) \bigotimes V$ (and, therefore, $\Gamma_{\mathfrak{s}_0}^0(V)$) can be seen as a space of functions from G into V. Let $X \in \mathfrak{sl}(1,2)$, $\psi \in \Gamma_{\mathfrak{s}_0}^0(V)$ and define $X\psi$ by

$$(\forall g \in G)\colon (X\psi)(g) = \operatorname{Ad}\left(g^{-1}\right)(X)(\psi(g)).$$

With these definitions, $\mathcal{L}_0^{\mathfrak{s}_0}$ and $\Gamma_{\mathfrak{s}_0}^0$ are, respectively, a right exact functor and a left exact functor.

Proposition 1.1. *Let $V \in \mathcal{HC}(\mathfrak{sl}(1,2), \mathfrak{s}_0)$ and suppose that, up to isomorphism, there are only a finite number of $\mathfrak{sl}(1,2)_0$-modules that are irreducible and isomorphic to $\mathfrak{sl}(1,2)_0$-submodules of $\mathcal{L}_0^{\mathfrak{s}_0}(V)$. Then there is a surjective homomorphism $\eta\colon V \twoheadrightarrow \mathcal{L}_0^{\mathfrak{s}_0}(V)$ and, furthermore, if W is a quotient of V that belongs to the category $\mathcal{HC}(\mathfrak{sl}(1,2), \mathfrak{sl}(1,2)_0)$ and $\pi\colon V \twoheadrightarrow W$ is the projection of V onto W, then there is a surjective homomorphism $\psi\colon \mathcal{L}_0^{\mathfrak{s}_0}(V) \twoheadrightarrow W$ such that $\psi \circ \eta = \pi$. Up to isomorphism, $\mathcal{L}_0^{\mathfrak{s}_0}(V)$ is the only $\mathfrak{sl}(1,2)$-module in the category $\mathcal{HC}(\mathfrak{sl}(1,2), \mathfrak{sl}(1,2)_0)$ with this property.*

See [7, proposition 4.3] for a proof of this proposition (in a more general context).

Note that in the previous proposition the hypothesis that says that, up to isomorphism, there are only a finite number of $\mathfrak{sl}(1,2)_0$-modules irreducible and isomorphic to $\mathfrak{sl}(1,2)_0$-submodules of $\mathcal{L}_0^{\mathfrak{s}_0}(V)$ is certainly valid when $\mathcal{L}_0^{\mathfrak{s}_0}(V)$ is finite-dimensional. It happens that, according to [7, proposition 4.10], this is always the case when $V = M(\mathfrak{p}, E)$ for some finite-dimensional $\mathfrak{s}$-module E. This observation, together with proposition 1.1, shows that when E is a finite-dimensional $\mathfrak{s}$-module, $\mathcal{L}_0^{\mathfrak{s}_0}(M(\mathfrak{p}, E))$ is the greatest quotient of $M(\mathfrak{p}, E)$ in $\mathcal{HC}(\mathfrak{sl}(1,2), \mathfrak{sl}(1,2)_0)$, that is, the greatest quotient of $M(\mathfrak{p}, E)$ that, as an $\mathfrak{sl}(1,2)_0$-module, can be written as a direct sum of finite-dimensional irreducible modules.

It was also proved in [7, §4] that if V belongs to the category $\mathcal{HC}(\mathfrak{sl}(1,2), \mathfrak{s}_0)$, then $\Gamma_{\mathfrak{s}_0}^0(V)$ is the greatest submodule of V in $\mathcal{HC}(\mathfrak{sl}(1,2), \mathfrak{sl}(1,2)_0)$.

2. Borel subalgebras

Let

$$\mathfrak{h} = \left\{ \begin{pmatrix} h_1 + h_2 & 0 & 0 \\ 0 & h_1 & 0 \\ 0 & 0 & h_2 \end{pmatrix} \middle| \; h_1, h_2 \in \mathbb{C} \right\}.$$

Then $\mathfrak{h}$ is a Cartan subalgebra of $\mathfrak{sl}(1,2)$. Let $\alpha, \beta \in \mathfrak{h}^*$ be defined by

$$\alpha \begin{pmatrix} h_1 + h_2 & 0 & 0 \\ 0 & h_1 & 0 \\ 0 & 0 & h_2 \end{pmatrix} = h_1 \quad \text{and} \quad \beta \begin{pmatrix} h_1 + h_2 & 0 & 0 \\ 0 & h_1 & 0 \\ 0 & 0 & h_2 \end{pmatrix} = h_2$$

and let

$$h_\alpha = \begin{pmatrix} 1 & 0 & 0 \\ 0 & 0 & 0 \\ 0 & 0 & 1 \end{pmatrix} \quad \text{and} \quad h_\beta = \begin{pmatrix} 1 & 0 & 0 \\ 0 & 1 & 0 \\ 0 & 0 & 0 \end{pmatrix}.$$

The even roots of $(\mathfrak{sl}(1,2), \mathfrak{h})$ are $\pm(\alpha - \beta)$ and the odd roots are $\pm\alpha$ and $\pm\beta$. If $\lambda \in \mathfrak{h}^* \setminus \{0\}$, then we define

$$\mathfrak{sl}(1,2)_\lambda = \{X \in \mathfrak{sl}(1,2) \mid (\forall H \in \mathfrak{h})\colon [H, X] = \lambda(H)X\}.$$

Note that this is done for $\lambda \neq 0$ only; the reason for this is that the notation $\mathfrak{sl}(1,2)_0$ is reserved for the even part of $\mathfrak{sl}(1,2)$. Let

$$\mathfrak{b}_+ = \mathfrak{h} \oplus \mathfrak{sl}(1,2)_{\alpha-\beta} \oplus \mathfrak{sl}(1,2)_\alpha \oplus \mathfrak{sl}(1,2)_\beta,$$
$$\mathfrak{b}_- = \mathfrak{h} \oplus \mathfrak{sl}(1,2)_{\alpha-\beta} \oplus \mathfrak{sl}(1,2)_{-\alpha} \oplus \mathfrak{sl}(1,2)_{-\beta}$$

and

$$\mathfrak{b}_\pm = \mathfrak{h} \oplus \mathfrak{sl}(1,2)_{\alpha-\beta} \oplus \mathfrak{sl}(1,2)_\alpha \oplus \mathfrak{sl}(1,2)_{-\beta};$$

then $\mathfrak{b}_+$, $\mathfrak{b}_-$ and $\mathfrak{b}_\pm$ are Borel subalgebras of $\mathfrak{g}$ and, up to conjugacy, these are the only ones, as can be easily seen (or deduced from [2, §2.5.4]). They all have the same even part, namely $\mathfrak{h} \oplus \mathfrak{sl}(1,2)_{\alpha-\beta}$. If $\lambda \in \mathfrak{h}^*$, then λ is a weight if and only if $\lambda = n(\alpha - \beta)/2 + t(\alpha + \beta)$ with $n \in \mathbb{Z}$ and $t \in \mathbb{C}$. It will be said that λ is a dominant weight when $n \in \mathbb{Z}_+$ and that λ is a regular weight when $\lambda(h_\alpha - h_\beta) \neq 0 (\Longleftrightarrow (\lambda, \alpha - \beta) \neq 0)$.

In this article, when we speak of "dominant weight" or "highest weight" this is meant to correspond to the choice of $\{\alpha - \beta\}$ as a set of positive roots of $\mathfrak{sl}(1,2)_0$.

If $\lambda \in \mathfrak{h}^*$ is a dominant weight, then let $L_0(\lambda)$ denote the finite-dimensional irreducible $\mathfrak{sl}(1,2)_0$-module whose highest weight is λ. If

$$\mathfrak{sl}(1,2)_+ = \mathfrak{sl}(1,2)_\alpha \oplus \mathfrak{sl}(1,2)_\beta \quad \text{and} \quad \mathfrak{sl}(1,2)_- = \mathfrak{sl}(1,2)_{-\alpha} \oplus \mathfrak{sl}(1,2)_{-\beta},$$

then $\mathfrak{sl}(1,2)_+$ and $\mathfrak{sl}(1,2)_-$ are supercommutative ideals of $\mathfrak{sl}(1,2)_0 \oplus \mathfrak{sl}(1,2)_+$ and $\mathfrak{sl}(1,2)_0 \oplus \mathfrak{sl}(1,2)_-$ respectively and therefore every $\mathfrak{sl}(1,2)_0$-module M becomes an $\mathfrak{sl}(1,2)_0 \oplus \mathfrak{sl}(1,2)_+$-module (resp. an $\mathfrak{sl}(1,2)_0 \oplus \mathfrak{sl}(1,2)_-$-module) with $\mathfrak{sl}(1,2)_+$ (resp. $\mathfrak{sl}(1,2)_-$) acting trivially in M. Let

$$K_+(\lambda) = \mathcal{U}(\mathfrak{sl}(1,2)) \bigotimes_{\mathcal{U}(\mathfrak{sl}(1,2)_0 \oplus \mathfrak{sl}(1,2)_+)} L_0(\lambda)$$

and

$$K_-(\lambda) = \mathcal{U}(\mathfrak{sl}(1,2)) \bigotimes_{\mathcal{U}(\mathfrak{sl}(1,2)_0 \oplus \mathfrak{sl}(1,2)_-)} L_0(\lambda);$$

these are the Kac modules in the specific case of $\mathfrak{sl}(1,2)$ (cf. [3, §2.2]).

Note that the functors $\mathcal{L}_i^{\mathfrak{s}0}$ and $\Gamma_{\mathfrak{s}0}^i$ can be defined for every basic classical Lie superalgebra. It will be useful for what will be done later to see now in more detail what is $\mathcal{L}_0^{\mathfrak{s}0}(V)$ and $\mathcal{L}_1^{\mathfrak{s}0}(V)$ in the specific case of the Lie algebra $\mathfrak{sl}(1,2)_0$ when $\mathfrak{s}(=\mathfrak{s}_0)$ is the standard Cartan subalgebra $\mathfrak{h}$ (that is, the subalgebra whose elements are the diagonal matrices), $\mathfrak{b}$ is the set of upper triangular matrices and V is a Verma module $M(\mathfrak{b}, \lambda)$ (where $\lambda \in \mathfrak{h}^*$ is a weight). In this situation, the functors $\mathcal{L}_0^{\mathfrak{s}0}$ and $\mathcal{L}_1^{\mathfrak{s}0}$ will be represented by $\mathcal{L}_0^0$ and $\mathcal{L}_1^0$ respectively. Notice that the Weyl group of $(\mathfrak{sl}(1,2)_0, \mathfrak{h})$ has only two elements: the identity and the automorphism $w\colon \mathfrak{h} \to \mathfrak{h}$ that exchanges α and β. With this notation, the description is quite simple:

$$\mathcal{L}_0^0(M(\mathfrak{b}, \lambda)) \simeq \begin{cases} L(\mathfrak{b}, \lambda) & \text{if } \lambda \text{ is dominant} \\ \{0\} & \text{otherwise} \end{cases}$$

and

$$\mathcal{L}_1^0(M(\mathfrak{b}, \lambda)) \simeq \begin{cases} L(\mathfrak{b}, w(\lambda) - \alpha + \beta) & \text{if } w(\lambda) - \alpha + \beta \text{ is dominant} \\ \{0\} & \text{otherwise.} \end{cases}$$

This is just a particular case of the Borel-Weil-Bott theorem (see [4, §IV.11]).

In what follows, $\mathcal{L}_i^{\mathfrak{s}0}$ and $\Gamma_{\mathfrak{s}0}^i$ will be replaced by $\mathcal{L}_i$ and Γ^i whenever it is clear from the context what subalgebra of $\mathfrak{g}_0$ is being taken. It was proved in [7, §4] that, for any basic classical Lie superalgebra $\mathfrak{g}$, any parabolic subalgebra $\mathfrak{p}$ and any finite-dimensional $\mathfrak{s}$-module E, $\mathcal{L}_i(M(\mathfrak{p}, E)) = \{0\}$ when $i > \dim(\mathfrak{g}_0/\mathfrak{s}_0)/2$. Therefore, $(\forall \lambda \in \mathfrak{h}^*)\colon \mathcal{L}_i(M(\mathfrak{b}, \lambda)) = \{0\}$ for any Borel subalgebra $\mathfrak{b}$ of $\mathfrak{sl}(1,2)$ and any $i > 1$.

We are ready to state and prove the first of the three theorems which will describe $\mathcal{L}_0(M(\mathfrak{p}, E))$ and $\mathcal{L}_1(M(\mathfrak{p}, E))$ when E is an irreducible finite-dimensional $\mathfrak{s}$-module. However, when $\mathcal{L}_i(M(\mathfrak{p}, E))$ ($i \in \{0,1\}$) is isomorphic to $\{0\}$ or to $\mathbb{C}$, the method that will be used to prove that assertion will always be the same and we will explain it now. The generalized Verma modules $M(\mathfrak{p}, E)$ are always, as $\mathfrak{sl}(1,2)_0$-modules, direct sum of three modules, two of which are Verma modules whereas the third one admits a filtration by Verma modules. To be more precise, we shall denote the $\mathfrak{sl}(1,2)_0$-Verma module with highest weight λ by $M_0(\mathfrak{b}_0, \lambda)$. Then, for instance, in the case when $\mathfrak{p} = \mathfrak{b}_+$ and $E = \mathbb{C}_\lambda$, we have an $\mathfrak{sl}(1,2)_0$-module isomorphism

$$M(\mathfrak{b}_+, \lambda) \simeq M_0(\mathfrak{b}_0, \lambda) \bigoplus M_0(\mathfrak{b}_0, \lambda - \alpha - \beta) \bigoplus M, \tag{1}$$

where M is such that there is a short exact sequence

$$\{0\} \to M_0(\mathfrak{b}_0, \lambda - \beta) \hookrightarrow M \twoheadrightarrow M_0(\mathfrak{b}_0, \lambda - \alpha) \to \{0\}. \tag{2}$$

On the other hand, if V is an $\mathfrak{sl}(1,2)$-module from the category $\mathcal{HC}(\mathfrak{sl}(1,2), \mathfrak{h})$ then, as an $\mathfrak{sl}(1,2)_0$-module, $\mathcal{L}_i(V) \simeq \mathcal{L}_i^0(V)$ where $\mathcal{L}_i^0$ is the functor from the category $\mathcal{HC}(\mathfrak{sl}(1,2)_0, \mathfrak{h})$ to the category $\mathcal{HC}(\mathfrak{sl}(1,2)_0, \mathfrak{sl}(1,2)_0)$ built the same way as $\mathcal{L}_i$ (see [7, proposition 4.4]). Therefore, it follows from (1) that we have

$$\mathcal{L}_i^0(M_0(\mathfrak{b}_0, \lambda)) \simeq \mathcal{L}_i^0(M_0(\mathfrak{b}_0, \lambda)) \bigoplus \mathcal{L}_i^0(M_0(\mathfrak{b}_0, \lambda - \alpha - \beta)) \bigoplus \mathcal{L}_i^0(M) \tag{3}$$

and it is a consequence of (2) that there is an exact sequence

$$\{0\} \to \mathcal{L}_1^0(M_0(\mathfrak{b}_0, \lambda - \beta)) \to \mathcal{L}_1^0(M) \to \mathcal{L}_1^0(M_0(\mathfrak{b}_0, \lambda - \alpha)) \to$$
$$\to \mathcal{L}_0^0(M_0(\mathfrak{b}_0, \lambda - \beta)) \to \mathcal{L}_0^0(M) \to \mathcal{L}_0^0(M_0(\mathfrak{b}_0, \lambda - \alpha)) \to \{0\}. \tag{4}$$

So, to prove that some $\mathcal{L}_i(M(\mathfrak{p}, E))$ is isomorphic to $\{0\}$ or to $\mathbb{C}$ it will be enough to use (3), (4) and the description made above of the $\mathfrak{sl}(1,2)_0$-modules of the type $\mathcal{L}_i^0(M_0(\mathfrak{b}, \lambda))$ ($i \in \{0, 1\}$).

Theorem 2.1. *Let $\lambda \in \mathfrak{h}^*$ be a weight.*

(1) If λ is dominant, then

    (a) $\mathcal{L}_0(M(\mathfrak{b}_+, \lambda)) \simeq K_+(\lambda)$;
    (b) $\mathcal{L}_1(M(\mathfrak{b}_+, \lambda)) \simeq \{0\}$;
    (c) $\mathcal{L}_0(M(\mathfrak{b}_-, \lambda)) \simeq K_-(\lambda)$;
    (d) $\mathcal{L}_1(M(\mathfrak{b}_-, \lambda)) \simeq \{0\}$.

(2) If λ is not dominant and $\lambda + (\alpha - \beta)/2$ is regular, then,

    (a) $\mathcal{L}_0(M(\mathfrak{b}_+, \lambda)) \simeq \{0\}$;
    (b) $\mathcal{L}_1(M(\mathfrak{b}_+, \lambda)) \simeq K_+(w(\lambda) - \alpha + \beta)$;
    (c) $\mathcal{L}_0(M(\mathfrak{b}_-, \lambda)) \simeq \{0\}$;
    (d) $\mathcal{L}_1(M(\mathfrak{b}_-, \lambda)) \simeq K_-(w(\lambda) - \alpha + \beta)$.

(3) If λ is not dominant and $\lambda + (\alpha - \beta)/2$ is not regular, $\mathcal{L}_0(M(\mathfrak{b}_+, \lambda))$, $\mathcal{L}_1(M(\mathfrak{b}_+, \lambda))$, $\mathcal{L}_0(M(\mathfrak{b}_-, \lambda))$ and $\mathcal{L}_1(M(\mathfrak{b}_-, \lambda))$ are all isomorphic to $\{0\}$.

(4) One has:

    (a) $\lambda(h_\beta) \neq 0 \Longrightarrow M(\mathfrak{b}_\pm, \lambda) \simeq M(\mathfrak{b}_+, \lambda + \beta)$;
    (b) $\lambda(h_\alpha) \neq 0 \Longrightarrow M(\mathfrak{b}_\pm, \lambda) \simeq M(\mathfrak{b}_-, \lambda - \alpha)$.

(5) $\mathcal{L}_0(M(\mathfrak{b}_\pm, 0)) \simeq \mathcal{L}_1(M(\mathfrak{b}_\pm, 0) \simeq \mathbb{C}$.

338

Proof. Among these assertions, those which state that some $\mathcal{L}_i(M(\mathfrak{b}, \lambda))$ is isomorphic to $\{0\}$ or to $\mathbb{C}$ (where $\mathfrak{b}$ can be either $\mathfrak{b}_+$ or $\mathfrak{b}_-$) can be proved using the method describe before the statement of the theorem. For instance, in order to prove 1b all that has to be proved is that $\mathcal{L}_1^0(M_0(\mathfrak{b}, \eta)) = \{0\}$, for each $\eta \in \{\lambda, \lambda - \alpha, \lambda - \beta, \lambda - \alpha - \beta\}$, but this is a consequence of the fact that, for each such η, $w(\eta) - \alpha + \beta$ is not dominant. Besides this, since there is an automorphism φ of $\mathfrak{sl}(1, 2)$ such that $\varphi(\mathfrak{b}_+) = \mathfrak{b}_-$, it is clear that for any result concerning $\mathfrak{b}_+$ there is a similar result concerning $\mathfrak{b}_-$. Hence, the assertions that will be proved are 1a, 2b and 4a.

It follows from proposition 1.1 and from the fact that $\mathcal{L}_0(M(\mathfrak{b}_+, \lambda))$ is finite-dimensional (see [7, proposition 4.10]) that, to prove that $\mathcal{L}_0(M(\mathfrak{b}_+, \lambda))$ and $K_+(\lambda)$ are isomorphic, it will be enough to prove that when V is a quotient of $M(\mathfrak{b}_+, \lambda)$ that, as an $\mathfrak{sl}(1, 2)_0$-module, can be written as a direct sum of finite-dimensional irreducible modules, then the projection $\pi \colon M(\mathfrak{b}_+, \lambda) \to V$ factors through $K_+(\lambda)$. But since λ is a maximal weight of V (unless $V = \{0\}$, but in this case there is nothing to prove), $L_0(\lambda)$ is isomorphic to a submodule $V(\lambda)$ of V. On the other hand, every weight of $M(\mathfrak{b}_+, \lambda)$ has the form $\omega - n(\alpha - \beta)$ with $\omega \in \{\lambda, \lambda - \alpha, \lambda - \beta, \lambda - \alpha - \beta\}$ and n a non-positive integer; therefore, the weights of V have the same form. It follows that $\mathfrak{sl}(1, 2)_\alpha \bigoplus \mathfrak{sl}(1, 2)_\beta$ acts trivially in $V(\lambda)$ and this proves that the inclusion $V(\lambda) \hookrightarrow V$ induces an $\mathfrak{sl}(1, 2)$-morphism

$$\eta \colon K_+(\lambda)\left(= \mathcal{U}(\mathfrak{sl}(1, 2)) \bigotimes_{\mathcal{U}(\mathfrak{sl}(1,2)_0 \bigoplus \mathfrak{sl}(1,2)_+)} V(\lambda)\right) \to V.$$

It is clear that, if Π is the projection of $M(\mathfrak{b}_+, \lambda)$ onto $K_+(\lambda)$, then $\eta \circ \Pi = \pi$. This proves the assertion 1a.

To prove the assertion 2b, take $\eta = w(\lambda) - \alpha + \beta$. Then η is a dominant weight and if π is a projection from $M(\mathfrak{b}_+, \eta)$ onto $K_+(\eta)$, then its kernel is isomorphic to $M(\mathfrak{b}_+, \lambda)$. In other words, one has a short exact sequence

$$\{0\} \to M(\mathfrak{b}_+, \lambda) \overset{\iota}{\hookrightarrow} M(\mathfrak{b}_+, \eta) \overset{\pi}{\twoheadrightarrow} K_+(\eta) \to \{0\}$$

which induces an exact sequence

$$\mathcal{L}_2(M(\mathfrak{b}_+, \eta)) \to \mathcal{L}_2(K_+(\eta)) \to \mathcal{L}_1(M(\mathfrak{b}_+, \lambda)) \to \mathcal{L}_1(M(\mathfrak{b}_+, \eta)).$$

But, since $M(\mathfrak{b}_+, \eta)$ is a Verma module, $\mathcal{L}_2(M(\mathfrak{b}_+, \eta)) \simeq \{0\}$ and, since η is dominant, the assertion 1a shows that $\mathcal{L}_1(M(\mathfrak{b}_+, \eta)) \simeq \{0\}$. Therefore,

$$\mathcal{L}_1(M(\mathfrak{b}_+, \lambda)) \simeq \mathcal{L}_2(K_+(\eta)).$$

But, according to [7, proposition 4.8], $\mathcal{L}_2(K_+(\eta))$ is isomorphic to $\Gamma^0(K_+(\eta))$. Since $K_+(\eta)$ belongs to the category $\mathcal{HC}(\mathfrak{sl}(1,2),\mathfrak{sl}(1,2)_0)$, it follows that $\Gamma^0(K_+(\eta)) \simeq K_+(\eta)$, because, as it was stated at the introduction, when V is an $\mathfrak{sl}(1,2)$-module that belongs to the category $\mathcal{HC}(\mathfrak{sl}(1,2),\mathfrak{s}_0)$, $\Gamma^0(V)$ is the greatest submodule of V in the category $\mathcal{HC}(\mathfrak{sl}(1,2),\mathfrak{sl}(1,2)_0)$.

Finally to prove the assertion 4a simply take $X_\beta \in \mathfrak{sl}(1,2)_\beta \setminus \{0\}$. It is easy to see that

$$M(\mathfrak{b}_+,\lambda) \to M(\mathfrak{b}_\pm,\lambda+\beta)$$
$$u \otimes c \rightsquigarrow uX_\beta \otimes c$$

is an isomorphism.

Note that assertion 4 of the theorem is only a particular case of a much more general result; cf. [5, §0.1.5] or [6, p. 23].

3. Irreducible representations of $\mathfrak{gl}(1,1)$

Theorem 2.1 proves that an $\mathfrak{sl}(1,2)$-module of the form $\mathcal{L}_i(M(\mathfrak{b},\lambda))$, where $\mathfrak{b}$ is a Borel subalgebra of $\mathfrak{sl}(1,2)$, is never an atypical finite-dimensional irreducible representation of $\mathfrak{sl}(1,2)$. It will be seen that such representations may be obtained by homological induction if one uses a parabolic subalgebra of $\mathfrak{sl}(1,2)$. To be more precise, it will be enough to use the parabolic subalgebras associated with one of the following sets of roots: $\{\pm\alpha,\beta\}$, $\{\pm\alpha,-\beta\}$, $\{\alpha,\pm\beta\}$ and $\{-\alpha,\pm\beta\}$. In each case, $\mathfrak{s} \simeq \mathfrak{gl}(1,1)$.

In order to study the finite-dimensional irreducible $\mathfrak{gl}(1,1)$-modules, we shall fix the notation. Put $X_\beta = \left(\begin{smallmatrix} 0 & 1 \\ 0 & 0 \end{smallmatrix}\right)$, $X_{-\beta} = \left(\begin{smallmatrix} 0 & 0 \\ 1 & 0 \end{smallmatrix}\right)$ and $h_\beta = [X_\beta, X_{-\beta}] = \left(\begin{smallmatrix} 1 & 0 \\ 0 & 1 \end{smallmatrix}\right)$. If E is a finite-dimensional irreducible $\mathfrak{gl}(1,1)$-module, then the usual argument shows that E is $\mathfrak{h}$-semisimple and that, for some weight $\lambda \in \mathfrak{h}^*$, every weight of E has the form $\lambda - n\beta$, for some non negative integer n. It will be said then that λ is the highest weight of E.

If E is a super vector space, $\dim E = \dim E_0 + \epsilon \dim E_1$.

Proposition 3.1. *Let E be a finite-dimensional irreducible representation of the Lie superalgebra $\mathfrak{gl}(1,1)$ and let λ be its highest weight. Then the dimension of E is 1 or ϵ if $\lambda(h_\beta) = 0$ and $1+\epsilon$ otherwise.*

Proof. Let $v_\lambda \in E_\lambda \setminus \{0\}$. Then $\mathbb{C}X_{-\beta}v_\lambda \oplus \mathbb{C}v_\lambda \oplus \mathbb{C}X_\beta v_\lambda$ is a submodule of E, since $X_\beta^2 v = X_{-\beta}^2 v = 0$ for every $v \in E$. In fact, since $\lambda + \beta$ is not a weight of E, $E = \mathbb{C}X_{-\beta}v_\lambda \oplus \mathbb{C}v_\lambda$. There are now two possibilities.

- $\lambda(h_\beta) \neq 0$. In this case, $X_{-\beta}v_\lambda \neq 0$ since

$$X_\beta(X_{-\beta}v_\lambda) = X_\beta(X_{-\beta}v_\lambda) + X_{-\beta}(X_\beta v_\lambda) = h_\beta v_\lambda = \lambda(h_\beta)v_\lambda \neq 0.$$

- $\lambda(h_\beta) = 0$. The same argument as above shows that $\mathbb{C}X_{-\beta}v_\lambda$ is a submodule of E; therefore, $E = \mathbb{C}X_{-\beta}v_\lambda$ or $X_{-\beta}v_\lambda = 0$. But, since $v_\lambda \in E$, $E \neq \mathbb{C}X_{-\beta}v_\lambda$ and this implies that $E = \mathbb{C}v_\lambda$.

4. Parabolic subalgebras

We will deal now with the case where $\mathfrak{p} = \mathfrak{h} \oplus (\oplus_{\eta \in \Psi}\mathfrak{sl}(1,2)_\eta)$, where Ψ is one among the following four sets: $\{\pm\alpha, \beta, -\alpha + \beta\}$, $\{\pm\alpha, -\beta, \alpha - \beta\}$, $\{\alpha, \pm\beta, \alpha-\beta\}$ and $\{-\alpha, \pm\beta, -\alpha+\beta\}$. Since all these cases are similar, it will be enough to do things in detail for one of these cases; this will be done with $\Psi = \{\alpha, \pm\beta, \alpha - \beta\}$. Therefore $\mathfrak{s} = \mathfrak{h} \oplus \mathfrak{sl}(1,2)_\beta \oplus \mathfrak{sl}(1,2)_{-\beta}(\simeq \mathfrak{gl}(1,1))$.

Theorem 4.1. *Let E be an irreducible finite-dimensional $\mathfrak{s}$-module and let λ be its highest weight.*

(1) If λ is dominant, then

    (a) if $\lambda(h_\beta) \neq 0$, $\mathcal{L}_0(M(\mathfrak{p}, E)) \simeq K_+(\lambda)$;
    (b) if $\lambda(h_\beta) = 0$, $\mathcal{L}_0(M(\mathfrak{p}, E))$ is irreducible with highest weight λ (in other words, $\mathcal{L}_0(M(\mathfrak{p}, E)) \simeq L(\mathfrak{b}_+, \lambda)$);
    (c) $\mathcal{L}_1(M(\mathfrak{p}, E)) \simeq \{0\}$.

(2) If λ is not dominant, then

    (a) $\mathcal{L}_0(M(\mathfrak{p}, E)) \simeq \{0\}$;
    (b) if $\lambda(h_\beta) \neq 0$, $\mathcal{L}_1(M(\mathfrak{p}, E)) \simeq K_+(w(\lambda) - \alpha + \beta)$, unless $w(\lambda) - \alpha + \beta$ is not dominant, in which case $\mathcal{L}_1(M(\mathfrak{p}, E)) \simeq \{0\}$;
    (c) if $\lambda(h_\beta) = 0$, $\mathcal{L}_1(M(\mathfrak{p}, E))$ is irreducible with highest weight $w(\lambda) - \alpha + \beta$ (that is, it is isomorphic with $L(\mathfrak{b}_+, w(\lambda) - \alpha + \beta)$), unless $\lambda = \beta$, in which case $\mathcal{L}_1(M(\mathfrak{p}, E)) \simeq \mathbb{C}$.

Proof. By the argument presented before the statement of theorem 2.1, the cases where it is stated that $\mathcal{L}_i(M(\mathfrak{p}, E))$ is isomorphic either to $\{0\}$ or to $\mathbb{C}$ are easy to establish.

Take $X_{\pm\alpha} \in \mathfrak{sl}(1,2)_{\pm\alpha}$ and $X_{\pm\beta} \in \mathfrak{sl}(1,2)_{\pm\beta}$ such that $h_\alpha = [X_\alpha, X_{-\alpha}]$ and $h_\beta = [X_\beta, X_{-\beta}]$. Suppose that λ is dominant and that $\lambda(h_\beta) \neq 0$. If V is a quotient of $M(\mathfrak{p}, E)$ that belongs to the category $\mathcal{HC}(\mathfrak{sl}(1,2), \mathfrak{sl}(1,2)_0)$ and if π is the projection of $M(\mathfrak{p}, E)$ onto V, then, in order to be able to use proposition 1.1 (and the remark made after its statement), it must be proved that π factors through $K_+(\lambda)$. Let $v \in E_\lambda \setminus \{0\}$ and let $w = \pi(1 \otimes v)$; it can (and will) be assumed that $w \neq 0$. Since $w \in V_\lambda$ and since V belongs to the category $\mathcal{HC}(\mathfrak{sl}(1,2), \mathfrak{sl}(1,2)_0)$, there is an $\mathfrak{sl}(1,2)_0$-submodule of V isomorphic to $L_0(\lambda)$ and the inclusion of this module into V induces an $\mathfrak{sl}(1,2)$-morphism $F\colon K_+(\lambda) \to V$ such that, for some $v^* \in L_0(\lambda)_\lambda \setminus \{0\}$, $F(1 \otimes v^*) = w$. Since $\lambda(h_\beta) \neq 0$, $\mathbb{C}(1 \otimes v^*) \bigoplus \mathbb{C}(X_{-\beta} \otimes v^*) \simeq E$ (as a $\mathfrak{gl}(1,1)$-module) and X_α acts trivially on $\mathbb{C}(1 \otimes v^*) \bigoplus \mathbb{C}(X_{-\beta} \otimes v^*)$, there is an $\mathfrak{sl}(1,2)$-morphism $\eta\colon M(\mathfrak{p}, E) \to K_+(\lambda)$ such that $\eta(v) = 1 \otimes v^*$. Therefore $(F \circ \eta)(v) = w = \pi(v)$; since v generates $M(\mathfrak{p}, E)$, $F \circ \eta = \pi$. This proves the assertion 1a.

Suppose now that λ is dominant and that $\lambda(h_\beta) = 0$. In this case, and since proposition 3.1 tells us that $E = \mathbb{C}v$ for some $v \in E$ whose weight is λ, the $\mathfrak{sl}(1,2)_0$-module $M(\mathfrak{p}, E)$ is the direct sum of two Verma modules whose highest weights are λ and $\lambda - \alpha$. Observe that, since $\lambda(h_\beta) = 0$ and λ is dominant, $\lambda = n\alpha$ for some non-negative integer n. There are two possibilities.

- $\lambda \neq 0$. Then λ and $\lambda - \alpha$ are both dominant weights and therefore, as an $\mathfrak{sl}(1,2)_0$-module, $\mathcal{L}_0(M(\mathfrak{p}, E))$ is isomorphic to the direct sum

$$L_0(\lambda) \bigoplus L_0(\lambda - \alpha).$$

  Let ω_1 be an element of $\mathcal{L}_0(M(\mathfrak{p}, E))_\lambda$ different from 0 and define $\omega_2 = X_{-\alpha}\omega_1$. Then the weight of ω_2 is $\lambda - \alpha$ and $\omega_2 \neq 0$ since

$$X_\alpha \cdot \omega_2 = X_\alpha \cdot (X_{-\alpha} \cdot \omega_1) = -X_{-\alpha} \cdot (X_\alpha \cdot \omega_1) + h_\alpha \cdot \omega_1 = \lambda(h_\alpha)\omega_1$$

  and both $\lambda(h_\alpha)$ and ω_1 are different from 0. Since, as an $\mathfrak{sl}(1,2)_0$-module, $\mathcal{L}_0(M(\mathfrak{p}, E))$ is the direct sum of two irreducible modules, generated by ω_1 and ω_2, $\omega_2 = X_{-\alpha}\omega_1$, and $X_\alpha \cdot \omega_2 = \lambda(h_\alpha)\omega_1$, the $\mathfrak{sl}(1,2)$-module $\mathcal{L}_0(M(\mathfrak{p}, E))$ is irreducible and generated by ω_1; it is therefore isomorphic to $L(\mathfrak{b}_+, \lambda)$.
- $\lambda = 0$. Then $\lambda - \alpha = -\alpha$ and therefore it is not a dominant weight. In this case then, as a $\mathfrak{sl}(1,2)_0$-module, $\mathcal{L}_0(M(\mathfrak{p}, E))$ is simply $L_0(\lambda)(= L_0(0))$, which is isomorphic to $\mathbb{C}$. Therefore, $\mathcal{L}_0(M(\mathfrak{p}, E)) \simeq \mathbb{C} \simeq L(\mathfrak{b}_+, 0) = L(\mathfrak{b}_+, \lambda)$.

Finally, the statements concerning $\mathcal{L}_1(M(\mathfrak{p}, E))$ when λ is not dominant can be proved in the same way as in theorem 2.1.

There are only two parabolic subalgebras $\mathfrak{p}$ left (distinct from $\mathfrak{sl}(1,2)$), namely

$$\mathfrak{p} = \mathfrak{sl}(1,2)_0 \bigoplus \mathfrak{sl}(1,2)_+ \quad \text{and} \quad \mathfrak{p} = \mathfrak{sl}(1,2)_0 \bigoplus \mathfrak{sl}(1,2)_-.$$

We shall simply describe the $\mathfrak{sl}(1,2)$-modules homologically induced that can be obtained from the first of these two parabolic subalgebras; the proof is similar (in fact, easier) to the proofs of theorems 2.1 and 4.1.

Theorem 4.2. *Let* $\mathfrak{p} = \mathfrak{sl}(1,2)_0 \bigoplus \mathfrak{sl}(1,2)_+$ *and let* E *be an irreducible finite-dimensional* $\mathfrak{sl}(1,2)_0$-*module with highest weight* $\lambda \in \mathfrak{h}^*$. *Then* $\mathcal{L}_0(E) \simeq K_+(\lambda)$ *and* $\mathcal{L}_1(E) \simeq \{0\}$.

5. Final remarks

The description made in sections 2 and 4 of the $\mathfrak{sl}(1,2)$-modules that are homologically induced show that

(1) no single parabolic subalgebra of $\mathfrak{sl}(1,2)$ is enough to obtain every irreducible finite-dimensional $\mathfrak{sl}(1,2)$-module by homological induction;

(2) every homologically induced $\mathfrak{sl}(1,2)$-module is indecomposable and, furthermore, every indecomposable $\mathfrak{sl}(1,2)$-module is homologically induced.

Remember that it was stated at the beginning of the article that all our modules are $\mathfrak{h}$-semisimple. In fact, there are $\mathfrak{sl}(1,2)$-modules which are indecomposable but that are not $\mathfrak{h}$-semisimple (see [1]).

References

1. J. Germoni, *Indecomposable representations of special linear Lie superalgebras*, J. Algebra **209** (1998), 367–401.
2. V. G. Kac, *Lie superalgebras*, Adv. Math. **26** (1977), 8–96.
3. V. G. Kac, *Representations of classical Lie superalgebras*, Differential Geometrical Methods in Mathematical Physics, Lecture Notes in Math., vol. 676, Springer-Verlag, 1978, pp. 597–626.
4. A. W. Knapp and D. A. Vogan, *Cohomological induction and unitary representations*, Princeton Mathematical Series, no. 45, Princeton University Press, 1995.
5. I. Penkov and V. Serganova, *Representations of classical Lie superalgebras of type I*, Indag. Math. (N.S.) **3** (1992), 419–466.
6. J. C. Santos, *Induction homologique dans les super algèbres de Lie basiques classiques*, Thèse de Doctorat, Université Paris VII, 1996.
7. J. C. Santos, *Foncteurs de Zuckerman pour les superalgèbres de Lie*, J. Lie Theory **9** (1999), 69–112.

INJECTIVE DIMENSION RELATIVE TO A TORSION THEORY

PATRICK F. SMITH

Department of Mathematics
University of Glasgow
Glasgow, G12 8QW, Scotland, UK
E-mail: pfs@maths.gla.ac.uk

Let τ be a hereditary torsion theory on a ring R. If $\tau_G \leq \tau$, where τ_G is the Goldie torsion theory on R, then the right global τ-injective dimension, r. gl. τ-id R, of R is precisely the right global dimension, r. gl. dim R, of R. In general, r. gl. τ-id $R \leq 1 + $ r. gl. dim R, with equality for certain rings R and hereditary torsion theories τ on R. For a stable hereditary torsion theory τ on a ring R we have r. gl. dim $R \leq$ r. gl. τ-id $R \leq 1 + $ r. gl. dim R.

Throughout this paper all rings are associative with identity, all modules are unitary right modules and all torsion theories are hereditary. This paper continues the discussion in [2] concerning projective and injective R-modules relative to a torsion theory τ on R. Any unexplained terminology can be found in [1], [2], [6], [8] or [9].

Let τ be a torsion theory on a ring R. Recall that an R-module P is called τ-*projective* provided for every τ-torsion-free submodule K of an R-module M, every homomorphism $\phi\colon P \to M/K$ can be lifted to a homomorphism $\theta\colon P \to M$. On the other hand, an R-module E is called τ-*injective* if, for every monomorphism $\alpha\colon A \to B$ of R-modules A, B such that $B/\alpha(A)$ is τ-torsion and every homomorphism $\phi\colon A \to E$, there exists a homomorphism $\theta\colon B \to E$ such that $\theta\alpha = \phi$. Basic properties of τ-projective modules and of τ-injective modules can be found in [1], [2] or [6].

In [2] we introduced and discussed the τ-projective dimension τ-$\mathrm{pd}_R(M)$ of an R-module M and proved in [2, Theorem 4.2] that τ-$\mathrm{pd}_R(M)$ equals the projective dimension $\mathrm{pd}_{R/T}(M/MT)$ of the (R/T)-module M/MT, where T is the τ-torsion submodule of the right R-module R (so that T is a two-sided ideal of R). We also introduced in [2] the right global τ-projective dimension of R which we shall denote the r. gl. τ-pd R in this paper. We proved in [2, Corollary 4.3] that r. gl. τ-pd R equals the right global dimension of the ring R/T.

In this paper we define and discuss the τ-injective dimension, τ-id$_R(M)$, of an R-module M and the right global τ-injective dimension, r. gl. τ-id R, of the ring R. We prove that, for any torsion theory τ on a ring R and for any R-module M with τ-torsion submodule T, τ-id$_R(T)$ is at most the injective dimension, id$_R(M)$, of M (Theorem 1.8) and, in consequence, r. gl. τ-id $R \leq 1 + $ r. gl. dim R (Corollary 1.9 (ii)). In case the torsion theory τ is stable we have τ-id$_R(M) = $ id$_R(M)$ for every τ-torsion R-module M (Theorem 2.1). On the other hand, if $\tau_G \leq \tau$, where τ_G is the Goldie torsion theory on R, then r. gl. τ-id $R = $ r. gl. dim R (Theorem 2.2). Next, if I is an idempotent ideal of R and τ_I denotes the I-adic torsion theory on R then τ_I-id$_R(M) = $ id$_{R/I}(M)$ for every τ_I-torsion R-module M (Theorem 2.4) and, in consequence, r. gl. dim$(R/I) \leq $ r. gl. τ_I-id $R \leq 1 + $ r. gl. dim(R/I) (Corollary 2.5). Finally if R is a regular local ring of dimension $d > 1$ and τ_s the torsion theory generated by all simple R-modules, then r. gl. τ_s-id $R = 1 + $ r. gl. dim R (Theorem 2.6). These theorems are illustrated by a number of examples at the end.

1. Relative injective dimension

Let R be any ring. Following [2] and [8, §5], for any R-module M, pd$_R(M)$ and id$_R(M)$ will denote the projective dimension and the injective dimension of M, respectively. The right global dimension of R will be denoted by r. gl. dim R, as in [8, 5.13]. Let τ be any torsion theory on R. For any R-module M, $E(M)$ will denote the injective envelope of M. We begin with a well known fact but will give its proof for completeness.

Lemma 1.1. *Let N be a submodule of a τ-injective R-module M. Let L be the submodule of M containing N such that L/N is the τ-torsion submodule of M/N. Then the R-module L is τ-injective.*

Proof. Let $\alpha\colon A \to B$ be a monomorphism of R-modules such that $B/\alpha(A)$ is τ-torsion and let $\phi\colon A \to L$ be any homomorphism. Let $i\colon L \to M$ denote the inclusion mapping. Since M is τ-injective it follows that there exists a homomorphism $\theta\colon B \to M$ such that $\theta\alpha = i\phi$. Let $b \in B$. Then $bF \subseteq \alpha(A)$ for some τ-dense right ideal F of R. Hence

$$\theta(b)F = \theta(bF) \subseteq \theta\alpha(A) = i\phi(A) \subseteq L.$$

But M/L is τ-torsion-free, so that $\theta(b) \in L$. Thus $\theta(B) \subseteq L$. It follows that L is τ-injective. $\qquad\square$

Lemma 1.2. *The following statements are equivalent for an R-module M with injective envelope E.*

(i) M *is τ-injective.*

(ii) *Whenever M is a τ-dense submodule of an R-module M', then M is a direct summand of M'.*

(iii) E/M *is a τ-torsion-free module.*

Proof. (i) $\Rightarrow$ (ii) Suppose that M is a τ-dense submodule of an R-module M'. If $i\colon M \to M'$ denotes inclusion then, because M'/M is τ-torsion, it follows that $\theta i = 1$ for some homomorphism $\theta\colon M' \to M$. Then $M' = M \oplus \ker\theta$.

(ii) $\Rightarrow$ (iii) Clear because M is an essential submodule of E.

(iii) $\Rightarrow$ (i) By Lemma 1.1. $\qquad\square$

Let M be an R-module with injective envelope E. Following [1, p. 102], we define $E_\tau(M)$ to be the set of elements e in E such that $eF \subseteq M$ for some τ-dense right ideal F of R, i.e. $E_\tau(M)$ contains M and $E_\tau(M)/M$ is the τ-torsion submodule of E/M. By Lemma 1.1, $E_\tau(M)$ is a τ-injective R-module. Note that $E_\tau(M)/M$ is τ-torsion. Thus there exists an exact sequence

$$0 \to M \xrightarrow{\theta_0} E_0 \xrightarrow{\phi_0} L_0 \to 0$$

of R-modules with E_0 τ-injective and L_0 τ-torsion.

Similarly there exists an exact sequence $0 \to L_0 \xrightarrow{\theta_1} E_1 \xrightarrow{\phi_1} L_1 \to 0$ of R-modules with E_1 τ-injective and L_1 τ-torsion. Repeat this process, so that for every integer $n \geq 0$ there is an exact sequence $0 \to L_n \xrightarrow{\theta_{n+1}} E_{n+1} \xrightarrow{\phi_{n+1}} L_{n+1} \to 0$ of R-modules such that E_{n+1} is τ-injective and L_{n+1} is τ-torsion. Combining these exact sequences together we can produce, for any integer $n \geq 0$, the following exact sequence

$$0 \to M \xrightarrow{\theta_0} E_0 \xrightarrow{\chi_0} E_1 \xrightarrow{\chi_1} \ldots \to E_n \xrightarrow{\phi_n} L_n \to 0$$

where $\chi_i = \theta_{i+1}\phi_i$ $(0 \leq i \leq n-1)$. Note further that $\operatorname{coker}\theta_0$ and $\operatorname{coker}\chi_i$ $(0 \leq i \leq n-1)$ are all τ-torsion. We have proved the following result.

Lemma 1.3. *Let M be an R-module. Then for any integer $n \geq 0$, there exists an exact sequence*

$$0 \to M \xrightarrow{\chi_0} E_0 \xrightarrow{\chi_1} E_1 \xrightarrow{\chi_2} \ldots \xrightarrow{\chi_n} E_n \xrightarrow{\chi_{n+1}} L_n \to 0$$

of R-modules such that E_i is τ-injective and $\operatorname{coker}\chi_i$ *is τ-torsion for all* $0 \le i \le n$.

The next result is Schanuel's Lemma for τ-injective modules. The proof is an easy adaptation of the proof of Schanuel's Lemma for injective modules (see, for example, [8, 5.40]).

Lemma 1.4. *Let M_1 and M_2 be isomorphic R-modules and let $0 \to M_1 \to E_1 \to L_1 \to 0$ and $0 \to M_2 \to E_2 \to L_2 \to 0$ be exact sequences of R-modules such that E_1 and E_2 are τ-injective and L_1 and L_2 are τ-torsion. Then $E_1 \oplus L_2 \cong E_2 \oplus L_1$.*

Corollary 1.5. *Let M_1 and M_2 be isomorphic R-modules, let n be a non-negative integer and let*

$$0 \to M_1 \xrightarrow{\phi_0} E_0 \xrightarrow{\phi_1} E_1 \to \ldots \to E_n \xrightarrow{\phi_{n+1}} E_{n+1} \to 0$$

and

$$0 \to M_2 \xrightarrow{\theta_0} F_0 \xrightarrow{\theta_1} F_1 \to \ldots \to F_n \xrightarrow{\theta_{n+1}} F_{n+1} \to 0$$

be exact sequences of R-modules such that E_i and F_i are τ-injective and $\operatorname{coker}\phi_i$ and $\operatorname{coker}\theta_i$ are τ-torsion for all $0 \le i \le n$. Then

$$E_0 \oplus F_1 \oplus E_2 \ldots \cong F_0 \oplus E_1 \oplus F_2 \oplus \ldots.$$

Proof. By induction on n using Lemma 1.4, adapting the standard proof. $\square$

Let M be an R-module. If M is τ-injective then we say that M has *τ-injective dimension* 0 and write $\tau\text{-id}_R(M) = 0$. More generally, for any integer $n \ge 0$, we say that M has *τ-injective dimension n*, and write $\tau\text{-id}_R(M) = n$, provided there exists an exact sequence

$$0 \to M \xrightarrow{\phi_0} E_0 \xrightarrow{\phi_1} E_1 \xrightarrow{\phi_2} \ldots \xrightarrow{\phi_n} E_n \to 0$$

of R-modules such that E_i is τ-injective and $\operatorname{coker}\phi_i$ is τ-torsion for all $0 \le i \le n$, but there does not exist a shorter such exact sequence. If M does not have τ-injective dimension n, for some integer $n \ge 0$, then we write $\tau\text{-id}_R(M) = \infty$. The *right global τ-injective dimension* of R, denoted by r. gl. τ-id R, is defined by

$$\text{r. gl. }\tau\text{-id } R = \sup\{\tau\text{-id}_R(M) : M \text{ is an } R\text{-module}\}.$$

This is the analogue of the *right global τ-projective dimension* of R, denoted by r. gl. τ-pd R and defined by r. gl. τ-pd $R = \sup\{\tau\text{-pd}_R(M) : M$ is an R-module$\}$ which is discussed in [2].

We look at two special torsion theories next. Let τ_0 denote the torsion theory τ on R for which every R-module is τ-torsion free. In this case, $\tau_0\text{-pd}_R(M) = \text{pd}_R(M)$ for every R-module M by [2, Theorem 4.2] and, consequently, $\text{r. gl.}\,\tau_0\text{-pd}\,R = \text{r. gl. dim}\,R$. Note that every R-module is τ_0-injective and hence $\tau_0\text{-id}_R(M) = 0$ for every R-module M. It follows that $\text{r. gl.}\,\tau_0\text{-id}\,R = 0$. Thus $\text{r. gl.}\,\tau_0\text{-id}\,R < \text{r. gl.}\,\tau_0\text{-pd}\,R$ for any ring R which is not semiprime Artinian by [8, 5.14].

Next, let τ_1 denote the torsion theory τ on R for which every R-module is τ-torsion. Then every τ_1-injective R-module is injective. Clearly this implies that $\text{id}_R(M) \leq \tau_1\text{-id}_R(M)$ for every R-module M and hence $\text{r. gl. dim}\,R \leq \text{r. gl.}\,\tau_1\text{-id}\,R$ by [8, 5.45]. However, by [2, Theorem 4.2], $\tau_1\text{-pd}_R(M) = 0$ for every R-module M and hence $\text{r. gl.}\,\tau_1\text{-pd}\,R = 0$. Thus $\text{r. gl.}\,\tau_1\text{-pd}\,R < \text{r. gl.}\,\tau_1\text{-id}\,R$ for any ring R which is not semiprime Artinian. Therefore, for a particular ring R and torsion theory τ on R it may well happen that $\text{r. gl.}\,\tau\text{-pd}\,R \neq \text{r. gl.}\,\tau\text{-id}\,R$, in contrast to [8, 5.45].

The next result is the key result of this paper.

Theorem 1.6. *Let τ be any torsion theory on a ring R. Let $0 \to M \to E \to L \to 0$ be an exact sequence of R-modules such that E is τ-injective and L is τ-torsion. Then M is τ-injective or $\tau\text{-id}_R(M) = 1 + \tau\text{-id}_R(L)$.*

Proof. Let $\theta\colon M \to E$ and $\phi\colon E \to L$ denote the homomorphisms in the given exact sequence. Suppose that $\tau\text{-id}_R(L) = n$, for some integer $n \geq 0$. Then there exists an exact sequence $0 \to L \xrightarrow{\phi_0} E_0 \xrightarrow{\phi_1} E_1 \to \ldots \xrightarrow{\phi_n} E_n \to 0$ of R-modules such that E_i is τ-injective and $\text{coker}\,\phi_i$ is τ-torsion for all $0 \leq i \leq n$. This yields the exact sequence

$$0 \to M \xrightarrow{\theta} E \xrightarrow{\phi_0\phi} E_0 \xrightarrow{\phi_1} E_1 \to \ldots \xrightarrow{\phi_n} E_n \to 0$$

where $\text{coker}\,\theta$, $\text{coker}\,\phi_0\phi$ and $\text{coker}\,\phi_i$ $(1 \leq i \leq n)$ are all τ-torsion. It follows that $\tau\text{-id}_R(M) \leq n + 1$. Thus $\tau\text{-id}_R(M) \leq 1 + \tau\text{-id}_R(L)$.

Now suppose that $\tau\text{-id}_R(M) = m < 1 + \tau\text{-id}_R(L)$, for some positive integer m. Then there exists an exact sequence

$$0 \to M \xrightarrow{\theta_0} F_0 \xrightarrow{\theta_1} F_1 \to \ldots \xrightarrow{\theta_m} F_m \to 0$$

of R-modules such that F_i is τ-injective and $\text{coker}\,\theta_i$ is τ-torsion for all $0 \leq i \leq m$.

Suppose that $m = 1$. Then the exact sequences $0 \to M \to E \to L \to 0$ and $0 \to M \to F_0 \to F_1 \to 0$ together give, by Lemma 1.4, $E \oplus F_1 \cong F_0 \oplus L$ and hence L is τ-injective. Thus $\tau\text{-id}_R(L) = 0 = \tau\text{-id}_R(M) - 1$. Now suppose that $m \geq 2$. Consider the exact sequence

$$0 \to L \xrightarrow{\phi_0} E_0 \xrightarrow{\phi_1} E_1 \to \ldots \xrightarrow{\phi_{m-1}} E_{m-2} \to \operatorname{coker} \phi_{m-2} \to 0.$$

This yields the exact sequence

$$0 \to M \xrightarrow{\theta} E \xrightarrow{\phi_0\phi} E_0 \xrightarrow{\phi_1} \ldots \xrightarrow{\phi_{m-1}} E_{m-2} \to \operatorname{coker} \phi_{m-2} \to 0.$$

By Corollary 1.5, we have

$$F_0 \oplus E_0 \oplus F_2 \oplus \ldots \cong E \oplus F_1 \oplus E_1 \oplus \ldots.$$

Since F_i $(0 \leq i \leq m)$ and E_i $(0 \leq i \leq m-1)$ are all τ-injective it follows that $\operatorname{coker} \phi_{m-2}$ is τ-injective. Hence $\operatorname{id}_R(L) \leq m - 1 < \tau\text{-id}_R(L)$, a contradiction. Thus $\tau\text{-id}_R(M) = 1 + \tau\text{-id}_R(L)$. $\qquad\square$

For any torsion theory τ on the ring R we define $\mathrm{r.\,gl.}\,\tau\text{-id}^* R = \sup\{\tau\text{-id}_R(M) : M$ is a τ-torsion R-module$\}$. Theorem 1.6 has the following immediate consequence.

Corollary 1.7. *For any torsion theory τ on a ring R, $\mathrm{r.\,gl.}\,\tau\text{-id}^* R \leq \mathrm{r.\,gl.}\,\tau\text{-id}\, R \leq 1 + \mathrm{r.\,gl.}\,\tau\text{-id}^* R$.*

Let τ be a torsion theory on a ring R. For any R-module M, the τ-torsion submodule of M will be denoted by $t_\tau(M)$.

Theorem 1.8. *Let τ be any torsion theory on a ring R. Then $\tau\text{-id}_R(t_\tau(M)) \leq \operatorname{id}_R(M)$ for every R-module M.*

Proof. Let M be any R-module. If $\operatorname{id}_R(M) = 0$ then M is an injective R-module and $t_\tau(M)$ is a τ-injective R-module by Lemma 1.1 so that $\tau\text{-id}_R(t_\tau(M)) = 0$. Let n be a positive integer and suppose that $\operatorname{id}_R(M) = n$. Let E denote the injective envelope of M. Consider the exact sequence $0 \to M \to E \to E/M \to 0$ and note that $t_\tau(E/M) = E_\tau(M)/M$. It is well known that $\operatorname{id}_R(E/M) = n - 1$. By induction on n, $\tau\text{-id}_R(E_\tau(M)/M) \leq n - 1$. Now consider the exact sequence $0 \to M \to E_\tau(M) \to E_\tau(M)/M \to 0$. By Theorem 1.6, $\tau\text{-id}_R(M) = 0$ or $\tau\text{-id}_R(M) = 1 + \tau\text{-id}_R(E_\tau(M)/M)$. In any case, $\tau\text{-id}_R(M) \leq n$. The result follows by induction. $\qquad\square$

Corollary 1.9. *Let τ be any torsion theory on a ring R. Then*

(i) $\tau\text{-id}_R(M) \leq \operatorname{id}_R(M)$ *for every τ-torsion R-module M.*
(ii) $\mathrm{r.\,gl.}\,\tau\text{-id}\, R \leq 1 + \mathrm{r.\,gl.}\,\tau\text{-id}^* R \leq 1 + \mathrm{r.\,gl.}\,\dim R$.

Proof. (i) By Theorem 1.8.

(ii) By (i) and Corollary 1.7. $\qquad\square$

Let R be any ring with $\operatorname{r.gl.dim} R = \infty$. We have seen above that $\operatorname{r.gl.} \tau_0\text{-id} R = 0$. This fact puts Corollary 1.9 (ii) into perspective.

2. Special torsion theories

In §1 we considered a general torsion theory τ on a ring R. In this section we shall restrict τ to be a torsion theory with a particular property or to be a particular type of torsion theory. Recall that a torsion theory τ on a ring R is called *stable* if the injective envelope $E(M)$ is τ-torsion for every τ-torsion R-module M. For example, τ_0 and τ_1 are both stable torsion theories on R, and so too is the Goldie torsion theory on R (see [12, p. 153 Proposition 7.3]). In case, R is a commutative Noetherian ring, every torsion theory on R is stable by [12, p. 170 Proposition 4.5]. We can improve Corollary 1.9 (i) in case τ is a stable torsion theory.

Theorem 2.1. *Let τ be a stable torsion theory on a ring R. Then $\tau\text{-id}_R(M) = \text{id}_R(M)$ for every τ-torsion R-module M.*

Proof. Let M be any τ-torsion R-module. By Corollary 1.9 (i) we have $\tau\text{-id}_R(M) \leq \text{id}_R(M)$. Now suppose that $\tau\text{-id}_R(M) = n$, for some non-negative integer n. If $n = 0$ then M is τ-injective and hence $E(M)/M$ is τ-torsion free by Lemma 1.2. Because τ is stable we have $M = E(M)$, i.e. $\text{id}_R(M) = 0$. Now suppose that $n \geq 1$. Then there exists an exact sequence

$$0 \to M \xrightarrow{\phi_0} E_0 \xrightarrow{\phi_1} E_1 \to \ldots \xrightarrow{\phi_n} E_n \to 0$$

of R-modules such that E_i is τ-injective and $\operatorname{coker}\phi_i$ is τ-torsion for all $0 \leq i \leq n$. For each $0 \leq i \leq n$, an easy argument will show that E_i is τ-torsion and hence injective by the above argument. It follows that $\text{id}_R(M) \leq n = \tau\text{-id}_R(M)$. Thus $\tau\text{-id}_R(M) = \text{id}_R(M)$. $\qquad\square$

Corollary 1.9 (ii) shows that $\operatorname{r.gl.} \tau_G\text{-id} R \leq 1 + \operatorname{r.gl.dim} R$, where τ_G is the Goldie torsion theory on R. Our next aim is to show that $\operatorname{r.gl.dim} R = \operatorname{r.gl.} \tau_G\text{-id} R$. Let σ and τ be torsion theories on a ring R. Then we write $\sigma \leq \tau$ if every σ-torsion R-module is τ-torsion, equivalently $t_\sigma(M) \subseteq t_\tau(M)$ for every R-module M.

Theorem 2.2. *The following statements are equivalent for a torsion theory τ on a ring R.*

350

(i) $\tau_G \leq \tau$.

(ii) *Every τ-injective module is injective.*

(iii) $\mathrm{id}_R(M) = \tau\text{-}\mathrm{id}_R(M)$ *for every R-module M.*

Moreover, in this case, $\mathrm{r.\,gl.}\,\tau\text{-}\mathrm{id}\,R = \mathrm{r.\,gl.\,dim}\,R$.

Proof. (i) $\Leftrightarrow$ (ii) By [2, Theorem 2.1]. (see also [6, Example 8.1 and Propositions 8.2 and 8.3].)

(iii) $\Rightarrow$ (ii) Clear.

(ii) $\Rightarrow$ (iii) Let M be any R-module. It is easy to see that (ii) implies that $\mathrm{id}_R(M) \leq \tau\text{-}\mathrm{id}_R(M)$. Suppose that $\mathrm{id}_R(M) = n$, for some non-negative integer n. If $n = 0$ then M is injective, whence τ-injective, and we have $\tau\text{-}\mathrm{id}_R(M) = 0$. Suppose that $n \geq 1$. Consider the exact sequence

$$0 \to M \to E \to E/M \to 0,$$

where $E = E(M)$. Note that E/M is τ_G-torsion and hence τ-torsion. By Corollary 1.9 (i), $\tau\text{-}\mathrm{id}_R(E/M) \leq \mathrm{id}_R(E/M)$. Now applying Theorem 1.6 for τ and for τ_1, we have

$$\tau\text{-}\mathrm{id}_R(M) = 1 + \tau\text{-}\mathrm{id}_R(E/M) \leq 1 + \mathrm{id}_R(E/M) = \mathrm{id}_R(M).$$

Thus $\tau\text{-}\mathrm{id}_R(M) \leq \mathrm{id}_R(M)$. This proves (iii).

The last part is clear. $\qquad\square$

Later we shall give an example of a torsion theory τ on a ring R such that $\mathrm{r.\,gl.}\,\tau\text{-}\mathrm{id}\,R = \mathrm{r.\,gl.\,dim}\,R$ but $\tau_G \not\leq \tau$ (see Example 3.6). Next we look at another type of torsion theory. For any idempotent ideal I of a ring R let τ_I denote the I-adic torsion theory on R. In this case an R-module M is τ_I-torsion if and only if $MI = 0$, equivalently M is an (R/I)-module. We next prove a simple lemma.

Lemma 2.3. *Let I be any idempotent ideal of a ring R. Then an (R/I)-module M is injective if and only if the R-module M is τ_I-injective.*

Proof. Suppose that the R-module M is τ_I-injective. Let Y be a submodule of an $\overline{R}$-module X, where $\overline{R}$ denotes the ring R/I, and let $\phi\colon X \to M$ be an $\overline{R}$-homomorphism. Then ϕ is an R-homomorphism and the R-module X/Y is τ_I-torsion. By hypothesis, ϕ lifts to an R- (and hence $\overline{R}$-) homomorphism $\theta\colon X \to M$. It follows that the $\overline{R}$-module M is injective. Conversely, suppose that the $\overline{R}$-module M is injective. Let A be a submodule of an R-module B such that B/A is τ_I-torsion and let $\alpha\colon A \to M$ be a homomorphism. Define $\overline{\alpha}\colon A/AI \to M$ by $\overline{\alpha}(a + AI) = \alpha(a)$ for each $a \in A$. Note that $\overline{\alpha}$ is well defined because $MI = 0$ and hence $\overline{\alpha}$ is an

$\overline{R}$-homomorphism. Note that $BI \subseteq A$ so that $BI = AI$ and hence B/AI is an $\overline{R}$-module. By hypothesis there exists a homomorphism $\overline{\beta}\colon B/AI \to M$ such that $\overline{\beta}(a + AI) = \alpha(a)$ for each $a \in A$. Define a homomorphism $\beta\colon B \to M$ by $\beta(b) = \overline{\beta}(b + AI)$ for each $b \in B$. Clearly β lifts α to B. It follows that M is τ_I-injective. $\qquad\square$

Compare the next result with Theorem 1.8.

Theorem 2.4. *Let I be an idempotent ideal of a ring R. Then $\tau_I\text{-id}_R(M) = \text{id}_{R/I}(M)$ for every τ_I-torsion R-module M. Moreover,*

$$\text{r. gl. } \tau_I\text{-id}^* R = \text{r. gl. dim}(R/I).$$

Proof. Let M be an (R/I)-module. Suppose that $\text{id}_{R/I}(M) = n$ for some non-negative integer n. Then there exists an exact sequence

$$0 \to M \to E_0 \to E_1 \to \ldots \to E_n \to 0$$

where E_i is an injective (R/I)-module for each $0 \leq i \leq n$. By Lemma 2.3, the R-module E_i is τ_I-torsion and τ_I-injective for each $0 \leq i \leq n$. Hence $\tau_I\text{-id}(M) \leq n$. We have proved that $\tau_I\text{-id}_R(L) \leq \text{id}_{R/I}(L)$ for every (R/I)-module L. Suppose that $\tau_I\text{-id}_R(N) < \text{id}_{R/I}(N)$ for some (R/I)-module N. Let k be the least non-negative integer such that there exists an (R/I)-module X with $k = \tau_I\text{-id}_R(X) < \text{id}_{R/I}(X)$. By Lemma 2.3, $k \geq 1$. Let E denote the injective envelope of X. Then $E_\tau(X) = \{e \in E : eI \subseteq X\} = \{e \in E : eI = 0\}$ which is the injective envelope of the (R/I)-module X by [11, Proposition 2.27]. Consider the exact sequence of (R/I)-modules:

$$0 \to X \to E_\tau(X) \to Y \to 0 \tag{$*$}$$

where $Y = E_\tau(X)/X$. By Theorem 1.6 applied to $(*)$ with respect to the torsion theory τ_1 for the ring R/I we have $\text{id}_{R/I}(X) = 1 + \text{id}_{R/I}(Y)$. On the other hand, Theorem 1.6 applied to $(*)$ with respect to the torsion theory τ_I for R gives $\tau_I\text{-id}_R(X) = 1 + \tau_I\text{-id}_R(Y)$. But by the choice of k, $\text{id}_{R/I}(Y) = \tau_I\text{-id}_R(Y)$ and hence $\text{id}_{R/I}(X) = \tau_I\text{-id}_R(X)$, a contradiction Thus $\tau_I\text{-id}_R(V) = \text{id}_{R/I}(V)$ for every (R/I)-module V. The last part is obvious by [8, 5.45]. $\qquad\square$

Corollary 2.5. *Let I be an idempotent ideal of a ring R. Then $\text{r. gl. dim}(R/I) \leq \text{r. gl. } \tau_I\text{-id}(R) \leq 1 + \text{r. gl. dim}(R/I)$.*

Proof. By Corollary 1.7 and Theorem 2.4. $\qquad\square$

Let τ_s denote the hereditary torsion theory on Mod-R generated by all simple R-modules (see [12, p. 182]). Note that a non-zero R-module M is τ_s-torsion if and only if every non-zero factor module of M has non-zero socle. The final theorem of this section concerns (commutative) regular local rings R. See [4] or [7] or [8] for the definition and properties of regular local rings. In Corollary 1.9 (ii) we saw that r. gl. τ-id $R \leq 1 +$ r. gl. dim R for any torsion theory τ on an arbitrary ring R. Now we show that r. gl. τ_s -id $R = 1 +$ r. gl. dim R for any regular local ring R which is not one-dimensional. (Note that regular local rings have finite global dimension by [8, 5.84].)

Theorem 2.6. *Let R be any regular local ring of dimension $d > 1$. Then* r. gl. τ_s -id $R = d + 1$.

Proof. By Corollary 1.9 (ii), gl. τ_s -id $R \leq d + 1$. By [4, Corollary 10.14] R is an integral domain. Let Q denote the field of fractions of R. Let P denote the unique maximal ideal of R and let $P^* = \{q \in Q : qP \subseteq R\}$. Then P^* is an R-submodule of Q, $R \subseteq P^*$, P^*P is an ideal of R and $P \subseteq P^*P$. If $P^*P = R$ then P is invertible and by [7, Theorem 59], $d = 1$, a contradiction. Hence $P^*P = P$ and hence $P^* = R$ by [7, Theorem 12], because R is a UFD by the Auslander-Buchsbaum Theorem (see [4, Theorem 19.19] or [7, Theorem 184]). It follows that the R-module Q/R has zero socle and hence the R-module R is τ_s-injective by Lemma 1.2. By Lemma 1.2, P is not τ_s-injective.

Consider the exact sequence $0 \to P \to R \to R/P \to 0$ and note that the R-module R/P is τ_s-torsion. By Theorem 1.6, τ_s -id$_R P = 1 + \tau_s$ -id$_R(R/P)$. But τ_s -id$_R(R/P) =$ id$_R(R/P)$ by Theorem 2.1 because τ_s is stable (see [12, p. 170 Proposition 4.5]). Next id$_R(R/P) = d$ by [7, Theorem 214] (or [8, (5.83)]). Thus τ_s -id$_R P = d + 1$ and hence gl. τ_s -id $R \geq d + 1$. We have now proved that gl. τ_s -id $R = d + 1$. $\square$

3. Examples

In this section we shall look at some particular examples. If R is a semiprime Artinian ring then every R-module is both projective and injective and hence both τ-projective and τ-injective for every torsion theory τ on R. In other words, r. gl. dim $R = 0$ implies that r. gl. τ-pd $R = 0$ and r. gl. τ-id $R = 0$ for every torsion theory τ on R. First of all in this section we shall give an example of a ring R which is neither semiprime nor Artinian but which has the property that r. gl. τ-pd $R =$ r. gl. τ-id $R = 0$ for a certain torsion theory τ.

Let S and T be rings and let M be a left S-, right T-bimodule. We shall use $[S, M; 0, T]$ to denote the ring of all "matrices" of the form

$$\begin{bmatrix} s & m \\ 0 & t \end{bmatrix}$$

with $s \in S$, $m \in M, t \in T$, and with the usual matrix addition and multiplication. We shall denote the above "matrix" by $[s, m; 0, t]$.

Lemma 3.1 (See [5] or [9, 7.5.1]). *Let S and T be rings, let M be a left S-, right T-bimodule and let $R = [S, M; 0, T]$. Then $\sup\{\mathrm{r.\,gl.\,dim}\, S, \mathrm{r.\,gl.\,dim}\, T] \leq \mathrm{r.\,gl.\,dim}\, R \leq \sup\{\mathrm{r.\,gl.\,dim}\, T, \mathrm{r.\,gl.\,dim}\, S + \mathrm{pd}_T(M) + 1\}$.*

Corollary 3.2 (See [3, Lemma 3.8] or [5]). *With the notation of Lemma 3.1, let S be a semiprime Artinian ring and let $M \neq 0$. Then*

$$\mathrm{r.\,gl.\,dim}\, R = \sup\{\mathrm{r.\,gl.\,dim}\, T, \mathrm{pd}_T(M) + 1\}.$$

Lemma 3.3 (See [2, Corollary 4.3(a)]). *Let τ be any torsion theory on a ring R. Then $\mathrm{r.\,gl.}\,\tau\text{-pd}\,R = \mathrm{r.\,gl.\,dim}(R/t_\tau(R_R))$.*

Example 3.4. *Let F be any field and let V be any non-zero vector space over F. Let R denote the ring $[F, V; 0, F]$. Let $I = [F, V; 0, 0]$. Then*

(i) *the ring R is not semiprime,*
(ii) *the ring R is right (and left) Artinian if and only if V is finite dimensional over F,*
(iii) *I is an idempotent ideal of R,*
(iv) *$\mathrm{r.\,gl.}\,\tau\text{-pd}\,R = \mathrm{r.\,gl.}\,\tau_I\text{-id}\,R = 0$, and*
(v) *$\mathrm{r.\,gl.\,dim}\, R = 1$ and $\mathrm{r.\,gl.\,dim}(R/I) = 0$.*

Proof. (i), (ii), (iii) Clear.

(iv) Note that if $A = \{r \in R : rI = 0\}$ then $A = [0, V; 0, F]$. By Lemma 3.3, $\mathrm{r.\,gl.}\,\tau\text{-pd}\,R = \mathrm{r.\,gl.\,dim}(R/A) = \mathrm{r.\,gl.\,dim}\, F = 0$. Next let B be a right ideal of R such that B is a τ_I- dense submodule of R. Then $I \subseteq B$ and hence $B = I$ or $B = R$. If $B = I$ then $R = B \oplus C$ where C is the right ideal $[0, 0; 0, F]$ of R. By [1, Proposition 4.1.5], $\mathrm{r.\,gl.}\,\tau_I\text{-id}\,R = 0$.

(v) By Corollary 3.2, $\mathrm{r.\,gl.\,dim}\, R = 1$. Moreover, $R/I \cong F$ so that $\mathrm{r.\,gl.\,dim}(R/I) = 0$. $\qquad\square$

Note that the rings R in Example 3.4 are all right (and left) hereditary. Note further that for *any* right hereditary ring R we have $0 \leq \mathrm{r.\,gl.}\,\tau\text{-pd}\,R \leq$

354

1 and $0 \leq \mathrm{r.\,gl.\,}\tau\text{-id}\,R \leq 1$ for every torsion theory τ on R by [2, Theorem 3.1].

For any ring R, let $\mathrm{Soc}(R_R)$ denote the right socle of R. It is a well known fact (see, for example, [8, p. 242 ex 12 (2)]) that $\mathrm{Soc}(R_R)$ is the intersection of all the essential right ideals of R.

Lemma 3.5. *Let I be an idempotent ideal of a ring R. Then $\tau_G \leq \tau_I$ if and only if $I \leq \mathrm{Soc}(R_R)$.*

Proof. Suppose that $\tau_G \leq \tau_I$. Let E be any essential right ideal of R. Then the R-module R/E is τ_G-torsion so that $(R/E)I = 0$, i.e. $I \subseteq E$. It follows that $I \subseteq \mathrm{Soc}(R_R)$. Conversely, if $I \subseteq \mathrm{Soc}(R_R)$ then it is clear that every τ_G-torsion R-module is τ_I-torsion, i.e. $\tau_G \leq \tau_I$. $\qquad\square$

Next we give (as promised) an example of a torsion theory τ on a ring R such that $\mathrm{r.\,gl.\,}\tau\text{-id}\,R = \mathrm{r.\,gl.\,dim}\,R$ but it is not the case that $\tau_G \leq \tau$.

Example 3.6. *Let F be any field, let $F[x]$ denote the polynomial ring in an indeterminate x over F and let $R = [F, F[x]; 0, F[x]]$. Let I denote the ideal $[0, F[x]; 0, F[x]]$ of R. Then*

   (i) *R is a right Noetherian PI-algebra and I is an idempotent ideal of R.*
  (ii) *$\mathrm{r.\,gl.\,}\tau_I\text{-id}\,R = \mathrm{r.\,gl.\,dim}\,R = 1$.*
(iii) *$\tau_G \nleq \tau_I$.*

Proof. (i) It is very easy to see that R is a right (but not left) Noetherian PI-algebra. That I is idempotent is a consequence of the fact that $I = Re$ where e is the idempotent $[0, 0; 0, 1]$ of R.

(ii) By Corollary 3.2, $\mathrm{r.\,gl.\,dim}\,R = 1$. Next, Corollary 2.5 gives $0 \leq \mathrm{r.\,gl.\,}\tau_I\text{-id}\,R \leq 1$. Suppose that $\mathrm{r.\,gl.\,}\tau_I\text{-id}\,R = 0$. By [1, Proposition 4.1.5], $I = fR$ for some idempotent f of R and it is easy to check that this cannot be the case. Thus $\mathrm{r.\,gl.\,}\tau_I\text{-id}\,R = 1$.

(iii) It is easy to check that the ring R has zero right socle and hence (iii) follows by Lemma 3.5. $\qquad\square$

Finally we give two examples which highlight Corollary 1.9 (ii) by showing that in general there is no relationship between $\mathrm{r.\,gl.\,}\tau\text{-pd}\,R$ and $\mathrm{r.\,gl.\,dim}\,R$, for a given torsion theory τ on a ring R. The first example is of a torsion theory τ on a ring R such that $\mathrm{r.\,gl.\,}\tau\text{-pd}\,R = 0$ but $\mathrm{r.\,gl.\,dim}\,R = \infty$. The second example is of a torsion theory τ on a ring R such that $\mathrm{r.\,gl.\,}\tau\text{-pd}\,R = \infty$ but $\mathrm{r.\,gl.\,dim}\,R = 2$.

Example 3.7. *Let F be any field of characteristic $p \neq 0$, let G be any finite group whose order is divisible by p, let $F[G]$ denote the group algebra of G over F and let $R = [F, F[G]; 0, F[G]]$. Let I denote the ideal $[F, F[G]; 0, 0]$ of R. Then I is an idempotent ideal of R such that* r. gl. τ_I-pd $R = 0$ *but* r. gl. dim $R =$ r. gl. τ_I-id $R = \infty$.

Proof. Clearly $I = eR$ where e is the idempotent $[1, 0; 0, 0]$ of R, so that $I = I^2$. If $A = \{r \in R : rI = 0\}$ then $A = [0, F[G]; 0, F[G]]$ so that r. gl. τ_I-pd $R =$ r. gl. dim$(R/A) =$ r. gl. dim $F = 0$ by Lemma 3.3. However, by Corollary 3.2 and [10, Corollary 10.3.7 (iii)], r. gl. dim $R = \infty$. Finally $R/I \cong F[G]$ so that r. gl. τ_I-id $R = \infty$ by Corollary 2.5 and [10, Corollary 10.3.7 (iii)]. $\qquad\square$

Note that, with our earlier notation, r. gl. τ_1-pd $R = 0$ but r. gl. dim $R =$ r. gl. τ_1-id $R = \infty$ for the ring R in Example 3.7. Moreover for this same ring $\tau_S = \tau_1$. Earlier in this section we noted that if R is a right hereditary ring then r. gl. τ-pd $R \leq 1$ for every torsion theory τ on R. To complete this paper we give an example of a ring R with r. gl. dim $R = 2$ and an idempotent ideal I of R such that r. gl. τ_I-pd $R = \infty$. (Compare [9, Theorem 7.3.10 (ii)].)

Example 3.8. *Let F be a field of characteristic $p \neq 0$, let G be a finite group whose order is divisible by p and let H be a free group such that there exists an epimorphism $\phi \colon H \to G$. Let $\theta \colon F[H] \to F[G]$ denote the ring epimorphism of group algebras induced by ϕ and let A denote the kernel of θ. Let $S = F[H]$, let R denote the ring $[S, S/A; 0, F]$ and let I denote the idempotent ideal $[0, S/A; 0, F]$ of R. Then* r. gl. dim $R = 2$ *but* r. gl. τ_I-pd $R = \infty$.

Proof. It is clear that I is idempotent because $I = Re$ where e is the idempotent element $[0, 0; 0, 1]$ of R. By [10, Corollary 10.3.7 (iv)], r. gl. dim $S = 1$ and by Lemma 3.1, $1 \leq$ r. gl. dim $R \leq 2$. Let $B = \{r \in R : rI = 0\}$. Then $B = [A, S/A; 0, 0]$ and hence $R/B \cong (S/A) \oplus F$. Now Lemma 3.3 gives r. gl. τ_I-pd $R =$ r. gl. dim$(R/B) =$ r. gl. dim$(S/A) =$ r. gl. dim $F[G] = \infty$ by [10, Corollary 10.3.7 (iii)]. By [9, Theorem 7.3.10 (ii)], I is not a projective right ideal of R and hence r. gl. dim $R = 2$. $\qquad\square$

Note that in Example 3.8, $t_{\tau_s}(R) = [0, S/A; 0, F]$ so that r. gl. τ_S-pd$(R) =$ r. gl. dim $S = 1$. Note further that $1 \leq$ r. gl. τ_I-id $R \leq 2$ by Corollary 2.5.

Acknowledgment

The author would like to thank Professor Paul E. Bland of Eastern Kentucky University for introducing him to this topic and for many helpful and stimulating discussions during Professor Bland's visit to the University of Glasgow in 2001.

References

1. P. E. Bland, Topics in Torsion Theory, Wiley-VCH, Mathematics Research 103 (1998).
2. P. E. Bland and P. F. Smith, Injective and projective modules relative to a torsion theory, New Zealand J. Math. 32 (2003), 1-11.
3. A. W. Chatters and P. F. Smith, A note on hereditary rings, J. Algebra 44 (1977), 181-190.
4. D. Eisenbud, Commutative Algebra with a view towards Algebraic Geometry (Springer, New York 1995).
5. K. L. Fields, On the global dimension of residue rings, Pacific J. Math. 32 (1970), 345-349.
6. J. S. Golan, Torsion Theories (Longman, Harlow 1986).
7. I. Kaplansky, Commutative Rings (Allyn and Bacon, London 1974).
8. T. Y. Lam, Lectures in Rings and Modules (Springer, New York 1999).
9. J. C. McConnell and J. C. Robson, Noncommutative Noetherian Rings (Wiley, Chichester 1987).
10. D. S. Passman, The Algebraic Structure of Group Rings (Wiley, New York 1977).
11. D. W. Sharpe and P. Vamos, Injective Modules (Cambridge Univ. Press, Cambridge 1972).
12. B. Stenström, Rings of Quotients (Springer-Verlag, Berlin 1975).

STRUCTURE THEOREMS ON COUNTABLY COMPACT RINGS

M. URSUL

Department of Mathematics, University of Oradea
Armatei Române 5, Oradea, România

1. Introduction

The Wedderburn-Mal'cev Theorem is a subtle fact of the theory of finite-dimensional algebras over a field (see [CR], Chapter X). D. Zelinsky [Z] proved the following result: *Let R be a compact ring of prime characteristic. Then $R = S + J(R)$ is a direct (group) sum for some compact subring S of R.* K. Numakura [N] found necessary and sufficient conditions for decomposability of a compact ring. See also related results in [AAU] and [Ec].

The author has proved Wedderburn-Mal'cev Theorem for commutative countably compact rings of prime characteristic (see [U1, Theorem I.8.41]). In this paper we will prove the Wedderburn-Mal'cev Theorem for a countably compact ring R of prime characteristic with the metrizable factor ring $R/J(R)$. We will outline here a new proof of Zelinsky's Theorem.

We will also construct an example of a countably compact ring of prime characteristic which does not admit a Wedderburn decomposition. This answers in the negative the Problem VII.25 posed in [U]. Furthermore, we will give some structure theorems for countably compact rings.

2. Notation and conventions

All spaces considered here are assumed to be completely regular and Hausdorff (i.e., Tychonoff spaces). The closure of a subset A of a topological space X is denoted by $\overline{A}$. If $f : X \to Y$ is a mapping, $A \subseteq X$ a subset, then $f \,|\, A$ stands for the restriction of f on A. All topological rings and groups are assumed to be Hausdorff. Rings are assumed to be associative not necessarily with identity. The Jacobson radical of a ring R will be denoted by $J(R)$. Recall that a ring R with identity is called *local* provided $R/J(R)$ is a division ring. The subring (subgroup) of a ring R generated by a subset S is denoted by $\langle S \rangle (\langle S \rangle^+$, respectively). For a ring R and a natural number n, $[n]R = \{x_1 + \cdots + x_n : x_1, \ldots, x_n \in R\}$ and $R^{[n]} = \{x_1 \ldots x_n : x_1, \ldots, x_n \in R\}$. An ideal I of a ring R is said to be *cofinite* provided the factor ring R/I is finite. The group of units of a ring R with identity is denoted by $U(R)$. If $\{M_\alpha : \alpha \in \Omega\}$ is a collection of subgroups of an Abelian group M, then $\Sigma_{\alpha \in \Omega} M_\alpha = \bigcup \{M_{\alpha_1} + \cdots + M_{\alpha_n} : \alpha_1, \ldots, \alpha_n \in \Omega \ (n = 1, 2, \ldots)\}$. By $\mathbb{T}$ is denoted the unit circle group $\mathbb{R}/\mathbb{Z}$ in additive notation. $A \supset B$ means that B is a subset of A and $A \neq B$. The symbol ω denotes the set of all natural numbers, ω_1 denotes the set of all countable ordinals and ω_2 is the set of all ordinal numbers of cardinality $\leq \aleph_1$. By $\mathbb{N}$ is denoted the set of numbers $1, 2, 3, \ldots$. We use the terminology on topological rings as it is given in [U2].

Recall [En, Chapter 3, paragraph 10] that a topological space X is said to be *countably compact* provided every countable open cover has a finite subcover.

Theorem 2.1 ([En, Theorem 3.10.2]). *For every Hausdorff space X the following conditions are equivalent:*

(i) *The space X is countably compact.*

(ii) *Every countable family of closed subsets of X which has the finite intersection property has non-empty intersection.*

(iii) *For every decreasing sequence $F_1 \supseteq F_2 \supseteq \cdots$ of non-empty closed subsets of X the intersection $\bigcap_{i=1}^\infty F_i$ is non-empty.*

Recall [ComRo, page 5] that a Tychonoff space is called ω-*compact* provided every its countable subset lies in a compact subspace.

Recall [U1, p. 39] that a topological ring R is said to be *left countably linearly compact* if:

(i) R has a fundamental system of neighborhoods of zero consisting of left ideals;

(ii) $\bigcap \{F : F \in \mathfrak{F}\} \neq \varnothing$ for every countable filter base $\mathfrak{F}$ consisting of cosets with respect to closed left ideals.

Recall that a topological group G is called *precompact* if for every neighborhood V of identity there exists a finite subset F such that $G = F \cdot V = \{xy : x \in F, y \in V\}$.

A topological ring R is called (see [K, p. 162]) *topologically nilpotent* if for any neighborhood U of 0 there exists N such that $R^n \subseteq U$ for $n > N$.

The class of topologically nilpotent rings is closed under operations I, S, and C, where I denotes a continuous homomorphic image, S a (not necessarily closed) subring, and C an arbitrary Cartesian product with the Tychonoff product topology. This class is not closed under extensions.

3. Compact and countably compact rings

Theorem 3.1. *Let R be a countably compact ring. Assume that there exists a filter base $\{I_n : n \in \omega\}$ of closed two-sided ideals such that $\bigcap I_n = 0$. Then $R \cong_{top} \lim_{\leftarrow} R/I_n$.*

Proof. By easy facts about inverse limits, there is no loss in generality in assuming that the sequence (I_n) is decreasing.

Since $\{R/I_n : n \in \omega\}$ is an inverse system of topological rings, we can consider the diagonal mapping $\alpha : R \to \lim_{\leftarrow} R/I_n$ which is continuous. It is injective because $\bigcap_n I_n = 0$.

We claim that α is surjective. Indeed, an element of the inverse limit is a sequence $x = (x_n + I_n)$ of cosets such that, for $m \geq p$, $x_m + I_p = x_p + I_p$. This means that these cosets are a decreasing sequence of closed subsets of R, so their intersection is non-empty. If $r \in \bigcap(x_n + I_n)$, it is plain that $\alpha(r) = x$.

Let us prove that α is open. First of all we prove that, for every neighborhood V of 0 in R, there exists $m \in \omega$ such that $I_m \subseteq V$.

Consider the open cover $\{V\} \cup \{R/I_n : n \in \omega\}$. Since R is countably compact, there exists $m \in \omega$ such that $R = V \cup (R \setminus I_m)$. Then $I_m \subseteq V$.

Now, let U be any neighborhood of 0; take an open neighborhood V of 0 and $m \in \omega$ such that $V + V \subseteq U$ and $I_m \subseteq V$. The image V_m of V in R/I_m is open; we want to prove that $\alpha(V) \supseteq \mathrm{pr}_m^{-1}(V_m)$. Let $\alpha(r) \in \mathrm{pr}_m^{-1}(V_m)$. Then $r + I_m = v + I_m$, for some $v \in V$, so that $r \in V + V \subseteq U$. $\square$

Remark 3.1. The method of the proof of the last Theorem led to the following known result [ComRo, Theorem 3.2]: Let $(G, \mathfrak{T})$ be a countably compact group and let $(X, \mathfrak{T}_0)$ be a Tychonoff space in which each point is an intersection of a countably set of neighborhoods. If there exists a continuous surjective injective mapping of $(G, \mathfrak{T})$ on $(X, \mathfrak{T}_0)$ then $(G, \mathfrak{T})$ is a metrizable compact group.

Corollary 3.2 ([U2, Theorem 27.8]). *Every topologically nilpotent countably compact ring R is an inverse limit of nilpotent countably compact rings.*

Since each quasiregular countably compact ring is topologically nilpotent, we obtain the following result:

Corollary 3.3. *Every quasiregular countably compact ring R is an inverse limit of nilpotent countably compact rings.*

Corollary 3.4. *Every ω-compact topologically nilpotent ring is an inverse limit of nilpotent ω-compact rings.*

It follows from the proof of Theorem 3.1 that a left countably linearly compact topologically nilpotent ring is an inverse limit of left countably linearly compact nilpotent rings. In particular, every left countably linearly compact topologically nilpotent ring is quasiregular.

Lemma 3.5. *Let R be a topological ring and I a compact ideal of R. If $e' + I$ is an idempotent in R/I, then there exists an idempotent $e \in R$ such that $e + I = e' + I$.*

Proof. Since $e' + I$ is an idempotent of R/I, the subset $e' + I$ of R is a compact subsemigroup of the multiplicative semigroup of R. Since every compact semigroup contains an idempotent (see, e.g., [U2, Proposition 4.19]), there exists an idempotent $e \in e' + I$. $\qquad\square$

Lemma 3.6. *Let R be a precompact ring topologically generated by a family $\{e_\alpha : \alpha \in \Omega\}$ of orthogonal idempotents. Then R has a fundamental system of neighborhoods of zero consisting of two-sided ideals.*

Proof. Evidently, the subring S generated by the subset $\{e_\alpha : \alpha \in \Omega\}$ is commutative, hence R is commutative. The completion $\widehat{R}$ of R is commutative and S is dense in $\widehat{R}$. It suffices to show that $\widehat{R}$ has has a fundamental system of neighborhoods of zero consisting of two-sided ideals. Therefore we may consider without loss of generality that R is compact and commutative.

Put for every $\alpha \in \Omega, F(\alpha) = \{x \in R : xe_\alpha = e_\alpha\}$; evidently, $F(\alpha)$ is a closed subset of R. If $\alpha_1, \ldots, \alpha_n$ are different elements from Ω, then $e_{\alpha_1} + \cdots + e_{\alpha_n} \in \bigcap_{i=1}^n F(\alpha_i)$. By compactness of R, $\bigcap_{\alpha \in \Omega} F(\alpha) \neq \varnothing$. If $e \in \bigcap_{\alpha \in \Omega} F(\alpha)$, then $\forall_{\alpha \in \Omega}[ee_\alpha = e_\alpha]$. This implies that $\forall_{x \in S}[ex = x]$ and by continuity, $\forall_{x \in R}[ex = x]$, i.e., e is the identity of R. By Kaplansky's Theorem [K, Theorem 8] R has a fundamental system of neighborhoods of zero consisting of ideals. $\qquad\square$

Lemma 3.7. *Let A be a countably compact Abelian group with the property that for every $a \in A$ the subgroup $\overline{\langle a \rangle^+}$ is compact and totally disconnected.*

Then A has a fundamental system of neighborhoods of zero consisting of subgroups.

Proof. Let f be any continuous homomorphism of A in the group $\mathbb{T}$. Then $f(A)$ is a countably compact metrizable group each element of which lies in a compact totally disconnected group. It is well known that $f(A)$ is finite.

By [ComRoss, Theorem 1.1] A is a precompact topological group. Therefore there exists a cardinal number $\mathfrak{m}$ such that A is topologically isomorphic to a subgroup of $\mathbb{T}^{\mathfrak{m}}$. This finishes the proof. $\qquad\square$

Lemma 3.8. *Let A be a countably compact Abelian group for which there exists a continuous injective homomorphism $f : A \to B$ in a compact Abelian totally disconnected group B. Then A has a fundamental system of neighborhoods of zero consisting of subgroups.*

Proof. For every $a \in A, f(\overline{\langle a\rangle^+})$ is a countably compact subgroup of B, topologically generated by one element. Since B is totally disconnected, $f(\overline{\langle a\rangle^+})$ is a compact metrizable group. Then by Remark 3.1 $\overline{\langle a\rangle}$ is a compact metrizable group. Obviously, $\overline{\langle a\rangle}$ is totally disconnected. According to Lemma 3.7 A has a fundamental system of neighborhoods of zero consisting of subgroups. $\qquad\square$

Lemma 3.9. *Let R be a topological ring and I a closed two-sided countably compact ideal of R. Then there exists a closed two-sided ideal K of R such that $K \subseteq I, KI = IK = 0$ and I/K is a closed countably compact ideal of R/K having a fundamental system of neighborhoods of zero consisting of ideals.*

Proof. Denote by $\widehat{R}$ the completion of R and by L the closure of I in $\widehat{R}$. By [ComRoss, Theorem 1.1] L is a compact ideal of $\widehat{R}$. Denote by C the maximal connected subring of L. According to Kaplansky's Theorem [K, Theorem 8] $CL = LC = 0$. Put $K = C \cap I$ and denote by f the canonical homomorphism of $\widehat{R}$ on $\widehat{R}/C$. The homomorphism f induces a continuous injective homomorphism g of I/K in L/C. By Lemma 3.8 I/K has a fundamental system of neighborhoods of zero consisting of ideals. $\qquad\square$

Corollary 3.10. *Let R be a countably compact ring. Then there exists a closed two-sided ideal K of R such that $KR = RK = 0$ and R/K is a countably compact ring having a fundamental system of neighborhoods of zero consisting of ideals.*

Lemma 3.11 ([U1, 1.6.27]). *Let R be a topological ring and I a closed countably compact ideal of R. If $x + I$ is an idempotent of the factor ring R/I, then there exists an idempotent $e \in R$ such that $e + I = x + I$.*

Proof. By Lemma 3.9 there exists a closed ideal K of R such that $K \subseteq I$ and $KI = IK = 0$; then $K^2 = 0$. Denote by π the canonical homomorphism of R on R/K and by γ the canonical isomorphism of R/K on $R/K/I/K$. Then $\gamma\pi(x) + \gamma\pi(I)$ is an idempotent of the ring $R/K/I/K$.

Since $K^2 = 0$, every idempotent of R/K can be lifted modulo K [J, Chapter VII, Proposition 3]. Therefore it suffices to find an idempotent of R/K in the coset $\pi(e') + \pi(I)$. We have reduced the proof to the case when I is a closed countably compact ideal having a fundamental system of neighborhoods of zero consisting of ideals.

We have that $x^2 = x + i$, where $i \in I$. It is easy to prove by induction on n that $x^n \in x + \langle i, ix, xi, xix \rangle$ for every $n \in \mathbb{N}$. Therefore $\overline{\{x^n : n \in \mathbb{N}\}} \subseteq x + A$, where $A = \overline{\langle i, ix, xi, xix \rangle}$. By [U2, Corollary 27.31], A is compact, therefore, [U2, Proposition 4.19], $\overline{\{x^n : n \in \mathbb{N}\}}$ contains an idempotent. $\square$

Lemma 3.12. *Let R be a countably compact ring topologically generated by a system $\{e_n : n \in \omega\}$ of orthogonal idempotents. Then R is a compact metrizable ring.*

Proof. By Lemma 3.6 R has a fundamental system of neighborhoods of zero consisting of two-sided ideals. The cardinality of the set of all cofinite ideals of the subring $S = \langle e_n : n \in \omega \rangle$ is $\leq \aleph_0$. Therefore S is metrizable and so is R. It follows that R is compact. $\square$

We will outline the proof of the following result [U1, Lemma 1.6.31]: *Any family $\{e_\alpha : \alpha \in \Omega\}$ of orthogonal idempotents of a compact ring R is summable.*

The family $\{e_\alpha : \alpha \in \Omega\}$ is summable in R if and only if it is summable in the subring $\overline{\langle \{e_\alpha : \alpha \in \Omega\} \rangle}$. This implies that there is no loss in generality in assuming that $R = \overline{\langle \{e_\alpha : \alpha \in \Omega\} \rangle}$.

By Lemma 3.6 R has a fundamental system neighborhoods of zero consisting of ideals. Let V be an arbitrary open ideal of R. Denote by f the canonical homomorphism of R on R/V. Since R/V is finite, there exists a finite subset $\Omega_0 \subseteq \Omega$ such that $\forall_{\alpha \notin \Omega_0}[f(e_\alpha) = 0]$, i.e., $\forall_{\alpha \notin \Omega_0}[e_\alpha \in V]$. This implies that for each finite subset $\Omega_1 \subseteq \Omega$, $\Omega_0 \cap \Omega_1 = \varnothing$, $\Sigma_{\alpha \in \Omega_1} e_\alpha \in V$. By Cauchy's Criterion [B, Chapter III, paragraph 5, Theorem 1] the family $\{e_\alpha : \alpha \in \Omega\}$ is summable.

Theorem 3.13. *Let R be a countably compact ring and $\{e'_\alpha + J(R) : \alpha \in \omega_1\}$ be a system of orthogonal idempotents of $R/J(R)$. Then there exists a system $\{e_\alpha : \alpha \in \omega_1\}$ of orthogonal idempotents of R such that $e_\alpha + J(R) = e'_\alpha + J(R)$ for each $\alpha \in \Omega$.*

Proof. According to [U2, Corollary 13.8, page 182] $J(R)$ is a closed ideal of R.

We shall construct the needed system of orthogonal idempotents in R by recursion of length ω_1. By Lemma 3.11 there exists an idempotent $e_0 \in R$ such that $e_0 + J(R) = e'_0 + J(R)$. Assume that we have constructed a system $\{e_\beta : \beta < \alpha\}$ of orthogonal idempotents such that $e_\beta + J(R) = e'_\beta + J(R), \beta < \alpha$. By Lemma 3.12 the subring $\overline{\langle e_\beta : \beta < \alpha \rangle}$ is compact, therefore exists the sum $\Sigma_{\beta < \alpha} e_\beta = e$ and is an idempotent. By Lemma 3.11 there exists an idempotent $g_\alpha \in R$, $g_\alpha + J(R) = e'_\alpha + J(R)$. Denote $p = (1 - e)g_\alpha(1 - e) = g_\alpha - g_\alpha e - eg_\alpha + eg_\alpha e$. Then $p + J(R) = g_\alpha - g_\alpha e - eg_\alpha + eg_\alpha e + J(R) = e'_\alpha + J(R)$. Evidently, $ep = pe = 0$.

There is an idempotent and $e_\alpha \in \overline{\langle p \rangle}$, $e_\alpha + J(R) = e'_\alpha + J(R)$. Evidently, $e_\alpha e_\beta = e_\beta e_\alpha = 0$ for every $\beta < \alpha$. $\qquad\square$

Theorem 3.14. *If a countably compact ring R satisfies A.C.C. for two-sided ideals then R is a compact, totally disconnected and metrizable ring with open radical.*

Proof. By Theorem of Comfort and Ross [ComRoss, Theorem 1.1] R is a precompact topological ring. Let $\widehat{R}$ be the completion of R and C the maximal connected subring of $\widehat{R}$. By Kaplansky's Theorem [K, Theorem 8] $C\widehat{R} = \widehat{R}C = 0$. Therefore $C \cap R$ is a countably compact finitely generated Abelian group. It follows that $C \cap R$ is finite. The ring $R/C \cap R$ is mapped continuously and isomorphic on the ring $R + C/C$. It suffices to show that the ring $R + C/C$ is compact and metrizable. Therefore we may assume that R has a fundamental system of neighborhoods of zero consisting of two-sided ideals.

Assume that $J(R) = 0$. According to [U2, Theorem 27.11] and [U2, Corollary 27.6] we may consider that R/V is a finite semisimple ring for every open ideal V. We affirm that under these conditions R is finite. Indeed, on the contrary, there exists a chain $R \supset V_1 \supset V_2 \supset \cdots \supset V_n \supset \cdots$ of open ideals. Then $R/\bigcap V_n$ is a compact semisimple infinite ring with A.C.C. for ideals, a contradiction.

Therefore, if R is an arbitrary ring satisfying the conditions of Theorem then J is open. The ring $J/\overline{J^2}$ is a $(R/J, R/J)$-bimodule. Every its submodule has the form $H/\overline{J^2}$, where H is a subgroup of R containing $\overline{J^2}$ and $RH \subseteq H$, $HR \subseteq H$. This means that H is a two-sided ideal of R. It follows that $J/\overline{J^2}$ is a finitely generated $(R/J, R/J)$-bimodule. This implies that $J/\overline{J^2}$ is finite. Since J is topologically nilpotent, it has a finite number of topological generators. It follows that R is metrizable. $\qquad\square$

4. Wedderburn-Mal'cev decomposition of countably compact rings

Theorem 4.1. *Let R be a countably compact ring of prime characteristic. If $R/J(R)$ is metrizable, then there exists a compact subring S of R such that $R = S + J(R), S \cap J(R) = 0$. If S, S' are two compact subrings such that $R = S + J(R) = S' + J(R), S \cap J(R) = 0 = S' \cap J(R)$, then there exists an element $a \in J(R)$ such that $S' = (1 + a)S(1 + a)^{-1}$.*

Proof. By Kaplansky's Theorem [K, Theorem 16] there exists a family R_n, $n \in \omega$ of finite simple rings such that R/J is topologically isomorphic to $\prod_{n \in \omega} R_n$. Let $\phi : R \to \prod_{n \in \omega} R_n$ be a continuous surjective homomorphism such that $\ker \phi = J(R)$. Let e'_n be the identity of R_n (identified with $R_n \times \prod_{m \neq n}\{0_m\}$). By Theorem 3.13 there exists a system $\{e_n : n \in \omega\}$ of orthogonal idempotents of R such that $\phi(e_n) = e'_n$ for each $n \in \omega$. Choose for each $n \in \omega$ a finite subset $X_n \subseteq e_n R e_n$, $\phi(X_n) = R_n$. Let $P_n = \overline{\langle X_n \rangle}$. Then P_n is a compact metrizable subring.

Affirmation. $J(P_n) = P_n \cap J(R)$. Since $\phi(P_n) = R_n, J(P_n) \subseteq J(R) \cap P_n = \ker(\phi \,|\, P_n)$. Since $J(R) \cap P_n$ is a topologically nilpotent ideal of a compact ring P_n, it is contained in $J(P_n)$. We have proved that $J(P_n) = P_n \cap J(R)$. By Zelinsky's Theorem [Z, Theorem E] there exists a finite semisimple subring S_n of P_n such that $P_n = S_n + J(P_n)$. Consider the subring $S = \sum_{n \in \omega} S_n$. Then the cardinality of all cofinite ideals of S is $\leq \aleph_0$. Therefore $\overline{S}$ is metrizable and so it is compact.

Affirmation. $R = \overline{S} + J(R)$.

Indeed, let $r \in R$, $\phi(r) = \{r_n\}$, $r_n \in R_n$. For each $n \in \omega$ there exists $s_n \in S_n$ such that $\phi(s_n) = r_n$. Then the family $\{s_n : n \in \omega\}$ is summable and $\phi(\sum s_n) = \{r_n\}$.

Claim. $\overline{S} \cap J(R) = 0$. Indeed, $\overline{S}$ is a compact ring containing a dense regular in the sense of von Neumann ring. It follows that $\overline{S}$ is regular in the sense of von Neumann ring, hence $\overline{S} \cap J(R) = 0$.

Mal'cev's part.

Let $R = S + J(R) = S' + J(R)$ be a representation of R where S, S' are two compact semisimple subrings. Then $S = \overline{\sum_{n \in \omega} S_n}, S' =$

$\overline{\sum_{m\in\omega} S'_m}$, where S_n, S'_m, $n,m \in \omega$ are finite simple rings. Denote for each $n \in \omega$ by $e_n(e'_n)$ the identity of S_n (S'_n, respectively) and put $B = \overline{\langle \bigcup_{n,m\in\omega}(S_n \cup S'_m)\rangle}$.

Claim. B is metrizable.

It suffices to show that the subring $\langle \bigcup_{n,m\in\omega}(S_n \cup S'_m)\rangle$ is metrizable. Let V be an open two-sided ideal of $\langle \bigcup_{n,m\in\omega}(S_n \cup S'_m)\rangle$. There exists $n \in \omega$ such that $S_k \cup S'_k \subseteq V$ for $k \geq n$. By Lewin's Theorem [L, Theorem 1], the ring $\langle \bigcup_{n,m\in\omega}(S_n \cup S'_m)\rangle$ has at most $\aleph_0$ cofinite ideals. It follows that B is metrizable, hence it is compact.

Claim. $J(B) = J(R) \cap B$.

Indeed, the inclusion $J(R) \cap B \subseteq J(B)$ is obvious. By construction, $\phi(B) = \prod_{n\in\omega} R_n$, hence $J(B) \subseteq J(R) \cap B$.

It follows that $B = S + J(B) = S' + J(B)$. By Eckstein's Theorem [Ec], [U2, Theorem 10.3, page 170] there exists an inner automorphism i_b, $b \in J(B)$ of B such that $S' = i_b(S)$. But i_b can be considered as an inner automorphism of R. $\qquad\square$

Theorem 4.2. *Let R be a compact local ring of prime characteristic p. Then there exists a finite subring F which is a field such that $R = F + J(R)$.*

Proof. Denote by ϕ be the canonical homomorphism of R on $R/J(R)$ and let $\mathrm{card}(R/J(R)) = p^n$. Let θ_1 be any generator of the multiplicative group of $R/J(R)$ and let $\theta \in \phi^{-1}(\theta_1)$. The subring $\overline{\langle \theta \rangle}$ is compact and metrizable. There exists a sequence (s_k) of natural numbers divisible by n such that the sequence $(\theta^{p^{s_k}})$ is convergent; let $\lim_{k\to\infty} \theta^{p^{s_k}} = \lambda$.

Then $\lambda^{p^n} = \lambda$: Let V be any open ideal of R. Then there exists $k \in \omega$ such that $(\theta^{p^n} - \theta)^{p^{s_k}} \in V, \theta^{p^{s_k}} - \lambda \in V$. Write $\theta^{p^{s_k}} = \lambda + v$, where $v \in V$. Then $\theta^{p^n p^{s_k}} - \theta^{p^{s_k}} \in V$, hence $(\theta^{p^{s_k}})^{p^n} - \theta^{p^{s_k}} \in V$, or $(\lambda+v)^{p^n} - \lambda - v \in V$. This implies $\lambda^{p^n} - \lambda \in V$ and so $\lambda^{p^n} = \lambda$.

Let $s_k = q_k n$. Then $\lim_{k\to\infty} \theta^{p^{q_k n}} = \lambda$, hence $\lim_{k\to\infty} \phi(\theta^{p^{q_k n}}) = \phi(\lambda)$ or $\lim_{k\to\infty} \theta_1^{p^{q_k n}} = \phi(\lambda) \Rightarrow \theta_1 = \phi(\lambda)$.

Evidently, $R = \langle \lambda \rangle + J(R)$. Since $\langle \lambda \rangle$ satisfies the identity $x^{p^n} = x$, $\langle \lambda \rangle \cap J(R) = 0$. $\qquad\square$

Note. By using of Theorem 4.2 we can give another proof of Zelinsky's Theorem [Z, Theorem E].

Example 4.1. Example of a countably compact ring which does not admit Wedderburn decomposition.

This example is a modification of the example of D. Zelinsky [Z, page 316]. Consider any finite field k and the set $\{e_\alpha, n_{\alpha\beta} : \alpha,\beta \in \omega_2\}$

of symbols with the following multiplication table: $e_\alpha^2 = e_\alpha$; $e_\alpha e_\beta = n_{\alpha\beta}$ if $\alpha \neq \beta$; $e_\alpha n_{\beta\gamma} = \delta_{\alpha\beta} n_{\beta\gamma}$; $n_{\alpha\beta} e_\gamma = \delta_{\beta\gamma} n_{\alpha\beta}$; $n_{\alpha\beta} n_{\gamma\delta} = 0$ (Kronecker deltas).

Let $R = \prod_{\alpha\in\omega_2} ke_\alpha \times \prod_{\beta,\gamma\in\omega_2} kn_{\beta\gamma}$ be the topological product of discrete groups ke_α, $kn_{\beta\gamma}$, $\alpha, \beta, \gamma \in \omega_2$. Then R becomes a compact Abelian group.

The multiplication on R is defined as follows: If $x, y \in R$, $\alpha \in \omega_2$, then put:

$$\operatorname{pr}_\alpha(xy) = \operatorname{pr}_\alpha(x)\operatorname{pr}_\alpha(y),$$

$$\operatorname{pr}_{\alpha\beta}(xy) = (1 - \delta_{\alpha\beta})\operatorname{pr}_\alpha(x)\operatorname{pr}_\beta(y) + \operatorname{pr}_\alpha(x)\operatorname{pr}_{\alpha\beta}(y) + \operatorname{pr}_{\alpha\beta}(x)\operatorname{pr}_\beta(y).$$

It is easy to prove that R becomes an associative ring.

We claim that R is a compact ring. Indeed, let W be a neighborhood of zero. We can assume without loss of generality that

$$W = \{0\}_{\alpha_1} \times \cdots \times \{0\}_{\alpha_m} \times \prod_{\alpha\neq\alpha_1,\ldots,\alpha_m} ke_\alpha \times$$

$$\times \{0\}_{\beta_1\gamma_1} \times \cdots \times \{0\}_{\beta_n\gamma_n} \times \prod_{(\beta,\gamma)\neq(\beta_1,\gamma_1),\ldots,(\beta_n,\gamma_n)} kn_{\beta\gamma}$$

and $\{\beta_1, \ldots, \beta_m, \gamma_1, \ldots, \gamma_n\} \subseteq \{\alpha_1, \ldots, \alpha_m\}$. We affirm that W is a two-sided ideal of R. Indeed, let $w \in W$, and $x \in R$. If $\alpha \in \{\alpha_1, \ldots, \alpha_m\}$, then $\operatorname{pr}_\alpha(xw) = \operatorname{pr}_\alpha(wx) = 0$. Let $(\beta, \gamma) \in \{(\beta_1, \gamma_1), \ldots, (\beta_n, \gamma_n)\}$; if $\beta = \gamma$, then

$$\operatorname{pr}_{\beta\beta}(xw) = \operatorname{pr}_\beta(x)\operatorname{pr}_{\beta\beta}(w) + \operatorname{pr}_{\beta\beta}(x)\operatorname{pr}_\beta(w) = 0 =$$
$$= \operatorname{pr}_{\beta\beta}(wx) = \operatorname{pr}_\beta(w)\operatorname{pr}_{\beta\beta}(x) + \operatorname{pr}_{\beta\beta}(w)\operatorname{pr}_\beta(x).$$

If $\beta \neq \gamma$, then

$$\operatorname{pr}_{\beta\gamma}(xw) = \operatorname{pr}_\beta(x)\operatorname{pr}_\gamma(w) + \operatorname{pr}_\beta(x)\operatorname{pr}_{\beta\gamma}(w) + \operatorname{pr}_{\beta\gamma}(x)\operatorname{pr}_\gamma(w) = 0 =$$
$$= \operatorname{pr}_{\beta\gamma}(wx) = \operatorname{pr}_\beta(w)\operatorname{pr}_\gamma(x) + \operatorname{pr}_\beta(w)\operatorname{pr}_{\beta\gamma}(x) + \operatorname{pr}_{\beta\gamma}(w)\operatorname{pr}_\gamma(x).$$

We proved that W is an open ideal of R.

Consider the Σ-product S of $ke_\alpha, kn_{\beta\gamma}$. We will write for simplicity the elements of S as $\Sigma_{\alpha\in\omega_2} a_\alpha e_\alpha + \Sigma_{\beta,\gamma\in\omega_2} a_{\beta\gamma} n_{\beta\gamma}$, where only a countable number of $a_\alpha, a_{\beta\gamma}$ are non-zero. Assume that there exists orthogonal idempotents e_α^* in S, $e_\alpha^* - e_\alpha \in J(S)$. Write $e_\alpha^* - e_\alpha = x_\alpha$. Denote for every $\alpha \in \omega_2$, $A_\alpha = \{\beta : e_\alpha x_\beta \neq 0\}$, $B_\alpha = \{\beta : e_\beta x_\alpha \neq 0\}$, $C_\alpha = \{\beta : x_\alpha e_\beta \neq 0\}$.

Claim. $\omega_2 \setminus A_\alpha \subseteq C_\alpha \cup \{\alpha\}$.

Indeed, let $\beta \in \omega_2 \setminus A_\alpha, \beta \neq \alpha$. Then $(e_\alpha + x_\alpha)(e_\beta + x_\beta) = 0$, hence $n_{\alpha\beta} + e_\alpha x_\beta + x_\alpha e_\beta = 0$. Since $e_\alpha x_\beta = 0$, we obtain that $x_\alpha e_\beta \neq 0$, i.e., $\beta \in C_\alpha$.

Obviously, $|C_\alpha| \leq \aleph_0$, hence $|\omega_2 \setminus A_\alpha| \leq \aleph_0$.

Fix an element $\beta \in \omega_2$. This element cannot lie in a uncountable family of sets A_α. Therefore if Ω_0 is an uncountable subset of ω_2, then $\bigcap_{\alpha \in \Omega_0} A_\alpha = \emptyset$. Choose Ω_0 such that $|\Omega_0| = \aleph_1$. Then $\omega_2 = \omega_2 \setminus \bigcap_{\alpha \in \Omega_0} A_\alpha = \bigcup_{\alpha \in \Omega_0} (\omega_2 \setminus A_\alpha)$, hence $|\omega_2| \leq \aleph_1$, a contradiction. The proof is complete.

We will give another example of a compact ring R for which R^2 is not closed. Let p be a prime number, $\mathbb{F}_p$ a finite field consisting of p elements and n a natural number. Consider the ring $\mathbb{F}_p[x_1, \ldots, x_n]$ of polynomials and the ideal I generated by polynomials having the lower degree ≥ 3.

Let R_n be the image in the factor ring $\mathbb{F}_p[x_1, \ldots, x_n]/I$ of the subring of $\mathbb{F}_p[x_1, \ldots, x_n]$ consisting of polynomials without terms of degree 0.

Lemma 4.3. *The set R_{3n}^2 is not the sum of n copies of the subset $R_{3n}^{[2]}$.*

Proof. Assume the contrary. The cardinality of $R_{3n}^{[2]} = (\mathbb{F}_p x_1 + \cdots + \mathbb{F}_p x_{3n}) \cdot (\mathbb{F}_p x_1 + \cdots + \mathbb{F}_p x_{3n})$ does not exceed $p^{3n} \cdot p^{3n} = p^{6n}$. Therefore the cardinality of $[n]R_{3n}^2$ does not exceed p^{3n^2}. The set $\{x_i x_j : i, j \in [1, 3n]\}$ is a linear base of R_{3n} over $\mathbb{F}_p$ and has $3n(3n+1)/2$ elements. Since $3n(3n+1)/2 > 3n^2$ the Lemma is proved. $\qquad\square$

The ring $R = \prod_{n \in \omega} R_{3n}$ satisfies the desired properties.

Recall that for any ring R the circular multiplication $\circ$ is defined as follows: $x \circ y = x + y + xy$ for all $x, y \in R$.

Corollary 4.4. *There exists a metabelian compact group G of exponent three whose commutator is not closed.*

Proof. Consider the ring $R = \prod_{n \in \omega} R_{3n}$ over the field $\mathbb{F}_3$. The subset $G = \{xE_{12} + yE_{13} + zE_{23} : x, y, z \in R\}$ of the ring of 3×3-matrices over R with the circular multiplication $\circ$ is a group (here E_{12}, E_{13}, E_{23} are the matrix units).

The inverse $(xE_{12} + yE_{13} + zE_{23})^{(-1)}$ of $xE_{12} + yE_{13} + zE_{23}$ is $-xE_{12} + (-y + xz)E_{13} - zE_{23}$. Let $u = xE_{12} + yE_{13} + zE_{23}, v = aE_{12} + bE_{13} + cE_{23}$. We will calculate the commutator $u^{(-1)} \circ v^{(-1)} \circ u \circ v$:

$$(-xE_{12} + (-y + xz)E_{13} - zE_{23}) \circ (-aE_{12} + (-b + ac)E_{13} - cE_{23}) \circ$$
$$\circ (xE_{12} + yE_{13} + zE_{23}) \circ (aE_{12} + bE_{13} + cE_{23})$$
$$= [-(x + a)E_{12} + (-y - b + xz + ac + xc)E_{13} - (z + c)E_{23}] \circ$$
$$\circ [(x + a)E_{12} + (y + b + xc)E_{13} + (z + c)E_{23}]$$
$$= (xz + ac + xc + xc)E_{13} - (x + a)(z + c)E_{13} = (xc - az)E_{13}.$$

Furthermore, for each $t, x, y, z \in R$,

$$tE_{13} \circ (xE_{12} + yE_{13} + zE_{23}) = xE_{12} + (t + y)E_{13} + zE_{23};$$
$$(xE_{12} + yE_{13} + zE_{23}) \circ tE_{13} = xE_{12} + (y + t)E_{13} + zE_{23}.$$

We proved that G is metabelian.

The group G has exponent three: $(xE_{12} + yE_{13} + zE_{23})^2 = 2xE_{12} + (2y + xz)E_{13} + 2zE_{23}$; $(xE_{12} + yE_{13} + zE_{23})^3 = [2xE_{12} + (2y + xz)E_{13} + 2zE_{23}] \circ (xE_{12} + yE_{13} + zE_{23}) = 0$. $\qquad\square$

We will give below a new characterization of the Jacobson radical of a compact ring.

Lemma 4.5. *A subring $S \subseteq \prod_{\alpha \in \Omega} R_\alpha$ of a topological product of finite simple rings R_α, $\alpha \in \Omega$ is maximal topologically nilpotent if and only if:*

(i) $S = \prod_{\alpha \in \Omega} \mathrm{pr}_\alpha(S)$;
(ii) *for every $\alpha \in \Omega$ the subring $\mathrm{pr}_\alpha(S)$ is a maximal nilpotent subring of R_α.*

Proof. ($\Rightarrow$) (i) Let $x = \{x_\alpha\} \in \prod_{\alpha \in \Omega} \mathrm{pr}_\alpha(S)$. Then the subring $\langle S \cup \{x\}\rangle$ is topologically nilpotent, hence $x \in S$.

(ii) Assume the contrary, i.e., that there exists $\beta \in \Omega$ such that $\mathrm{pr}_\beta(S)$ is not a maximally nilpotent subring. Let T a nilpotent subring of R_β strongly containing $\mathrm{pr}_\beta(S)$. Then $T \times \prod_{\alpha \neq \beta} \mathrm{pr}_\alpha(S)$ is a topologically nilpotent subring strongly containing S, a contradiction.

($\Leftarrow$) Assume the contrary and let P be a topologically nilpotent subring of R, $P \supset S$. Then there exist $\{x_\alpha\} \in P$ and $\beta \in \Omega$ such that $x_\beta \notin S_\beta = \mathrm{pr}_\beta(S)$. It follows that $\mathrm{pr}_\beta(P)$ is a nilpotent subring of R_β strongly containing S_β, a contradiction. $\qquad\square$

Corollary 4.6. *If A, A' are two maximal topologically nilpotent of a semisimple compact ring R then there exists an element $x \in U(R)$ such that $A' = x^{-1}Ax$.*

Proof. We may consider that $R = \prod_{\alpha \in \Omega} R_\alpha$ is a topological product of finite simple rings R_α, $\alpha \in \Omega$. Then $A = \prod_{\alpha \in \Omega} A_\alpha$, $A' = \prod_{\alpha \in \Omega} A'_\alpha$, where A_α, A'_α are maximal nilpotent subrings of R_α, $\alpha \in \Omega$. Corollary follows from [S, Corollary 1, paragraph 13]. $\qquad\square$

Recall that a topological ring R is called *topologically nil* if $x^n \to 0$ for any $x \in R$.

Note that if R is a precompact ring then each its topologically nil subring is contained in a maximal topologically nilpotent subring.

We will extend the Koethe Theorem [Ko] (see notes at the end of the Chapter 18 of [F]):

Theorem 4.7. *The Jacobson radical of a compact ring R is the intersection of all its maximal topologically nilpotent subrings.*

Proof. Let A be a maximal topologically nilpotent subring of R; $A + J(R)/J(R)$ is topologically isomorphic to $A/A \cap J(R)$. Since $A \cap J(R)$ is quasiregular, A is a quasiregular subring of R, hence it is topologically nilpotent. It follows that $A + J(R) = A$, hence $J(R) \subseteq \bigcap_{\alpha \in \Omega} A_\alpha$, where A_α runs all maximal topologically nilpotent subrings of R.

Denote by p the canonical homomorphism of R on $R/J(R)$. We note that if B is a maximal topologically nilpotent subring of $R/J(R)$ then $p^{-1}(B)$ is a maximal topologically nilpotent subring of R. Indeed, p induces a continuous homomorphism p_1 of $p^{-1}(B)$ on B and $\ker p_1 = J(R)$. It follows that $p^{-1}(B)$ is topologically nilpotent. If C is a closed topologically nilpotent subring of R containing $p^{-1}(B)$, then C contains $J(R)$. Since $p(C) = B$, we obtain that $p^{-1}(B) = p^{-1}(C) = C + J(R) = C$, therefore $p^{-1}(B)$ is a maximal topologically nilpotent subring of R.

By Lemma 4.5, $0 = \bigcap B$ where B runs all maximal topologically nilpotent subrings of $R/J(R)$. Then $J(R) = \bigcap p^{-1}(B)$ where B runs all maximal topologically nilpotent subrings of $R/J(R)$. Therefore $J(R) \supseteq \bigcap_{\alpha \in \Omega} A_\alpha$ where A_α runs all maximal topologically nilpotent subrings of R. $\qquad\square$

Theorem 4.8. *If A, B are two maximal topologically nilpotent subrings of a compact ring R with identity then there exists $x \in U(R)$ such that $B = x^{-1}Ax$.*

Proof. By Corollary 4.6 there exists an element $x \in U(R)$ such that $x^{-1}Ax \subseteq B + J(R) = B$. Since $x^{-1}Ax$ is a maximal topologically nilpotent subring too, $x^{-1}Ax = B$. $\qquad\square$

5. Open questions

Question 1. Let R be a countably compact ring of prime characteristic of weight $\aleph_1$. Does R admit Wedderburn decomposition?

Question 2. Let R be a countably compact ring of prime characteristic having a subring S such that $R = S \oplus J(R)$ (a direct group sum). Does R admit Wedderburn decomposition?

Question 3. Let R be a countably compact ring which admits Wedderburn decomposition. Is true the Mal'cev Theorem for R?

Question 4. Is the Jacobson radical of an arbitrary countably compact ring the intersection of all its maximal topologically nilpotent subrings?

Acknowledgments

The author is grateful to the members of the algebraic seminar of the Institute of Mathematics of Hungary where the first version of this paper was reported. I am also grateful to the referee for her/his very careful reading and substantially helpful suggestions.

Bibliography

AAU.   V. A. Andrunakievich, V. I. Arnautov, M. I. Ursul, Wedderburn Decomposition of Hereditarily Linearly Compact Rings, Doklady Akademii Nauk SSSR, 211 (1973), 15–18.

B.   N. Bourbaki, General Topology. Topological Groups. Numbers and Related Groups and Spaces. Chapters III-VIII. Nauka, Moscow, 1969 (in Russian).

ComRo.   W. W. Comfort and L. C. Robertson, Extremal phenomena in certain classes of totally bounded groups, Dissertationes Math. (Rozprawy Mat.) 272 (1988), 1–48.

ComRoss.   W. W. Comfort, K. A. Ross, Pseudocompactness and uniform continuity in topological groups, Pacific J. Math., 16(3) (1966), 483–496.

CR.   C. W. Curtis, I. Reiner, Representation Theory of Finite Groups and Associative Algebras, Interscience Publishers, New York, London, 1962.

Ec.   F. Eckstein, On the Mal'cev Theorem, J. Algebra, 12 (1969), 372–385.

En.   R. Engelking, General Topology, Polska Akademia Nauk, Monografie Matematyczne, vol. 60, Panstwowe Wydawnictwo Naukowe — Polish Scientific Publishers, Warszawa 1977.

F.   C. Faith, Algebra II. Ring Theory, Grundlehren der Mathematischen Wissenschaften 191, Springer-Verlag, Berlin-Heidelberg-New York, 1976.

J.   N. Jacobson, Structure of Rings, Amer. Math. Soc. Colloquium Publ., vol. 37, Providence, RI, 1964 (revised edition).

K.   I. Kaplansky, Topological Rings, Amer. J. Math., 69 (1947), 153–183.

Ko.   G. Koethe, Über maximale nilpotente Unterringe und Nilringe, Math. Ann., 103 (1930), 359–363.

L.   J. Lewin, Subrings of Finite Index in Finitely Generated Rings, J. Algebra, 8 (1967), 84–88.

N.   K. Numakura, A note on Wedderburn Decomposition of Compact Rings, Nihon Gakusin. Proceedings 35 (1959), 313–315.

S.   D. A. Suprunenko, Matrix groups, Nauka, Moscow, 1972.

U.   M. I. Ursul, Unsolved problems of topological algebra (preprint), Chisinau, Stiintsa, 1985.

U1.   M. I. Ursul, Compact Rings and Their Generalizations, Chisinau, Stiinta, 1991.

U2.   M. I. Ursul, Topological Rings Satisfying Compactness Conditions, Kluwer Academic Publishers, 2002.

Z.   D. Zelinsky, Raising Idempotents, Duke Math. J., 21 (1954), 315–322.